# Microbial Interactions at Nanobiotechnology Interfaces

Molecular Mechanisms and Applications

*Edited by*

*R. Navanietha Krishnaraj*

*South Dakota School of Mines and Technology*
*Rapid City, South Dakota*

*and*

*Rajesh K. Sani*

*Department of Chemical and Biological Engineering*
*South Dakota School of Mines and Technology*
*Rapid City, South Dakota*

WILEY

This edition first published 2022


*Registered Office*
John Wiley & Sons, Inc., 111 River Street, Hoboken, NJ 07030, USA

*Editorial Office*
111 River Street, Hoboken, NJ 07030, USA

For details of our global editorial offices, customer services, and more information about Wiley products visit us at www.wiley.com.

Wiley also publishes its books in a variety of electronic formats and by print-on-demand. Some content that appears in standard print versions of this book may not be available in other formats.



*Library of Congress Cataloging-in-Publication Data applied for:*

ISBN: 9781119617198

Cover Design: Wiley
Cover Image: © iStock\MF3d

MIX
Paper from responsible sources
FSC® C013604
FSC www.fsc.org

# Contents

# Preface

Nanomaterials have been realized as potential agents for several biological applications including therapy and diagnostics. Realizing the biological activity of nanomaterials and their importance for nanomedicine, the field of nanobiotechnology emerged to emphasize the toxic effects of nanomaterials and biological applications. The field of nanobiotechnology covers concepts on the biological synthesis of nanomaterials, characterization of biologically synthesized nanomaterials, and applications in the biological sector. The unique physicochemical characteristics of the nanomaterials offer interesting activities including antibacterial, antifungal, antiviral, and anticancer. Recent decades have evidenced tremendous advancements in nanobiotechnology research with applications in diagnosis, drug delivery, and therapy.

Considering the importance of the field, the book aims to provide a basic understanding about the about the interaction of nanomaterials with living cells/proteins/enzymes at a molecular level and factors influencing the interactions. This book covers specialized topics on protein nanoparticle interactions and factors influencing these interactions, shape- and size-dependent activity of the nanomaterials, and surface functionalization of nanoparticles for stability/activity in biological systems. An interesting chapter covering the molecular factors and mechanisms that drives the anticancer activity of the nanomaterials. Understanding the surface factors of nanomaterials influencing the biological activity will help in reverse engineering of nanomaterials with tailored surface characteristics.

Overall this would be an ideal textbook for nanobiotechnology courses for chemical, biochemical, and biomedical engineering students. Glossary and reasoning type questions have also been included at the end of each chapter. This book will also help scientists to understand advanced concepts in nanobiotechnology and nano-biomanufacturing with specific characteristics. This book discusses concepts on factors influencing better microbial interactions and surface engineering approaches specifically focused on surface engineering of nanomaterials for biomedical applications.

The editors would like to thank all the authors for their valuable contribution and the Wiley editorial team for their support.

| | | |
|---|---|---|
| 1 | Shape- and Size-Dependent Antibacterial Activity of Nanomaterials | Prashant Mishra |
| 2 | Size- and Shape-Selective Synthesis of DNA-Based Nanomaterials and Their Application in Surface-Enhanced Raman Scattering | Subrata Kundu |
| 3 | Surface Modification Strategies to Control the Nanomaterial–Microbe Interplay | Vignesh MuthuVijayan |
| 4 | Surface Functionalization of Nanoparticles for Stability in Biological Systems | D. Sakthi Kumar |
| 5 | Molecular Mechanisms Behind Nano-Cancer Therapeutics | Aravind Kumar Rengan |
| 6 | Protein Nanoparticle Interactions and Factors Influencing These Interactions | Mala R. |
| 7 | Interaction Effects of Nanoparticles with Microorganisms Employed in the Remediation of Nitrogen-Rich Wastewater | Sheeja Jagadevan |
| 8 | Silver-Based Nanoparticles for Antibacterial Activity | Chittaranjan Patra |
| 9 | Microbial Gold Nanoparticles and Their Potential Biomedical Applications | Dindyal Mandal |
| 10 | Nano-Bio Interactions and Their Practical Implications in Agriculture | Achintya N. Bezbaruah |
| 11 | Biogeochemical Interactions of Bioreduced Uranium Nanoparticles | S. Sevinç Şengör |
| 12 | Characterization and Quantification of Mobile Bioreduced Uranium Phases | S. Sevinç Şengör |

# List of Contributors

***W. Aadinath***
Department of Biotechnology
Bhupat and Jyoti Mehta School of Biosciences
Indian Institute of Technology Madras
Chennai
India

***Srishti Agarwal***
Bio-Nano Electronics Research Center
Graduate School of Interdisciplinary New Science
Toyo University
Kawagoe
Saitama
Japan

***R. Akhil***
Department of Biotechnology
Bhupat and Jyoti Mehta School of Biosciences
Indian Institute of Technology Madras
Chennai
India

***Papia Basuthakur***
Department of Applied Biology
CSIR-Indian Institute of Chemical Technology
Hyderabad
Telangana State
India
Academy of Scientific and Innovative Research (AcSIR)
Ghaziabad
Uttar Pradesh
India

***Achintya N. Bezbaruah***
Department of Civil and Environmental Engineering
North Dakota State University
Fargo
ND
USA

***Parmita Chawley***
Department of Environmental Science and Engineering
Indian Institute of Technology (Indian School of Mines)
Dhanbad
Jharkhand
India

***Ann-Marie Fortuna***
United States Department of Agriculture

Grazingland Research Laboratory
Agricultural Research Service
El Reno
OK
USA

***Shagufta Haque***
Department of Applied Biology
CSIR-Indian Institute of Chemical Technology
Hyderabad
Telangana State
India
Academy of Scientific and Innovative Research (AcSIR)
Ghaziabad
Uttar Pradesh
India

***Sheeja Jagadevan***
Department of Environmental Science and Engineering
Indian Institute of Technology (Indian School of Mines)
Dhanbad
Jharkhand
India

***K. Karthick***
Electrochemical Process Engineering (EPE) Division
CSIR-Central Electrochemical Research Institute (CECRI)
Karaikudi
Tamil Nadu
India

***Senthilguru Kulanthaivel***
Department of Biochemical Engineering and Biotechnology
Indian Institute of Technology Delhi
New Delhi
India

***Subrata Kundu***
Electrochemical Process Engineering (EPE) Division
CSIR-Central Electrochemical Research Institute (CECRI)
Karaikudi
Tamil Nadu
India

***Uday Suryakant Maharana***
School of Biotechnology
KIIT University
Bhubaneswar
Odisha
India

***R. Mala***
Department of Biotechnology
MepcoSchlenk Engineering College
Sivakasi
Tamil Nadu
India

***Dindyal Mandal***
School of Biotechnology
KIIT University
Bhubaneswar
Odisha
India

***Prashant Mishra***
Department of Biochemical Engineering and Biotechnology
Indian Institute of Technology Delhi
New Delhi
India

*Sourav Mishra*
School of Biotechnology
KIIT University
Bhubaneswar
Odisha
India

*Vignesh Muthuvijayan*
Department of Biotechnology
Bhupat and Jyoti Mehta School of Biosciences
Indian Institute of Technology Madras
Chennai
India

*Bijayananda Panigrahi*
School of Biotechnology
KIIT University
Bhubaneswar
Odisha
India

*Chitta Ranjan Patra*
Department of Applied Biology
CSIR-Indian Institute of Chemical Technology
Hyderabad
Telangana State
India
Academy of Scientific and Innovative Research (AcSIR)
Ghaziabad
Uttar Pradesh
India

*Aravind Kumar Rengan*
Department of Biomedical Engineering
Indian Institute of Technology
Kandi, Sangareddy
Hyderabad
India

*Arpita Roy*
Department of Applied Biology
CSIR-Indian Institute of Chemical Technology
Hyderabad
Telangana State
India
Academy of Scientific and Innovative Research (AcSIR)
Ghaziabad
Uttar Pradesh
India

*D. Sakthi Kumar*
Bio-Nano Electronics Research Center
Graduate School of Interdisciplinary New Science
Toyo University
Kawagoe
Saitama
Japan

*Rajesh K. Sani*
Department of Chemical and Biological Engineering
South Dakota School of Mines and Technology
Rapid City
SD
USA

*S. Sevinç Şengör*
Bobby B. Lyle School of Engineering
Southern Methodist University
Dallas
TX
USA
Department of Environmental Engineering
Middle East Technical University
Ankara, Turkey

***Rohit Kumar Singh***
School of Biotechnology
KIIT University
Bhubaneswar
Odisha
India

***Surya Prakash Singh***
Department of Biomedical Engineering
Indian Institute of Technology
Kandi, Sangareddy
Hyderabad
India

Academy of Scientific and Innovative Research (AcSIR)
Ghaziabad
Uttar Pradesh
India

***T. K. Vasudha***
Department of Biotechnology
Bhupat and Jyoti Mehta School of Biosciences
Indian Institute of Technology Madras
Chennai
India

# 1

# Shape- and Size-Dependent Antibacterial Activity of Nanomaterials

*Senthilguru Kulanthaivel and Prashant Mishra*

*Department of Biochemical Engineering and Biotechnology, Indian Institute of Technology Delhi, New Delhi, India*

## Objectives

- To understand the importance of studying nanomaterials-bio interface and the factors that dictate it.
- To understand the current situation of antibiotic resistance of microbes and their effect on worldwide healthcare.
- To know the strategies to overcome the antibiotic resistance mechanism and role of nanomaterials in combatting it.
- To understand the mechanism of action of antimicrobial nanomaterials and the factors that influence their antimicrobial properties.
- To understand the key role of size and shape of the nanomaterials on the antimicrobial properties of nanomaterials.

## 1.1 Introduction

Over the past three decades "Nanotechnology" has emerged as a promising strategy to overcome impasses that have accumulated in various fields of science and technology (Albanese, Tang, & Chan, 2012). Nanomaterials (NMs) are defined as minuscule structures having at least one of their dimensions equal to or between 1 and 100 nm. Since there is no single universally accepted definition, so far various organizations have given their own definition to the term "NMs" (Boverhof et al., 2015). US Food and Drug Administration (USFDA) defines nanomaterial as "materials that have at least one dimension in the range of approximately 1–100 nm

*Microbial Interactions at Nanobiotechnology Interfaces: Molecular Mechanisms and Applications*, First Edition. Edited by R. Navanietha Krishnaraj and Rajesh K. Sani.

and exhibit dimension dependent phenomena" (Bleeker et al., 2012). As per Environmental Protection Agency (EPA) definition, "NMs can exhibit unique properties dissimilar than the equivalent chemical compound in a larger dimension." Similarly, the International Organization for Standardization (ISO) defines it as "material with any external nanoscale dimension or having internal nanoscale surface structure." EU Commission has described it as "a manufactured or natural material that possesses unbound, aggregated or agglomerated particles where external dimensions are between 1–100 nm size range" (Jeevanandam et al., 2018).

NMs generally exist in the shape of spheres, cubes, rods, tubes, flowers, and platelets (Machado et al., 2015). NMs in the nanoscale dimensions possess a high surface area-to-volume ratio and also a high number of atoms/molecules present on the surface rather than the interior of the materials. These are the properties that majorly contribute to the unique functionality of the nanoscale materials, which vary from the bulk of the same material. Therefore, modulation in their structural properties such as change of size or shape will significantly affect their optical, electrical, magnetic, and biological activity (He et al., 2010; Machado et al., 2015). This is one of the distinct advantages of nanotechnology where by engineering the design or production parameters we can modulate the functionality of the NMs specific to particular application (Machado et al., 2015). Hence, the recent studies in nanotechnology are majorly focused on understanding the effect of physicochemical properties such as size, shape, and surface chemistry of the material on the optical, electrical, magnetic, and biological activities.

Considering the aforementioned advantages, NMs have found enormous applications in various fields and products such as cosmetics, catalysts, fillers, biomedical devices, and semiconductors. As per data report from 2015, approximately more than 1800 products from 622 companies in 32 countries contain engineered NMs. In summary, 762 (i.e. around 42%) of the total products are used in the health and fitness category where silver is the most predominantly used NM, in almost 435 products, which are around 24% of the total. Further, about 528 products (i.e. 29% of total) contain NMs as liquid suspension where dermal contact is highly possible. Hence, the abundant application of these materials is leading towards a long-term co-existence of such NMs with living systems which may result in adverse toxicological effects to the living bodies. In this context, it is necessary to study the effect of these materials on biological entities such as proteins, DNA, RNA, cell membranes, cell organelles, cells, tissues, and organs. The interactions between the biological system and NMs strongly depend upon the environment and the biophysico-chemical property of the nano-bio interface (Nel et al., 2009). As discussed, size, shape, and surface chemistry are the most important factors that govern the physicochemical properties of the NM that in turn throws light at the nano-bio interface. Understanding the effect of these physicochemical factors and extrapolating them toward the interactions at the nano-bio interface would help us to design or engineer NMs for specific applications with an added advantage of minimal toxicity to living bodies. There are a number of analytical tools to study the interaction of nano-biomolecules/proteins. Among them the most employed are mass spectroscopy, Fourier transform infrared spectroscopy, circular dichroism spectroscopy, Raman spectroscopy, nuclear magnetic resonance, UV–vis spectroscopy, surface plasma resonance, quartz crystal balance, atomic force microscopy, fluorescence correlation spectroscopy, fluorescence spectroscopy, and isothermal calorimetry (Saptarshi, Duschl, & Lopata, 2013). Among the different strategies, mass spectroscopy-based proteomics is the most preferred. Even though the technique is a qualitative measure of proteins bound to NMs, such as nanoparticles (NPs), it can be applied over a wide range of NMs. UV–vis spectroscopy is employed to measure changes observed in the adsorption spectrum caused by NM–protein interaction. Similar to UV–vis spectroscopy, fluorescence spectroscopy is employed to measure changes in the fluorescence spectra caused by the binding of protein on to NMs. Surface plasma resonance is used to study changes in electrons' oscillation on the surface of metal NMs as a result of protein interaction with NM. Isothermal calorimetry analysis is employed to determine the binding constant and other thermodynamic parameters of the nano/bio interface (Saptarshi et al., 2013). Quartz crystal balance is used to measure changes in mass on the surface of oscillating quartz caused by the NM–protein interaction. In a study, adsorption of proteins myoglobin, bovine serum albumin, and cytochrome over the surface of gold NPs was studied using quartz crystal balance (Kaufman et al., 2007). Confocal Raman spectroscopy and confocal spectroscopy can be employed to study and visualize NM–protein interaction and intake of NMs into cells by fluorescent labeling of NPs. In recent times, a combination of these techniques has been strategically employed to study the different aspects of NM–protein/biomolecule interaction. NMs can be synthesized through different routes such as chemical, physical, and green methods. Changes in the synthesis methods, concentration of reactants, and conditions can definitely modulate the morphological parameters (size and shape) of NMs. Taking this into account, the selection of synthesis route also plays an important role in governing NMs' morphological features and their functions. Keeping the aforementioned perspectives in mind, this chapter describes the effects of size and shape of NMs on their biological activity.

## 1.2 Synthesis of Nanomaterials

NMs can be synthesized by a number of methods that are grouped into two different categories: (i) bottom-up and (ii) top-down method. Schematics of typical methods for synthesis of NMs are given in Figure 1.1 as described by Ealias and Saravanakumar (2017).

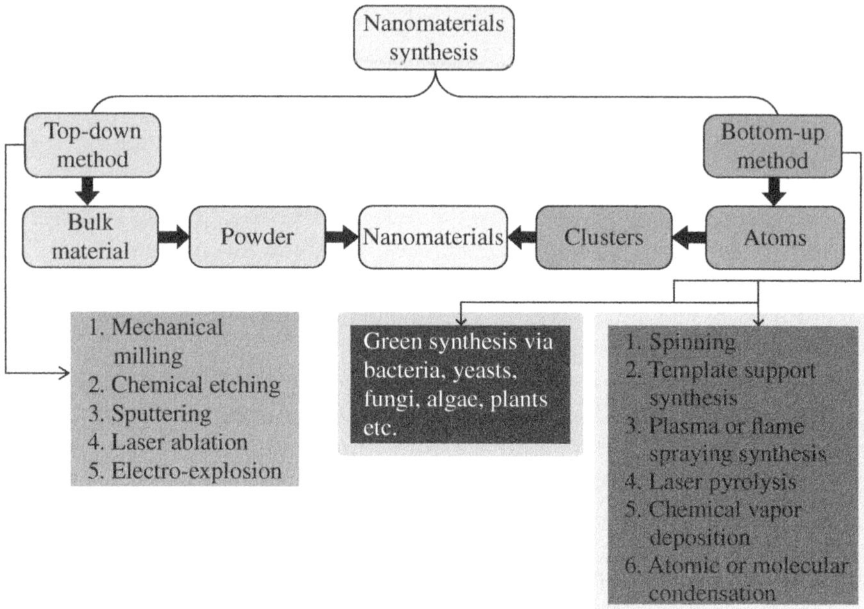


**Figure 1.1** Schematics of typical methods for synthesis of NMs.

i) *Bottom-up method:* It is a constructive method wherein the atoms build up clusters that in turn form the NMs. This category includes methods such as sol–gel, spinning, chemical vapor deposition, pyrolysis, and biosynthesis.
ii) *Top-down method:* On the contrary, the top-down method is a destructive method where the bulk materials are reduced into nanoscale materials. It includes methods such as mechanical milling, nanolithography, laser ablation, sputtering, and thermal decomposition. The typical method of synthesis for various NMs is given in Table 1.1.

## 1.3 Classification of NMs

In nature, NMs are built by nanoscale or submicron-sized blocks that exhibit size-dependent effects. Over the last two decades, a number of NMs or nanostructured materials have been developed and a lot of new developments are underway. The abundant increase in the number of NMs has set forth the need for the classification of these materials. An understanding of the classification would give insight into the interaction of NMs with various surfaces and resultant functionality of the NMs. The first classification of NMs was given by Gleiter (2000) in which the

**Table 1.1** Various approaches for NM synthesis (Ealias & Saravanakumar, 2017).

| S. No | Category | Method | NMs |
|---|---|---|---|
| 1 | Bottom-up | • Sol–gel | • Metal and metal oxide and carbon NMs |
| | | • Spinning | • Organic polymers |
| | | • Chemical vapor deposition | • Carbon and metal NMs |
| | | • Pyrolysis | • Metal oxide and carbon NMs |
| | | • Biosynthesis | • Metal and organic polymer NMs |
| 2 | Top-down | • Mechanical milling | • Metal, metal oxide, and polymeric NMs |
| | | • Nanolithography | • Metal NMs |
| | | • Laser ablation | • Carbon and metal oxide NMs |
| | | • Sputtering | • Metal NMs |
| | | • Thermal decomposition | • Metal oxide and carbon NMs |

materials were classified on the basis of crystalline forms and chemical compositions. The scheme subdivided the materials into three classes where each class has four kinds of materials (Gleiter, 2000). However, the list of Gleiter was not considered complete as it failed to take account of 0D and 1D NMs such as fullerenes and nanotubes into the classification.

### 1.3.1 Classification Based on Dimensions

Later, Pokropivny and Skorokhod (2007) proposed a new scheme of NM classification where the dimensionality (shape and size or form) of the NMs was considered as a primary criterion. In general, nanostructures are structures with at least one dimension $d$ equal to or less than 100 nm, which is considered as $d^*$. The value $d^*$ is always dictated by physical phenomena such as path length of phonons and electrons, diffusional length, length of de Broglie wave, penetration length, and correction length. According to the scheme, NMs were classified into four major categories: 0D, 1D, 2D, and 3D (Pokropivny & Skorokhod, 2007).

#### 1.3.1.1 Zero-Dimensional NMs

Zero-dimensional NMs are defined as materials where all the three dimensions are confined to the nanoscale (1–100 nm). The same definition could also be stated on basis of the movement of electrons along the dimensions of the NMs. Zero-dimensional materials are materials where the electrons are merely entrapped in a dimensionless space without any possible movement (Jeevanandam et al., 2018). The best examples for 0D NMs are NPs and quantum dots. Over the past decades, 0D materials have gained a lot of interest where a number of methods have been designed to fabricate 0D NMs with precise dimensions. 0D NMs can be crystalline or amorphous in nature. They may be mono- or polycrystalline, single- or multi-element, and may exist in various forms (shapes and sizes). These materials have found application in a number of fields such as solar cells (Lee et al., 2009), light-emitting diodes (Stouwdam & Janssen, 2008), single-electron transistors (Mokerov et al., 2001), lasers, therapeutics and diagnosis (Azzazy, Mansour, & Kazmierczak, 2007).

#### 1.3.1.2 One-Dimensional NMs

One-dimensional NM are the materials where one of the dimensions is in macroscale with other two dimensions confined to the nanoscale (<100 nm) (Xia et al., 2003). Herein, the electrons can move across one axis freely whereas they entrapped in other two dimensions of the NMs (Jeevanandam et al., 2018). These 1D NMs are ideal choice for studying the dimension-dependent activity of the materials. Similar to 0D NMs they also can be amorphous or crystalline, mono- or polycrystalline, ceramic, polymeric or composite materials of different shapes and sizes. 1D materials such as nanotubes, nanowires, and nanofibers have attracted a lot of interest in the development of hierarchal nanostructures such as nanofilms, nanosheets, and nanoribbons with profound applications in the field of optoelectronics and nanoelectronics (Cui et al., 2001; Kong et al., 2000).

#### 1.3.1.3 Two-Dimensional NMs

Materials with one of the dimensions in the nanoscale (≤100 nm) and the other two dimensions in macroscale are called 2D materials. Here the electrons are confined in one direction whereas they can move across in other two axes freely (Jeevanandam et al., 2018). Similar to 0D and 1D, 2D materials can also be amorphous or crystalline, poly or monocrystalline, single- or multi-element, which also exist in different forms. 2D materials such as nanosheets, nanofilms, and nanoribbons have shown promising applications in the fields of optoelectronics, sensors, and biomedicine (Weaver et al., 2014).

#### 1.3.1.4 Three-Dimensional NMs

Herein, the materials have all the three dimensions in macroscale but are comprised of uniformly distributed nanometer-sized grains. Hence, the movement of the electrons can be free across all the three dimensions without any confinement (Jeevanandam et al., 2018). 3D NMs also called bulk NMs are widely used in catalysis, electrodes, and magnetic materials. Nano balls, nano coils, and nanoflowers are typical 3D NMs that have high surface area and can provide maximum adsorption sites for all the molecules in a small-area framework (Shen et al., 2008).

### 1.3.2 Classification Based on Chemical Compositions

Similar to dimension, the composition of NMs also plays a vital role in deciding their activities and application. On the basis of composition, NMs are classified into four subcategories, namely: (i) carbon-based NMs, (ii) organic NMs, (iii) inorganic NMs, and (iv) composite NMs.

#### 1.3.2.1 Carbon-Based NMs

The NMs with carbon atoms as their backbone are called carbon-based NMs. They can exist in different forms such as 0D (fullerenes), 1D (carbon nanotubes), 2D (graphene sheets), and 3D (diamond crystal and graphite). General methods to prepare these NM include chemical vapor deposition, arc discharge, and laser ablation. Carbon-based NMs exist in different forms with multiple shapes such as hollow spheres, nanotubes, and ellipsoids (Jeevanandam et al., 2018). Fullerenes are carbon materials with spherical morphology where the carbon atoms are held by $sp^2$ hybridization. A unique advantage of the fullerenes is their high symmetric property (Astefanei, Núñez, & Galceran, 2015). In general, fullerenes contains 28–980 carbon atoms where the diameter of single layer is up to 8.2 nm and for multilayered fullerenes it is about 4–36 nm (Ealias & Saravanakumar, 2017). Carbon nanotubes are 1D carbon NMs where carbon atoms are wound up to form hollow cylinders, which can also be described as an extension of fullerenes or buckyball. Carbon nanotubes can be single-walled, double- or multi-walled with thickness varying from 0.7 nm for single-walled to 100 nm for multi-walled CNTs. The length of CNTs generally varies from few micrometers to several millimeters (Ealias & Saravanakumar, 2017). CNTs have been exploited in various fields owing to their versatile properties such as elasticity, strength, rigidity, field emission, and electrical conductivity (Saeed & Khan, 2014, 2016). Graphene is one of the 2D carbon-based materials formed by $sp^2$ hybridized carbon atoms. It is a hexagonal network of carbon atoms with honeycomb atomic structure that is confined to a two-dimensional planar surface. Graphene elucidates commendable physical, chemical, optical, and mechanical properties owing to their unique honeycomb atomic structure. These unique properties altogether make them remarkable materials that are extensively applied in the fields of electronics, optics, storage, thermal applications, photovoltaics, and composite materials (Goenka, Sant, & Sant, 2014; Pumera, 2010).

#### 1.3.2.2 Organic-Based NMs

The NMs formed from proteins, lipids, carbohydrates, and other organic substances are termed as organic-based NMs, which are generally 10 nm to 1 μm in size. Commonly exploited organic NMs are dendrimers, liposomes, micelles, and polymeric NPs. The superior advantage of these systems over the other NM systems is because of their biodegradable and nontoxic nature. Some of these materials like micelles, liposomes, and polymeric NPs have hollow core also called as nanocapsules, which are being exploited for loading and delivery of drug molecules (Biswas et al., 2013; Tiwari, Behari, & Sen, 2008). Apart from that, these materials are sensitive to heat and light, which can be used as platform for responsive and targeted drug delivery system. Similarly, the surface of dendrimers has many chain ends that can also be engineered for specific chemical functions and targeted delivery. Owing to the aforementioned properties coupled with their structural stability, structural integrity, and controlled release profile, organic NMs have emerged as a promising drug delivery system (Wei et al., 2015).

#### 1.3.2.3 Inorganic-Based NMs

The NMs that are based on metal, metal oxide, and ceramic are called inorganic NMs.

##### *1.3.2.3.1 Metal-Based NMs*

Nanometer-sized particles that are synthesized from the metal either by constructive or destructive routes are metal-based NMs. Most of the metals can be synthesized in form of NMs (Salavati-Niasari, Davar, & Mir, 2008); however, the most extensively studied metal-based NMs include cadmium, aluminum, silver (Kim et al., 2007), iron, gold (Sun & Xia, 2002), copper (Ramyadevi et al., 2012; Ruparelia et al., 2008), and lead-based NMs. The size of these materials varies from 10 to 100 nm with high surface area-to-volume ratio, unique surface charge, and pore size. Further, they can be either amorphous or crystalline, which can exist in different sizes and shapes such as spheres, and cylinders.

##### *1.3.2.3.2 Metal Oxide-Based NMs*

Metal-based NMs are sensitive to environmental factors such as heat, sunlight, moisture, and air. In order to overcome the demerits of metal NMs, metal oxide-based NMs were synthesized. One of the most common examples of metal oxide NPs are iron oxide NPs, which are synthesized from the oxidation of iron particles at room temperature. The metal oxide NPs are preferred over metal NPs due to their increased reactivity and efficiency (Tai et al., 2007). Routinely employed metal oxides include: magnetite (Sun & Zeng, 2002), iron oxide, aluminum oxide (Mukherjee et al., 2011), silicon dioxide, titanium dioxide, zinc oxide (Sharma, Jandaik, Kumar, Chitkara, & Sandhu, 2016), cerium oxide, and copper oxide (Ren et al., 2009).

##### *1.3.2.3.3 Composite-Based NMs*

In general, composite materials are described as materials with two or more different materials combined to blend the properties of all the constituent materials. In the same way, the composite-based NM is a multiphase material with at least one of the dimensions in nanoscale, which is obtained by combining one NM with other or blending one NM with bulk or larger material to form a NM (Vollath, 2013). Bones and the eggshells are the best examples of naturally occurring composite NMs. These NMs generally possess highly improved physical, chemical, mechanical, and biological properties in comparison to their constituent materials. Nanocomposite materials can be of different combinations such as metal/metal, metal/ceramic, carbon/carbon, and ceramic/ceramic (Jeevanandam et al., 2018).

### 1.3.3 Classification Based on Origin

Based on their origin, NMs are classified as natural and synthetic NMs. The NMs that are produced by biological species or anthropogenic activities in nature without human intervention are called natural NMs. The NMs formed in nature are present throughout earth's atmosphere, hydrosphere, and lithosphere. This may include the NMs present in whole troposphere, oceans, sea, rivers, lake, groundwater, rocks, lava, soils, even microorganism, and higher organisms (Hochella, Spencer, & Jones, 2015; Sharma et al., 2015). Synthetic NMs are the NMs that are synthesized through physical, chemical, biological, or hybrid methods besides the materials that are produced from engine exhaust, smoke, and mechanical grinding (Wagner et al., 2014). Even though synthetic NMs are more advantageous as aforementioned, the major problem is predicting the fate and behavior of the materials in the environment. Currently there are a lot of strategies to perform the risk assessment of the synthetic NP in various environmental conditions. Still extrapolating the behavior of synthetic NMs from existing knowledge is a major challenge.

## 1.4 Application of NMs

NMs have found broad applications in various fields such as nanofluids, medical sectors, in cutting tools, automotive sector, wear, and corrosion-resistant coatings. The engineering of NMs to form lighter, as well as extremely stronger materials has found its application in making hard and strong surface coating over material as resistive coatings, faster acting switches, medicines, storage devices with enhanced storage capacity and in building materials.

### 1.4.1 Advanced Application of NMs as Antimicrobial Agents

Apart from the above applications, NMs have been employed in the medical field as both theragnostic and diagnostic agents. Gold NPs are well-known for their application in the medical field whereas silver NPs have found applications as antimicrobial materials. Apart from silver, in recent times, a number of NMs have gained attention as antimicrobial agents in the healthcare sector because of the development of resistant bacteria caused by the uncontrolled usage of antibiotics. Antimicrobial-resistance is considered one of the critical issues that need an immediate solution. In this regard NMs with unique features and specific functionality have gained interest in combating antimicrobial resistance. In the following sections, we will briefly explain bacterial resistance and the role of NMs in bacterial resistance.

## 1.5 Bacterial Resistance to Antibiotics

The most serious threat to public health are infectious diseases and mortalities that have resulted from chronic infections. The common causative agents for most infectious diseases are bacteria. Before the discovery of antibiotics, the old treatment modalities involved the use of synthetic compounds such as sulfa drugs, quinolones, and salvarsan as chemotherapeutic agents (Aminov, 2010). Later on, in the twentieth century antibiotics emerged as wonder drugs. However, the wild use of antibiotics with uncontrolled measures led to the emergence of antibiotic-resistant pathogens and the foremost dangerous multidrug-resistant strains.

The first antibiotic resistance was reported with the enzyme called penicillinase produced from pathogenic *Escherichia coli* (Abraham & Chain, 1940). In nature, the organism that produces antibiotics has self-resistance against its own antibiotic. Most of them have more than one simultaneous mechanism to protect the cells completely from their own bioactive molecules. The most common mechanism of self-resistance involves antibiotic modification or degradation, antibiotic efflux, antibiotic sequestration, and target modification. In the producer organisms, the genetic code for the self-resistances are clustered with the antibiotic synthesis gene and hence their expression is co-regulated. The widespread use of antibiotics and coexistence of antibiotic producer organism with nonproducers led to the origin of antibiotic resistance (Kaur & Peterson, 2018). Since NMs have shown potential to deal with antibiotic resistance, a brief discussion on the mechanism of antibiotic resistance is included in this section.

### 1.5.1 Mechanism of Antibiotic Resistance

The mechanism of bacterial antibiotic resistance can be categorized into intrinsic and extrinsic. The antibiotic resistance mechanism that fixed in the genetic core of the organisms is an intrinsic mechanism encoded in chromosomes. This may include the enzyme system which inactivates antibiotics, nonspecific efflux pump systems, and permeability barrier mechanisms (Cox & Wright, 2013; Fajardo et al., 2008). AcrAB/TolC efflux pump in *E. coli* is one of the well-studied intrinsic resistance systems. These efflux systems are generally very nonspecific and help in exporting different antibiotics, detergents, dyes, and disinfectants (Nikaido & Takatsuka, 2009). Similarly, a mechanism involving permeability barrier in *E. coli* and other Gram-negative bacteria for vancomycin is also an intrinsic resistance system where the outer membrane acts as a permeability barrier (Arthur & Courvalin, 1993). On the other hand, the resistance system that is obtained from other organisms such as producers by horizontal gene transfer is called the acquired resistance system. Unlike the intrinsic resistance system, the resistance elements of the acquired systems are generally embedded in plasmids and transposons. Acquired resistance system includes the plasmid-encoded specific efflux pumps and enzymes that can alter or modify the antibiotics or the target of antibiotics (Bismuth et al., 1990).

According to Wang, Hu, and Shao (2017), the resistance mechanism can be categorized into different subdivisions on the basis of the biochemistry at the protein level target alterations, passive or inactive enzyme generation, active efflux pumps, permeability barrier, biofilm formation, elimination and emergence of certain specific protein. It has been noted that in the same bacterium there may exist two or more simultaneous mechanisms from the aforementioned categories as resistance mechanism such as antagonist induction through metabolic pathway and production of competitive inhibitor to counteract the antibiotics. In general, the molecular mechanisms of antibiotic resistance are divided into three types: (i) antibiotic modification, (ii) antibiotic efflux, and (iii) target modification or bypass or protection mechanisms (Wang et al., 2017).

#### 1.5.1.1 Antibiotics Modification

Antibiotics modification is the common resistance mechanism of pathogenic bacteria against antibiotics of aminoglycosides class. So far, multiple types of aminoglycosides modifying enzymes (AMEs) have been identified in both Gram-negative and Gram-positive bacteria (Ramirez and Tolmasky, 2010; Schwarz et al., 2004). The genetic code for these systems is embedded in the mobile genetic elements (MGEs) of pathogenic or resistant bacteria (Ramirez & Tolmasky, 2010). The chromosomal determinants of the aminoglycosides modifying enzymes have been found in the large number of bacteria present in the environment such as *Acinetobacter* and *Providencia*. These chromosomal determinants are the sources from where pathogenic strains acquired the genetic codes onto their mobile genetic elements (Schwarz et al., 2004). A well-known AME is the *N*-acetyl transferase, which acetylates the aminoglycosides. Apart from AMEs, chloramphenicol acetyltransferase (CAT) and antibiotic hydrolyzing enzyme β-lactamases belong to the same group of enzymes that acts on the antibiotics and modifies them (Martinez, 2018; Schwarz et al., 2004).

#### 1.5.1.2 Antibiotic Efflux

The second most common mechanism of antibiotic resistance is antibiotic efflux and permeability barrier. As we discussed earlier, the permeability barrier mechanism is mostly availed by the greatest number of Gram-negative bacteria. The presence of an extra outer membrane in Gram-negative bacteria exhibits a barrier against hydrophilic antimicrobial agents and antibiotics such as vancomycin. However, a mutation in the genes related to outer membrane such as porin or even change in their expression level makes them vulnerable to hydrophilic antibiotics (Li et al., 2012).

The antibiotic efflux pumps in bacteria are categorized into five different families: ATP-binding cassette (ABC), major facilitator superfamily (MFS), resistance–nodulation–division (RND), small multidrug resistance (SMR), and multidrug and toxin extrusion (MATE) (Sun, Deng, & Yan, 2014). Among these, only ABC family proteins use ATP as an energy source for efflux whereas the rest couple the export of their substrate with ion gradients. The acquired determinants of the efflux system are generally located on the plasmids in the pathogenic bacteria such as Tet genes. At least 22 genes have been identified in both Gram-positive and Gram-negative bacteria (Roberts, 2005). In pathogenic bacteria, the resistance–nodulation–division (RND) pump systems are operative synergistically with the Tet pump systems. The simple Tet protein effluxes the tetracycline into periplasm where RND captures and exports it outside. This is the plausible reason for increase in the minimum inhibitory concentration of tetracycline against pathogenic bacteria (Lee et al., 2000).

#### 1.5.1.3 Target Modification or Bypass or Protection

The resistant mechanism involving target modification and protection has been also observed in various clinically resistant strains of bacteria. A typical example of target modification is found in methicillin-resistant *Staphylococcus aureus* (MRSA) strains. In MRSA strains, the resistance mechanism to β-lactams is conferred by the exogenous penicillin-binding protein (PBP) called PBP2a. The acquired PBP2a is devoid of trans-glycosylase activity; hence, it acts along with native PBP2 to confer the β-lactams. PBP2a coded by mecA gene is located in large mobile genetic elements called staphylococcal chromosomal cassette (Fishovitz et al., 2014; Liu et al., 2016). Another example of target modification is vancomycin resistance in enterococci. The acquired vancomycin resistance genes called "van gene clusters" are located on the mobile genetic elements. Among the different types of van clusters, vanA and vanB are the most effective resistant systems as they are found in critical clinical strains (Miller, Munita, & Arias, 2014).

## 1.6 Microbial Resistance: Role of NMs

Every year millions of people suffer from severe illness due to antibacterial resistance and the mortality rate is expected to reach 10 million by 2050 (Gupta et al., 2019). For patients having such infections, antibiotic therapy is the most common treatment used with removal of infected tissue in few cases. There are many disadvantages to such treatments. Along with the high cost, these treatments make the condition worse by increasing antibiotic tolerance of the surviving bacteria. Therefore, it becomes important to find new ways to tackle such problem. Recent advancement in the field of nanotechnology provides new prospects to develop novel NMs with antimicrobial properties. Various NMs have been synthesized to defend against multidrug resistance (MDR) and microbial resistance. NMs have the ability to act not only against bacteria but also as carrier for synthetic and natural antimicrobial compounds. Different NMs have different mechanisms to combat bacteria.

### 1.6.1 Overcoming the Existing Antibiotic Resistance Mechanisms

Most kinds of NMs can defeat a minimum of one among the prevalent resistance mechanisms mentioned in Section 1.5.1. These impacts are the consequences of NMs' bactericidal mode, which is based on their physicochemical properties. Unlike traditional antibiotics, the characteristic sizes of NPs are 1–100 nm, which gives them novel properties like better interactions with cells (Huh & Kwon, 2011). How the NMs interact with the cell barriers and disrupt the bacterial cell membrane is discussed in this section. Since few mutations cannot change bacterial cell membrane, it further reduces the chance of drug resistance.

Not only bacterial membrane but the hindrance of biofilm formation is also an important mechanism as biofilms develop bacterial resistance by providing shelter to microorganisms, thus escaping most of the antibiotics (Peulen & Wilkinson, 2011); also, they are breeding grounds for frequent resistant mutations (Khameneh et al., 2016). NMs play a vital role in the prevention of this biofilm formation and the size of these NMs determines the level of their effectiveness in the destruction of these biofilms.

#### 1.6.1.1 Combating Microbes Using Multiple Mechanisms Simultaneously

The simple mechanism of action of traditional antibiotics is the main reason why bacterial resistance occurred in the first place. On contrary, NMs have different action mechanisms and can be designed to have multiple mechanisms that act simultaneously against microbes. Hence, it becomes difficult for microbes to develop resistance against NMs as it is unlikely to have many mutated genes.

#### 1.6.1.2 Acting as Good Carriers of Antibiotics

As mentioned above, NMs can also act as a carrier for various antibiotics. The delivery mechanism of NMs is different from that of other available drug delivery systems. Various NMs that have attracted attention and are currently in common use for drug delivery are polymeric NMs, metallic NMs, ceramic-based NMs, micelles, liposomal NMs (Daeihamed et al., 2017), and carbon-based NMs (Ranghar et al., 2014). For being an efficient antibiotics delivery system NMs possess following advantages:

i) *Size:* The tunable ultrasmall size of NMs makes them a suitable delivery system as they can act against intracellular bacteria. Antibiotics have poor membrane transport, which makes it difficult for them to kill intracellular and drug-resistant microbes. On the other hand, drug-loaded NMs can easily pass cell membranes and act. Further, NMs can also enter host cells by phagocytosis and get released inside by endocytosis (Andrade et al., 2013).
ii) *Protection:* When a direct drug goes inside the body, the chemical present inside could deteriorate drug molecules or microbes could develop resistance against them. Also, uptake of antibiotics in bacterial cells is very slow and less. NMs-based carriers maintain the drug potency and protect it from resistance by microbes. The uptake rate of NMs can be manipulated as needed. For example, in gastrointestinal (GI) tract, dendrimers inhibit glycoprotein-mediated efflux of drug (Liu, Tee, & Chiu, 2015).
iii) *Precision and security:* Side effects of antibiotics and their targeted delivery is an important concern, which is difficult to achieve with conventional drug delivery. The targeted delivery of antibiotics to the infection site minimizes systemic side effects. NM-based drug delivery helps in delivering the drug to targeted site and reduces the risk of side effects. With this, higher dose can be applied to the site of infection directly. This targeted drug delivery can be either active or passive. In active targeting, the NM's surface is modified to selectively recognize the signals on the target infected site, whereas, in passive targeting, permeation, and retention of drug-loaded NMs is increased at the infection site. Drugs like vancomycin that have kidney toxicity but are good for Gram-positive infections can be delivered to the desired location by loading in mesoporous silica NMs (Qi et al., 2013). The prerequisite for effective target therapy is to target macrophages with NMs, make macrophages active, and then release the drug (Xiong et al., 2012).
iv) *Controllability:* As mentioned above, the controlled release of any drug is crucial for its action mechanism. Conventional delivery methods failed to maintain controlled and sustained release of many drugs. This resulted in either high drug levels for short time periods or very low drug levels. Thus, repeated dosage is given, which has its own side effects. NMs' ability to slowly release the drugs at therapeutic concentration results in reduced frequency of dosing and pain. The prolonged drug release from NM-based delivery systems provides better inhibitory effects on microbial growth (Liu, Zhang, Li, Yang, Pan, Kong, & Zhang, 2016).

Further, the drug release from NMs can also be controlled by making them responsive to various stimulatory factors like temperature, pH, light, chemicals, or magnetic field (Lim, Chung, & Chung, 2018; Wu et al., 2016). Similar systems are already being explored for better controlled drug release. For example, drugs like levofloxacin are delivered using solid–lipid NMs, which prolong the retention for ocular applications (Baig et al., 2016).

v) *Combination:* Another important advantage of NM-based drug delivery is its ability to deliver multiple drugs at the same time through the same channel. In many cases, the targeted site of infection has multiple microbes present, requiring the delivery of specific drugs for each type. In such cases, multiple drugs can be packed and delivered using same NMs. Also, for single-cell type, it is difficult to develop resistance to multiple antibiotics that are delivered through same NM at the same time. Multiple action mechanisms make this kind of system more efficient.

On the other hand, two or more types of NMs can also be combined to overcome the disadvantages of a single type of NM. The much explored liposome-based drug delivery system has a short shelf life, less stability, and low encapsulation efficacy. This system can be combined with other delivery systems like solid lipid NMs to obtain hybrid NM with improved properties. Another example of a hybrid NM delivery system are lipid–polymer NMs with better efficacy (Hadinoto, Sundaresan, & Cheow, 2013).

## 1.7 Antibacterial Application of NMs

Generally, NMs have been used as an antimicrobial coating material for implantable medical devices such as heart valves, dental implants, and catheters (intravenous catheters or neurosurgical catheters). One of the best examples is the coating of drug-loaded mesoporous $TiO_2$ on the implants, which inhibits the growth and even the adhesion of *E. coli* (Della Valle et al., 2012; Xia et al., 2012). The skin is part of the primary immune defense system in our body where trauma or burns or chronic ulcers damage skin and compromise its functions. Common microbial infections associated with skin wounds are *Staphylococcus, Streptococcus, E. coli,* and *Klebsiella* species. Here, the infections involve multiple bacteria including antibiotic-resistant bacteria where the antimicrobial agent needs to have a broad spectrum of antimicrobial property. In this regard, NMs have been exploited in the preparation of wound dressing materials. Silver NPs along with polyvinyl alcohol and chitosan have been used in the synthesis of fiber mat for wound healing. Nanosilver with high antimicrobial property significantly inhibited the growth of bacteria and along with mat fiber enhanced the wound healing rate

(Li et al., 2013). In addition, NMs with antimicrobial properties have been also used in bone and dental implants. Bone cement containing polymethyl methacrylate (PMMA) with silver-doped silica glass powder significantly inhibited the formation of biofilm over the implants (Miola et al., 2015). $TiO_2$ mixed with prosthesis retarded bacterial growth and prevented biofilm formation over light exposure (Aboelzahab et al., 2012). Among various NMs, the most commonly exploited antibacterial NMs include nanometals, metal oxides, carbonaceous materials, and cationic polymers.

### 1.7.1 Nanometals

Metals over a threshold concentration exhibit toxicity on the biological system. Among them, heavy metals such as arsenic, cadmium, chromium, lead, and mercury are highly toxic to the living system (Tchounwou et al., 2012). The property of metal toxicity has been exploited to prepare antibacterial nanometals that are toxic to bacterial cells. An additional advantage of nanometals is the possibility of tuning the properties of the system during the synthesis phase.

Metal NMs include silver NPs, copper NPs, and iron NPs. Silver NPs are one of the well-known antimicrobial materials that have been used in various healthcare, industrial, and medicinal sectors, their mechanism of action being the toxic effect of $Ag^+$ ions exerted by interacting with sulfhydryl groups in proteins. Further, silver ions also inhibit DNA replication and also affect the cell membrane integrity, thus compromising the permeability (Feng et al., 2000). Wigginton et al. (2010) showed that the Ag NPs have high affinity for tryptophanase protein of *E. coli*. It is evident from the literature that the enzyme tryptophanase is crucial for the production of indole from tryptophan amino acid where indole plays a crucial role behind the multidrug exporters and also acts as a signaling molecule in biofilm formation. It was clearly evident from the study that the binding of Ag NPs deactivated the enzyme which could make the bacteria prone to effect of drugs and other antimicrobial materials (Wigginton et al., 2010). Ag NPs exhibited bactericidal activity by inactivating the enzyme phosphomannose isomerase in bacteria, which converts mannose 6 phosphate to fructose 6 phosphate. This conversion is one of the crucial steps in glycolysis of bacterial sugar metabolism (Dakal et al., 2016; Sundar & Kumar Prajapati, 2012). In another study, Elechiguerra et al. (2005) showed that the interaction of Ag NPs with HIV-1 was not random but based on the structure of virus envelope. The envelope consists of a protruding gp120 glycoprotein connected to intracellular matrix protein p17 through a transmembrane glycoprotein gp41. Importantly, HIV-1 binds to CD4 receptor sites on the host cell using gp120 protein knobs. It was observed in this study that Ag NPs attach to the HIV-1 surface by sulfur-containing residues of the gp120 protein (Elechiguerra et al., 2005). Next to silver NPs, copper NPs have gained attention in the past few years as cost-effective alternatives to the former. Copper NPs also exert a mechanism similar to that of silver. They act by damaging the bacterial cell wall and disrupting the DNA structure (Raffi et al., 2010).

### 1.7.2 Metal Oxides

Similar to metal NMs, oxides of metal also exhibit antimicrobial activity through the metal ions released from the system. The mechanism of action is the ROS production induced by metal ions or by the accumulation of NPs inside the cells. Most metal oxides are semiconductors in nature with larger bandgap energy. Among the metal oxides, ZnO (3.2 eV) has the highest bandgap energy similar to $TiO_2$, which gets activated at a wavelength of about 390 nm to induce ROS production. CdO has a bandgap energy of about 2.1 eV with activation wavelength at 590 nm (Table 1.2).

Among the metal oxides, the most widely used are $TiO_2$, ZnO, CuO, and MgO for antibacterial activity. A vastly studied metal oxide is $TiO_2$, which is used in various industrial and environmental applications such as in sunscreen preparation, implant coatings, and removal of water and air contaminants. The antimicrobial efficiency of a system depends on the wavelength of light used for activation, the intensity of light, concentration, interaction time, temperature, and target microbes (Markowska-Szczupak, Ulfig, & Morawski, 2011). Efficacy of the $TiO_2$ system depends on the photoinduced generation of ROS, which happens effectively at the oxide anatase phase. When light of energy equal to or higher than the bandgap energy (3.22 eV, anatase phase) is exposed, photocatalytic activation produces an energy-rich electron–hole pair. The electron produced is transferred to reducible species such oxygen to generate free radicals like superoxides $O_2^-$. In a similar way, it induces the production of hydroxyl radicals and hydrogen peroxide in acidic conditions (Kühn et al., 2003). A schematic representation of photoactivated ROS generation and antimicrobial property of NMs is given in Figure 1.2 as described by Gardini et al. (2018)

**Table 1.2** Bandgap energy and activation wavelength for various metal oxide NMs (Gardini et al., 2018).

| S. No | Material | Bandgap energy (eV) | Activation wavelength (nm) |
|---|---|---|---|
| 1 | CdO | 2.1 | 590 |
| 2 | $Fe_2O_3$ | 2.2 | 565 |
| 3 | $WO_3$ | 2.8 | 443 |
| 4 | $TiO_2$ | 3.2 | 387 |
| 5 | ZnO | 3.2 | 390 |

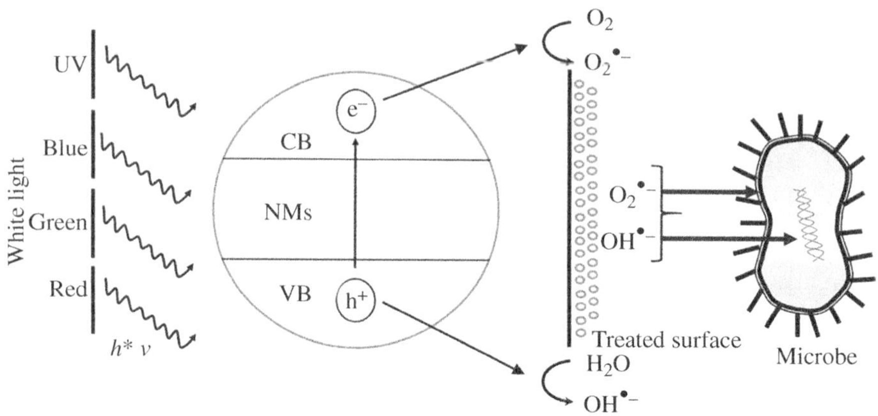


**Figure 1.2** Schematic representation of photoactivated ROS generation and antimicrobial property of NMs.

In a similar way, ZnO, a semiconductor with larger bandgap energy, is applied in coatings, paints, and sunscreens. Upon exposure of UV light, photocatalysis induces the ROS production, which is responsible for its antimicrobial effect. Here, the roughness of the system depends on the surface defects. Efficacy of the ZnO NP system increases with decrease in size where the roughness of the particle along with its high surface area causes the disruption of the microbial cell wall (Padmavathy & Vijayaraghavan, 2008). ZnO NMs have also been reported to interact with some disease target proteins. Chatterjee et al. (2010) studied the effect of ZnO NPs over periplasmic domain structure of ToxR protein of *Vibrio cholerae*. ToxR protein plays a critical role in the regulation of expression of many virulence factors of the bacteria. It was observed that the binding of the protein ToxR to ZnO NPs' surface reduced the stability of protein where it was more susceptible to denaturation. Further, significant change in the structure of the protein was also observed (Chatterjee et al., 2010).

### 1.7.3 Carbonaceous NMs

At the nanoscale, carbon forms different allotropes, which include graphene, carbon nanotubes, fullerenes, and nanodiamonds where each of them exerts inherent unique properties. The antimicrobial effect of carbon NMs such as graphene, graphene oxide, reduced graphene oxide, and carbon nanotubes depends on the level of cell wall disruption and amount of ROS induced in the microbial cells (Liu et al., 2011). Carbon materials hold the advantages of commercial viability and environmental safety in comparison to conventional NPs such as silver and other metal NPs.

### 1.7.4 Cationic Polymer NMs

Polymers with an inherent positive surface charge or added with positively charged moieties are called cationic polymer NMs. The most commonly used cationic polymer in antimicrobial application is chitosan (Rekha Deka, Kumar Sharma, & Kumar, 2015). Chitosan is a cationic polysaccharide derived from chitin by partial deacetylation. The positively charged surface moieties aid in the interaction of NM with negatively charged bacterial cell wall, which causes the rupture of membrane and subsequent cytoplasmic content leakage (Qi et al., 2004). The efficacy of the system relies on the pH, degree of deacetylation, molecular weight, and presence of other substances such as proteins, lipids, and metal ions. The major disadvantage of chitosan-based nanosystems is their poor solubility at neutral pH causing them to precipitate in culture medium.

In general, the efficacy of the antimicrobial NMs depends on the interaction of the material with bacteria and the mechanism of action. Further, the interaction of NMs with bacteria depends on a few crucial factors such as electrostatic attraction, van der Waals forces, receptor–ligand interaction, and hydrophobic interaction. Understanding the basics of the NMs' interaction with the microbial cell is likely to pave way for the design of novel antimicrobial agents with crucial insight into their toxicity and mechanism of action. In the following sections, we discuss the interaction of NMs with microbial cells and the mode of action.

## 1.8 Interaction of NMs with Bacteria

Based on the cell wall structure, bacteria are divided into two categories as Gram-positive bacteria and Gram-negative bacteria. Gram-positive bacteria have a thick peptidoglycan layer ranging from 15 to 100 nm. Further, they contain a phosphate-containing polymeric chain of teichoic acid that is responsible for the negative charge of the bacteria (Neu, 1992). In the case of Gram-negative bacteria, an extra hydrophobic lipid bilayer is present over a thin peptidoglycan layer (20–50 nm). The presence of an extra lipid layer limits the permeability of several hydrophilic antimicrobial agents, which is the reason for the high resistant nature of Gram-negative bacteria (Gupta, Landis, & Rotello, 2016). The negative surface charge of bacteria is due to the lipids and carbohydrates of the lipopolysaccharide layer. Hence, the structure of the bacterial cell wall determines the interaction of bacteria with NMs. Schematics of cell wall structure of Gram-positive and Gram-negative bacteria are given in Figure 1.3 as detailed by Hajipour et al. (2012)

In an earlier study, a homogenous distribution of cetyltrimethylammonium bromide (CTAB) coated gold NPs was observed over the *Bacillus cereus*. This was explained on the basis of electrostatic interaction between negatively charged teichoic

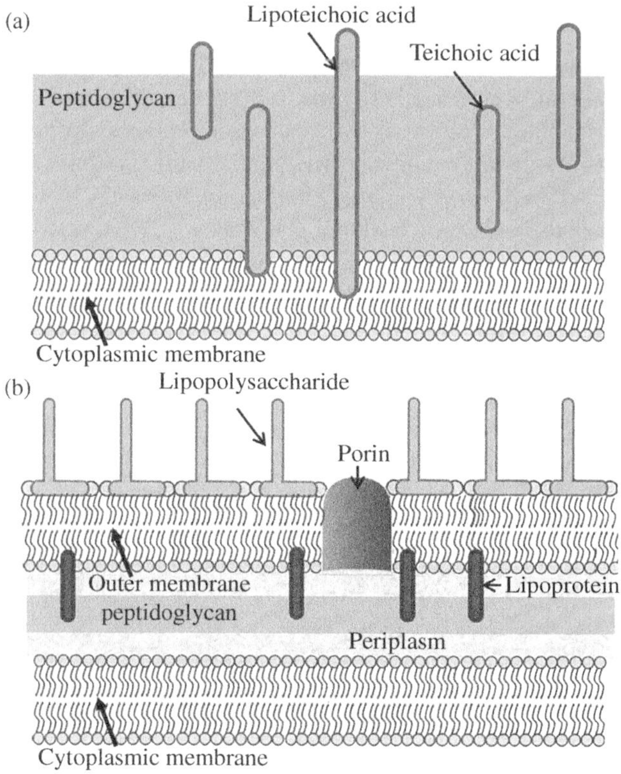


**Figure 1.3** Bacterial cell wall structure of (a) Gram-positive bacteria (b) Gram-negative bacteria.

acid moieties on the bacteria and positively charged gold NPs (Berry et al., 2005). In another study, it was reported that mannose functionalized gold NPs bound to the pili of Gram-negative *E. coli*. It was attributed to the receptor–ligand interaction between mannose and lectin containing pili (Lin et al., 2002). Later, Hayden et al. (2012) suggested that the positively charged NMs exhibited high toxicity over bacteria. This electrostatic interaction might be a plausible reason for the spatialized aggregation of cationic and hydrophobic gold NPs on the negatively charged bacterial membrane (Hayden et al., 2012). The interaction of NPs with the membrane generally effects in membrane blebbing, tubule formation, and other membrane damage.

## 1.9 Antibacterial Mechanism of NMs

NMs exert different mechanisms of action against microbes as represented in Figure 1.4. Most of the time, the antimicrobial systems employ multiple mechanisms to take over the multiple resistance mechanisms developed by microorganisms,

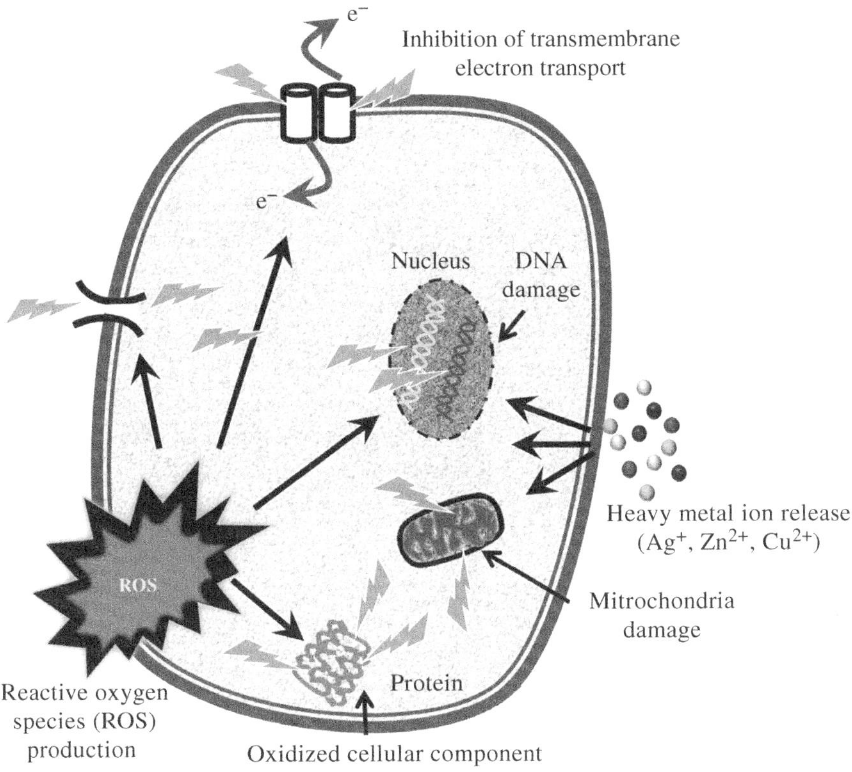


**Figure 1.4** Schematic of the antibacterial mechanism of NMs.

thereby increasing the antimicrobial efficiency. In brief, different mechanisms of action of NMs are given below:

**Cell membrane disruption**: Interaction of the NMs with the surface of the microbes causes mechanical damage to cell wall, which in turn leads to cell wall disruption followed by leakage of the cytoplasmic content and subsequent cell death (Pal, Tak, & Song, 2007).

**DNA damage**: NMs or metal ions from NMs diffuse through the cell wall and effectively interact with DNA and affect or modify the morphology of the DNA. This, in turn, interrupts the duplication or replication of DNA, leading to cell death (Feng et al., 2000).

**Release of metal ions**: Metal ions released from the NM diffuse into the cells and bind to thiol-containing proteins including enzymes and compromise their function. In addition to that, metal ions generally function as cofactors for a number of enzymes. Hence homeostasis of metal ions is very important for the survival of the microorganism. Metal ions released from the NM diffuse in excess into bacterial cells, affecting the homeostasis leading to dysfunction of proteins and enzymes (Feng et al., 2000).

**Interrupted transmembrane electron transport**: Often the interaction of NM with cell wall damages the cell wall and hampers the electron transport chain leading to the disruption of cellular respiration and eventual cell death (Lemire, Harrison, & Turner, 2013).

**Oxidative stress**: Few of the NMs such as metals and metal oxides prominently induce production of ROS, which includes hydroxyl free radicals, superoxides, and hydrogen peroxide. The produced free radicals further oxidize cell wall components causing their disruption and act on the internal components of cells. In few cases, the oxidation continues till the entire cell is oxidized into $CO_2$ and water. In some specific NMs such as carbon NMs, oxidative stress is induced without the involvement of ROS, which acts by electron transfer mechanism (Jacoby et al., 1998). Schematic of the antibacterial mechanism of NMs is given in the Figure 1.4 as described by Hajipour et al. (2012).

## 1.10 Factors Affecting the Antibacterial Activity of NMs

As discussed earlier, the activity of any NM system depends on its physicochemical properties such as size, shape, zeta potential, crystal structure, charge, and other factors. The physicochemical factors' influence on the surface area, surface energy, and atomic ligand deficiency dictates the behavior of NMs, which in turn affects the activity of NMs. Hence it is very important to study the effect of these factors on the activity of the NMs. Schematic of the factors influencing the antimicrobial activity of the NMs is given the Figure 1.5 as detailed by Daima and Bansal (2015).

### 1.10.1 Size

A key advantage of any NM system is its higher surface area-to-volume ratio in comparison to the micro and macro structures. Practically it is possible to contain a significantly high number of smaller NMs in comparison to bigger particles in the same volume. High surface area-to-volume ratio owing to an increased number of particles results in exposure of greater numbers of atoms, which could increase the activity. Considering the case of antibacterial system, the formation of biofilm is a key event in the development of resistant bacteria. Exposure of a greater number of atoms on the surface results in increased interaction of NM surface with bacteria, increasing the number of reactive oxygen species at a faster rate followed by inhibition or elimination of the bacteria. Supporting this fact,

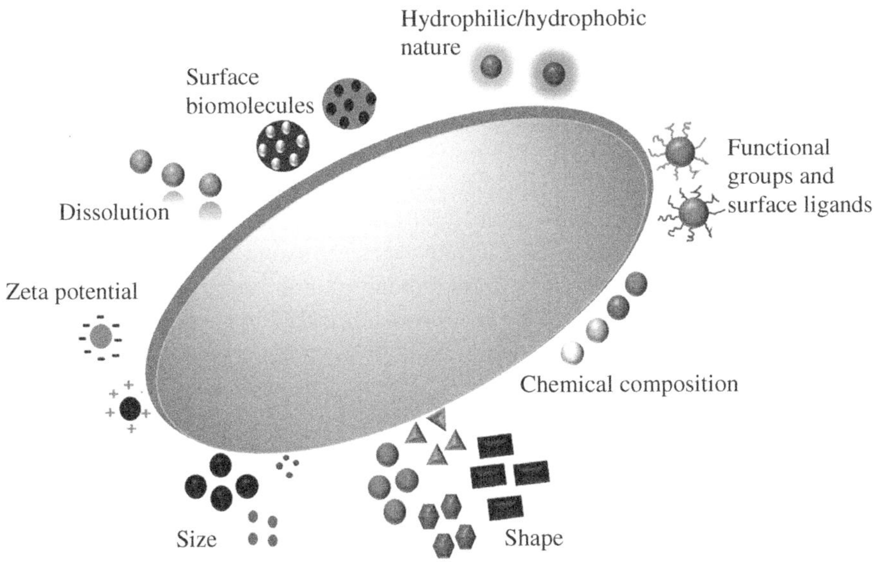


**Figure 1.5** Schematic of the factors influencing the antimicrobial activity of the NMs.

several recent studies have shown that the size of NMs plays a critical role in dictating the antimicrobial property of the NM.

Generally, the smaller the size of NM the higher will be the surface area with a high chance of prolonged interaction with microbial system and high diffusion through the cell membrane in comparison to bigger NMs with smaller surface area (Gurunathan et al., 2014). In the case of silver NPs, the size of the NPs clearly influences the surface area exposure, release rate of silver ions, and the antimicrobial efficacy of the particles. Similarly, ZnO NPs of smaller size (12 nm) had better antimicrobial activity in comparison to larger particles of size 45 nm due to its high cell permeability (Padmavathy & Vijayaraghavan, 2008). In another study involving the $TiO_2$ and silica NPs, the antimicrobial property and the mechanism of action of the system were influenced by the size of titanium nanotubes (Çalışkan et al., 2014). In contrast, a study involving three different sizes of $Mg(OH)_2$, the smallest NPs had the least antibacterial effect (Pan et al., 2013). Thus, it is necessary to consider the effect of other factors also with size in determining the mechanism action of NMs.

### 1.10.2 Shape

Next, to size, shape is also an important factor that affects the structural behavior and antimicrobial activity of the NMs. NMs of different shapes have shown to cause varying degrees of bacterial cell damage, mechanism of action, and antibacterial property (Cha et al., 2015). Hong et al. (2016) studied the antimicrobial property of sphere-, wire-, and cube-shaped silver NPs of the same diameter. Notably, the nanoparticle with cubic shape exhibited the highest antimicrobial effect in comparison to other shapes. This was attributed to the specific facet reactivity and surface area of the cube-shaped NPs (Hong et al., 2016). Another study was done to compare the enzyme inhibition and antimicrobial effects of the sphere-, plate-, and pyramid-shaped nano-ZnO particles. It was reported that the particles of pyramid shape showed a better β-galactosidase enzyme inhibition and greater antimicrobial property. The system was a reversible inhibition system and worked by obstructing the enzyme similar to natural inhibitors unlike by the degradation of enzymes as reported earlier (Cha et al., 2015). In another study involving $Y_2O_3$ NPs, the particles of prismatic shape showed better antibacterial activity against *Pseudomonas desmolyticum* and *S. aureus*. The antimicrobial activity was due to the direct interaction of prismatic-shaped particles with cell membrane followed by the membrane damage and bacterial kill (Prasannakumar et al., 2015).

### 1.10.3 Zeta Potential

In addition to size and shape, zeta potential (surface charge) of the NMs is also known to affect the behavior of NMs. It is clear from the literature that the surface charge of the NMs has a strong influence on the adhesion of bacteria. Since bacterial cell surface is negatively charged, NMs with positive charge exert electrostatic attraction, which helps in the adsorption of bacteria onto the surface. This is the reason behind the enhanced ROS production by positively charged NM in comparison to neutral and negatively charged NMs. However, negatively charged NMs exhibit antimicrobial property at a higher concentration through molecular crowding leading to the interaction of NM surface with bacteria.

Pan et al. (2013) studied the antibacterial activity of $Mg(OH)_2$ prepared using different precursors ($MgCl_2$, MgO, and $MgSO_4$). It has been reported that positively charged $Mg(OH)_2$ NPs prepared with $MgCl_2$ exhibited greater antimicrobial property against *E. coli* in comparison to negatively charged particles prepared with MgO. This was due to electrostatic interaction between the positively charged $Mg(OH)_2$ with the negatively charged bacterial cell membrane resulting in damage to bacterial cell (Pan et al., 2013).

### 1.10.4 Roughness

Although majority of the studies were done by analyzing the effect of size, shape, and surface charge of NMs on their activity, a few studies have been performed to understand the effect of roughness. In one of them, Sukhorukova et al. (2015) prepared silver ion-doped titanium calcium phosphate films. Here, it was found that the surface with high roughness promoted the retention *E. coli* bacteria onto the surface (Sukhorukova et al., 2015). Longer retention time enhanced antibacterial property of the films with higher roughness.

### 1.10.5 Synthesis Methods and Stabilizing Agents

The choice of synthesis methods and stabilizing agents is very crucial in the fabrication of antimicrobial NMs, since these factors also can influence the properties of the NMs to a major extent. As stated, during synthesis, NMs are synthesized by different methods such as laser ablation, mechanical milling, chemical etching, melt mixing, sputtering, and other chemical methods such as thermolysis, microemulsion, and sol–gel. However, the NMs that are synthesized through the chemical or physical methods are unstable, have surface-attached toxic materials, and are formed along with toxic by-products. Considering Ag NPs' synthesis, the process involves a reducing agent such as sodium borohydride or sodium citrate with capping agent such as polyethyl glycol. On the other hand, the biological synthesis methods employ biological sources such as microorganisms and plants. In the case of microbial biosynthesis, the microbes exert a bio-reduction process to reduce and accumulate the metallic ions to avoid the metal-related toxicity. The mechanism involves the reduction of metal ions inside the cell through intracellular reducing species and outside using their different extracellular metabolites. Plants also contain a number of reducing agents such as proteins, flavonoids, and other water-soluble biomolecules (Singh et al., 2018). Green synthesis methods improve the stability of NMs with no hazardous by-products. Further, they provide a biocompatible coating over NMs, which not only improves the biocompatibility but also increases surface area with reactive groups, which can improve the interaction with biological environment (Singh, Garg, Pandit, Mokkapati, & Mijakovic, 2018). For example, Sudhasree et al. (2014) showed that nickel NPs prepared from *Desmodium gangeticum* were monodispersed. The green synthesized NPs were found to possess high antibacterial activity against *Klebsiella pneumonia, Pseudomonas aeruginosa,* and *Proteus vulgaris* whereas the chemically synthesized nickel NPs had the least effect on the same microbes. Apart from enhancing the antimicrobial property, it also improved the biocompatibility as observed from biocompatibility studies using LLC PK1 (epithelial cell lines) (Sudhasree et al., 2014). However, the choice of a particular synthesis method depends on the nature of NM required and the type of applications.

Antimicrobial property is directly proportional to the surface area, surface charge, and the extent of interaction or contact of NMs with bacterial cell. In this regard, understanding the effect of stabilizing agents on NMs is also important. The stabilizing agents reduce agglomeration and provide net charge over NMs. Hence, these stabilizing agents to an extent determine the toxicity or antimicrobial property of the NMs. It was observed from a past study that the Ag NPs stabilized with chitosan and citrate enhanced the antimicrobial property of Ag NPs against multidrug-resistant bacteria (*S. aureus* and *K. pneumonia*). The study also showed that effect of stabilizing agents chitosan and citrate over the microbes was at basal level (Cavassin et al., 2015). Similarly, Ag NPs coated with 11-mercaptoundecanoic acid exhibited higher toxicity over *P. aeruginosa* in comparison to Ag NPs coated with citrate. Citrate capping provides net negative charge to the Ag NPs, which results in electrostatic repulsion between negatively charged bacterial cell and Ag NPs. In the case of 11-mercaptoundecanoic acid capped Ag NPS, the particles agglomerated over the surface of the hydrophilic *P. aeruginosa,* which facilitated the release of $Ag^+$ ions near the proximity of cell and improved the toxicity of the NPs toward the bacteria (Dorobantu et al., 2015). In another study, El Badawy et al. (2010) studied the toxicity of Ag NPs coated with different capping agents such as citrate, polyvinylpyrrolidone, and branched polyethyleneimine. The coating provided Ag NPs with a range of surface charge from highly negative to highly positive. Among the different coatings, citrate coating showed the least toxicity against *Bacillus* species. The surface potential of the citrate capped Ag NPs was found to be −38 mV, which was in line with the surface charge of *Bacillus* species (−37 mV). The electrostatic repulsion between the negatively charged citrate capped Ag NPs and bacteria was the probable reason for the least toxic effect of citrate capped NPs. In accordance with that, highly positively charged branched polyethyleneimine-coated Ag NPs (+40 mV) showed the highest toxicity whereas uncoated Ag NPs (−22 mV) and polyvinylpyrrolidone-coated Ag NPs (−10 mV) exerted toxicity above the citrate capped Ag NPs (El Badawy et al., 2010). This clearly reveals that the nature and the structure of stabilizing agents affect the toxicity and the bactericidal potential of the NMs, which should be taken into account while fabricating NMs for antimicrobial properties.

### 1.10.6 Environmental Conditions

Various environmental factors have shown influence over the antimicrobial property of NMs. Among different environmental conditions, temperature causes significant effect on the antimicrobial property because of its influence on ROS production rate. The stimulation of ZnO NMs with temperature resulted in the generation of electrons at the active site. Later the electrons interact with oxygen to generate ROS, which enhances the antimicrobial efficiency of ZnO NMs (Saliani, Jalal, & Goharshadi, 2015). Next to temperature, pH is another very common factor that has a strong influence over the antimicrobial efficacy of NMs. In the case of ZnO NMs, with decrease in pH the dissolution rate of the NMs increases, thereby increasing its antimicrobial properties. Adding to pH, the osmotic pressure of the medium also influences the behavior of NMs, such as the aggregation, surface charge, and solubility of NMs (Saliani et al., 2015). A study of

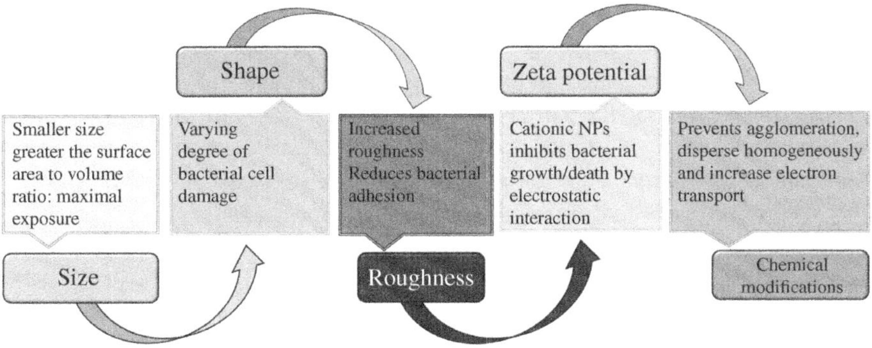


**Figure 1.6** Schematic of the key factors that contribute to the antimicrobial property of NMs.

ZnO in five different mediums suggested that the antimicrobial activity of NMs is mainly caused by the $Zn^{2+}$ ions and the complexes of zinc. Additionally, by supplying the nutrients and other substances, the medium improves the tolerance of bacteria to the NMs (Li, Zhu, & Lin, 2011). Another study showed that stirring conditions of ZnO NPs during the synthesis phase has influence over their antimicrobial property against Gram-positive (*Bacillus subtilis*) bacteria, Gram-negative bacteria (*E. coli*), and fungus (*Candida albicans*) (Khan et al., 2016). Schematic of the key factors that contribute to the antimicrobial property of NMs is given in Figure 1.6 as described by Jagadeeshan and Parsanathan (2019).

## 1.11 Influence of Size on the Antibacterial Activity and Mechanism of Action of Nanomaterials

It is understood from the definition and previous discussions of NMs that size is the predominant factor of any NM, which lies between the atomic and bulk zone of the same composition. Along with size, ion composition and active biomolecules or functional group on the surface also contribute to the interaction. As have discussed, the properties of the material at the nanoscale differ significantly from the bulk material, which affects its interaction with the biological system. With the development of NMs, several new opportunities have emerged due to their reduced size with increased number of particles contributing to the high surface area-to-volume ratio. The reduced size of the NMs may promote the interaction of bacterial cells with the surface of material and cause membrane damage with subsequent bacterial cell death. Similarly, in the case of in vivo; applications, the size of the materials plays a crucial role in the kinetics of adsorption, distribution, metabolism, and excretion of the NMs. In order to study the effect of size on the antibacterial activity and toxicity of the material, several studies have been conducted.

In one of the studies, Raghupathi, Koodali, and Manna (2011) showed that the antibacterial activity of the ZnO NPs varied significantly with the particle size. The authors studied the antibacterial activity of NPs' size ranging from 12 to 307 nm against *S. aureus*. They found that the particles with size more than 100 nm at a concentration of 6 mM were merely acting as bacteriostatic whereas particles of size 12 nm at the same concentration not only limited the growth of the bacteria but also killed them completely. Here, the mechanism of action involved ROS production and the accumulation of nano-sized particles in the cytoplasm of *S. aureus* (Raghupathi et al., 2011).

In another study, size-dependent antimicrobial activity of cobalt ferrite core/shell NPs was demonstrated. Antimicrobial property of three different sizes of NPs (1.65, 5, and 15 nm) was studied against *Saccharomyces cerevisiae* and *Candida parapsilosis*. Notably, against *S. cerevisiae* 1.65 nm exhibited 12 and 25% higher killing than 5 and 15 nm particles, respectively. Similarly, in the case of *C. parapsilosis,* the same trend was reported whereas 1.6 nm particles showed 15 and 44% higher killing efficiency in comparison to 5 and 15 nm particles, respectively. The antimicrobial activity of the cobalt ferrite NPs at size of 7–8 nm was suggested to be due to intracellular diffusion with subsequent interaction with cell membrane causing oxidative stress and finally DNA damage. It was also suggested that with decrease in size, the cobalt content of the shell might have increased, which in turn improved the interaction or binding efficiency of particle with bacterial cell (Žalnėravičius et al., 2016).

Morones et al. (2005) investigated the effect of size of silver NPs in the range of 1–100 nm against four different Gram-negative bacteria (*E. coli, V. cholera, P. aeruginosa,* and *Scrub typhus*). High-angle annular dark-field (HAADF) scanning transmission electron microscopy (STEM) technique showed that silver NP in range of 1–10 nm was able to attach to bacterial cell membrane, which altered its permeability and respiration. Further the NPs that have penetrated caused intracellular damage by interacting with sulfur and phosphorous-containing substances such as proteins and DNA. Through this study the author confirmed that the size of silver NPs does play a crucial role in the antibacterial effect (Morones et al., 2005).

Later, Adams et al. (2014) demonstrated a size-dependent antimicrobial activity of the palladium NPs with a small difference in the particle size (<1 nm). Three different self-contained sub-10 nm particles ($2.0 \pm 0.1$, $2.5 \pm 0.2$, and $3.0 \pm 0.2$ nm) were tested against Gram-negative and Gram-positive bacteria (i.e. *E. coli* and *S. aureus*). In the case of *E. coli* it was observed that smallest particles had higher bacterial killing effect followed by medium-sized and larger particles. Interestingly, in the case of *S. aureus,* middle-sized particles exhibited higher effect than the smallest and larger particles. This study clearly suggested that even a small difference in size (i.e. <1 nm) affects the antimicrobial property of the NPs, which also depends on the strains used (Adams et al., 2014).

In a similar study, the effect of size was explained in terms of change in diameter of carbon nanotubes: a well-known antibacterial material. In this study, the antibacterial activity of single-walled (SWCNTs) and multi-walled carbon nanotubes (MWCNTs) with outer diameter of about 0.9 and 30 nm respectively was considered for assessing the effect of size. Scanning electron microscopy studies showed that *E. coli* cells attached to SWCNTs exhibited higher degree of cellular damage than those attached to MWCNTs. It was also observed that the *E. coli* cells treated with SWCNTs got inactivated ($80 \pm 10\%$) at a higher percentage than those treated with MWCNTs ($24 \pm 4\%$). Similarly, a metabolic activity study also suggested that the cells attached to SWCNTs had lesser metabolic activity than the cells with MWCNTs. Further, the measurement of cytoplasmic content efflux and gene expression of stress and DNA-related products of CNTs-treated bacterial cells confirmed the superior toxicity of SWCNTs in comparison to MWCNTs. Overall these results clearly suggested that the SWCNTs exhibited a greater antimicrobial property than MWCNTs. The mechanism of action involved the partial penetration of CNTs and subsequent membrane damage. These effects of SWCNTs are attributed to the diameter (size) of the nanotubes where the smaller diameter aided in better penetration of CNTs into bacterial cells. Penetration was followed by membrane damage affecting the metabolic activity and altered stress-related gene expressions (Kang et al., 2008).

Zhang et al. (2008) prepared different metallic silver and gold NPs by in situ reduction and stabilized with poly(amidoamine) with terminal dimethylamine groups [HPAMAM-$N(CH_3)_2$]. The size and dispersity of the Ag (7.1–1 nm) and Au (7.7–3.9 nm) NMs can be changed by changing the molar ratio of metal with stabilizer. The antimicrobial property of these series of NMs was tested against Gram-positive bacteria, Gram-negative bacteria, and fungi. In these cases, the smallest particles with high surface-to-volume ratio exhibited the maximum antimicrobial activity against bacteria and fungi. Along with the size, the cationic terminal groups on surface contributed to a certain amount through interaction with the negative bacterial surface (Zhang et al., 2008).

Apart from individual particle size, experimental or physiological size of the materials does matter in terms of antimicrobial activity. In general, NPs tend to aggregate in experimental and physiological conditions due to their high reactivity. There is a greater chance that NMs that are exposed to bacterial cells at physiological conditions will aggregate rather than existing as individual particles. In such cases, the antibacterial activity obtained may be attributed to agglomerated units rather than the specified size. Neglecting this factor generally results in the misinterpretations of the data. In order to address this issue, a study was conducted with three photosensitive materials such as $TiO_2$, $SiO_2$, and ZnO to analyze their antimicrobial properties in water suspension. The authors reported that the experimental size of NPs was not the same as the true particle size as resulted from the potential aggregation of the NMs. Even though the antibacterial activity of the agglomerated system was similar to that of the material at the same concentration (Zhang et al., 2008). However, antimicrobial activity would have significantly higher in the case of individual units in comparison to the aggregates. It suggests that the size of NM is a very important factor that dictates the physicochemical property of NMs; however, it cannot be considered a general phenomenon in all the cases.

## 1.12 Influence of Shape on the Antibacterial Activity and Mechanism of Action of Nanomaterials

In addition to size, surface chemistry, composition, and shape also affect the functionality or activity of NMs. Shape plays a crucial role with regard to the interaction and the toxic effects on bacterial cell. Notably, the shape and size of the NMs dictate the physicochemical characteristics such as optical, electromagnetic, catalytic, and the crucial biological properties of the NMs. Taking the aforementioned factors into consideration, researchers attempted to develop various synthesis processes to gain precise control over the physicochemical factors such as size and shape (Bansal et al., 2010; Mulvaney, 1996; Narayanan & El-Sayed, 2004). In addition to size, the shape of the NM also determines the surface area of the material where even same materials with the same size will have different surface areas because of a change in shape. Next to shape of NM, crystalline nature of the nanostructures also plays an important role. It is generally defined as the relative abundance of particular crystallographic planes where each of them shows specific properties and reactivity. Conclusively, the shape and crystalline nature are important parameters, next to size, that play a significant role in the nano-bio interaction. Various studies have documented that the shape and crystallinity of NM have a great influence over the behavior of NMs and their biological activity such as antibacterial activity, and their uptake rate. In an earlier study, it was observed that spherical NPs had higher cellular uptake than nanorods (Chithrani, Ghazani, & Chan, 2006). Yang et al. (2016) showed that gold nanorods of different aspect ratio illustrated a significant variation in the cellular uptake rate. A significant increase in the internalization rate was observed with increase in the aspect ratio from 1 to 2 (AR2) where further increase did decrease the cellular internalization rate (Yang et al., 2016).

A well-known antibacterial material, $TiO_2$ is generally constituted by three crystalline phases such as anatase, rutile, and brookite. In this case, the anatase phase possesses the highest photocatalytic activity, which is due to its high charge transport properties and the presence of highly reactive (001) crystal facet. In comparison to the less-reactive crystal facet (101), (001) facet produced more efficient electron–hole pairs and reduced their recombination rates. Hence, the antimicrobial property of $TiO_2$ can be enhanced by increasing the ratio of exposed anatase (001) crystal facet. The anatase crystal of $TiO_2$ is generally present in a truncated octahedral bipyramid shape, which is composed of eight less-reactive (101) facets in the sides added with two highly reactive (001) facets in the top and bottom. However, during synthesis, the highly reactive facets tend to reduce their surface area to minimize the surface free energy. The use of capping agents such as hydrofluoric acid in the synthesis can bind and stabilize the reactive facets (Ong et al., 2014).

Tong et al. (2013) prepared different shapes of $TiO_2$ NMs such as nanorods, nanotubes, and nanosheets with exposed high-reactive (001) facets. All the nanostructures with more exposed (001) facets produced high hydroxyl radicals in comparison to classical $TiO_2$ NPs P25 (25 nm). Though it enhanced the photocatalytic activity, the antimicrobial property of nanostructures did not follow the same trend where P25 exhibited the highest antimicrobial activity followed by nanorods, nanosheets, and nanotubes, respectively. This has been attributed to the aspect ratio of the nanostructures where the interaction of bacteria and NMs depends on the surface area of the NM. The low antimicrobial profile of the elongated structures such as nanotubes, nanorods, and nanosheets could be attributed to reduced or limited exposure of ROS producing surface to bacterial cells. Since, the elongated nanostructures generally tend to stack over each other due to their strong van der Waals attraction forces (Tong et al., 2013).

Such a direct correlation of the active facets and antimicrobial property has also been found true in the case of silver NPs. In the case of silver, facet (111) is a highly atomic dense lattice that interacts with bacterial cell surface directly and causes membrane damage in comparison to less atomic dense (100) facets. Pal et al. (2007) studied the antibacterial activity of the silver NPs of different shapes against *E. coli*. Silver NMs with truncated triangular nanoplates exhibited higher antibacterial activity in comparison to spherical and rod-shaped NMs. Bacterial cell treated with the triangle-shaped nanoplates having (111) lattice plane showed drastic changes in the membrane, which caused rupture and cell death. This study clearly indicated in addition to the nano-size of the material, the morphology of NM having (111) lattice plane enhanced the antimicrobial property of silver NMs (Pal et al., 2007).

Gilbertson et al. (2016) studied the antibacterial activity of CuO NMs as a function of their shape. In this study, the author synthesized nanopowders (<50 nm) and nanosheets (~250–1000 $nm^2 \times 15$ nm thick) of CuO and compared their antimicrobial property with bulk of CuO material (500 nm–3 μm). The nanosheets of CuO illustrated the highest antimicrobial activity against the tested *E. coli* followed by nanopowders and bulk CuO. The difference in their level of antimicrobial activity was clearly attributed to their shape. As we discussed in the mechanism of action of NMs, the mechanism of action of NMs can be physical or chemical. The TEM studies revealed that CuO nanosheets oriented parallel to bacterial surface similar to other 1D NMs. This could have led to better interaction of CuO nanosheets with *E. coli*. Similarly, biochemical reactivity of NMs was evaluated using glutathione oxidation assay. It was observed from the study that the nanosheets exhibited higher oxidation of glutathione than the other two materials. Other studies such electrochemical and catalytic surface reactivity assay also revealed that CuO nanosheets had higher reactivity in comparison to other samples. The above results pertained to the high catalytic reactivity of CuO nanosheets, which produces oxidative stress-related species and activates the pathways for cellular death (Gilbertson et al., 2016).

Raza et al. (2016) showed that spherical silver NPs of smaller size (15–50 nm) had higher antibacterial activity against *E. coli* and *P. aeruginosa* in comparison to bigger spherical NPs (30–200 nm) and triangular NMs (edge length 150 nm). This observation was in contrast with previous studies of Pal et al. (2007) and Van Dong Ha, Binh, and Kasbohm (2012), where the triangle-shaped silver NMs exhibited better antimicrobial property than spherical NMs. The explanation of enhanced antibacterial effect of triangular silver nanostructures was based on the presence of highly reactive and dense atomic crystal facets (111). But here the XRD study revealed that the spherical silver NPs had a strong diffraction peak at $2\theta$ 38.5° from (111) facets. This suggested that spherical NPs were made up of top basal plane with the reactive (111) crystal facets, which could have enhanced the ROS production in bacterial cell and so their antimicrobial property (Raza et al., 2016). In a recent study Cheon et al. (2019) showed a shape-dependent antimicrobial property of Ag NPs. Antimicrobial property of differently shaped Ag NPs was studied using *S. aureus, E. coli,* and *P. aeruginosa*. The zone of inhibition studies showed that Ag spherical NP exhibited highest antibacterial property followed by Ag NM disks and triangular plate Ag NMs (Cheon et al., 2019). The difference in the antimicrobial property was attributed to the release of $Ag^+$ ions from NMs. Considering the surface area, spheres had the highest surface area ($1307 \pm 5\,cm^2$) followed by disk ($1104 \pm 109\,cm^2$) and triangular plate ($1028 \pm 35\,cm^2$). The higher the surface area the higher the release of $Ag^+$ ions, which could be the plausible reason for the highest antimicrobial property of spherical nanospheres followed by disk and triangular plate. The released $Ag^+$ ions interact with bacterial proteins or enzymes through sulfhydryl groups, thereby inactivating or destabilizing the cellular components. $Ag^+$ ions also bind to the membrane proteins that are involved in the ATP generation and ion transport across the cell membrane. Further, the released $Ag^+$ ions interact with nucleic acids and disrupt the H-bonds of DNA strands, thereby preventing the division and growth of the bacteria. Additionally, they induce the production of ROS, which in turn oxidizes the cellular components such as proteins and DNA. Gao et al. (2013) showed that the antibacterial activity of the Ag nanospheres is higher than that of nanoplates. The results suggested that nanospheres with higher surface area might have had greater contact with bacterial surface in comparison to nanoplates (Gao et al., 2013). In another recent study, Acharya et al. (2018) compared the antibacterial property of Ag nanorods with spherical NPs against *E. coli, P. aeruginosa, S. aureus,* and *B. subtilis.* The study revealed that both nanospheres and nanorods were very effective against both the Gram-positive and Gram-negative bacteria. The enhanced antibacterial property of both spheres and rods has been attributed to the presence of (111) plane. In general, plane (111) possesses high atomic density, which is one of the factors that determines the antibacterial activity of the Ag NMs (Acharya et al., 2018).

In a similar study, Sharma, Agarwal, and Balani (2016) studied the effect of shapes of ZnO nanostructures on their bactericidal property. Here the author synthesized ZnO microrods and microdisks from ZnO NPs of size (<100 nm) using hydrothermal route. The antimicrobial activity of ZnO structures against Gram-positive *S. aureus* and *Staphylococcus epidermidis* revealed that minimum inhibitory concentration (MIC) of all the three materials was within 0.5 μg/ml, whereas the same for Gram-negative bacteria *E. coli* was in the range of 70–76 μg/ml. The mechanism of action involved the production of $H_2O_2$, $Zn^{2+}$ ions release, and the presence of surface oxygen vacancies. In general, the release of metal ion from a material depends on the plane area of the material. Notably, the ZnO microdisks had the highest basal plane area, which led to the highest release of $Zn^{2+}$ ions from the surface ($75.3 \pm 14.6$ μg/l in MilliQ water and $631.3 \pm 17.3$ μg/l in LB medium) in 48 hours at 37 °C. Hence the mechanism of action of ZnO microdisks and microrods were suggested to be through release of $Zn^{2+}$ ions whereas the release of $H_2O_2$ and $Zn^{2+}$ along with cellular internalization was in case of ZnO particles (Sharma et al., 2016). Cha et al. (2015) studied the effect of ZnO NMs over a model enzyme β-galactosidase (GAL), which can be extrapolated to similar bacterial enzymes. The inhibition mechanisms of three different shapes of ZnO NMs – pyramid, plates, and spheres – were studied using: Michaelis Menten, Lineweaver Burk, and Eadie–Hofstee kinetics models. Among the three different shapes, nanoplates exhibited competitive inhibition over GAL whereas the ZnO NMs of pyramid shape exhibited noncompetitive mechanism. Such variation in the inhibitory effects has been explained by the ability of particular shapes to partially enter the grooves of the active site and inhibit the catalytic reaction by interfering in the reconfiguration of enzyme during substrate binding (Cha et al., 2015).

The sharp edges and the apexes of nanopyramids could have been a better geometrical match to the surface of enzyme. These factors further determine the association of proteins with NMs and hence their inhibition mechanism. This clearly showed that the antimicrobial property and mechanism of action strongly depend on the shape of NMs. However, the antimicrobial property of different NMs or same NMs synthesized by different methods could not be generalized on a particular shape. Since, along with shape, the organization of active facets present in the particular NM also plays a crucial role in determining its antimicrobial property. Hence, along with shape other sub-parameters such as facets organization and crystallinity should also be taken into account. Apart from size and shape, surface chemistry is the third most important factor that dictates the antimicrobial property, which we discuss briefly in the following section.

## 1.13 Effects of Functionalization on the Antimicrobial Property of Nanomaterials

Although several antimicrobial agents have been developed so far, they are still not able to meet the required therapeutic index. Even though NMs are well-known for their renowned antibacterial activities, their application is still limited due to their certain nonspecific toxicity. In order to improve antimicrobial therapeutic index and reduce the nonspecific toxicity, biofunctionalization or chemical modification of NPs with bioactive molecules has emerged as a plausible and promising solution. The selection of a NM along with a rational biomolecule is likely to improve the applicability of the composite NM.

Several techniques have been employed to functionalize NMs such as covalent bonding (Veerapandian et al., 2010), non-covalent bonding (Knopp, Tang, & Niessner, 2009), simple coating or deposition (Bunker et al., 2007), stober technique (Luckarift et al., 2007), coupling reaction-assisted immobilization (Wang et al., 2004), reverse micelle and sol–gel technique (Yang et al., 2004). Photo-Fenton oxidation, radiofrequency plasma, and vacuum-UV radiation methods have been employed for click chemistry (Mazille et al., 2010). Protein and peptides, especially those with cationic nature, have been found to be toxic against many drug-resistant microbes. The antimicrobial property of these proteins or peptides depends on the ability to form α-helical or β-sheets or α-helical bundles because of the interaction with anionic bacterial cell wall and self-association in solution state (Fernandez-Lopez et al., 2001; Oren et al., 2002). In an earlier study, the antibacterial activity of hen egg lysozyme-conjugated polystyrene latex NPs against *Micrococcus lysodeikticus* was studied. It was observed from the study that antimicrobial property of cationic NP-conjugated enzyme was twice that of free enzyme (Satishkumar & Vertegel, 2008). Similar to proteins, carbohydrate also enhances the antimicrobial property of NMs. Veerapandian et al. (2010) reported that the glucosamine (amino sugar) functionalization of silver NPs improved the antimicrobial property against eight Gram-positive and Gram-negative bacteria. The functionalization of glucosamine over silver NPs enhanced the interaction and penetration of NPs into bacterial cell, which improved the antibacterial activity of glucosamine-functionalized silver NPs (Veerapandian et al., 2010). Next to proteins and carbohydrates, lipids also possess antimicrobial property and they are part of innate immune system. The common antimicrobial lipids found in skin cells of human involve sphingosine, dihydro-sphingosine, 6-hydrosphingosine, sapienic– acid, and lauric acid (Drake et al., 2008). A study reported that oleic acid-stabilized silver NPs exhibited highest antimicrobial property against *E. coli* and *S. aureus*. It was observed that the hybrid material produced a quick response over *E. coli* than *S. aureus*. The stabilization of oleic acid improved the permeability of the NP inside the bacterial cell, which inhibited or altered the cellular transport across the bacterial cell and resulted in bacterial cell death (Le et al., 2010). Apart from other biomolecules like proteins/peptides, carbohydrates, lipids, and DNA, the NMs have been functionalized with antibiotics to improve their antibacterial effect through synergistic effect. It was reported earlier that the introduction of antibiotics such as kanamycin, erythromycin, ampicillin, and chloramphenicol along with silver NPs has enhanced the antibacterial property of silver NPs (Fayaz et al., 2010). The ampicillin silver NP complex has shown the highest antimicrobial effect over other complexes. Here, the strong van der Waals attraction force caused the interaction of NP with the bacterial cell surface. This interaction led to the lysis of cell wall and subsequent penetration of NP into cell where it intervened with the DNA unwinding and effected in the death of bacterial cell (Fayaz et al., 2010). In another study, Jaiswal and Mishra (2018) showed that the functionalization of silver NPs with curcumin improved the antimicrobial properties and also reduced the cytotoxicity of the silver NPs against human keratinocytes (Jaiswal & Mishra, 2018). However, in recent years a lot of work is going on with functionalization of NMs and the basic understanding of the interaction of the bacterial system with functionalized NPs. Better understanding of mechanism of functionalized NMs along with suitable nano-bio interface phenomenon will guide us to develop more standard design criteria to develop advanced materials with peculiar and desired properties.

## 1.14 Conclusion and Future Perspectives

This section delineates an overview of NMs, their role in microbial resistance, and the effect of physicochemical factors on their antimicrobial property. The development of microbial resistance to antibiotics and other common disinfectants has driven researchers to look for novel strategies to treat infections. Pertaining to this issue, NMs have emerged as a promising solution to microbial resistance due to their broad spectrum of antimicrobial property along with ease of integration with other products for diverse applications. It is clear from the literature that the antimicrobial property of the material depends on certain crucial physicochemical properties such as size, shape, and surface chemistry. Further, the antimicrobial property of the system can be tuned by controlling those crucial physicochemical properties of nanostructures. Understanding of this phenomenon can be exploited to tailor NMs of interest with reduced nonspecific toxicity where the NMs can be engineered to be specifically active against microbial cells rather than mammalian cell systems. The discussion over the mechanism of action of metal and metal oxide NMs suggested that the mechanisms of action of these NMs are not merely dependent on the release of metal ions from the NMs but also depend on the nanostructures (size and shape) of the NMs, which contribute directly to the antimicrobial property of the NMs. In the present scenario, the toxic effect of NM systems and the mechanism of action over human systems are not clearly understood. All these issues can be addressed if we develop standardized testing protocols and define the NMs' properties on an international level and enforce it. Most importantly, future research in this field should be directed to further understand the complex relationship between the various physicochemical factors over the antimicrobial property and mechanism of action of the NMs.

## Questions and Answers

**1** Why the study of the nano-bio interface is necessary?
Bio–nano interface hosts "the dynamic physicochemical interactions, kinetics and thermodynamic exchanges between nanomaterial surfaces and the surfaces of biological components." In the last three decades, there has been an exponential increase in the application of nanomaterials in various fields including the health sector. This is leading toward a long-term co-existence of such nanomaterials with living systems which may result in adverse toxicological effects to the living bodies. In this regard, it is necessary to study the effect of these materials on the biological entities such proteins, DNA, RNA, cell membrane, cell organelles, cells, tissues, and organs.

**2** Do nanomaterials occur in nature?
Yes, nanomaterials do occur in nature and are called "natural nanomaterials." They are produced by biological species or anthropogenic activities in nature without human intervention. The nanomaterials formed in nature are present throughout the earth's atmosphere, hydrosphere, and lithosphere, such as in volcanic ash, sea spray, and smoke.

**3** Explain the terms "antibiotic resistance" and "post-antibiotic era"?
"Antibiotic resistance" is the ability of microbes such as bacteria to resist the killing effects or overcoming the actions of the antibiotics. In recent times, a rapid increase in the level of microbial resistance to antibiotics is leading to an era called as "Post-antibiotic era" where the mortality rate caused because of microbial infections will be higher than that of cancer as stated by Centre for Disease Control and Prevention.

**4** How can nanomaterials combat antibiotic resistance and what are the different nanomaterials used as antimicrobial agents?
The traditional antibiotics exhibit their antimicrobial activity by one of these mechanisms: (i) inhibition of cell wall synthesis, (ii) inhibition of fatty acid biosynthesis, (iii) inhibition of protein synthesis, and (iv) compromising the cell membrane functions. Application of either one of the simple mechanisms by the traditional antibiotics is the main reason for the occurrence of bacterial resistance. On the other hand, nanomaterials have different mechanisms of action such as: (i) disputing cell membranes, (ii) causing DNA damage, (iii) interrupting the transmembrane electron transport, and (iv) inducing oxidative stress. Further, nanomaterials can be designed to have multiple mechanisms that act simultaneously against microbes. Hence, it becomes difficult for microbes to develop resistance against nanomaterials as they are unlikely to have many mutated genes. The most commonly exploited antibacterial nanomaterials include nanometals (silver, copper); metal oxides (Zinc oxide, titanium dioxide); carbonaceous materials (graphene, graphene oxide, carbon nanotubes); and cationic polymers (chitosan).

**5** Among Gram-negative and Gram-positive bacteria which has higher resistance over hydrophilic drugs and state the reason?
Among these two groups of bacteria, Gram-negative bacteria have higher resistance over hydrophilic drugs in comparison to Gram-positive since Gram-negative bacteria have an extra hydrophobic lipid bilayer over a thin peptidoglycan layer (20–50 nm). The presence of such an extra lipid layer limits the permeability of several hydrophilic antimicrobial agents, which is one of the reasons for high resistance of Gram-negative bacteria.

**6** In general, smaller sized nanomaterials exhibit higher antimicrobial property in comparison to the bigger sized particles of the same material. State the reason.
Generally, smaller sized nanomaterials have higher surface area and thus high chance of prolonged interaction with the microbial system and diffuse through the cell membrane in comparison to bigger nanomaterials with

smaller surface area. Additionally, it is possible to contain a significantly higher number of smaller NMs in comparison to bigger particles in the same volume. The higher surface area-to-volume ratio owing to increased number of particles results in exposure of greater numbers of atoms. Exposure of a greater number of atoms on the surface results in increased interaction of nanomaterial surface with bacteria increasing the number of reactive oxygen species at faster rate followed by inhibition or elimination of the bacteria.

**7** Does the shape of the nanomaterial affect the antimicrobial property? If yes explain why?
Yes, the shape is also one of the crucial factors that affect the structural behavior and antimicrobial activity of the nanomaterials. Apart from size, shape of the nanomaterials also has influence over the surface area where even same materials with the same size will have different surface area because of a change in shape. The surface area of the nanomaterials strongly determines the level of interaction of the nanomaterial with the microbial surface. Thus, shape plays a crucial role with regard to interaction and the toxic effects on bacterial cell where nanomaterials of different shapes would cause varying degrees of bacterial cell damage and antibacterial property.

## References


Aboelzahab, A., Azad, A. M., Dolan, S., & Goel, V. (2012). Mitigation of *Staphylococcus aureus*-mediated surgical site infections with IR photoactivated $TiO_2$ coatings on Ti implants. *Advanced Healthcare Materials*, 1(3), 285–291.

Abraham, E. P., & Chain, E. (1940). An enzyme from bacteria able to destroy penicillin. *Nature*, 146(3713), 837.

Acharya, D., Singha, K. M., Pandey, P., Mohanta, B., Rajkumari, J., & Singha, L. P. (2018). Shape dependent physical mutilation and lethal effects of silver nanoparticles on bacteria. *Scientific Reports*, 8(1), 201.

Adams, C. P., Walker, K. A., Obare, S. O., & Docherty, K. M. (2014). Size-dependent antimicrobial effects of novel palladium nanoparticles. *PLoS One*, 9(1), e85981.

Albanese, A., Tang, P. S., & Chan, W. C. (2012). The effect of nanoparticle size, shape, and surface chemistry on biological systems. *Annual Review of Biomedical Engineering*, 14, 1–16.

Aminov, R. I. (2010). A brief history of the antibiotic era: Lessons learned and challenges for the future. *Frontiers in Microbiology*, 1, 134.

Andrade, F., Rafael, D., Videira, M., Ferreira, D., Sosnik, A., & Sarmento, B. (2013). Nanotechnology and pulmonary delivery to overcome resistance in infectious diseases. *Advanced Drug Delivery Reviews*, 65(13–14), 1816–1827.

Arthur, M., & Courvalin, P. (1993). Genetics and mechanisms of glycopeptide resistance in enterococci. *Antimicrobial Agents and Chemotherapy*, 37(8), 1563.

Astefanei, A., Núñez, O., & Galceran, M. T. (2015). Characterisation and determination of fullerenes: A critical review. *Analytica Chimica Acta*, 882, 1–21.

Azzazy, H. M., Mansour, M. M., & Kazmierczak, S. C. (2007). From diagnostics to therapy: Prospects of quantum dots. *Clinical Biochemistry*, 40(13–14), 917–927.

Baig, M. S., Ahad, A., Aslam, M., Imam, S. S., Aqil, M., & Ali, A. (2016). Application of Box–Behnken design for preparation of levofloxacin-loaded stearic acid solid lipid nanoparticles for ocular delivery: Optimization, *in vitro;* release, ocular tolerance, and antibacterial activity. *International Journal of Biological Macromolecules*, 85, 258–270.

Bansal, V., Li, V., O'Mullane, A. P., & Bhargava, S. K. (2010). Shape dependent electrocatalytic behaviour of silver nanoparticles. *CrystEngComm*, 12(12), 4280–4286.

Berry, V., Gole, A., Kundu, S., Murphy, C. J., & Saraf, R. F. (2005). Deposition of CTAB-terminated nanorods on bacteria to form highly conducting hybrid systems. *Journal of the American Chemical Society*, 127(50), 17600–17601.

Bismuth, R., Zilhao, R., Sakamoto, H., Guesdon, J., & Courvalin, P. (1990). Gene heterogeneity for tetracycline resistance in *Staphylococcus* spp. *Antimicrobial Agents and Chemotherapy*, 34(8), 1611–1614.

Biswas, S., Deshpande, P. P., Navarro, G., Dodwadkar, N. S., & Torchilin, V. P. (2013). Lipid modified triblock PAMAM-based nanocarriers for siRNA drug co-delivery. *Biomaterials*, 34(4), 1289–1301.

Bleeker, E., Cassee, F., Geertsma, R., de Jong, W., Heugens, E., Koers-Jacquemijns, M., . . . Rietveld, A. (2012). Interpretation and implications of the European Commission's definition on nanomaterials. RIVM Letter Report 601358001, 43.

Boverhof, D. R., Bramante, C. M., Butala, J. H., Clancy, S. F., Lafranconi, M., West, J., & Gordon, S. C. (2015). Comparative assessment of nanomaterial definitions and safety evaluation considerations. *Regulatory Toxicology and Pharmacology*, 73(1), 137–150.

Bunker, C. E., Novak, K. C., Guliants, E. A., Harruff, B. A., Meziani, M. J., Lin, Y., & Sun, Y.-P. (2007). Formation of protein–metal oxide nanostructures by the sonochemical method: Observation of nanofibers and nanoneedles. *Langmuir*, 23(20), 10342–10347.

Çalışkan, N., Bayram, C., Erdal, E., Karahaliloğlu, Z., & Denkbaş, E. B. (2014). Titania nanotubes with adjustable dimensions for drug reservoir sites and enhanced cell adhesion. *Materials Science and Engineering C*, 35, 100–105.

Cavassin, E. D., de Figueiredo, L. F. P., Otoch, J. P., Seckler, M. M., de Oliveira, R. A., Franco, F. F., . . . Costa, S. F. (2015). Comparison of methods to detect the in vitro; activity of silver nanoparticles (AgNP) against multidrug resistant bacteria. *Journal of Nanobiotechnology*, 13(1), 64.

Cha, S.-H., Hong, J., McGuffie, M., Yeom, B., VanEpps, J. S., & Kotov, N. A. (2015). Shape-dependent biomimetic inhibition of enzyme by nanoparticles and their antibacterial activity. *ACS Nano*, 9(9), 9097–9105.

Chatterjee, T., Chakraborti, S., Joshi, P., Singh, S. P., Gupta, V., & Chakrabarti, P. (2010). The effect of zinc oxide nanoparticles on the structure of the periplasmic domain of the *Vibrio cholerae* ToxR protein. *The FEBS Journal*, 277(20), 4184–4194.

Cheon, J. Y., Kim, S. J., Rhee, Y. H., Kwon, O. H., & Park, W. H. (2019). Shape-dependent antimicrobial activities of silver nanoparticles. *International Journal of Nanomedicine*, 14, 2773.

Chithrani, B. D., Ghazani, A. A., & Chan, W. C. (2006). Determining the size and shape dependence of gold nanoparticle uptake into mammalian cells. *Nano Letters*, 6(4), 662–668.

Cox, G., & Wright, G. D. (2013). Intrinsic antibiotic resistance: Mechanisms, origins, challenges and solutions. *International Journal of Medical Microbiology*, 303(6–7), 287–292.

Cui, Y., Wei, Q., Park, H., & Lieber, C. M. (2001). Nanowire nanosensors for highly sensitive and selective detection of biological and chemical species. *Science*, 293(5533), 1289–1292.

Daeihamed, M., Dadashzadeh, S., Haeri, A., & Faghih Akhlaghi, M. (2017). Potential of liposomes for enhancement of oral drug absorption. *Current Drug Delivery*, 14(2), 289–303.

Daima, H. K., & Bansal, V. (2015). Influence of physicochemical properties of nanomaterials on their antibacterial applications. In M. Owais (Ed.), *Nanotechnology in Diagnosis, Treatment and Prophylaxis of Infectious Diseases* (pp. 151–166). Elsevier.

Dakal, T. C., Kumar, A., Majumdar, R. S., & Yadav, V. (2016). Mechanistic basis of antimicrobial actions of silver nanoparticles. *Frontiers in Microbiology*, 7, 1831–1831.

Della Valle, C., Visai, L., Santin, M., Cigada, A., Candiani, G., Pezzoli, D., . . . Chiesa, R. (2012). A novel antibacterial modification treatment of titanium capable to improve osseointegration. *The International Journal of Artificial Organs*, 35(10), 864–875.

Dorobantu, L. S., Fallone, C., Noble, A. J., Veinot, J., Ma, G., Goss, G. G., & Burrell, R. E. (2015). Toxicity of silver nanoparticles against bacteria, yeast, and algae. *Journal of Nanoparticle Research*, 17(4), 172.

Drake, D. R., Brogden, K. A., Dawson, D. V., & Wertz, P. W. (2008). Thematic review series: Skin lipids. Antimicrobial lipids at the skin surface. *Journal of Lipid Research*, 49(1), 4–11.

Ealias, A. M., & Saravanakumar, M. (2017). *A review on the classification, characterisation, synthesis of nanoparticles and their application*. Paper presented at the P Conf. Ser. Mater. Sci. Eng. 263 (2017) 032019, May 2–3 2017, VIT university, Vellore, Tamil Nadu, India.

El Badawy, A. M., Silva, R. G., Morris, B., Scheckel, K. G., Suidan, M. T., & Tolaymat, T. M. (2010). Surface charge-dependent toxicity of silver nanoparticles. *Environmental Science & Technology*, 45(1), 283–287.

Elechiguerra, J. L., Burt, J. L., Morones, J. R., Camacho-Bragado, A., Gao, X., Lara, H. H., & Yacaman, M. J. (2005). Interaction of silver nanoparticles with HIV-1. *Journal of Nanobiotechnology*, 3(1), 6.

Fajardo, A., Martínez-Martín, N., Mercadillo, M., Galán, J. C., Ghysels, B., Matthijs, S., . . . Baquero, F. (2008). The neglected intrinsic resistome of bacterial pathogens. *PLoS One*, 3(2), e1619.

Fayaz, A. M., Balaji, K., Girilal, M., Yadav, R., Kalaichelvan, P. T., & Venketesan, R. (2010). Biogenic synthesis of silver nanoparticles and their synergistic effect with antibiotics: A study against Gram-positive and Gram-negative bacteria. *Nanomedicine: Nanotechnology, Biology and Medicine*, 6(1), 103–109.

Feng, Q. L., Wu, J., Chen, G., Cui, F., Kim, T., & Kim, J. (2000). A mechanistic study of the antibacterial effect of silver ions on *Escherichia coli* and *Staphylococcus aureus*. *Journal of Biomedical Materials Research*, 52(4), 662–668.

Fernandez-Lopez, S., Kim, H.-S., Choi, E. C., Delgado, M., Granja, J. R., Khasanov, A., . . . Wilcoxen, K. M. (2001). Antibacterial agents based on the cyclic D, L-α-peptide architecture. *Nature*, 412(6845), 452.

Fishovitz, J., Hermoso, J. A., Chang, M., & Mobashery, S. (2014). Penicillin-binding protein 2a of methicillin-resistant *Staphylococcus aureus*. *IUBMB Life*, 66(8), 572–577.

Gao, M., Sun, L., Wang, Z., & Zhao, Y. (2013). Controlled synthesis of Ag nanoparticles with different morphologies and their antibacterial properties. *Materials Science and Engineering C*, 33(1), 397–404.

Gardini, D., Lüscher, C. J., Struve, C., & Krogfelt, K. A. (2018). Tailored nanomaterials for antimicrobial applications. In A. Barhoum & A. S. H. Makhlouf (Eds.), *Fundamentals of Nanoparticles* (pp. 71–104). Elsevier.

Gilbertson, L. M., Albalghiti, E. M., Fishman, Z. S., Perreault, F. O., Corredor, C., Posner, J. D., . . . Zimmerman, J. B. (2016). Shape-dependent surface reactivity and antimicrobial activity of nano-cupric oxide. *Environmental Science & Technology*, 50(7), 3975–3984.

Gleiter, H. (2000). Nanostructured materials: Basic concepts and microstructure. *Acta Materialia*, 48(1), 1–29.

Goenka, S., Sant, V., & Sant, S. (2014). Graphene-based nanomaterials for drug delivery and tissue engineering. *Journal of Controlled Release*, 173, 75–88.

Gupta, A., Landis, R. F., & Rotello, V. M. (2016). Nanoparticle-based antimicrobials: Surface functionality is critical. *F1000Research*, 5. doi:10.12688/f1000research.7595.1

Gupta, A., Mumtaz, S., Li, C.-H., Hussain, I., & Rotello, V. M. (2019). Combatting antibiotic-resistant bacteria using nanomaterials. *Chemical Society Reviews*, 48(2), 415–427.

Gurunathan, S., Han, J. W., Kwon, D.-N., & Kim, J.-H. (2014). Enhanced antibacterial and anti-biofilm activities of silver nanoparticles against Gram-negative and Gram-positive bacteria. *Nanoscale Research Letters*, 9(1), 373.

Hadinoto, K., Sundaresan, A., & Cheow, W. S. (2013). Lipid–polymer hybrid nanoparticles as a new generation therapeutic delivery platform: A review. *European Journal of Pharmaceutics and Biopharmaceutics*, 85(3), 427–443.

Hajipour, M. J., Fromm, K. M., Ashkarran, A. A., de Aberasturi, D. J., de Larramendi, I. R., Rojo, T., . . . Mahmoudi, M. (2012). Antibacterial properties of nanoparticles. *Trends in Biotechnology*, 30(10), 499–511.

Hayden, S. C., Zhao, G., Saha, K., Phillips, R. L., Li, X., Miranda, O. R., . . . Bunz, U. H. (2012). Aggregation and interaction of cationic nanoparticles on bacterial surfaces. *Journal of the American Chemical Society*, 134(16), 6920–6923.

He, C., Hu, Y., Yin, L., Tang, C., & Yin, C. (2010). Effects of particle size and surface charge on cellular uptake and biodistribution of polymeric nanoparticles. *Biomaterials*, 31(13), 3657–3666.

Hochella, M. F., Spencer, M. G., & Jones, K. L. (2015). Nanotechnology: Nature's gift or scientists' brainchild? *Environmental Science: Nano*, 2(2), 114–119.

Hong, X., Wen, J., Xiong, X., & Hu, Y. (2016). Shape effect on the antibacterial activity of silver nanoparticles synthesized via a microwave-assisted method. *Environmental Science and Pollution Research*, 23(5), 4489–4497.

Huh, A. J., & Kwon, Y. J. (2011). "Nanoantibiotics": A new paradigm for treating infectious diseases using nanomaterials in the antibiotics resistant era. *Journal of Controlled Release*, 156(2), 128–145.

Jacoby, W. A., Maness, P. C., Wolfrum, E. J., Blake, D. M., & Fennell, J. A. (1998). Mineralization of bacterial cell mass on a photocatalytic surface in air. *Environmental Science & Technology*, 32(17), 2650–2653.

Jagadeeshan, S., & Parsanathan, R. (2019). Nano-metal oxides for antibacterial activity. In M. Naushad, S. Rajendran, & F. Gracia (Eds.), *Advanced Nanostructured Materials for Environmental Remediation* (pp. 59–90). Springer.

Jaiswal, S., & Mishra, P. (2018). Antimicrobial and antibiofilm activity of curcumin-silver nanoparticles with improved stability and selective toxicity to bacteria over mammalian cells. *Medical Microbiology and Immunology*, 207(1), 39–53.

Jeevanandam, J., Barhoum, A., Chan, Y. S., Dufresne, A., & Danquah, M. K. (2018). Review on nanoparticles and nanostructured materials: History, sources, toxicity and regulations. *Beilstein Journal of Nanotechnology*, 9(1), 1050–1074.

Kang, S., Herzberg, M., Rodrigues, D. F., & Elimelech, M. (2008). Antibacterial effects of carbon nanotubes: Size does matter! *Langmuir*, 24(13), 6409–6413.

Kaufman, E. D., Belyea, J., Johnson, M. C., Nicholson, Z. M., Ricks, J. L., Shah, P. K., . . . Blomberg, E. (2007). Probing protein adsorption onto mercaptoundecanoic acid stabilized gold nanoparticles and surfaces by quartz crystal microbalance and $\zeta$-potential measurements. *Langmuir*, 23(11), 6053–6062.

Kaur, P., & Peterson, E. (2018). Antibiotic resistance mechanisms in bacteria: Relationships between resistance determinants of antibiotic producers, environmental bacteria, and clinical pathogens. *Frontiers in Microbiology*, 9, 2928.

Khameneh, B., Diab, R., Ghazvini, K., & Bazzaz, B. S. F. (2016). Breakthroughs in bacterial resistance mechanisms and the potential ways to combat them. *Microbial Pathogenesis*, 95, 32–42.

Khan, M. F., Ansari, A. H., Hameedullah, M., Ahmad, E., Husain, F. M., Zia, Q., . . . Khan, A. M. (2016). Sol-gel synthesis of thorn-like ZnO nanoparticles endorsing mechanical stirring effect and their antimicrobial activities: Potential role as nano-antibiotics. *Scientific Reports*, 6, 27689.

Kim, J. S., Kuk, E., Yu, K. N., Kim, J.-H., Park, S. J., Lee, H. J., . . . Hwang, C.-Y. (2007). Antimicrobial effects of silver nanoparticles. *Nanomedicine: Nanotechnology, Biology and Medicine*, 3(1), 95–101.

Knopp, D., Tang, D., & Niessner, R. (2009). Bioanalytical applications of biomolecule-functionalized nanometer-sized doped silica particles. *Analytica Chimica Acta*, 647(1), 14–30.

Kong, J., Franklin, N. R., Zhou, C., Chapline, M. G., Peng, S., Cho, K., & Dai, H. (2000). Nanotube molecular wires as chemical sensors. *Science*, 287(5453), 622–625.

Kühn, K. P., Chaberny, I. F., Massholder, K., Stickler, M., Benz, V. W., Sonntag, H.-G., & Erdinger, L. (2003). Disinfection of surfaces by photocatalytic oxidation with titanium dioxide and UVA light. *Chemosphere*, 53(1), 71–77.

Le, A.-T., Tam, P. D., Huy, P., Huy, T. Q., Van Hieu, N., Kudrinskiy, A., & Krutyakov, Y. A. (2010). Synthesis of oleic acid-stabilized silver nanoparticles and analysis of their antibacterial activity. *Materials Science and Engineering C*, 30(6), 910–916.

Lee, A., Mao, W., Warren, M. S., Mistry, A., Hoshino, K., Okumura, R., . . . Lomovskaya, O. (2000). Interplay between efflux pumps may provide either additive or multiplicative effects on drug resistance. *Journal of Bacteriology*, 182(11), 3142–3150.

Lee, W., Kang, S. H., Kim, J.-Y., Kolekar, G. B., Sung, Y.-E., & Han, S.-H. (2009). $TiO_2$ nanotubes with a ZnO thin energy barrier for improved current efficiency of CdSe quantum-dot-sensitized solar cells. *Nanotechnology*, 20(33), 335706.

Lemire, J. A., Harrison, J. J., & Turner, R. J. (2013). Antimicrobial activity of metals: Mechanisms, molecular targets and applications. *Nature Reviews Microbiology*, 11(6), 371.

Li, C., Fu, R., Yu, C., Li, Z., Guan, H., Hu, D., . . . Lu, L. (2013). Silver nanoparticle/chitosan oligosaccharide/poly (vinyl alcohol) nanofibers as wound dressings: A preclinical study. *International Journal of Nanomedicine*, 8, 4131.

Li, H., Luo, Y.-F., Williams, B. J., Blackwell, T. S., & Xie, C.-M. (2012). Structure and function of OprD protein in *Pseudomonas aeruginosa*: From antibiotic resistance to novel therapies. *International Journal of Medical Microbiology*, 302(2), 63–68.

Li, M., Zhu, L., & Lin, D. (2011). Toxicity of ZnO nanoparticles to *Escherichia coli*: Mechanism and the influence of medium components. *Environmental Science & Technology*, 45(5), 1977–1983.

Lim, E.-K., Chung, B. H., & Chung, S. J. (2018). Recent advances in pH-sensitive polymeric nanoparticles for smart drug delivery in cancer therapy. *Current Drug Targets*, 19(4), 300–317.

Lin, C.-C., Yeh, Y.-C., Yang, C.-Y., Chen, C.-L., Chen, G.-F., Chen, C.-C., & Wu, Y.-C. (2002). Selective binding of mannose-encapsulated gold nanoparticles to type 1 pili in *Escherichia coli*. *Journal of the American Chemical Society*, 124(14), 3508–3509.

Liu, J., Chen, D., Peters, B. M., Li, L., Li, B., Xu, Z., & Shirliff, M. E. (2016). Staphylococcal chromosomal cassettes mec (SCCmec): A mobile genetic element in methicillin-resistant *Staphylococcus aureus*. *Microbial Pathogenesis*, 101, 56–67.

Liu, J.-L., Zhang, W.-J., Li, X.-D., Yang, N., Pan, W.-S., Kong, J., & Zhang, J.-S. (2016). Sustained-release genistein from nanostructured lipid carrier suppresses human lens epithelial cell growth. *International Journal of Ophthalmology*, 9(5), 643.

Liu, S., Zeng, T. H., Hofmann, M., Burcombe, E., Wei, J., Jiang, R., . . . Chen, Y. (2011). Antibacterial activity of graphite, graphite oxide, graphene oxide, and reduced graphene oxide: Membrane and oxidative stress. *ACS Nano*, 5(9), 6971–6980.

Liu, Y., Tee, J. K., & Chiu, G. N. C. (2015). Dendrimers in oral drug delivery application: Current explorations, toxicity issues and strategies for improvement. *Current Pharmaceutical Design*, 21(19), 2629–2642.

Luckarift, H. R., Balasubramanian, S., Paliwal, S., Johnson, G. R., & Simonian, A. L. (2007). Enzyme-encapsulated silica monolayers for rapid functionalization of a gold surface. *Colloids and Surfaces B: Biointerfaces*, 58(1), 28–33.

Machado, S., Pacheco, J., Nouws, H., Albergaria, J. T., & Delerue-Matos, C. (2015). Characterization of green zero-valent iron nanoparticles produced with tree leaf extracts. *Science of the Total Environment*, 533, 76–81.

Markowska-Szczupak, A., Ulfig, K., & Morawski, A. (2011). The application of titanium dioxide for deactivation of bioparticulates: An overview. *Catalysis Today*, 169(1), 249–257.

Martinez, J. (2018). Ecology and evolution of chromosomal gene transfer between environmental microorganisms and pathogens. *Microbiology Spectrum*, 6(1), 1–16.

Mazille, F., Moncayo-Lasso, A., Spuhler, D., Serra, A., Peral, J., Benítez, N., & Pulgarin, C. (2010). Comparative evaluation of polymer surface functionalization techniques before iron oxide deposition. Activity of the iron oxide-coated polymer films in the photo-assisted degradation of organic pollutants and inactivation of bacteria. *Chemical Engineering Journal*, 160(1), 176–184.

Miller, W. R., Munita, J. M., & Arias, C. A. (2014). Mechanisms of antibiotic resistance in *Enterococci*. *Expert Review of Anti-infective Therapy*, 12(10), 1221–1236.

Miola, M., Fucale, G., Maina, G., & Verné, E. (2015). Antibacterial and bioactive composite bone cements containing surface silver-doped glass particles. *Biomedical Materials*, 10(5), 055014.
Mokerov, V., Fedorov, Y. V., Velikovski, L., & Scherbakova, M. Y. (2001). New quantum dot transistor. *Nanotechnology*, 12(4), 552.
Morones, J. R., Elechiguerra, J. L., Camacho, A., Holt, K., Kouri, J. B., Ramírez, J. T., & Yacaman, M. J. (2005). The bactericidal effect of silver nanoparticles. *Nanotechnology*, 16(10), 2346.
Mukherjee, A., Mohammed Sadiq, I., Prathna, T., & Chandrasekaran, N. (2011). Antimicrobial activity of aluminium oxide nanoparticles for potential clinical applications. *Science Against Microbial Pathogens: Communicating Current Research and Technological Advances*, 1, 245–251.
Mulvaney, P. (1996). Surface plasmon spectroscopy of nanosized metal particles. *Langmuir*, 12(3), 788–800.
Narayanan, R., & El-Sayed, M. A. (2004). Shape-dependent catalytic activity of platinum nanoparticles in colloidal solution. *Nano Letters*, 4(7), 1343–1348.
Nel, A. E., Mädler, L., Velegol, D., Xia, T., Hoek, E. M., Somasundaran, P., . . . Thompson, M. (2009). Understanding biophysicochemical interactions at the nano–bio interface. *Nature Materials*, 8(7), 543.
Neu, H. C. (1992). The crisis in antibiotic resistance. *Science*, 257(5073), 1064–1073.
Nikaido, H., & Takatsuka, Y. (2009). Mechanisms of RND multidrug efflux pumps. *Biochimica et Biophysica Acta (BBA)-Proteins and Proteomics*, 1794(5), 769–781.
Ong, W. J., Tan, L. L., Chai, S. P., Yong, S. T., & Mohamed, A. R. (2014). Facet-dependent photocatalytic properties of $TiO_2$-based composites for energy conversion and environmental remediation. *ChemSusChem*, 7(3), 690–719.
Oren, Z., Ramesh, J., Avrahami, D., Suryaprakash, N., Shai, Y., & Jelinek, R. (2002). Structures and mode of membrane interaction of a short $\alpha$ helical lytic peptide and its diastereomer determined by NMR, FTIR, and fluorescence spectroscopy. *European Journal of Biochemistry*, 269(16), 3869–3880.
Padmavathy, N., & Vijayaraghavan, R. (2008). Enhanced bioactivity of ZnO nanoparticles – An antimicrobial study. *Science and Technology of Advanced Materials*, 9(3), 035004.
Pal, S., Tak, Y. K., & Song, J. M. (2007). Does the antibacterial activity of silver nanoparticles depend on the shape of the nanoparticle? A study of the gram-negative bacterium *Escherichia coli*. *Applied and Environmental Microbiolog*, 73(6), 1712–1720.
Pan, X., Wang, Y., Chen, Z., Pan, D., Cheng, Y., Liu, Z., . . . Guan, X. (2013). Investigation of antibacterial activity and related mechanism of a series of nano-$Mg(OH)_2$. *ACS Applied Materials & Interfaces*, 5(3), 1137–1142.
Peulen, T.-O., & Wilkinson, K. J. (2011). Diffusion of nanoparticles in a biofilm. *Environmental Science & Technology*, 45(8), 3367–3373.

Pokropivny, V., & Skorokhod, V. (2007). Classification of nanostructures by dimensionality and concept of surface forms engineering in nanomaterial science. *Materials Science and Engineering C*, 27(5-8), 990–993.

Prasannakumar, J., Vidya, Y., Anantharaju, K., Ramgopal, G., Nagabhushana, H., Sharma, S., . . . Rajanaik, H. (2015). Bio-mediated route for the synthesis of shape tunable $Y_2O_3$: $Tb^{3+}$ nanoparticles: Photoluminescence and antibacterial properties. *Spectrochimica Acta Part A: Molecular and Biomolecular Spectroscopy*, 151, 131–140.

Pumera, M. (2010). Graphene-based nanomaterials and their electrochemistry. *Chemical Society Reviews*, 39(11), 4146–4157.

Qi, G., Li, L., Yu, F., & Wang, H. (2013). Vancomycin-modified mesoporous silica nanoparticles for selective recognition and killing of pathogenic Gram-positive bacteria over macrophage-like cells. *ACS Applied Materials & Interfaces*, 5(21), 10874–10881.

Qi, L., Xu, Z., Jiang, X., Hu, C., & Zou, X. (2004). Preparation and antibacterial activity of chitosan nanoparticles. *Carbohydrate Research*, 339(16), 2693–2700.

Raffi, M., Mehrwan, S., Bhatti, T. M., Akhter, J. I., Hameed, A., Yawar, W., & ul Hasan, M. M. (2010). Investigations into the antibacterial behavior of copper nanoparticles against *Escherichia coli*. *Annals of Microbiology*, 60(1), 75–80.

Raghupathi, K. R., Koodali, R. T., & Manna, A. C. (2011). Size-dependent bacterial growth inhibition and mechanism of antibacterial activity of zinc oxide nanoparticles. *Langmuir*, 27(7), 4020–4028.

Ramirez, M. S., & Tolmasky, M. E. (2010). Aminoglycoside modifying enzymes. *Drug Resistance Updates*, 13(6), 151–171.

Ramyadevi, J., Jeyasubramanian, K., Marikani, A., Rajakumar, G., & Rahuman, A. A. (2012). Synthesis and antimicrobial activity of copper nanoparticles. *Materials Letters*, 71, 114–116.

Ranghar, S., Sirohi, P., Verma, P., & Agarwal, V. (2014). Nanoparticle-based drug delivery systems: Promising approaches against infections. *Brazilian Archives of Biology and Technology*, 57(2), 209–222.

Raza, M., Kanwal, Z., Rauf, A., Sabri, A., Riaz, S., & Naseem, S. (2016). Size-and shape-dependent antibacterial studies of silver nanoparticles synthesized by wet chemical routes. *Nanomaterials*, 6(4), 74.

Rekha Deka, S., Kumar Sharma, A., & Kumar, P. (2015). Cationic polymers and their self-assembly for antibacterial applications. *Current Topics in Medicinal Chemistry*, 15(13), 1179–1195.

Ren, G., Hu, D., Cheng, E. W., Vargas-Reus, M. A., Reip, P., & Allaker, R. P. (2009). Characterisation of copper oxide nanoparticles for antimicrobial applications. *International Journal of Antimicrobial Agents*, 33(6), 587–590.

Roberts, M. C. (2005). Update on acquired tetracycline resistance genes. *FEMS Microbiology Letters*, 245(2), 195–203.

Ruparelia, J. P., Chatterjee, A. K., Duttagupta, S. P., & Mukherji, S. (2008). Strain specificity in antimicrobial activity of silver and copper nanoparticles. *Acta Biomaterialia*, 4(3), 707–716.

Saeed, K., & Khan, I. (2014). Preparation and properties of single-walled carbon nanotubes/poly (butylene terephthalate) nanocomposites. *Iranian Polymer Journal*, 23(1), 53–58.

Saeed, K., & Khan, I. (2016). Preparation and characterization of single-walled carbon nanotube/nylon 6, 6 nanocomposites. *Instrumentation Science & Technology*, 44(4), 435–444.

Salavati-Niasari, M., Davar, F., & Mir, N. (2008). Synthesis and characterization of metallic copper nanoparticles via thermal decomposition. *Polyhedron*, 27(17), 3514–3518.

Saliani, M., Jalal, R., & Goharshadi, E. K. (2015). Effects of pH and temperature on antibacterial activity of zinc oxide nanofluid against *Escherichia coli O157: H7* and *Staphylococcus aureus*. *Jundishapur Journal of Microbiology*, 8(2), e17115.

Saptarshi, S. R., Duschl, A., & Lopata, A. L. (2013). Interaction of nanoparticles with proteins: Relation to bio-reactivity of the nanoparticle. *Journal of Nanobiotechnology*, 11(1), 26.

Satishkumar, R., & Vertegel, A. (2008). Charge-directed targeting of antimicrobial protein-nanoparticle conjugates. *Biotechnology and Bioengineering*, 100(3), 403–412.

Schwarz, S., Kehrenberg, C., Doublet, B., & Cloeckaert, A. (2004). Molecular basis of bacterial resistance to chloramphenicol and florfenicol. *FEMS Microbiology Reviews*, 28(5), 519–542.

Sharma, N., Jandaik, S., Kumar, S., Chitkara, M., & Sandhu, I. S. (2016). Synthesis, characterisation and antimicrobial activity of manganese-and iron-doped zinc oxide nanoparticles. *Journal of Experimental Nanoscience*, 11(1), 54–71.

Sharma, R. K., Agarwal, M., & Balani, K. (2016). Effect of ZnO morphology on affecting bactericidal property of ultra high molecular weight polyethylene biocomposite. *Materials Science and Engineering C*, 62, 843–851.

Sharma, V. K., Filip, J., Zboril, R., & Varma, R. S. (2015). Natural inorganic nanoparticles–formation, fate, and toxicity in the environment. *Chemical Society Reviews*, 44(23), 8410–8423.

Shen, Q., Jiang, L., Zhang, H., Min, Q., Hou, W., & Zhu, J.-J. (2008). Three-dimensional dendritic Pt nanostructures: Sonoelectrochemical synthesis and electrochemical applications. *The Journal of Physical Chemistry C*, 112(42), 16385–16392.

Singh, P., Garg, A., Pandit, S., Mokkapati, V., & Mijakovic, I. (2018). Antimicrobial effects of biogenic nanoparticles. *Nanomaterials*, 8(12), 1009.

Singh, P., Pandit, S., Garnæs, J., Tunjic, S., Mokkapati, V. R., Sultan, A., . . . Daugaard, A. E. (2018). Green synthesis of gold and silver nanoparticles from *Cannabis sativa* (industrial hemp) and their capacity for biofilm inhibition. *International Journal of Nanomedicine*, 13, 3571.

Stouwdam, J. W., & Janssen, R. A. (2008). Red, green, and blue quantum dot LEDs with solution processable ZnO nanocrystal electron injection layers. *Journal of Materials Chemistry*, 18(16), 1889–1894.

Sudhasree, S., Shakila Banu, A., Brindha, P., & Kurian, G. A. (2014). Synthesis of nickel nanoparticles by chemical and green route and their comparison in respect to biological effect and toxicity. *Toxicological & Environmental Chemistry*, 96(5), 743–754.

Sukhorukova, I., Sheveyko, A., Kiryukhantsev-Korneev, P. V., Zhitnyak, I., Gloushankova, N., Denisenko, E., . . . Shtansky, D. (2015). Toward bioactive yet antibacterial surfaces. *Colloids and Surfaces B: Biointerfaces*, 135, 158–165.

Sun, J., Deng, Z., & Yan, A. (2014). Bacterial multidrug efflux pumps: Mechanisms, physiology and pharmacological exploitations. *Biochemical and Biophysical Research Communications*, 453(2), 254–267.

Sun, S., & Zeng, H. (2002). Size-controlled synthesis of magnetite nanoparticles. *Journal of the American Chemical Society*, 124(28), 8204–8205.

Sun, Y., & Xia, Y. (2002). Shape-controlled synthesis of gold and silver nanoparticles. *Science*, 298(5601), 2176–2179.

Sundar, S., & Kumar Prajapati, V. (2012). Drug targeting to infectious diseases by nanoparticles surface functionalized with special biomolecules. *Current Medicinal Chemistry*, 19(19), 3196–3202.

Tai, C. Y., Tai, C.-T., Chang, M.-H., & Liu, H.-S. (2007). Synthesis of magnesium hydroxide and oxide nanoparticles using a spinning disk reactor. *Industrial & Engineering Chemistry Research*, 46(17), 5536–5541.

Tchounwou, P. B., Yedjou, C. G., Patlolla, A. K., & Sutton, D. J. (2012). Heavy metal toxicity and the environment. In A. Luch (Ed.), *Molecular, Clinical and Environmental Toxicology* (pp. 133–164). Springer.

Tiwari, D. K., Behari, J., & Sen, P. (2008). Application of nanoparticles in waste water treatment. *World Applied Science Journal*, 3(3), 417–433.

Tong, T., Shereef, A., Wu, J., Binh, C. T. T., Kelly, J. J., Gaillard, J. F., & Gray, K. A. (2013). Effects of material morphology on the phototoxicity of nano-$TiO_2$ to bacteria. *Environmental Science & Technology*, 47(21), 12486–12495.

Van Dong, P., Ha, C. H., Binh, L. T., & Kasbohm, J. (2012). Chemical synthesis and antibacterial activity of novel-shaped silver nanoparticles. *International Nano Letters*, 2, 9. doi:https://doi.org/10.1186/2228-5326-2-9

Veerapandian, M., Lim, S. K., Nam, H. M., Kuppannan, G., & Yun, K. S. (2010). Glucosamine-functionalized silver glyconanoparticles: Characterization and antibacterial activity. *Analytical and Bioanalytical Chemistry*, 398(2), 867–876.

Vollath, D. (2013). *Nanoparticles-Nanocomposites–Nanomaterials: An Introduction for Beginners*. Wiley.

Wagner, S., Gondikas, A., Neubauer, E., Hofmann, T., & von der Kammer, F. (2014). Spot the difference: Engineered and natural nanoparticles in the environment – Release, behavior, and fate. *Angewandte Chemie International Edition*, 53(46), 12398–12419.

Wang, H., Castner, D. G., Ratner, B. D., & Jiang, S. (2004). Probing the orientation of surface-immobilized immunoglobulin G by time-of-flight secondary ion mass spectrometry. *Langmuir*, 20(5), 1877–1887.

Wang, L., Hu, C., & Shao, L. (2017). The antimicrobial activity of nanoparticles: Present situation and prospects for the future. *International Journal of Nanomedicine*, 12, 1227.

Weaver, C. L., LaRosa, J. M., Luo, X., & Cui, X. T. (2014). Electrically controlled drug delivery from graphene oxide nanocomposite films. *ACS Nano*, 8(2), 1834–1843.

Wei, T., Chen, C., Liu, J., Liu, C., Posocco, P., Liu, X., . . . Fermeglia, M. (2015). Anticancer drug nanomicelles formed by self-assembling amphiphilic dendrimer to combat cancer drug resistance. *Proceedings of the National Academy of Sciences*, 112(10), 2978–2983.

Wigginton, N. S., Titta, A. D., Piccapietra, F., Dobias, J., Nesatyy, V. J., Suter, M. J., & Bernier-Latmani, R. (2010). Binding of silver nanoparticles to bacterial proteins depends on surface modifications and inhibits enzymatic activity. *Environmental Science & Technology*, 44(6), 2163–2168.

Wu, J., Shen, Y., Jiang, W., Jiang, W., & Shen, Y. (2016). Magnetic targeted drug delivery carriers encapsulated with pH-sensitive polymer: Synthesis, characterization and *in vitro;* doxorubicin release studies. *Journal of Biomaterials Science, Polymer Edition*, 27(13), 1303–1316.

Xia, W., Grandfield, K., Hoess, A., Ballo, A., Cai, Y., & Engqvist, H. (2012). Mesoporous titanium dioxide coating for metallic implants. *Journal of Biomedical Materials Research Part B: Applied Biomaterials*, 100(1), 82–93.

Xia, Y., Yang, P., Sun, Y., Wu, Y., Mayers, B., Gates, B., . . . Yan, H. (2003). One-dimensional nanostructures: Synthesis, characterization, and applications. *Advanced Materials*, 15(5), 353–389.

Xiong, M. H., Li, Y. J., Bao, Y., Yang, X. Z., Hu, B., & Wang, J. (2012). Bacteria-responsive multifunctional nanogel for targeted antibiotic delivery. *Advanced Materials*, 24(46), 6175–6180.

Yang, H., Chen, Z., Zhang, L., Yung, W. Y., Leung, K. C. F., Chan, H. Y. E., & Choi, C. H. J. (2016). Mechanism for the cellular uptake of targeted gold nanorods of defined aspect ratios. *Small*, 12(37), 5178–5189.

Yang, H.-H., Zhang, S.-Q., Chen, X.-L., Zhuang, Z.-X., Xu, J.-G., & Wang, X.-R. (2004). Magnetite-containing spherical silica nanoparticles for biocatalysis and bioseparations. *Analytical Chemistry*, 76(5), 1316–1321.

Žalnėravičius, R., Paškevičius, A., Kurtinaitiene, M., & Jagminas, A. (2016). Size-dependent antimicrobial properties of the cobalt ferrite nanoparticles. *Journal of Nanoparticle Research*, 18(10), 300.

Zhang, Y., Peng, H., Huang, W., Zhou, Y., & Yan, D. (2008). Facile preparation and characterization of highly antimicrobial colloid Ag or Au nanoparticles. *Journal of Colloid and Interface Science*, 325(2), 371–376.

## Keywords

Antibiotics
antibiotics carriers
antibiotic efflux
antibiotic resistance
antimicrobial
antimicrobial activity
application of nanomaterials
bandgap energy
biofilm
capping agent
carbon nano tubes
classification of nanomaterials
copper nanoparticles
functionalization of nanomaterials
Gram-negative bacteria
Gram-positive bacteria
inorganic nanomaterials
metallic nanoparticles
metal oxides
nano-bio interface
nanomaterials
nanostructures
nanoparticles
one dimensional nanomaterials
organic nanomaterials
oxidative stress
photocatalysis
reactive oxygen species
roughness
shape
silver nanoparticles
size
three-dimensional nanomaterials
titanium dioxide
two-dimensional nanomaterials
zero-dimensional nanomaterials
zeta potential
zinc oxide nanoparticles

## Glossary

**Antagonist:** substance that inhibits or interrupts the action or function of any another substance.

**Antibiotics:** antimicrobial drugs that inhibit the growth or kill bacteria such as penicillin and ciprofloxacin.

**Antimicrobials:** the substances that kill or inhibit the growth of microbes such as bacteria, viruses, and fungi. Antimicrobials include antibacterial, antifungal, and antiviral agents. Antibiotics are one type of antimicrobials that act as antibacterial agents.

**Bacteria:** living organisms that belong to the family of unicellular microbes that have cell wall but are devoid of an organized nucleus and cell organelles. Bacteria can be both helpful and can cause infections or illness such as pneumonia and strep throat.

**Bandgap energy:** the energy that is required by valence band electron to excite into conduction band.

**Cells:** small microscopic structures that form the basic structural and functional unit of every living organism. They consist of primarily two compartments nucleus and cytoplasm that are enclosed in a membrane.

**CNTs:** carbon nano tubes are a sheet of carbon atoms that are rolled up to form a tube, which generally measures in nanometers.

**Crystal:** solid material that has a regular or periodic arrangement of its constituent atoms or molecule to form a crystal lattice extending in all the three directions.

**Dendrimers:** nanosized, highly branched, radially symmetric molecules that form a tree-like structure with high homogeneity and monodispersity.

**DNA:** deoxyribonucleic acid: the substance or molecule that is present in nearly all living organisms which contains genetic code that constitutes the chromosomes with self-replicating potential.

**Facets:** flat faces on a geometric shape.

**Glycoproteins:** type of proteins where amino acid chain is covalently attached to carbohydrate (oligosaccharide chain). The oligosaccharide attachment occurs during post-translational modification of protein.

**Micelles:** an aggregate or organized auto-assembly of amphiphilic macromolecules in colloidal solution.

**Multidrug-resistant microbe:** microbe that is resistant to more than one kind of antibiotics.

**Nanofluids:** the fluids that consist of nanometer-sized particles. Base fluids that are added with colloidal suspension of nanosized particles.

**Nanomaterials:** minuscule structures having at least one of their dimensions equal to or between 1 and 100 nm

**Nanoscale:** having geometric dimensions less than 100 nm.

**Nanotechnology:** production, characterization, and application of structures, devices, and systems where their atoms or molecules are engineered to be the size of nanometer (1–100 nm).

**Periplasm:** the space consisting of concentrated gel-like matrix that lies between the outer cell membrane and inner cytoplasmic membrane of bacteria.

**Photocatalysis:** process of accelerating a chemical reaction or process by direct irradiation or by irradiating a catalyst, which in turn decreases the activation energy for the reaction to occur.

**Photo-Fenton oxidation:** generation of highly oxidizing species such as hydroxyl radicals from hydrogen peroxide and Fe II ions enhanced with irradiation of UV light to treat or degrade organic contents in waste water.

**Polycrystalline:** solid material where several crystalline units are oriented randomly with varying size. Most of the inorganic materials such as common metals and ceramics are polycrystalline in nature.

**Submicron:** on a scale of $10^{-6}$, smaller than one millionth of a meter.

**Therapeutics:** a branch of medicine that deals with disease treatment modalities and the action of the curing agents.

**Zeta potential:** the potential difference between the surface of solid material or particle dispersed in a conducting liquid and the bulk of the liquid.

# 2

# Size- and Shape-Selective Synthesis of DNA-Based Nanomaterials and Their Application in Surface-Enhanced Raman Scattering

*K. Karthick and Subrata Kundu*

*Electrochemical Process Engineering (EPE) Division, CSIR-Central Electrochemical Research Institute (CECRI), Karaikudi, Tamil Nadu, India*

## Objectives

- Understanding the formation of metal NPs utilizing DNA as a scaffold for metal–DNA nanoassembly.
- Understanding the mechanism behind the successful formation of metal NPs over DNA.
- Relating the concentration of DNA to the metal NPs and resultant morphological outcomes.
- Understanding the nature of "hot spots" with the assistance of DNA and the how it affects the SERS activity.
- Directing biomolecule DNA toward the preparation of different size shaped metal NPs in SERS.

## 2.1 Introduction

The formulation of nanomaterials with different approaches has recently gained much importance as the size and shape of them can find promising applications in multidisciplinary areas. In focusing the materials with varying sizes between bulk and nano levels, the different sizes and shapes formed with respect to the synthesis method adopted, will have exciting variance in the physicochemical properties (Anantharaj, Nithiyanantham, Ede, & Kundu, 2014). The prepared nanomaterials find extensive applications, mainly in areas of catalysis, sensors, medicinal, and nanotechnology. The influence of scaffolds in the preparation of nanomaterials results in differently shaped nanomaterials depending upon the concentration along with reaction time and other parameters. Different scaffolds can be adopted in directing the morphologies of the prepared nanomaterials such

*Microbial Interactions at Nanobiotechnology Interfaces: Molecular Mechanisms and Applications*, First Edition. Edited by R. Navanietha Krishnaraj and Rajesh K. Sani.

as sodium dodecyl sulfate (SDS), cetyl trimethyammonium bromide (CTAB), triton X-100, and a biomolecule like deoxyribonucleic acid (DNA). Among the above scaffolds, the first three fall under the category of surfactants and possess hydrophobic and hydrophilic natures. Moreover, based on the charges present, they are categorized as anionic, cationic, amphoteric, and nonionic surfactants with respect to the charge present over the hydrophilic part of the surfactant (Morsy, 2014). In the case of DNA, while considering the structure, it possesses a double-helical nature with adenine–thymine and guanine–cytosine pairs. These nucleobases with hydrogen bond between them form an assembly of DNA. DNA is a comparatively better scaffold than the abovementioned surfactants in anchoring nanomaterials due the presence of different binding sites like deoxyribose sugar unit, phosphates, and the nitrogen containing four nucleobases (Chen, Liu, Cao, Ren, & Qu, 2018). Therefore, it is easy to anchor different kinds of metal ions over the DNA and the same can be formed as NPs with respect to the method of reduction adopted. To prepare NPs with the aid of DNA, synthetic methods like microwave, chemical reduction, and photochemical reduction method can be adopted. Moreover, by varying the concentration of DNA, the morphology of the NPs can be fine-tuned and can be applied to different applications. Recently, our group explored the utilization of DNA-based nanomaterials in the applications of catalysis, electrocatalysis, and in surface-enhanced Raman scattering (SERS) (Anantharaj et al., 2014; Karthick, Anantharaj, Karthik, Subramanian, & Kundu, 2017). The presence of DNA along with metal NPs tends to give rise to enhanced activities with high rate constants in catalysis studies. Moreover, the biomolecule DNA itself acts as a binder in electrochemical studies with metal–DNA architectures that too with low loading. The exceptional application with metal NPs and DNA combination was observed with the SERS studies. In general, SERS is a tool to enhance the Raman signal over the metal surface for the probe molecules even when present in very low concentrations. When the localized electric field matches with the frequencies of incident laser, intense signals will arise due to plasmonic coupling between the laser and the localized metal surface. When the NPs have more sites for plasmonic coupling generally called as "hot spots," the SERS signal is observed even at pico-molar concentrations. Therefore, it is crucial to make sufficient "hot spots" by employing scaffolds like DNA. In the absence of effective scaffold like DNA, the formed NPs can get agglomerated and the reproducibility in finding the SERS enhancements is also a question. Moreover, the "hot spots" are limited and random in the absence of DNA. With the method of synthesis adopted and with the concentration of DNA, one can produce different morphological outcomes in such a way that the "hot spots" produced will be more with NPs arranged in closer proximities for the improvised SERS signal. Hence, the problem of reproducibility that restricts the application of NPs can be solved with the simple addition of biomolecule DNA. Making NPs with very low interparticle gap of ~2–5 nm is rather difficult and for that there are methods like electron beam lithography and optical methods but they are expensive and time-consuming processes. To form NPs with low interparticle gap below 5 nm, DNA provides a great platform with different morphological structures with respect to the concentration. Therefore, DNA will play a major role in giving enhanced signals at low concentrations of probes and therefore the mechanism of interactions of DNA with NPs, functional groups participating in anchoring NPs, and concentration-dependent morphological features need to be considered for improving SERS activities of the same at very low concentrations of probe molecules. This book chapter focuses on the aspects of synthesizing different NPs with biomolecular scaffold as DNA and their influence in the enhancement of SERS. The binding sites of DNA that participated in anchoring NPs and the evidence of the same with experimental results have been discussed. The subsequent use of those metal NP assemblies with DNA architectures in SERS application is elaborated in detail in this chapter.

## 2.2 Mechanism of Surface-Enhanced Raman Scattering (SERS)

The enhancement in Raman signal in SERS was first invented by Fleischmann et al. in the year of 1974 through the incorporation of rough Ag NPs with pyridine probe molecule (Fleischmann, Hendra, & McQuillan, 1974). The clear-cut mechanism behind such an enhancement is under debate to date. Electromagnetic enhancement mechanism and chemical enhancement mechanism explain the interactions and the enhancements of rough metal surfaces with probe molecules. Considering electromagnetic enhancement, it is based on the collective excitation of electrons from the rough surface of the metal and that matches with the incident laser beam for the increase in enhancement of the signal (Campion & Kambhampati, 1998). In general, the rough surface of metal is important for the localized plasmon excitation that interacts with the applied electromagnetic field (EM). The surface plasmon excitations from the close proximity NPs with low interparticle gap tend to enrich the local field experienced by the probe molecules. The localized fields of probe molecules with the surface plasmonic excitation enhance the light of EM field when incident on rough metal surface and also the scattered Raman signal. Further, the charge transfer mechanism between the HOMO and LUMO pair of the molecular probe adsorbates with the rough metal surfaces is explained earlier in detail (Campion & Kambhampati, 1998). Comparing both chemical and electromagnetic enhancements, the former follows the pure chemical interaction of the probe molecules with the metal surface and therefore it cannot be applied to all the systems under investigation and the latter is about the physisorption of the probe molecules to the rough metal surface. The intensity of the Raman signal can be enhanced depending upon the interparticle gap, surface plasmons of the conductive electrons from the rough surface, physisorption of the probe molecules, chemical bonds with the metal surface, and the distance between the metal and probe molecules.

### 2.2.1 Significance of Nano-Bio Interfaces and Role of DNA in Enhancing SERS Activity

Nanotechnology has emerged as a fascinating field of applications because of the variance in physicochemical properties at the nano level. Because of these advantages, it can be used in a wide variety of applications. Recently, the interactions at the interface of nano- and biomolecules show innovative area of research that needs to be improved for clinical applications. The metal NPs and bio-interfaces show mainly three kinds of surface interactions, namely interactions at: the surface of NPs, solid–liquid interface, and contact zone of the organic molecules present in the biomolecules (Bhaumik, Shearin, Delong, Wanekaya, & Ghosh, 2014). Gold-based NPs have been effectively studied with biomolecules and in different fields such as tissue imaging, drug delivery, and therapeutics (Alkilany, Lohse, & Murphy, 2012). The protein molecules with NPs show surface arranging at initial times to form soft-corona and after sometime it forms long-lived corona. The interactions between nano-biomolecules show the relative possibilities of physical forces such as weak van der Waal interactions, hydrophobic interactions, electrostatic interactions, and H-bonding interactions. Among the different kinds of interactions mentioned above, electrostatic interactions have been found to play an important role in interaction with surface of NPs (Mahmoudi, Bertrand, Zope, & Farokhzad, 2016). Moreover, the hydrophobic interactions of aromatic groups show inhibition with surrounding water molecules. Also, when bringing weak forces with NPs, the conformations of biomolecules change and enhance the binding capabilities of NPs with bio-interfaces. Therefore, the utilization of nano-bio interfaces in different fields and improvising the same with fundamental understanding will have a huge impact in the near future. Here, in this chapter, the DNA as biomolecule with different metal NPs as nano-bio interface was considered for SERS applications. This interaction between DNA and metal NPs can be effectively understood with electronic and FT-IR spectral studies and also by microstructural studies. The use of nano-bio interface (metal NPs-DNA) here focused for SERS application as the production of "hot spots" from nano-biomolecule assembly ensured high enhancements in SERS at too low concentrations of probe molecules. The production of "hot spots" of DNA can be related to the close arrangement of NPs over the DNA assembly. In general, SERS is done to enhance the signal of probe molecules like MB at low concentrations. Metal NPs because of localized surface electrons will show surface plasmon resonance (SPR) with the applied laser beam and when there is matching of frequencies, strong intense signals will be observed even at nano-molar concentrations. This is enabled when the prepared NPs are in closer arrangement and hence the localized electrons between the two particles will show huge resonance with incident Raman signal. But here, the reproducibility will be difficult as the searching of closer proximity of NPs is rather struggling. It will be interesting to correlate the metal NPs with biomolecules to relate the SERS activity. In the case of DNA, with effective assembly of DNA with organic base pairs, phosphate backbones, and sugar moieties, the nano-bio interface is achieved with physical forces and electrostatic interactions. Here, the closely assembling of NPs will be highly probable and because of this, there are more "hot spots" available to effect the intense SERS signal at very low concentrations of probe molecules. Moreover, as finding "hot spots" is easier with DNA–metal interfaces, reproducibility is also achieved. Therefore, the use of DNA becomes significant in anchoring metal NPs for nano-bio interfaces and needs to be further improved for futuristic applications.

## 2.3 Size- and Shape-Selective Synthesis of Metal NPs with DNA for SERS Studies

The synthesis of metal–DNA as nano-bio interfaces gives different shapes and sizes based on the method adopted and the concentration of DNA used. The interactions of DNA with CNT ensured removing strong inter-tube interactions and hence agglomeration is restricted in CNT–DNA combination. Moreover, the hydrophobic CNT is water solubilized with DNA by means of aromatic nucleotides (Campbell, Tessmer, Thorp, & Erie, 2008). The integration of biomolecules like DNA and the nano interfaces like different metal NPs, CNT, and quantum dots gives combination of changes in catalytic functions of biomolecules and nanomaterials (Willner & Willner, 2010). The single-stranded DNA with CNT was further observed as monodispersed solution and by tuning them, it can be applied in sensors and drug delivery (Umemura, Izumi, & Oura, 2016). The focusing of only DNA as nanotubes from double helical structures showed variations in high aspect ratios as long and narrow channels. The focusing of only DNA as nanotubes from double helical structures showed variations in high aspect ratios and long and narrow channels (Rafati & Gill, 2016). For the shape-selective synthesis of metal NPs in DNA, methods like chemical reduction in aquesol and organosol, photochemical conditions, and microwaves can be used that result in metal NPs of varying particle sizes assembled in DNA chains. DNA in the presence of wide variety of functional groups, phosphate backbones, aromatic base pairs, and sugar moieties can assemble different NPs. As the agglomeration of NPs is restricted, it forms stable assemblies of metal NPs in DNA scaffold. DNA as biomolecular scaffold is utilized to prepare differently shaped NPs based on chemical reduction by employing strong reducing agents like sodium borohydride and tetrabutyl ammonium borohydride. Photochemical methods can be employed where UV light is irradiated to form NPs with different structures with the aid of DNA. The chemical reduction method is employed either in aqueous medium (aquesol) or in organic medium (organosol) with considerable variation in size and shape of the NPs and also by concentration variation in DNA. The microwave method can be adopted to facilitate the uniform heating of metal–DNA solution and, hence, the resultant assemblies and the diameter of the NPs in the DNA chains can be varied. Therefore, these processes can give stable metal NPs in DNA scaffolds with size and shape selectivity:

- Chemical reduction method by which metal ions are reduced to form metal NPs and assembled over the DNA chain: The same method can also be employed in both aqueous and organic conditions.
- Photochemical reduction method by which the metal ions anchored on DNA can be reduced with the aid of hydroxyl groups present in DNA that forms self-assemblies of nano-bio interfaces.
- The microwave method where the metal–DNA solution is getting uniformly heated to effect the variance in shape-selective formation of DNA chain.

The formed NPs with varying sizes and shapes by means of DNA and how they affect SERS enhancement are explained hereafter. This is elaborated under four subheadings as metal–DNA NPs prepared in photochemical conditions, metal–DNA prepared in aquesol, metal–DNA prepared in organosol, and metal–DNA prepared in microwave conditions for SERS respectively.

### 2.3.1 Metal NP Assemblies on DNA Using Photochemical Route for SERS Studies

The photochemical synthesis of metal NPs on DNA involves the irradiation of UV light with different times of exposures and by the use of hydroxyl groups present in DNA and 2,7-dihydroxy naphthalene (2,7-DHN). The use of DNA here assists the formed NPs to have very low interparticle gap and with varying morphologies such as nanowires and spherical particles at nano-size. In the preparation of Ag–DNA, $AgNO_3$ and DNA have been added together to which UV light with a wavelength of 260 nm was irradiated for three hours (Majumdar, Singha, Mondal, & Kundu, 2013). The confirmation of the nucleation of Ag NPs over DNA was first assessed by surface plasmon resonance (SPR) bands from the electronic spectral studies. Figure 2.1 shows the electronic spectral results and we can see that

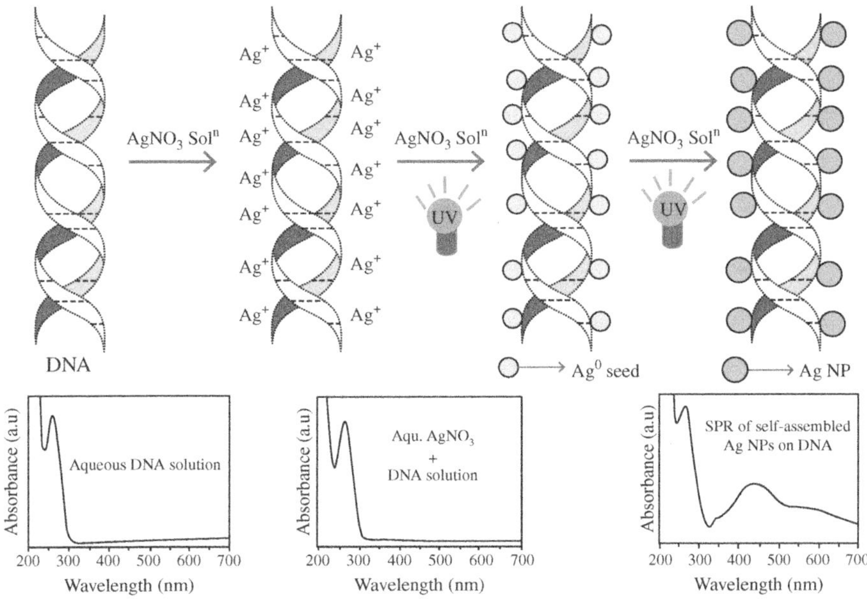


**Figure 2.1** Schematization of DNA with Ag for the formation of Ag–DNA. Lower panel shows the electronic spectral characteristics during DNA metallization. *Source*: Revised from Majumdar et al. (2013).

pristine DNA shows a peak around 260 nm, which is because of the presence of aromatic base pairs in DNA. Upon addition of $Ag^+$ to DNA, in addition to the parent peak, one more peak is observed near 420 nm because of the complex formation between them. After UV photoirradiation, finally a new peak is elevated around 440 nm that is responsible for the SPR band of the Ag NPs. This is easily understandable with the shown schematic, where $Ag^+$ is adsorbed over DNA chain and gets reduced further with UV irradiation. The morphologies at different magnifications of Ag–DNA system were analyzed with transmission electron microscopy (TEM) and observed to be nanowires as seen from Figure 2.2a–d. The formed Ag–DNA nanowires contained particle size of around 17 nm with an interparticle gap around 1.6 nm. The simulation study was performed through COSMOL multiphysics program to understand the electromagnetic enhancements. The Ag–DNA nanowire system was taken with the mentioned sizes. The dielectric constant of the medium surrounded by the metal surfaces increases the coupling of EM field. In this case, as shown in Figure 2.3a, two Ag NPs ($N = 2$) with mentioned radius and distance were taken with the dielectric constant of DNA around 2.53, which is higher than that of air, 1. The SPR bands were

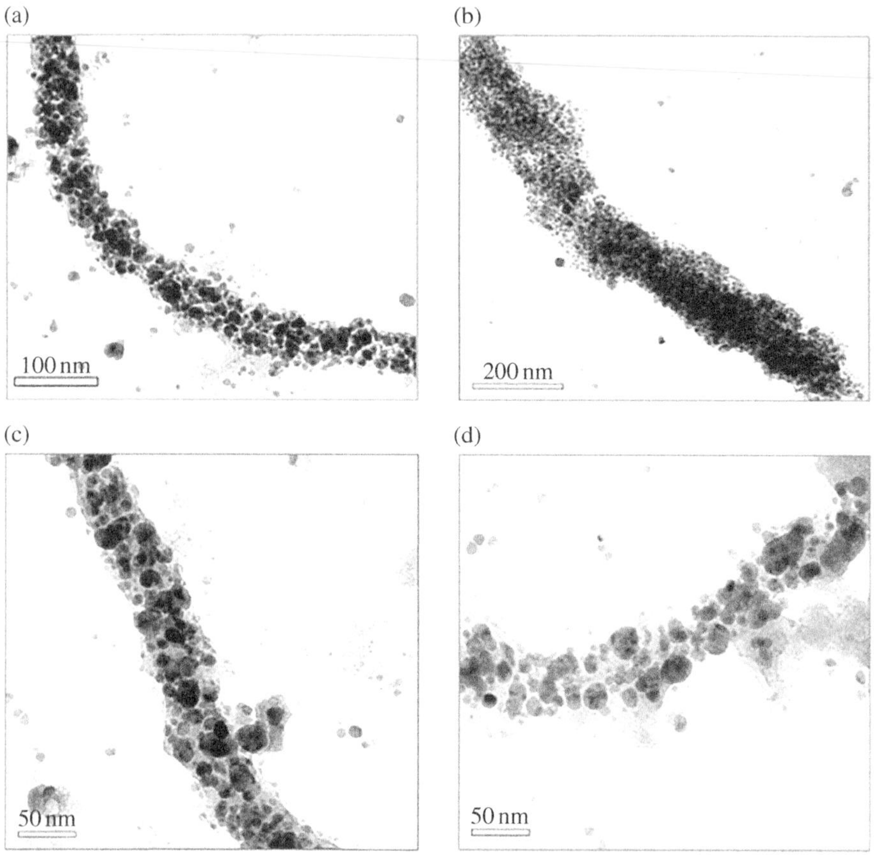


**Figure 2.2** (a–d) TEM images of Ag–DNA taken at different magnifications showing the wire-like structure with low interparticle gap. *Source*: Revised from Majumdar et al. (2013).

analyzed for an isolated Ag NP and two and twenty Ag NPs. The isolated Ag ($N = 1$) showed a band near 370 nm and the sample with two and twenty Ag NPs ($N = 20$) showed SPR band at 455 nm and confirmed the red shift that occurred when more than one Ag was present (Figure 2.3b). This meant the increase in EM field with high-order Ag NPs. Moreover, a threefold enhancement in SERS signal is observed when DNA is present, which could be attributed to the enhanced dielectric constant of the same. This dielectric field increases the dipole moment of metal surface and hence along with DNA medium, with low interparticle gap, the plasmonic coupling of Ag is greater with the applied EM field. This red shift can easily be seen in the right-hand side of Figure 2.3c and this shift is more prominent for resonance between plasmonic coupling of Ag NPs with EM field. Also,

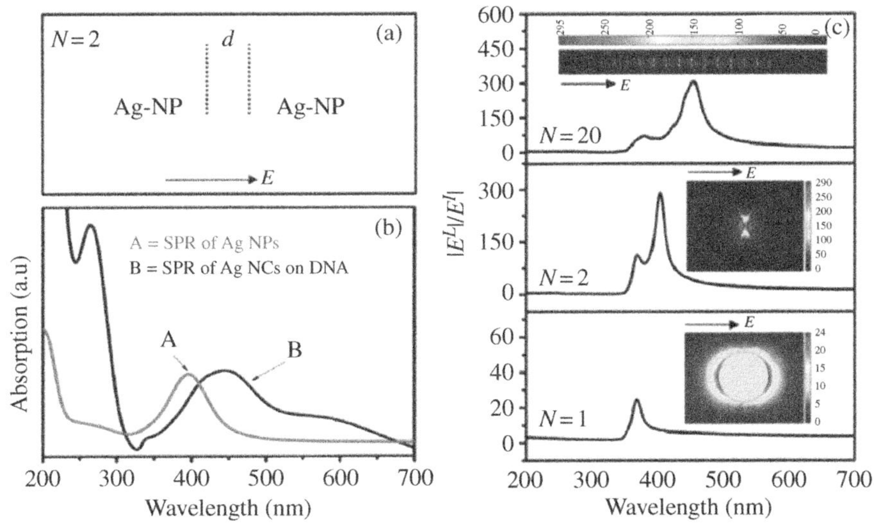


**Figure 2.3** (a) Model taken for calculating induced magnetic field. (b) SPR band for single and Ag–DNA. (c) EM field enhancements in $N = 1, 2$, and 20 Ag NPs. *Source*: Revised from Majumdar et al. (2013).

considerable induced EM field enhancements for $N = 2$ are depicted in Figure 2.4a. The induced EM field was much pronounced for $N = 20$ and also in a model 2D array of 14 Ag NPs as seen from Figure 2.4b and c, respectively. This 2D array is considered because of the Ag NPs' arrangements over the nanowires of DNA with low interparticle gaps. Figure 2.4d shows the linearity between the numbers of Ag NPs to the enhancement in the induced EM field and shows the saturation after 14 numbers. The close proximity ensures the enhancements in the presence of DNA and showed a high enhancement factor (EF) of $1 \times 10^8$ for $N = 20$ in Ag–DNA. Figure 2.5 summarizes the overall portrayal of Ag–DNA nanowires for SERS as mentioned in the upper panel. The probe used here is methylene blue (MB) and at low concentrations of MB; in the presence of Ag–DNA, intense peaks are observed due to the presence of enhanced "hot spots" with very low interparticle junctions. This is attributed to the Ag–DNA that allows identification of probe molecules even at pico-molar concentrations. The EF value is calculated using the formula

$$\text{EF} = \left(I_{\text{SERS}} / C_{\text{SERS}}\right) / \left(I_{\text{ref}} / C_{\text{ref}}\right)$$

Here, $I_{\text{SERS}}$ and $I_{\text{ref}}$ are the intensities of sample and reference probe molecule from SERS and $C_{\text{SERS}}$ and $C_{\text{ref}}$ are the concentrations of sample taken for analysis

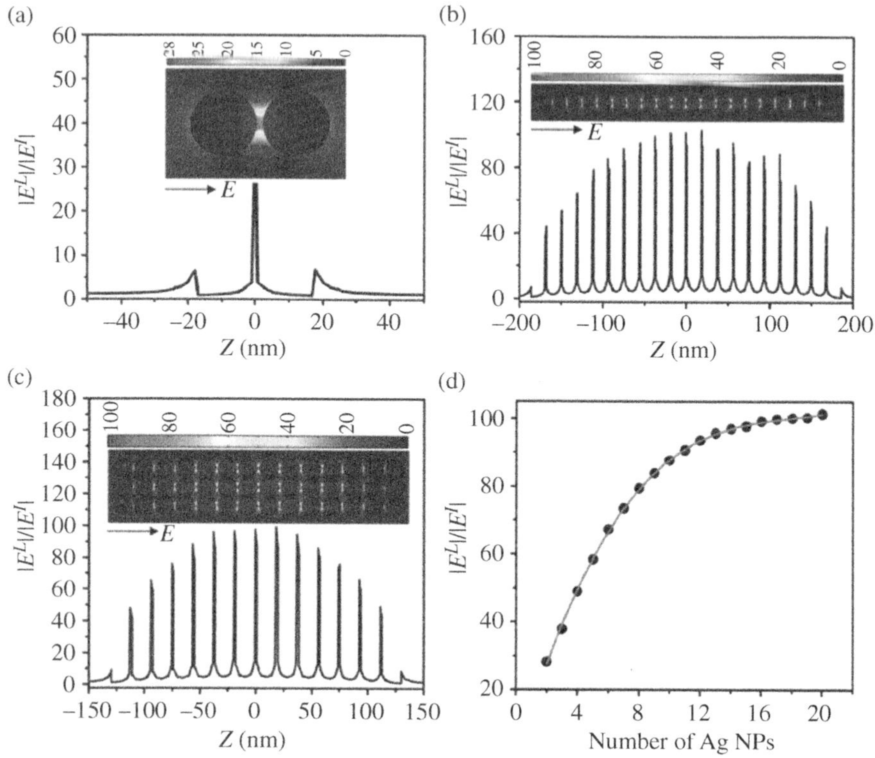


**Figure 2.4** (a) EM enhancement profile for $N = 2$. (b) $N = 14$. (c) 2D array using excitation wavelength 488 nm. (d) EM field enhancement with the presence of number of Ag NPs. *Source*: Revised from Majumdar et al. (2013).

and the reference probe molecules respectively. A closer look at Figure 2.5 from the lower panel shows that the probe molecule MB at $10^{-3}$ M concentration is observed to have very less intense peaks, but at pico-molar concentrations along with Ag–DNA shows highly intense peaks at 1625, 1595, 1376, and 1184 $cm^{-1}$ with EF values $10^8$ and $10^{10}$. This enhancement is related to the use of DNA, which ensured the increase in "hot spots" with enhanced plasmonic coupling.

In another work, Au NPs were prepared by UV irradiation for four hours with different concentrations of precursor solutions and ended in forming nanowires of Au in DNA (Kundu & Jayachandran, 2013). Here, the method of formation of $Au^0$ seeds is by UV photoirradiation where DNA helps to reduce $Au^{3+}$ ions to give Au0 NP seeds and finally generate Au–DNA nanowires. By varying concentrations, the resultant sizes of Au–DNA were also varied such as $12 \pm 3$ nm for Au–DNA nanowires with SPR band at 529, $65 \pm 5$ nm with SPR band at 540, and $80 \pm 10$ nm with SPR band at 548 nm. Therefore, from this it is easy to conclude

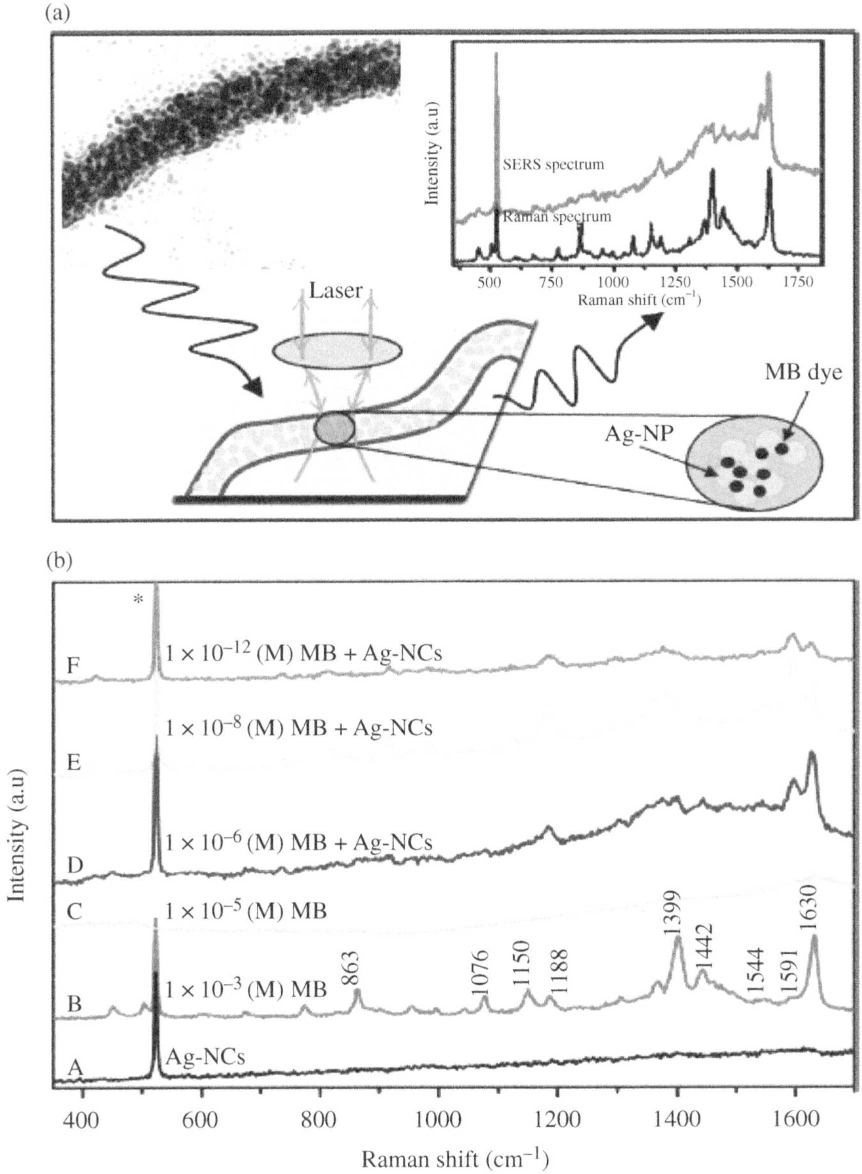


**Figure 2.5** (a) Upper panel shows the overall depiction of Ag–DNA for SERS and (b) the lower panel indicates the SERS enhancements at low concentrations of MB probe. *Source*: Revised from Majumdar et al. (2013).

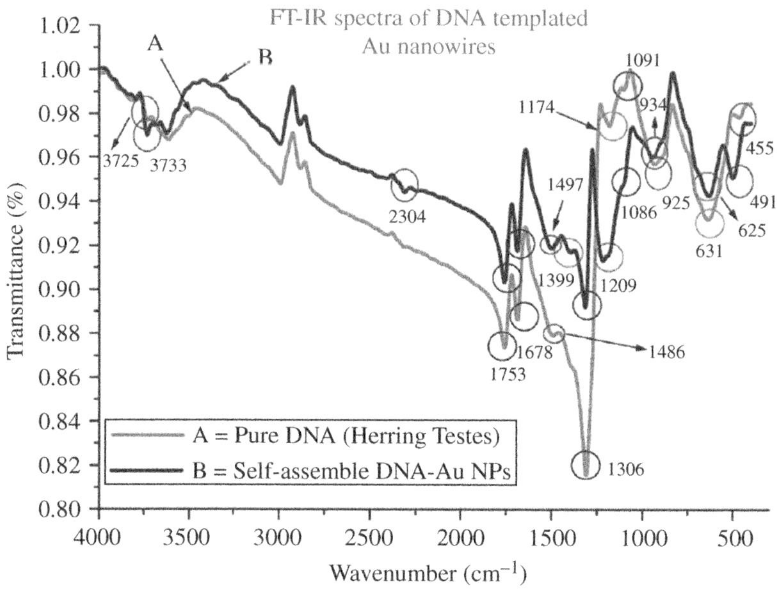


**Figure 2.6** FT-IR spectral characteristics of Au–DNA nanowires. *Source*: Revised from Kundu and Jayachandran (2013).

that the concentration of DNA plays a prominent role in designing the length of Au–DNA nanowires. This variation in wire length and the assembling of metal NPs also have a huge impact on the SPR bands as can be seen from the significant change in the absorbance wavelength. Moreover, this can influence the SERS activity with resonance of applied EM field and localized surface plasmons. The interaction of DNA with Au can initially be analyzed with electronic spectral studies with DNA and $Au^{3+}$ ions added where the peak position and intensity of pristine DNA vary while added with $Au^{3+}$. The shift in DNA peak from 258 nm for aromatic base pairs with $Au^{3+}$ ions implies that there is some interaction with DNA and $Au^{3+}$ ions. Fourier transform infrared spectroscopy (FT-IR) gives exact information on the functional groups that anchor the metal ions of interest. Here, as shown in Figure 2.6 for Au–DNA, the DNA shows different transmittance curves and the major ones are the hydroxyl, amides, phosphates, and carbon–nitrogen, hydrogen stretching and bending vibrations. These transmittance curves will get shifted to the higher or lower wavenumber regions when interacted with metal NPs. Therefore, it is apparent that the functional groups can be identified. Figure 2.6, which is the FT-IR spectra for DNA and Au–DNA, shows deviations in transmittance curve positions like 3615 (—OH), 1678(C=O, CONH), 1493 (C—H bending), 1306($PO_2$–stretching), 934(deoxyribose region), and 1091 $cm^{-1}$(C—O—C).

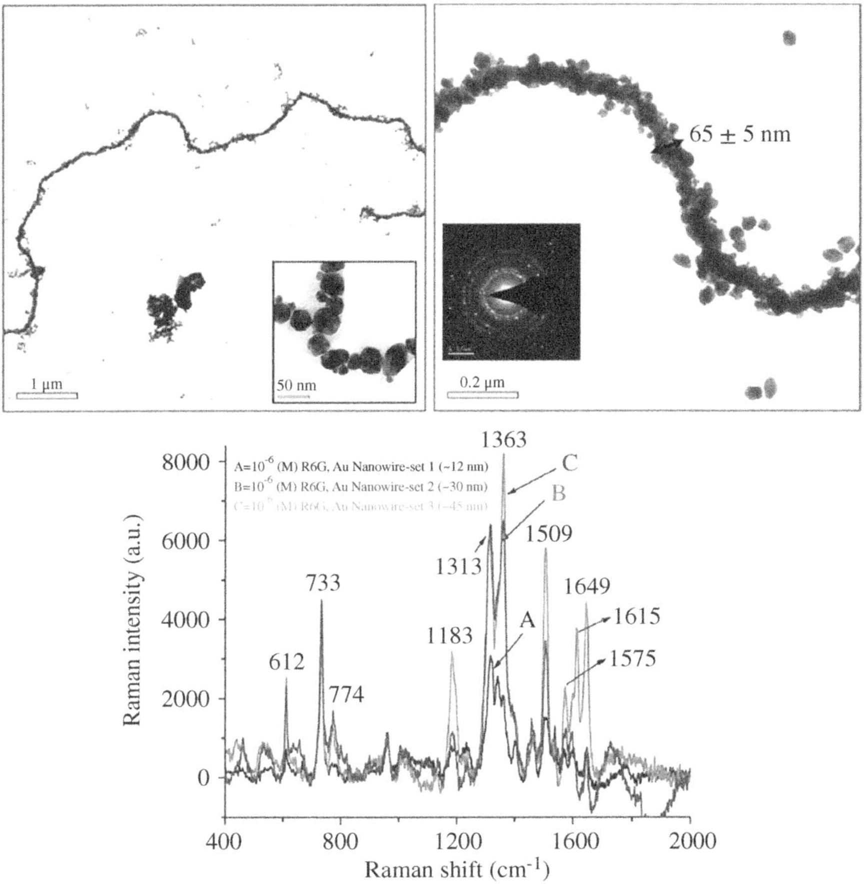


**Figure 2.7** Upper panel shows the TEM images of Au–DNA and the lower panel shows the SERS enhancements of Au–DNA with the probe molecule Rhodamine-6G at lower concentrations. *Source*: Revised from Kundu and Jayachandran (2013).

Therefore, the peak positions vary for the pristine DNA and Au–DNA for the particular functional groups of DNA and this in turn further ensures the nature of interactions between Au and DNA. The morphology of Au NPs over DNA nanowires was found to be spherical in nature and arranged so closely over DNA as seen from the upper panel of Figure 2.7. The SERS enhancements of Au–DNA were confirmed by using the Rhodamine-6G probe at micro-molar concentrations. The three differently sized Au–DNA NPs could deliver SERS activities with considerable intensities that showed the potentiality of DNA-embedded NPs like Au. Moreover, the calculated EF value was found to be in the range of $10^6$ and it was found that the larger Au NPs showed increased enhancement as observed from lower panel of Figure 2.7. This increased activity of Au NPs with increase in size is well observed as the SPR bands are also red shifted indicating the increase in induced localized surface plasmons for high probable resonance with the incident laser beam. Here, the Au NPs' size with 45 ± 5 nm and with DNA wire like morphology assist the closer arrangement of NPs and hence by both size and enhanced "hot spots," they can deliver considerable enhancements with lower probe molecule concentrations.

The size effect of NPs with DNA was studied in detail with the spherical Rh NPs over DNA by varying the concentration of DNA, where small-, medium-, and large-sized Rh NPs were formed over the DNA scaffold (Kundu, Chen, et al., 2017). UV irradiation was carried out for five hours with the wavelength of 260 nm. The Rh NPs' size was observed as 3 ± 0.5, 2 ± 0.6, and 0.9 ± 0.3 nm from larger to small size with DNA nanowires. Here, 2,7-DHN was used and ensured the fast and efficient reduction of Rh NPs over DNA and formed as too small sized Rh NPs. SERS activities of Au and Ag are well-explored and the Rh metal surfaces with DNA scaffold show induced electric field and hence that activity is enhanced with high EF values. The electronic spectra showed the interaction between the Rh and DNA moieties. The UV–vis spectra in Figure 2.8 show differences in the absorption intensities and shifts. The UV–vis spectra of RhCl3.H2O shows absorption peak at 519 nm, which is due to the LMCT and when it was added with the DNA

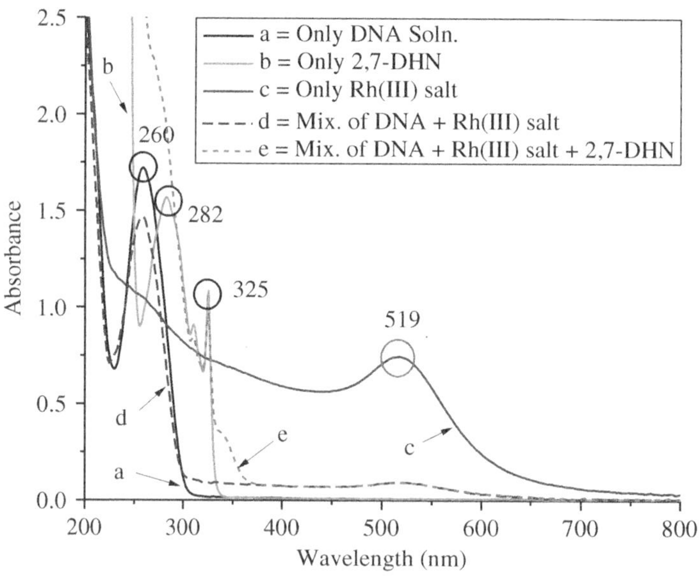


**Figure 2.8** Electronic spectral features of DNA aided Rh NPs. *Source*: Revised from Kundu et al. (2017).

which have peak at 260 nm is shifted and intensity is varied that gives the information about the probable interaction between them. The addition of both 2,7-DHN and NaOH to the reaction mixture produces peaks near 453 nm, which could be due to the interactions between the reagents. After washing, these peaks diminished, showing the presence of Rh NPs. This gives a clue that after the UV irradiation, Rh NPs' seed growth is initiated and gives Rh–DNA assemblies. The similar formation of Rh–DNA can provide more junctions between NPs and this will be more advantageous for enhancing the intensities of Raman signal like those observed in Ag–DNA and Au–DNA. The concentration differences of DNA could give small-, medium-, and large-sized NPs of Rh in DNA. The morphologies of Rh–DNA were observed to be spherical Rh NPs with DNA nanowires and the particles size was much smaller compared to the Ag–DNA and Au–DNA. This is attributed to the 2,7-DHN and NaOH used during the UV irradiation that enabled the fast and efficient reduction of Rh NPs with size below 5 nm. The schematization of formation of differently sized Rh NPs over DNA is shown in Figure 2.9. At a closer look, the particles' size can be varied with respect to the concentration of DNA, even though other parameters remain the same. This is really a huge

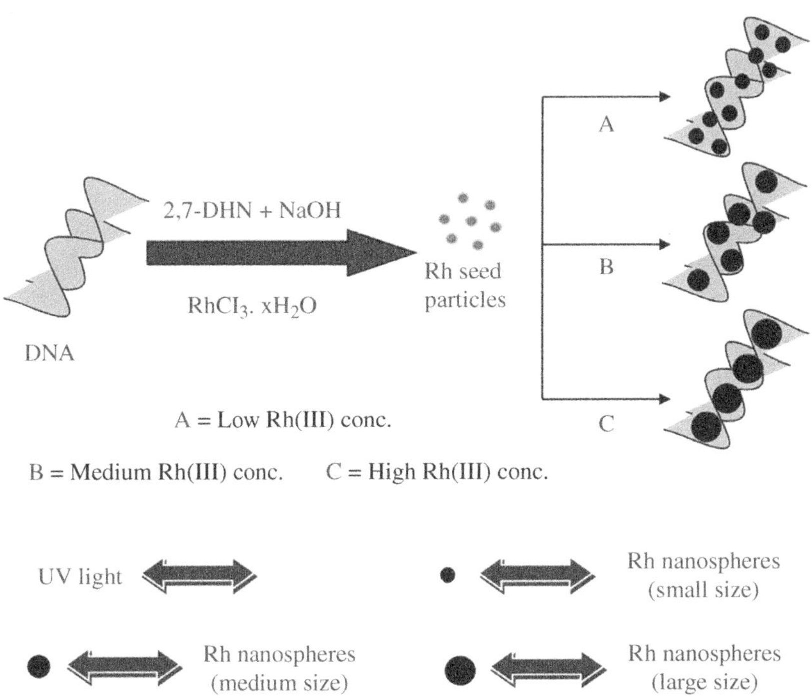


**Figure 2.9** Schematic illustration of small-, medium-, and large-sized Rh NPs over DNA scaffold. *Source*: Revised from Kundu et al. (2017).

advantage as only by varying the concentration of DNA, the outcome of the Rh NPs can be varied. This will have a big impact on the SERS signals of the probe molecules with increased EF values. The probe molecule MB is testified with the Rh–DNA for SERS applications with small, medium and large size Rh NPs. The Rh NPs with small size showed highly intense Raman signals at the MB peaks corresponding to the values of 443, 1392, and 1624 $cm^{-1}$. The high-intensity signals were observed with small-sized Rh–DNA with EF value of $1.53 \times 10^2$ compared to other medium- and large-sized Rh NPs. Compared to Au–DNA and Ag–DNA, here with Rh NPs, the EF value was lesser and this is because of the SPR bands lying in the range between 400 and 600 nm for Au or Ag NPs, matching with resonant frequency of incident laser with the wavelength of 532 nm, which is not observed for Rh NPs.

After exploring the prominent role of DNA in deciding EF values at low concentrations with size and shape, a comparative study was carried out with other constituents like cetyl trimethyl ammonium bromide (CTAB) and polyvinyl alcohol (PVA) (Kundu, Yi, et al., 2017). With UV photoirradiation, by changing the scaffolds, the morphologies of Pd have been varied as cubes and nanowires. The irradiation of UV light resulted in differently sized NPs and the nanocubes formed with CTAB show length of $37 \pm 5$ nm. The PVA-assisted Pd nanowires showed length of 3–4 μm and particle size of 25–35 nm. The DNA with Pd nanowires showed length of 3–5 μm and particle size of 3.5–5 nm. The schematic formation of Pd NPs with scaffolds like CTAB, PVA, and DNA is shown in Figure 2.10. The interactions of Pd with DNA found with Au, Ag, and Rh–DNA systems show prominent shifts in absorptions. Since the morphology is a main parameter in deciding the enhancements in SERS, here the morphological roles were studied. For SERS, the low interparticle gap is necessary and here the DNA only can provide such platform for particles to arrange in closer proximities as seen from the upper panel of Figure 2.11. With DNA, the advantageous presence of low interparticle gap and wire-like structure together assists the enhanced coupling of plasmons whereas in nanocubes of Pd from CTAB and nanowires of Pd from PVA could not do so. This shows the advantageous nature of DNA compared to other scaffolds for SERS applications. Therefore, it is apparent that the use of DNA in SERS will be highly useful for the detection of probe molecules at very low concentrations. The EF value for the Pd–DNA for the micro-molar concentrations of MB molecule was around $1.90 \times 10^5$ at a peak position of 1391 $cm^{-1}$ as shown in the lower panel of Figure 2.11. Pd NPs from CTAB and PVA showed $1.96 \times 10^3$ and $1.29 \times 10^2$, respectively which is very lesser comparing DNA scaffolds. In a trace detection of probe molecules at very lower concentrations, the appliance of DNA can be a fruitful and simple process along with rough metal surface of interest. This method of photoirradiation is advantageous as the process is simple and DNA itself reduces metal ions such as Au and Ag in a significant way.

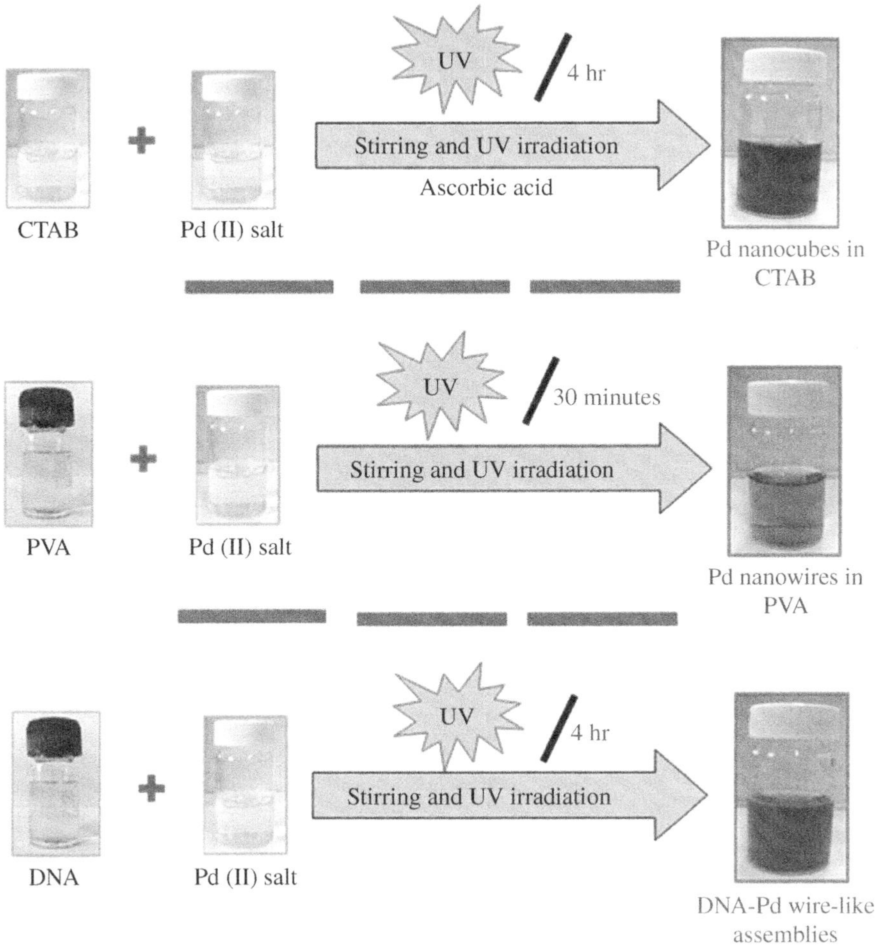


**Figure 2.10** Schematic illustration of Pd NPs over DNA, CTAB and PVA scaffolds for SERS. *Source*: Revised from Kundu et al. (2017).

### 2.3.2 Metal NP Assemblies on DNA Using Chemical Reduction Process as Aquasol for SERS Studies

The chemical reduction method of preparation of NPs is a fast and efficient method of synthesizing NPs and involves the addition of reducing agents like $NaBH_4$ that reduce metal ions to their corresponding metallic zero states. Moreover, these metal–DNA colloidal solutions in aquesol (prepared in aqueous medium) are highly stable even up to six months when kept in refrigerator. Therefore, the probability of using them for intensive applications in medical

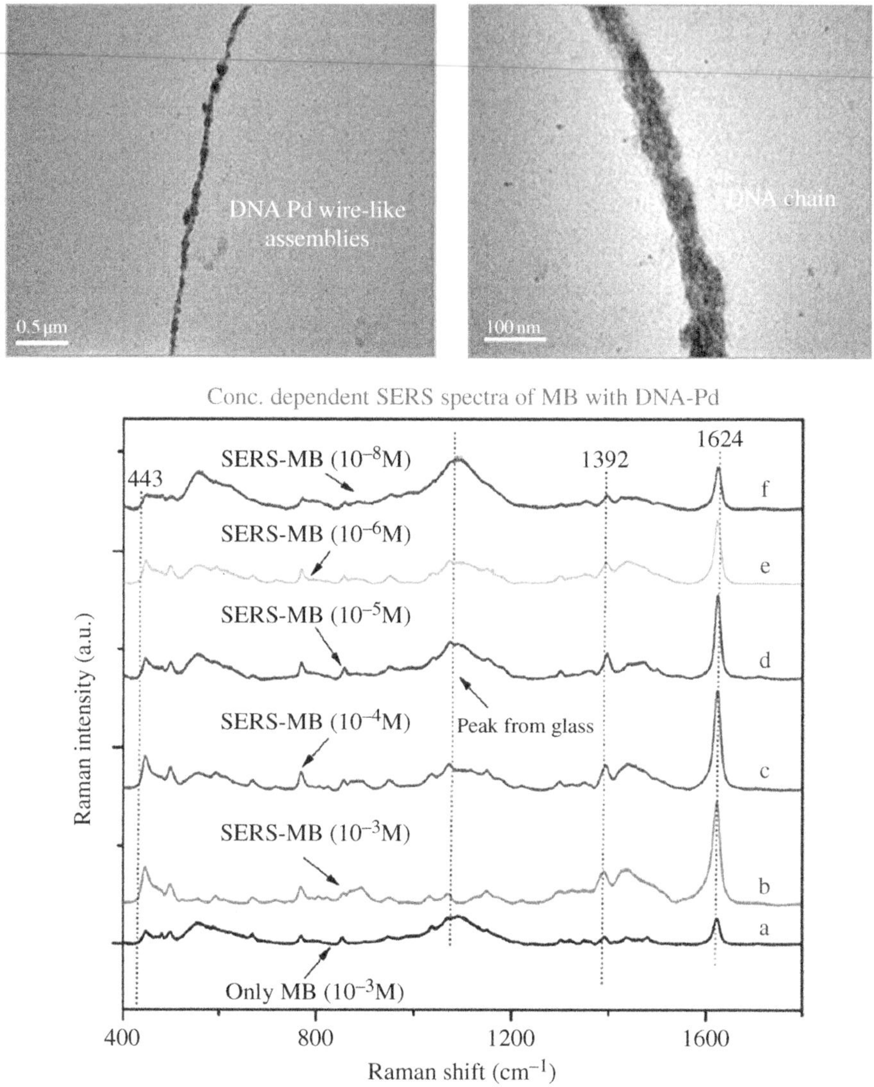


**Figure 2.11** Upper panel shows TEM images of Pd–DNA nanowires and lower panel shows the comparative SERS study of Pd–DNA with MB probe molecules. *Source*: Revised from Kundu et al. (2017).

fields for tracing probes by SERS is simple and fast way. The DNA scaffolds, even with fast reduction of metal ions to NPs tend to stag the NPs over it and forms assembly of metal–DNA structure provided the concentration of DNA is optimized. Hence, DNA acts as a versatile scaffold with slow and fast reduction of metal ions. In the formation of Pt NPs through chemical reduction method, DNA could stabilize Pt–DNA and formed as assemblies with stable nature (Sankar et al., 2018). The metal ions concentration was varied while the DNA concentration is fixed for Pt–DNA and formed stable solutions of Pt–DNA conferring the successful formation of Pt–DNA nano assembly. Here, $NaBH_4$ is added to the solution containing DNA and $Pt^{4+}$ ions. Initially, there is electrostatic interaction between the $Pt^{4+}$ ions and the negative ions from the aromatic molecules of DNA that forms an assembly. These ions, when $NaBH_4$ is added, simultaneously got reduced to metallic Pt and as such assembled over DNA chain and thus formed Pt–DNA nanostructures. The interactions of both Pt and DNA can again easily be identified with spectral studies like UV–vis studies and FT-IR. These two studies clearly confirm the functional groups' participation and interaction between Pt and DNA as similar to the observed earlier metal ions for DNA. The morphological outcomes of Pt–DNA are shown in Figure 2.12. The particle size of Pt is too small, below 5 nm, and this is the case normally observed when strong reducing agents like NaBH4 are used for the reduction of metal ions in DNA assembly. Figure 2.12a–d shows the small-sized Pt NPs with DNA chains and assembled closely, which is necessary for enhancing the activities in SERS. Figure 2.12e–f shows the fine lattice fringes of Pt with (111) and (200) planes and selected area diffraction (SAED) respectively. The Pt–DNA was used for detecting the probe

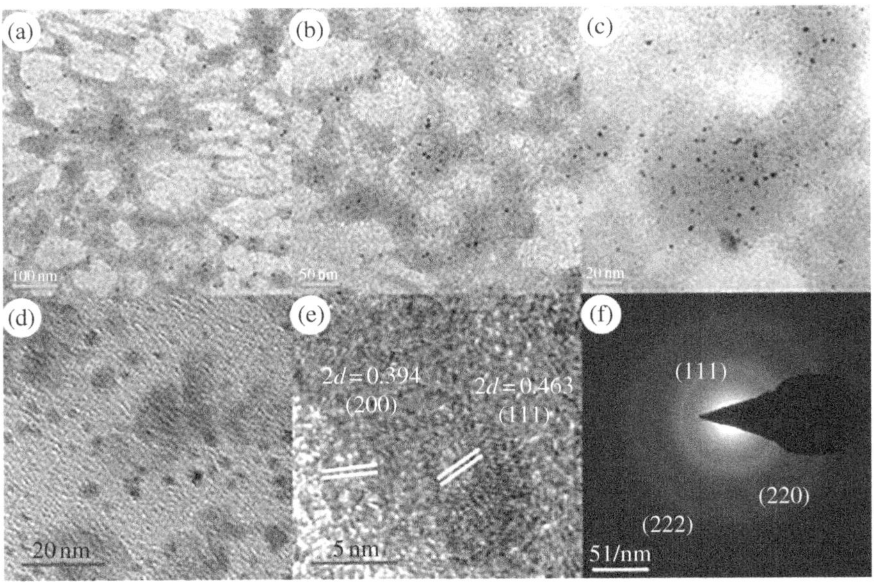


**Figure 2.12** (a–d) HR-TEM images of Pt–DNA with chain-like structure. (e, f) fine lattice fringes and SAED pattern respectively. *Source*: Revised from Sankar et al. (2018).

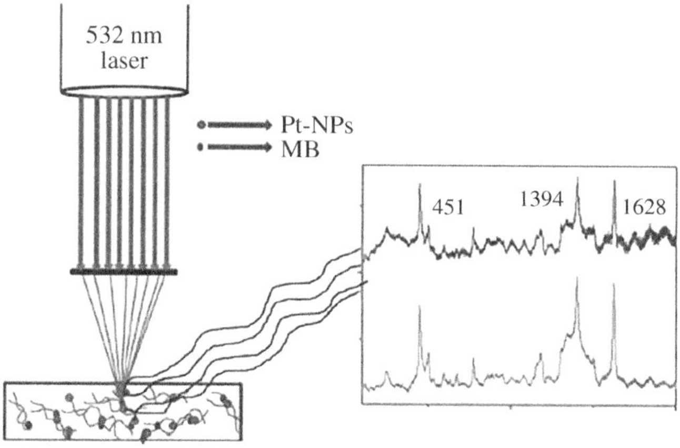


**Figure 2.13** Representation of 532 nm laser with Pt–DNA and MB probe for SERS applications. *Source*: Revised from Sankar et al. (2018).

molecule MB at lower concentrations. As expected, SERS activity was superior in Pt–DNA nanostructures. The close arrangement of Pt NPs with more "hot spots" is key in delivering high enhancement for the Raman signal even at very low concentrations. In chemical reductions, unlike photochemical reductions, fast reduction occurs and hence controlling the size and shape of metal NPs is rather difficult. Hence, in this case, the particle size is too small with irregular shapes of NPs but DNA strand joined and form a chain like and wire like structures as observed in photochemical conditions. Figure 2.13 shows the interaction of 532 nm laser with Pt–DNA and MB probe with enhanced peaks at 451, 1394, and 1628 $cm^{-1}$. The calculated EF value was very high, $2.52 \times 10^5$, for a Pt–DNA concentration of 0.003 M. This EF is comparable with other reports mentioned above like those on Rh–DNA prepared by photochemical conditions. Therefore, with the method of reduction, for different values of reaction time, concentration of DNA, and concentration of metal ions, it is possible to make different combinations and the same can be applied for SERS for obtaining improved intensities at nano- and pico-molar concentrations.

The role of chemical reduction in Rh NPs for SERS was monitored with $NaBH_4$ as a reducing agent. The DNA concentrations are varied to stabilize different Rh–DNA combinations (Sangeetha, Sankar, et al., 2019). The observed chain length of DNA was around 98 nm and the Rh NPs' size was below 5 nm. This is similar to the one observed in the Pt–DNA case for SERS. This study focuses on the effect of concentration of DNA in SERS while the former focused on the concentration of $Pt^{4+}$ in SERS with the same concentration of DNA. The schematic presentation

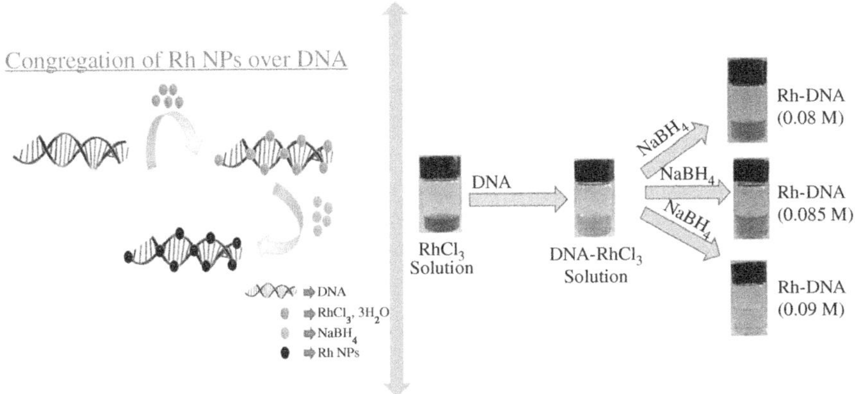


**Figure 2.14** Schematization of Rh–DNA shown on the left-hand side. The visual appearances are shown on right. *Source*: Revised from Sangeetha et al. (2019).

of Rh–DNA by chemical reduction is shown in Figure 2.14. The $Rh^{3+}$ ions first assembled over the DNA chain through aromatic molecules and once the reducing agent $NaBH_4$ was added, they got reduced rapidly and reassembled over the DNA chain. The same is depicted pictorially where $RhCl_3$ solution is in pink and once DNA, which is colorless, is added, the color fades a little. Once the reducing agent is added, fast nucleation is triggered and finally after forming Rh NPs, they became Rh–DNA, a wine-colored solution. This visual change with Rh–DNA combinations gives information on the formation of NPs. Moreover, the morphology observed here is chain-like nanostructures of Rh with DNA as depicted in the upper panel of Figure 2.15. From the figure, we can visualize the small NPs very clearly with the size below 5 nm over the DNA chain. The SERS activities of Rh–DNA with MB probe is testified and showed impressive enhancements with very less concentrations of MB. The peak positions of 445, 1392, and 1620 $cm^{-1}$ of MB probe become less intense at $10^{-3}$ M concentration. The same when added with Rh–DNA, because of the presence of excess "hot spots" with DNA architecture, could give high enhancements in the Raman signal even up to micro-molar concentrations as shown in the lower panel of Figure 2.15. The observed EF value was higher for Rh–DNA (0.08 M), $2.52 \times 10^5$. This is higher compared to that of the Rh–DNA prepared in the photochemical conditions, which was $1.53 \times 10^2$. From this, it is clear that the method of synthesis also show potentiality in SERS along with the concentration of DNA. Here, this much enhancement might be attributed to the closely arrangement of Rh NPs during the fast reduction method. The excess "hot spots" with Rh–DNA and facile DNA structure were able to deliver enhancements with the localized electric field with the incident laser.

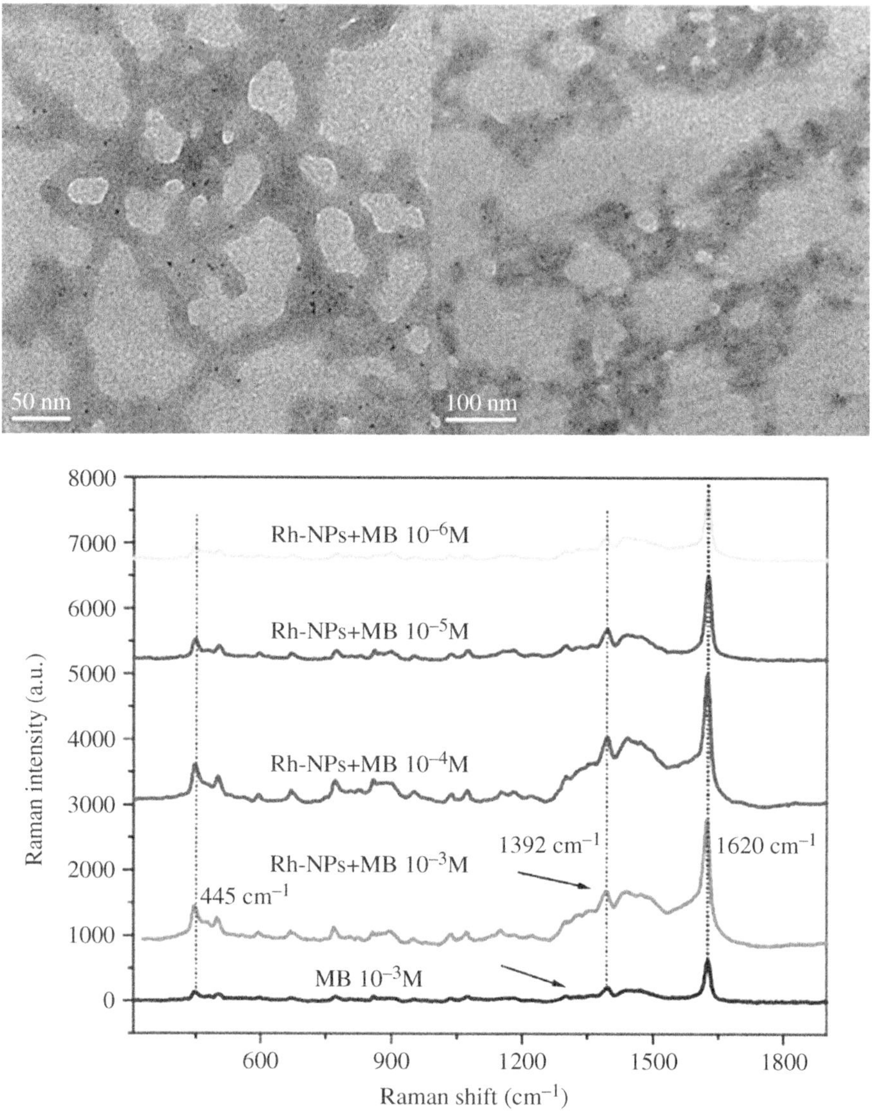


**Figure 2.15** Upper panel shows the morphology of Rh–DNA from TEM. Lower panel shows the SERS application of Rh–DNA with MB probe. *Source*: Revised from Sangeetha et al. (2019).

The Re–DNA nanostructures for SERS with fast chemical reduction showed high enhancements compared to Rh–DNA and Pt–DNA (Anantharaj et al., 2016). The schematic from Figure 2.16 shows the stepwise processes that occurred while forming metallic Re nanostructures over DNA scaffold. The $ReO_4^-$ ions get

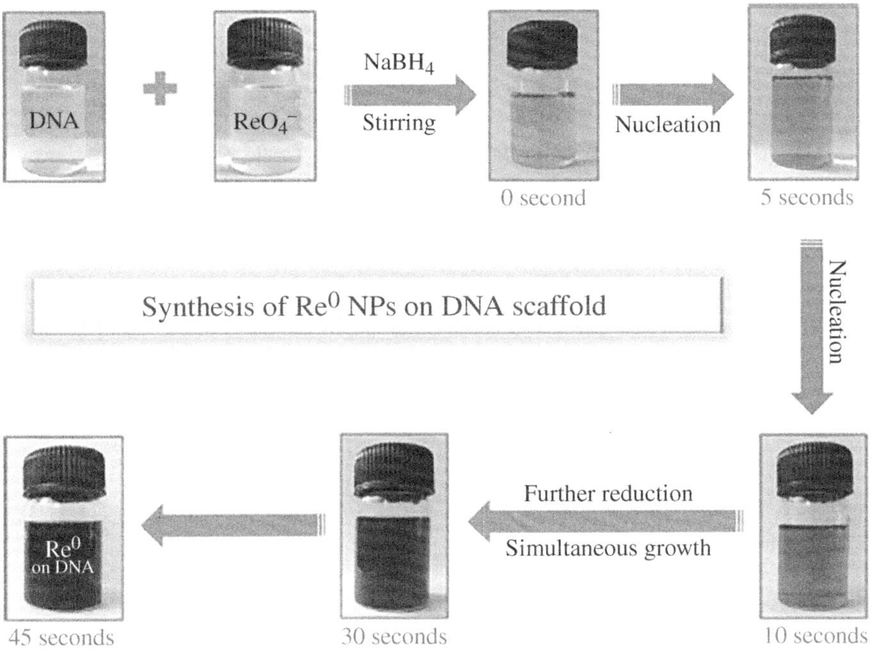


**Figure 2.16** Schematization of Re–DNA by a chemical reduction method. *Source*: Revised from Anantharaj et al. (2016).

assembled over the DNA chain. Here, unlike other metal ions studied so far where metal cations only anchored, in this case, $ReO_4$– is anchored which shows the potentiality of DNA as a scaffold. The DNA offers wide range of functional groups that can participate in anchoring neutral, anionic, and cationic ions over it and forms a perfect metal–DNA assembly. By varying the DNA concentration, particle sizes of 0.6–1.1 nm grain sizes of Re NPs were prepared. The morphology as depicted from the upper panel of Figure 2.17 shows tiny particles with size below 2 nm and diffused ring in SAED pattern. The SERS activity was much higher with the probe MB molecule and showed intense Raman signals with concentrations up to $10^{-7}$ M. The peaks at 448, 855, 1437, and 1619 $cm^{-1}$ show enhancements. The EF value was $2.07 \times 10^7$ for the Re NPs at a peak position of 1619 $cm^{-1}$ as shown in the lower panel of Figure 2.17. The overall picture of chemical reduction of metal–DNA nanostructures showed the fast reduction of NPs with very small size below 5 nm. These small-sized NPs with the assistance of DNA can have more accessible "hot spots" for SERS detection even up to micro- and nano-molar concentrations. By comparing the photochemical and chemical reduction methods, Rh–DNA showed higher EF values. This is attributed to the

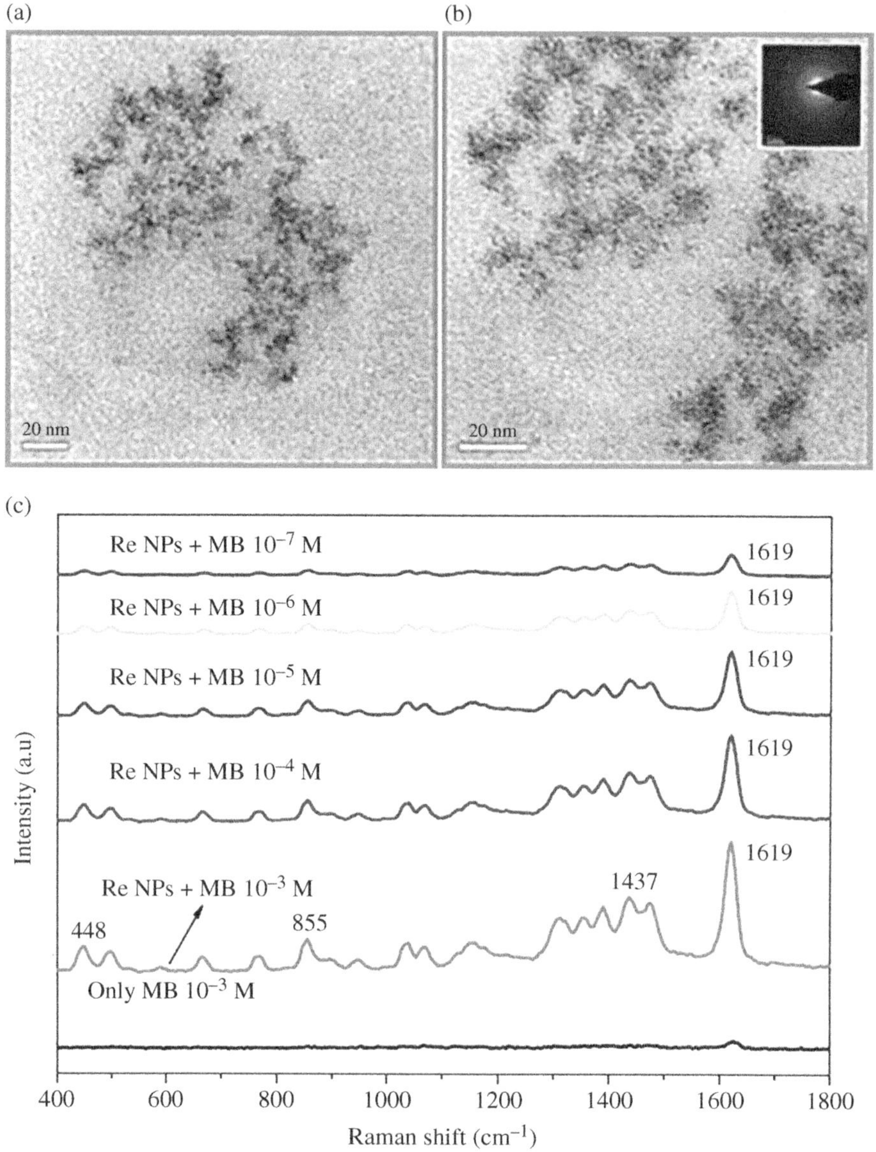


**Figure 2.17** (a) Upper panel shows the morphologies of Re–DNA and (b) lower panel shows the SERS enhancements with MB probe. *Source*: Revised from Anantharaj et al. (2016).

small-sized NPs with more active hot spots for the enhancements. Moreover, the chemical reduction method is a faster and more efficient way of preparing NPs. Recently, the effect of $RuO_2$–DNA catalyst in SERS was further identified by changing the concentrations of DNA (Sangeetha, Thriuvengetam, et al., 2019).

### 2.3.3 Metal NP Assemblies on DNA Using Chemical Reduction as Organosol for SERS Studies

The fine control of metal NPs over DNA can be achieved with the method of organosol (reaction done in organic medium) compared to aquesol where random arrangement of NPs is observed. The utilization of organosol is to control over the structure of NPs with monodispersity. As showed in Figure 2.18, the $ReO_4-$ ions and DNA have been kept in refrigerator with tetraoctyl ammonium bromide (TOAB) for getting transferred into the acetonitrile organic layer (Sakthikumar et al., 2016). Then after adding $NaBH_4$, monodispersed and spherical Re NPs were formed over the DNA scaffold. This method is advantageous to prepare controlled shapes of Re NPs over DNA. The morphology of Re–DNA organosol shows fine spherical particles of size $1.5 \pm 0.5$ and $6.5 \pm 0.5$ nm for different concentrations of DNA as observed from the upper panel of Figure 2.19. SERS was performed with the MB probe molecule as shown in the lower panel of Figure 2.19. The EF value was $3.35 \times 10^7$ for the Re–DNA organosol at a peak position of $1618\,cm^{-1}$. Considering the monodispersity, shape control, and along with DNA assembly, the Re–DNA combination could give high EF values for the MB probe. By the methods of organosol, as discussed above, the particle size and shape can be regulated and the same can be modified

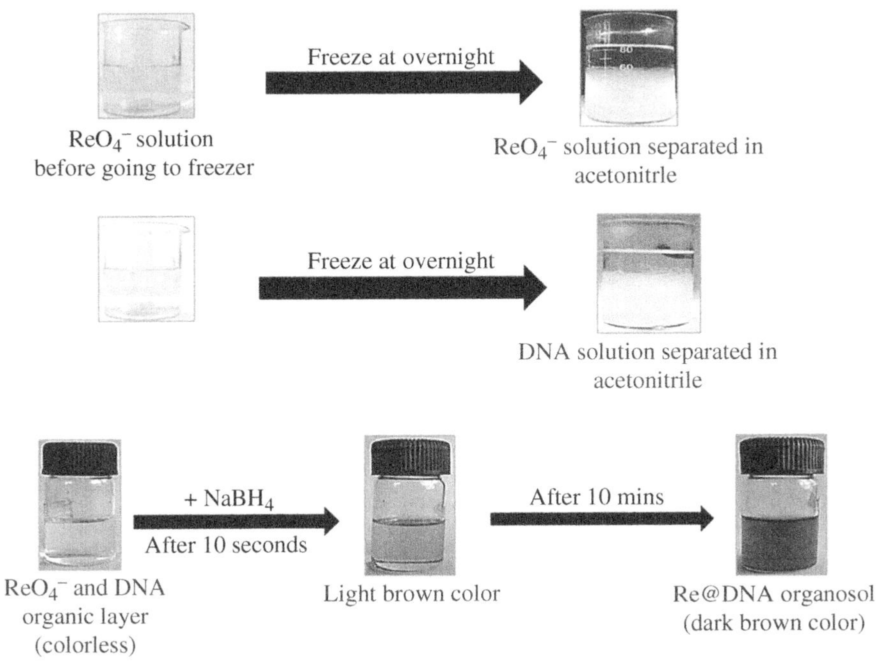


**Figure 2.18** Schematization of Re–DNA organosol by a chemical reduction method. *Source*: Revised from Sakthikumar, Anantharaj, Ede, Karthick, and Kundu (2016).

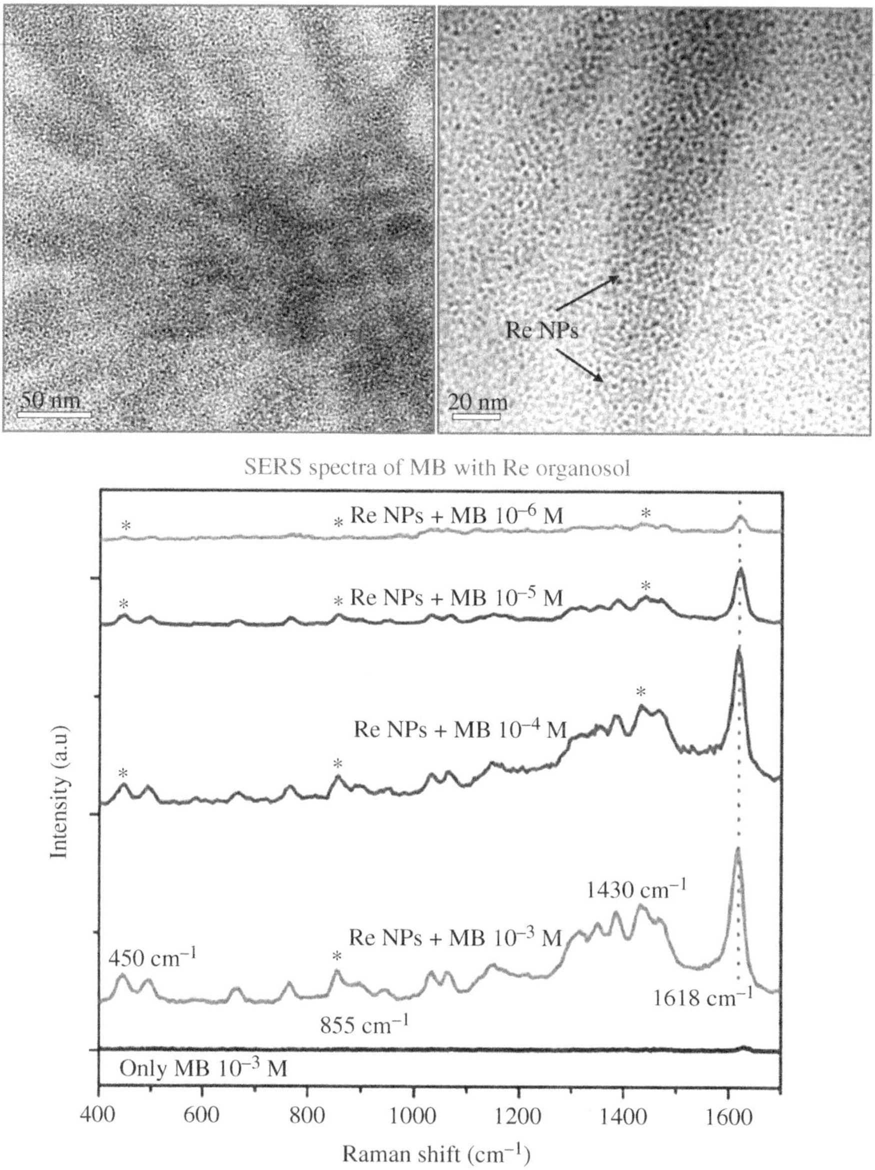


**Figure 2.19** SERS application of Re–DNA organosol by a chemical reduction method with MB probe. Upper panel shows TEM images and lower panel shows SERS data. *Source*: Revised from Sakthikumar et al. (2016).

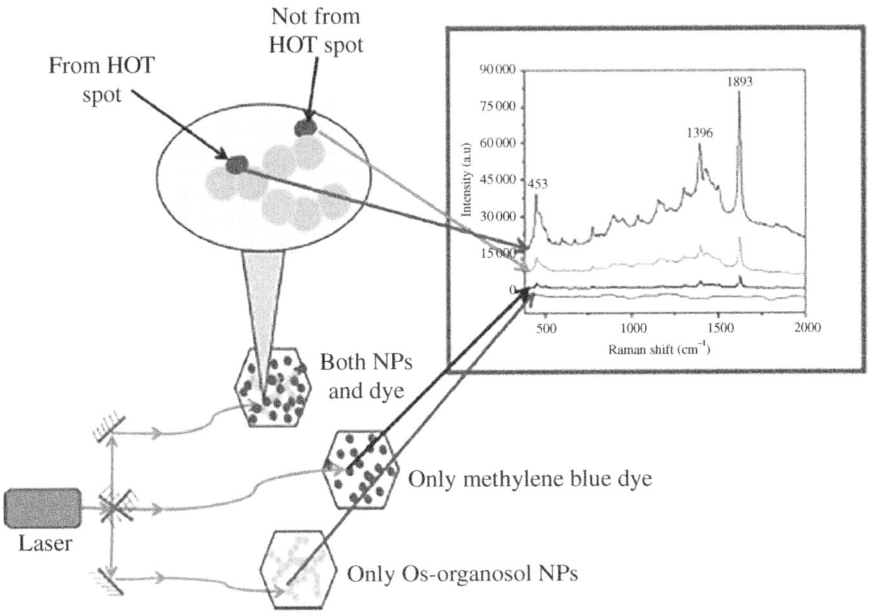


**Figure 2.20** Schematic of Os–DNA organosol for SERS application. *Source*: Revised from Anantharaj et al. (2014).

with the concentrations of DNA. In another work, the Os–DNA organosol showed higher EF values when applied for SERS (Anantharaj et al., 2014). The overall visualization of Os–DNA nanowires prepared in organosol for SERS is highlighted in Figure 2.20. The Os–DNA organosol showed small Os NPs with size around 3 nm and the nominal length of DNA was around 0.5–8 microns as seen from the upper panel of Figure 2.21. These Os NPs with more surface plasmons produced over the Os metallic surface interact more with the incident laser and give intense Raman signals for the MB molecule. The SERS signals of Os–DNA organosol with MB showed high enhancements with EF value of $5.22 \times 10^5$ as shown in the lower panel of Figure 2.21 at a peak position of 448 cm$^{-1}$. These activities observed in organosol with the appliance of DNA as a scaffold show higher shape selectivity and monodispersity of metal NPs prepared and moreover, the reproducibility in SERS enhancements is maintained while using DNA scaffold.

### 2.3.4 Metal NP Assemblies on DNA Prepared Using Microwave Heating for SERS Studies

The microwave method can also be adopted to design metal-DNA assemblies for SERS application. The Os–DNA nanowires and honeycomb structures were prepared in a microwave condition within just 30 secondss of reaction time

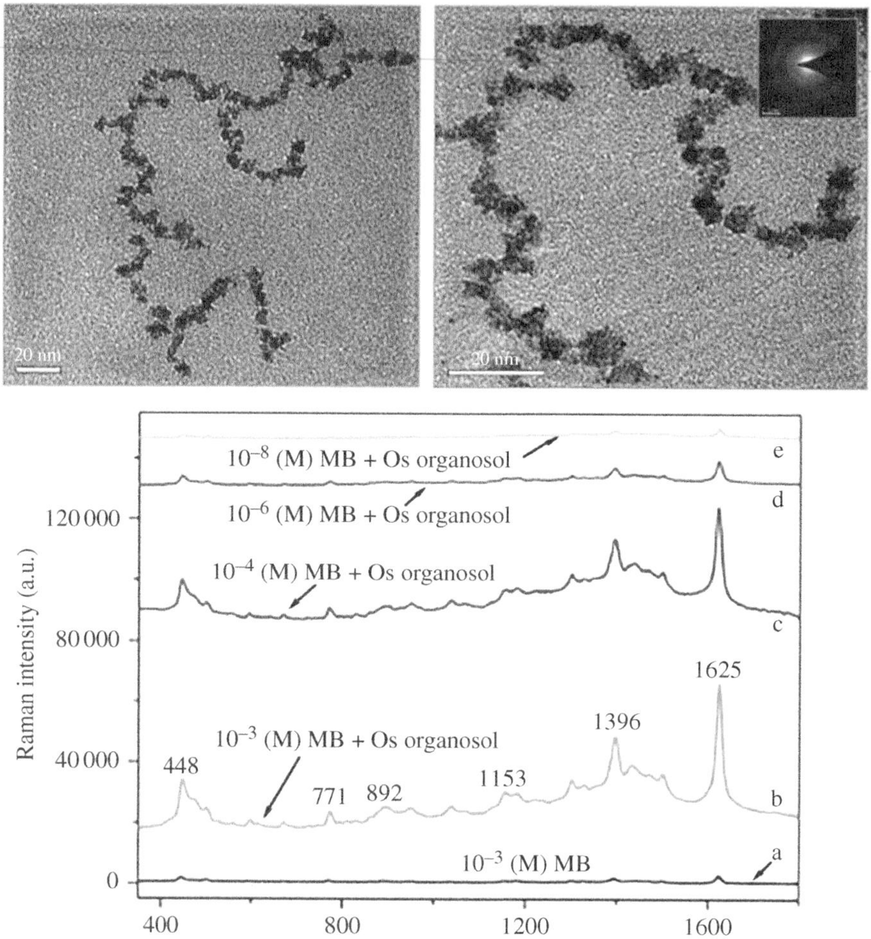


**Figure 2.21** Upper panel shows the morphologies of Os–DNA organosol and lower panel shows the SERS application with MB probe. *Source*: Revised from Anantharaj et al. (2014).

(Nithiyanantham, Ede, & Kundu, 2014). This microwave method of heating gives uniform and fast reduction and is even faster than both chemical and photochemical reductions. Here, with the presence of DNA, ethanol is used to reduce $OsO_4$–$Os^0$ by microwave heating. With the variation in the concentration of DNA, morphologies were observed to be Os–DNA nanowires and honeycomb structures. The confirmation of NPs is observed with the formation of SPR bands in the visible region of 600 nm. Electronic spectral studies give information on SPR

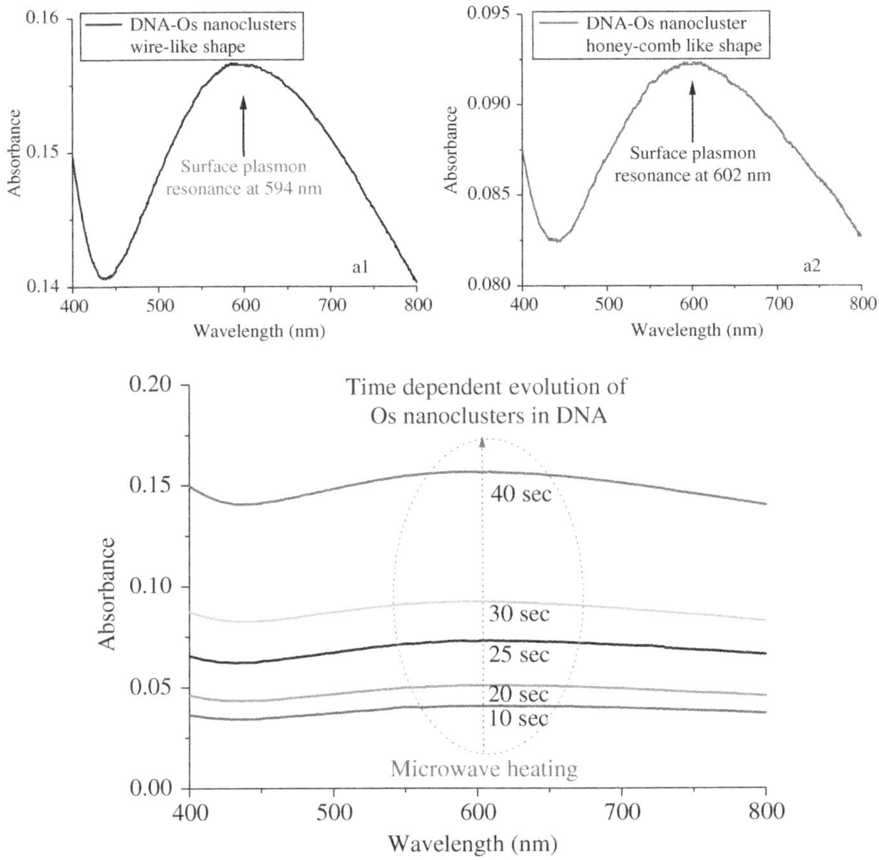


**Figure 2.22** Upper panel shows the SPR band of Os–DNA prepared by microwave conditions and the lower panel shows the time-dependent enhancement of SPR for Os–DNA. *Source*: Revised from Nithiyanantham et al. (2014).

bands for both wire-like and honeycomb-like structures as shown in the upper panel of Figure 2.22. The Os–DNA wire-like morphology shows an SPR band at 594 nm whereas Os–DNA honeycomb-like morphology shows an SPR band at 602 nm. The lower panel of Figure 2.22 shows the time-dependent enhancement in SPR band. The nanowire morphology of Os–DNA can be seen in the upper panel of Figure 2.23. The particle size was 1–3 nm with the DNA chain showing wire-like morphology. This method of microwave heating could give attractive assembling of Os NPs over DNA chain and forming wire-like and honeycomb-like morphologies within a short period of time. The potentiality of this Os–DNA was analyzed for SERS with MB probe molecule. The Os–DNA nanoclusters with the SPR band of 600 nm, which lies in the visible region, can be more advantageous when the applied laser beam comes at the same visible region and hence the Raman signal will be more intense even at very low concentrations. Here, the Os–DNA nanoclusters exhibited a high EF value of $1.34 \times 10^7$ at a peak position of $1400\,cm^{-1}$ as shown in the lower panel of Figure 2.23. This high enhancement again is similar to the apparent utilization of DNA that give sufficient room for arranging Os NPs with closer junctions with small size of below 3 nm. This set up

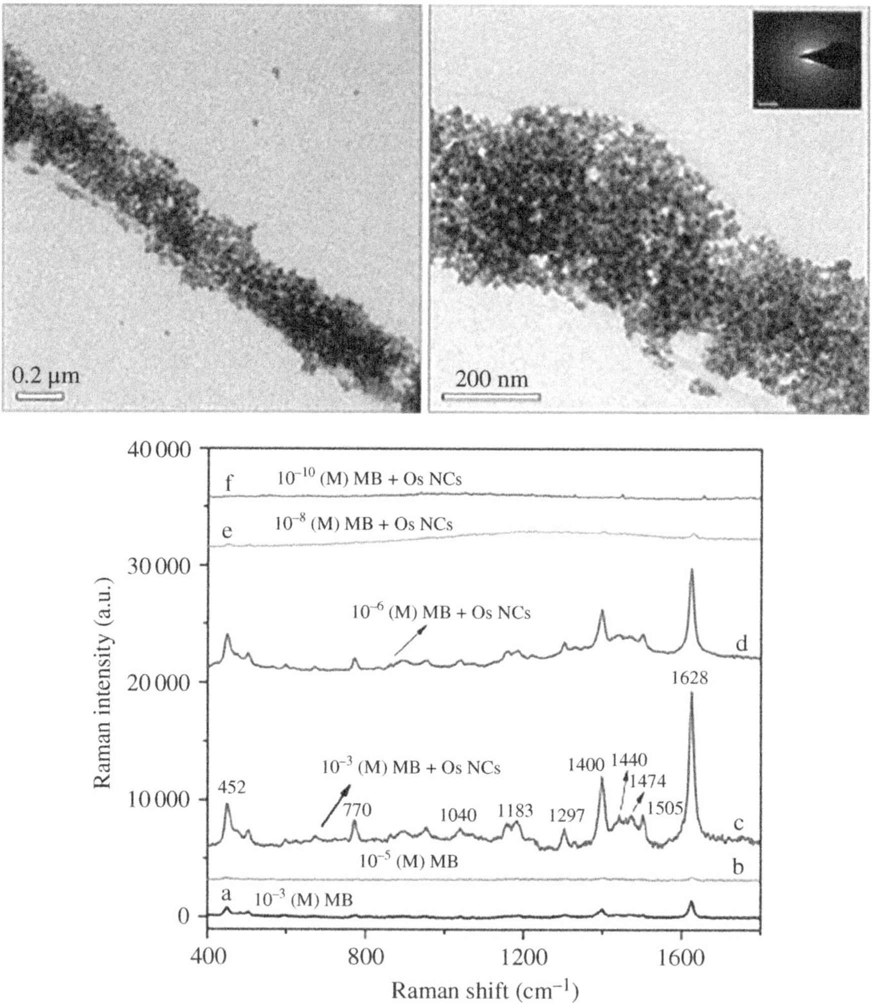


**Figure 2.23** Upper panel shows wire-like morphologies of Os–DNA and the lower panel shows the SERS activity of Os–DNA with MB. *Source*: Revised from Nithiyanantham et al. (2014).

of Os–DNA nanoclusters with SPR band in the visible region assured the pronounced "hot spots" and gave high EF values with $10^{-10}$ M concentrations of MB probe molecule. Therefore, as highlighted above with photochemical conditions, chemical reduction in both aquesol and organosol and in microwave conditions with the aid of DNA, it is possible to fine-tune the SERS activities of metal NPs in terms of size, shape, and monodispersity. A comparison of the overall depiction of morphology, reducing methods, and SERS enhancements is provided in Table 2.1.

### 2.3.5 Conclusions and Outcomes of DNA-Based Metal Nanostructures for SERS Studies

This chapter focuses on the fruitful utilization of biomolecule DNA for enhancing the SERS activity after successful scaffolding with metal NPs. The metal NPs and DNA interaction is possible due to the presence of aromatic base molecules along with sugar moieties and phosphate backbones can offer a wide variety of participants for anchoring both metal cations and metal anions. For application in SERS, the formation of localized electric field over the rough metal surface, which can produce surface plasmons, is important in judging the SERS signals even at low concentrations of probe molecules. The DNA is used for the formation of metal NPs, and this metal–DNA combination tends to produce more localized electric field because of the low interparticle gap between the NPs. Hence, when incident laser falls on the exposed "hot spots," intense Raman signals are observed even at pico-molar concentrations. Moreover, without the assistance of DNA, in SERS, reproducibility is questionable as the formation of NPs with low interparticle gap is rather difficult and the SERS activity cannot be reproduced to maximum extent. Based on the concentration of DNA, the particle size and structure of metal NPs-based DNA assemblies can be modified and the same will have influence on the SERS activities too. Similarly, the method of reduction of NPs over DNA, such as photochemical reduction, chemical reduction in aquesol and organosol and microwave conditions, will also play an influential role in directing the morphologies. Because of these advantages, DNA, a biomolecule, shows versatility in anchoring wide range of metal NPs and is able to detect probe molecules at pico-molar concentrations by creation of excess "hot spots." This chapter also highlights and studies the possible interactions of metal NPs-based DNA assemblies by electronic and FT-IR spectroscopic studies and morphological aspects by microscopic analyses. Moreover, the size and shape selectivity by various methods have been elaborated for finding best method for SERS detection. In futuristic, the readers can find many ways by simplifying the methods of synthesis of metal–DNA assemblies for SERS application. Hence, this kind of DNA-based metal NPs will have more impact in many interdisciplinary fields and particularly in medical

**Table 2.1** Comparison of metal–DNA assemblies and their corresponding SERS activities prepared in chemical, photochemical, and microwave conditions.

| S. No | Metal–DNA used | Method of reduction | Morphology | Size of metal NPs (nm) | EF value | References |
|---|---|---|---|---|---|---|
| 1 | Ag–DNA | Photochemical | Nanowires | $17 \pm 3$ | $10^8$–$10^{10}$ | Majumdar et al. (2013) |
| 2 | Au–DNA | Photochemical | Nanowires | 10–50 | $1.4 \times 10^6$ | Kundu and Jayachandran (2013) |
| 3 | Rh–DNA | Photochemical | Spherical | 1–3 | $1.53 \times 10^2$ | Kundu et al. (2017) |
| 4 | Pd–DNA | Photochemical | Nanowires | 3.5–5 | $1.59 \times 10^5$ | Kundu et al. (2017) |
| 5 | Pt–DNA | Chemical-aquesol | Nanowires | <5 | $2.52 \times 10^5$ | Sankar et al. (2018) |
| 6 | Rh–DNA | Chemical-aquesol | Nanowires | <5 | $1.19 \times 10^5$ | Sangeetha et al. (2019) |
| 7 | Re–DNA | Chemical-aquesol | Nanochains | $0.7 \pm 0.1$ | $2.07 \times 10^7$ | Anantharaj et al. (2016) |
| 8 | Re–DNA | Chemical-organosol | Spherical | 1.5–6.5 | $3.35 \times 10^6$ | Sakthikumar et al. (2016) |
| 9 | Os–DNA | Chemical-organosol | Wire-like | 1–3 | $2.07 \times 10^7$ | Anantharaj et al. (2014) |
| 10 | Os–DNA | Microwave | Honey-comb | 1–3 | $1.34 \times 10^7$ | Nithiyanantham et al. (2014) |

fields for the trace detections of probe molecules with the aid of metal NPs-based DNA assemblies.

### Take Home Message

- From this chapter, the influential role of DNA in anchoring metal ions and metal NPs and their interactions with DNA can be well-understood.
- The effect of concentration on DNA and its vital role for resultant particles' morphology and their sizes can be understood with this chapter.
- This chapter further deals with the methods of reduction for the formation of different NPs with DNA and the changes in SERS activities.
- The relation between the excess "hot spots" in the presence of DNA as scaffold and the enhancements in SERS activities are discussed clearly.
- This chapter also gives information on the effectiveness of DNA scaffold with different metal NPs prepared in different conditions for superior SERS activity.

## Questions and Answers

**1** How DNA can anchor metal ions and in what way the interactions between them can be identified?

The interaction of metal ions over DNA is possible because of the presence of different functional groups such as deoxyribose, aromatic N-molecules, and phosphates from the backbone of DNA. The interaction of DNA with metal ions can easily be monitored with electronic spectral studies. With FT-IR spectroscopy, it is possible to deduce the functional groups that interact with the metal ions.

**2** How the method of synthesis and concentration of DNA affect the morphological features of metal–DNA structures?

The morphological features of the formed metal–DNA assembly depend on the methods of reduction and the optimization of concentration of DNA. The conversion of metal ions to metal NPs is based on the nucleation of metal NPs from any method adopted like photochemical, chemical, and microwave conditions. This affects the substantial growth of NPs with varying concentrations of DNA. The DNA strands with the size of 2–3 nm will assemble together to form DNA chains with different nominal length and diameter and this in turn will have huge variation in the morphological outcomes with respect to both concentration and the method of reduction adopted.

**3** The stability of metal–DNA NPs is very high. Why?
By DNA assembly, with very less agglomeration of NPs, it is easy to have stabilized metal NPs. This extended stability is attributed to the self-assembling nature of DNA that strongly binds with metal NPs.

**4** The SERS activities of metal–DNA combinations are huge. Explain this.
In SERS, metal NPs with low interparticle gap is advantageous that shows excess "hot spots." These hot spots show enhanced plasmonic coupling because DNA is able to provide room for closer arrangement of NPs. Therefore, probe molecules with very low concentrations can be identified even in pico-molar concentrations.

**5** How DNA is more advantageous compared to other scaffolds for SERS of rough metal surfaces?
DNA is a biomolecular scaffold and has different functional groups that can assist the anchoring of anionic, cationic, and neutral molecules. The method of formation of assemblies of metal–DNA structure is too easy and fast method. Moreover, DNA itself can reduce metal ions by photochemical reductions using deoxyribose region. Likewise, it can act as both reducing agent and scaffold at a same time. Moreover, it can provide easy accessing of laser with more "hot spots" and so the probe molecules can be identified with intense Raman signals.

## Acknowledgements

This book chapter work was carried out by K. Karthick and Subrata Kundu. Authors wish to thank support from Council of Scientific and Industrial Research (CSIR)-Central Electrochemical Research Institute (CECRI), Karaikudi, India. K. Karthick also wishes to acknowledge University Grants Commission (UGC), New Delhi, India for the funding from the UGC-SRF.

## References

Alkilany, A. M., Lohse, S. E., & Murphy, C. J. (2012). The gold standard: Gold nanoparticle libraries to understand the nano–bio interface. *Accounts of Chemical Research*, 46(3), 650–661. doi:10.1021/ar300015b

Anantharaj, S., Nithiyanantham, U., Ede, S. R., & Kundu, S. (2014). Osmium organosol on DNA: Application in catalytic hydrogenation reaction and in SERS studies. *Industrial & Engineering Chemistry Research*, 53, 19228–19238. doi:10.1021/ie503667y

Anantharaj, S., Sakthikumar, K., Elangovan, A., Ravi, G., Karthik, T., & Kundu, S. (2016). Ultra-small rhenium nanoparticles immobilized on DNA scaffolds: An excellent material for surface enhanced Raman scattering and catalysis studies. *Journal of Colloid and Interface Science*, 483, 360–373. doi:10.1016/j.jcis.2016.08.046

Bhaumik, A., Shearin, A. M., Delong, R., Wanekaya, A., & Ghosh, K. (2014). Probing the interaction at the nano-bio interface using Raman spectroscopy: ZnO nanoparticles and adenosine triphosphate biomolecules. *The Journal of Physical Chemistry C*, 118(32), 18631–18639. doi:10.1021/jp506200a

Campbell, J. F., Tessmer, I., Thorp, H. H., & Erie, D. A. (2008). Atomic force microscopy studies of DNA-wrapped carbon nanotube structure and binding to quantum dots. *Journal of the American Chemical Society*, 130(32), 10648–10655. doi:10.1021/ja801720c

Campion, A., & Kambhampati, P. (1998). Surface-enhanced Raman scattering. *Chemical Society Reviews*, 27, 241–250. doi:10.1039/A827241Z

Chen, Z., Liu, C., Cao, F., Ren, J., & Qu, X. (2018). DNA metallization: Principles, methods, structures, and applications. *Chemical Society Reviews*, 47, 4017–4072. doi:10.1039/C8CS00011E

Fleischmann, M., Hendra, P. J., & McQuillan, A. J. (1974). Raman spectra of pyridine adsorbed at a silver electrode. *Chemical Physics Letters*, 26, 163–166. doi:10.1016/0009-2614(74)85388-1

Karthick, K., Anantharaj, S., Karthik, P. E., Subramanian, B., & Kundu, S. (2017). Self-assembled molecular hybrids of CoS-DNA for enhanced water oxidation with low cobalt content. *Inorganic Chemistry*, 56, 6734–6745. doi:10.1021/acs.inorgchem.7b00855

Kundu, S., Chen, Y., Dai, W., Ma, L., Sinyukov, A. M., & Liang, H. (2017). Enhanced catalytic and SERS activities of size-selective Rh NPs on DNA scaffolds. *Journal of Materials Chemistry C*, 5, 2577–2590. doi:10.1039/C6TC05529J

Kundu, S., & Jayachandran, M. (2013). The self-assembling of DNA-templated Au nanoparticles into nanowires and their enhanced SERS and catalytic applications. *RSC Advances*, 3, 16486–16498. doi:10.1039/C3RA42203H

Kundu, S., Yi, S. I., Ma, L., Chen, Y., Dai, W., Sinyukov, A. M., & Liang, H. (2017). Morphology dependent catalysis and surface enhanced Raman scattering (SERS) studies using Pd nanostructures in DNA, CTAB and PVA scaffolds. *Dalton Transactions*, 46, 9678–9691. doi:10.1039/C7DT01474K

Mahmoudi, M., Bertrand, N., Zope, H., & Farokhzad, O. C. (2016). Emerging understanding of the protein corona at the nano-bio interfaces. *Nano Today*, 11(6), 817–832. doi:10.1016/j.nantod.2016.10.005

Majumdar, D., Singha, A., Mondal, P. K., & Kundu, S. (2013). DNA-mediated wirelike clusters of silver nanoparticles: an ultrasensitive SERS substrate. *ACS Applied Materials & Interfaces*, 5, 7798–7807. doi:10.1021/am402448j

Morsy, S. M. (2014). Role of surfactants in nanotechnology and their applications. *International Journal of Current Microbiology and Applied Science*, 3, 237–260.

Nithiyanantham, U., Ede, S. R., & Kundu, S. (2014). Self-assembled wire-like and honeycomb-like osmium nanoclusters (NCs) in DNA with pronounced catalytic and SERS activities. *Journal of Materials Chemistry C*, 3782–3794. doi:10.1039/C4TC00049H

Rafati, A., & Gill, P. (2016). Ultrastructural characterizations of DNA nanotubes using scanning tunneling and atomic force microscopes. *Journal of Microscopy and Ultrastructure*, 4(1), 1–5. doi:10.1016/j.jmau.2015.08.001

Sakthikumar, K., Anantharaj, S., Ede, S. R., Karthick, K., & Kundu, S. (2016). A highly stable rhenium organosol on a DNA scaffold for catalytic and SERS applications. *Journal of Materials Chemistry C*, 4, 6309–6320. doi:10.1039/C6TC01250G

Sangeetha, K., Sankar, S. S., Karthick, K., Anantharaj, S., Ede, S. R., Wilson, S., & Kundu, S. (2019). Synthesis of ultra-small Rh nanoparticles congregated over DNA for catalysis and SERS applications. *Colloids and Surfaces B: Biointerfaces*, 173, 249–257. doi:10.1016/j.colsurfb.2018.09.052

Sangeetha, K., Thiruvengetam, P., Karthick, K., Sankar, S. S., & Kundu, S. (2019). Detection of lignin motifs with $RuO_2$-DNA as active catalyst via surface enhanced raman scattering studies. *ACS Sustainable Chemistry & Engineering.*, 7, 18463–18475. doi:10.1021/acssuschemeng.9b04414

Sankar, S. S., Sangeetha, K., Karthick, K., Anantharaj, S., Ede, S. R., & Kundu, S. (2018). Pt nanoparticle tethered DNA assemblies for enhanced catalysis and SERS applications. *New Journal of Chemistry*, 42, 15784–15792. doi:10.1039/C8NJ03940B

Umemura, K., Izumi, K., & Oura, S. (2016). Probe microscopic studies of DNA molecules on carbon nanotubes. *Nanomaterials*, 6(10), 180. doi:10.3390/nano6100180

Willner, I., & Willner, B. (2010). Biomolecule-based nanomaterials and nanostructures. *Nano Letters*, 10(10), 3805–3815. doi:10.1021/nl102083j

## Glossary

**Self-assembly:** Assembling metal NPs in a close proximity to generate uniform structure.

**SERS:** Surface-enhanced Raman scattering.

**SPR:** Surface plasmon resonance.

**Hot spots:** Localized surface electrons from metal NPs that matches with the frequency of incident LASER beam.
**Photochemical route:** A route where UV light is used to prepare solutions. Generally used for forming metal NPs via illuminating UV light to the precursor solution.
**Wet-chemical route:** Preparing metal particles in solution under ambient conditions.
**Aquesol:** Forming metal NPs in aqueous solutions.
**Organosol:** Forming metal NPs in organic solutions.
**Morphology:** Topographical view of metal NPs assembly.
**EF:** Enhancement factor that shows the order of increase in intensity of the parent Raman signal at low concentrations of probe molecule.
**Nanoparticles:** Particles with size between 1 and 100 nm.
**Electrostatic interaction:** The interactions between the positive ions and negative ions of elements etc.
**Interparticle gap:** The distance between two individual particles.
**DNA:** Deoxy ribonucleic acid that contains aromatic base pairs, nucleotides, and sugar moieties with A–T and G–C pairing.
**Probe molecule:** Large molecule having various functional groups and readily adsorbs charged ions/particles such as various dye molecules like MB (methylene blue) used for Raman scattering studies.
**HOMO:** Highest occupied molecular orbital.
**LUMO:** Lowest unoccupied molecular orbital.
**Nanowires:** Metal NPs assembly that forms wire-like structure at the nanoscale.
**Nanotubes:** Formation of nano-sized tube-like structure with hydrophobic environments like CNT.
**CNT:** Carbon nanotubes with water insoluble nature.
**Nanochains:** Assembling of metal NPs as chain-like structure such as NPs assembly in DNA molecule.
**MLCT:** Metal to ligand charge transfer.
**Honeycomb:** Materials assembly with a unique fashion that form honeycomb-like structure.
**Pico-molar concentration:** Concentration at $10^{-12}$ M concentration.
**Localized electrons:** Surface electrons from the formed metal NPs that give surface charging and frequencies.

## Academic Profile

**Karthick. K** completed his BSc in chemistry from Government Arts and Science College, Udumalpet, India and MSc in chemistry from The American College, Madurai, India. After getting qualified as junior research fellows (JRF) from UGC, New Delhi, India in December 2014, he joined Dr. Subrata Kundu's research group in CSIR-CECRI, Karaikudi in August 2015. He is currently working on his PhD thesis that is focused on the materials for electrocatalytic water splitting applications.

**Dr. Subrata Kundu** completed his PhD from the Indian Institute of Technology (IIT), Kharagpur, India in 2005, after which he pursued his postdoctoral work in the University of Nebraska, Lincoln, USA and in Texas A&M University, College Station, Texas, USA from 2005 to 2010. Dr. Kundu is now a senior scientist at CSIR-CECRI, Karaikudi, India. Dr. Kundu worked as visiting scientist in Texas A&M University, USA from July 2016 to May 2017 to avail of BASE (Bhaskara Advanced Solar Energy) fellowship given by DST and IUSSTF. He is serving as an editorial board member in several international journals including prestigious "Scientific Reports" from nature publishers. Dr. Kundu's research group is currently focusing on the forefront areas of material science for possible application in the areas of energy, environment, and in catalysis.

# 3

# Surface Modification Strategies to Control the Nanomaterial–Microbe Interplay

*T. K. Vasudha, R. Akhil, W. Aadinath, and Vignesh Muthuvijayan*

*Department of Biotechnology, Bhupat and Jyoti Mehta School of Biosciences, Indian Institute of Technology Madras, Chennai, India*

## Objectives

- Explore and understand the factors affecting NM–microbe interactions.
- Review the surface modification methods studied to control NM–microbe interactions.
- Study various analytical techniques used to characterize NM–microbe interactions.

## 3.1 Introduction

Nanotechnology is of increasing interest due to its wide-ranging applications in the fields of medicine and physiology (Bamrungsap et al., 2012), sensing (Munawar et al., 2019), agriculture (Achari & Kowshik, 2018), robotics (Hortelão, Patiño, Perez-Jiménez, Blanco, & Sánchez, 2018), electronics (S. J. Lee et al., 2019), cosmetics (Raj, Jose, Sumod, & Sabitha, 2012), etc. Nanoscale materials have desirable properties distinct from their bulk counterparts. NMs possess versatile mechanical, thermal, and physicochemical properties that make them indispensable in various consumer products and manufacturing processes. The structural stability, high surface-to-volume ratio and resistance to degradation make NMs useful for therapeutic and diagnostic applications (Bamrungsap et al., 2012).

NMs occur in nature in all spheres of the earth, i.e. the atmosphere, hydrosphere, lithosphere, and even biosphere. In nature, they are produced by mechanical (wind erosion, volcanic activity); thermal (such as combustion of biomass); or biological (microbial activity) processes. All forms of life, including microbes, plants, and animals have been continuously exposed to naturally occurring NMs. In the human body, the blood system and the gastrointestinal tract regularly

*Microbial Interactions at Nanobiotechnology Interfaces: Molecular Mechanisms and Applications*, First Edition. Edited by R. Navanietha Krishnaraj and Rajesh K. Sani.

encounter NMs that play an important role in (patho)physiology (Fakruddin, Hossain, & Afroz, 2012). Presence of natural silver nanoparticles in water of coastal areas (Akaighe et al., 2011) and production of nanocrystalline manganese and iron oxides by iron-oxidizing bacteria (*Leptothrix, Gallionella*) are representative examples of naturally occurring NMs (Emerson, Fleming, & McBeth, 2010).

With the advent of technology, NMs can be synthesized for a variety of applications. Improvement in analytical and synthesis procedures has enabled the visualization and large-scale manufacture of NMs (Bamrungsap et al., 2012; Hortelão et al., 2018; S. J. Lee et al., 2019; Munawar et al., 2019; Raj et al., 2012). As a result, engineered NMs are increasingly being developed for use in various areas of life sciences such as in pharmaceutical products, medical diagnostics, etc. This requires the study of nano-bio interactions to ensure the safe use of this technology.

The major focus of studies on NMs, concerning their interactions with microbes, has been on their antimicrobial activity. NMs act as excellent antimicrobial agents because they have varied mechanisms of action, and hence, less chance for development of resistance. NMs used as antimicrobial agents interact with microbes to effect killing or inhibition of microbes. Studying the interactions between the NM and the microbe is crucial for understanding the inhibitory mechanism of the NM on the microbes. The inhibitory mechanism depends on the physicochemical characteristics of NMs and the type of microbe. NMs act as antimicrobial agents by one of the following inhibitory mechanisms (Singh et al., 2019):

1) Release of metal ions.
2) Production of reactive oxygen species (ROS).
3) Physical damage of bacterial cell wall.
4) Inhibition of enzyme activity.
5) Inhibition of DNA synthesis.

NMs, owing to their large surface-to-volume ratio, have a greater number of functional sites for interaction with the bacterial cell surface (Sweet & Singleton, 2011). Moreover, NMs establish direct contact with the bacterial surface rather than by penetration through the cell wall, which makes several antibiotic resistance mechanisms in bacteria irrelevant (Beyth, Houri-Haddad, Domb, Khan, & Hazan, 2015). They are, therefore, less likely to provoke antibiotic resistance in bacteria (L. Wang, Hu, & Shao, 2017). Silver nanoparticles can inhibit the enzyme systems in the respiratory chain of some bacteria and also support antibacterial activity by release of silver ions (Salomoni, Léo, Montemor, Rinaldi, & Rodrigues, 2017). Zinc oxide nanoparticles damage the cell membrane and are also internalized to carry out their antibacterial activity (Siddiqi, Ur Rahman, Tajuddin, & Husen, 2018). Superparamagnetic iron oxide nanoparticles (SPIONs) also possess high antibacterial activity and exert low toxicity to host cells (Vallabani & Singh, 2018). Nanoengineered antibacterial peptide polymers act by membrane destabilization, upregulated ion movement across the cytoplasmic membrane, or induction of apoptosis (Bassous & Webster, 2019). Quantum dots (W. Wang et al., 2019), graphene (Xianjun Zhu, Zhu, Murali, Stoller, & Ruoff, 2011), gold nanoparticles (Mohamed, Fouad, Elshoky, Mohammed, & Salaheldin, 2017), and carbon nanotubes (Lohan et al., 2016) are more recent NMs explored as antimicrobial agents. Graphene-based NMs release heat upon near-infrared laser irradiation, and hence kill bacteria (Xianjun Zhu et al., 2011).

Apart from the antimicrobial activity of NMs, other NM–microbe interactions are less studied. NM–microbe interactions are exploited for various applications. NMs can be used for the rapid detection of microbial pathogens to overcome challenges in microbial diagnosis such as tedious sample preparation, expensive and time-consuming assays, and inability to distinguish live from dead bacteria (Kaushal et al., 2019). Fluorescence-based quantification of bacteria has been done using antibody-conjugated dye-doped silica nanoparticles (Qin, He, Wang, & Tan, 2008). The foodborne pathogen *Listeria monocytogenes* has been quantified with a detection limit of $6 \times 10^6$ bacteria by employing magnetic nanoparticles (MNPs) via superconducting quantum interference device (SQUID) (Chemla et al., 2000).

Surface functionalization of NMs can be done using antibodies or polymers to control NM–microbe interaction (Della Ventura et al., 2017; Pelaz et al., 2015). Gold nanoparticles have been surface functionalized with organic thiol molecules or thiol-containing polymers as the gold surface has a strong affinity to sulfhydryl groups (Corbierre & Lennox, 2005; Gao, Huang, Liu, Zan, & Ren, 2012). Silver nanoparticles have also been surface modified using small reducing agents, such as sodium citrate or by various functional polymers like polyvinyl alcohol, polyacrylic acid, and polyacrylamide. It has been reported that surface modifications can significantly impact the antimicrobial activity of NMs (Leung et al., 2012). Antimicrobial NMs are surface modified to strengthen the complex formation between NM and the microbe. Enhanced antibacterial activity has been observed by surface modification of titanium dioxide ($TiO_2$) with *Garcinia zeylanica* extract (Senarathna et al., 2017). Similar results have been observed in a study where $TiO_2/Ag^+$ nanoparticles grafted with ϒ-aminopropyltriethoxysilane (APS) displayed potent antibacterial activity (Cheng, Li, Pavlínek, Saha, & Wang, 2006). NMs can also be coupled to bacterial antibodies to achieve specific interaction between NM and bacteria. This can avoid unintended side effects. Selective laser killing of *Staphylococcus aureus* has been achieved with anti-protein A-conjugated gold nanoparticles. It was reported that direct irradiation of bacteria with laser did not damage the bacteria because of low absorption of natural endogenous cytochromes (Navarre & Schneewind, 1999; Zharov, Mercer, Galitovskaya, & Smeltzer, 2006). In another study, vancomycin-bound gold nanoparticles have been used as photothermal agents for the selective killing of bacteria under irradiation of near-infrared light (808 nm). Here, vancomycin binds to the terminal D-Ala-D Ala moieties present on the peptide units of the bacterial cell wall (Huang, Tsai, & Chen, 2007).

Apart from the control of NM–microbe interaction, surface functionalization can also improve the physicochemical properties, stability, and biocompatibility of NMs (N. Zhu et al., 2018). For instance, polymeric ligands attached to the NM surface can improve its stability by generating a physical barrier at the NM surface (steric stabilization) (Gref et al., 2000). An example is the coating of gold nanoparticles with polyethylene glycol (PEG) to improve the stability of the NM in solutions with high salt concentration (Rahme et al., 2013). Zwitterionic ligands (Breus, Heyes, Tron, & Nienhaus, 2009), lipid bilayers (Nyalosaso et al., 2016), proteins (Tebbe, Kuttner, Männel, Fery, & Chanana, 2015) or glycan (García et al., 2015), and aptamers (Koren, Apte, Jani, & Torchilin, 2012) are some examples of various ligands that can be used for surface functionalization of NMs.

In this chapter, we first introduce the reader to various types of interactions between NMs and microbes. Then, the difference in surface features of microbes and the physicochemical characteristics of NMs that can affect NM–microbe cross talk will be discussed. In the subsequent sections, we will explore various surface functionalization strategies to control the NM–microbe cross talk.

## 3.2 Factors Influencing NM–Microbe Cross talk

Various factors influence the interaction between NMs and microbes such as the surface features of the microbes and the physicochemical properties of the NMs. Different types of microbes interact differently with the same NM by virtue of their contrasting surface features. Gram-positive and Gram-negative bacteria have distinct cell wall composition, and this influences the way they interact with NMs. Similarly, fungi also have unique surface features that can play a role in NM–microbe interactions. Also, the physicochemical properties of NMs, such as their shape, size, surface charge and chemistry, and hydrophobicity play a role in the NM–microbe cross talk. The formation of a biomolecule corona at the NM surface also influences the interaction.

### 3.2.1 Surface Features of Microbes

Different types of bacteria and fungi differ in their surface morphology, which affects their interaction with NMs. Based on the difference in their cell wall composition, bacteria are classified as Gram-negative and Gram-positive, as shown in Figure 3.1. The Gram-positive bacteria have a large peptidoglycan structure. Surface expression of teichoic acids, polymers of glycerol and ribitol joined by

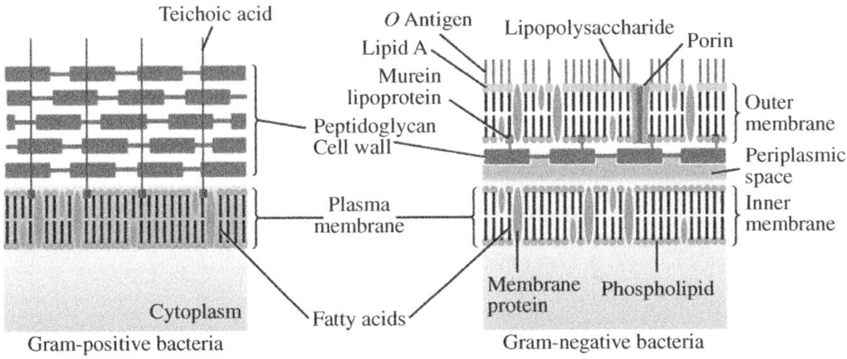


**Figure 3.1** Gram-positive and Gram-negative cell wall. *Source*: https://commons.wikimedia.org/wiki/File:Figure_22_02_08f.png.

phosphate groups, is unique to Gram-positive bacteria (Weidenmaier & Peschel, 2008). The Gram-negative bacteria have a thin peptidoglycan layer but possess an outer cytoplasmic membrane, mostly composed of lipopolysaccharides. These are also known as endotoxins that are toxic and elicit strong immune responses in the host system (Salton, 1953). Fungal cell walls are structurally unique and are comprised of glycoproteins and polysaccharides, mainly glucan and chitin. Several other minor cell wall components are present and vary among different species of fungi (Bowman & Free, 2006). In the case of biofilms, the extracellular polymeric substances (EPSs) that are secreted result in the high affinity of NMs to biofilms (Harper et al., 2019; Nevius, Chen, Ferry, & Decho, 2012).

### 3.2.2 Physicochemical Properties of NMs

The physicochemical properties of NMs play a crucial role in their antimicrobial activity. The large surface-to-volume ratio of NMs can improve antibacterial activity due to an increase in the number of functional sites (Sweet & Singleton, 2011). However, this also increases the chances of toxicity due to increased interaction with biomolecules in the host system. The shape, size, surface charge, composition, hydrophilicity, oxidative dissolution, presence of functional groups or ligands or biomolecules on the surface, formation of biomolecule corona, etc. influence the interaction of NMs with the bacteria (Daima & Bansal, 2015). Figure 3.2 shows the various physicochemical properties of a nanomaterial that can influence NM–microbe interaction.

Different physicochemical properties like electromagnetic, optical, and catalytic properties of NMs are strongly influenced by their shape and size. The shape and size of the NM affect the effective surface area. For example, a tetragonal

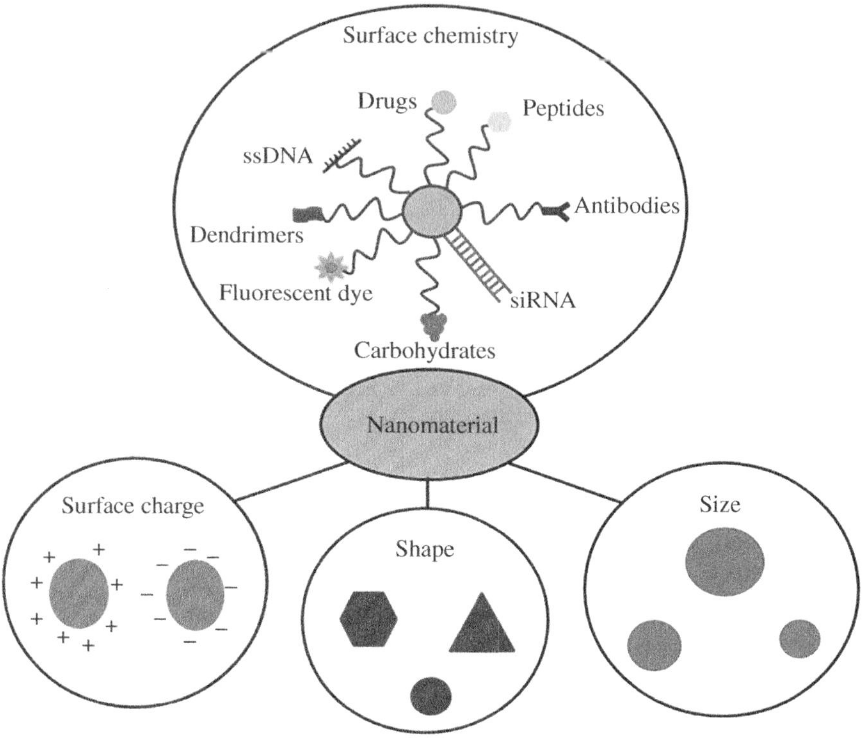


**Figure 3.2** Physicochemical properties of NMs that affect NM–microbe interaction.

particle has a different area compared to spherical ones (Boonrungsiman et al., 2017). This affects NM–microbe interaction. It has been observed that the shape of the NM affects its ability for uptake by cells (Acharya et al., 2018; Raza et al., 2016). For example, titanate nanofibers and titanate nanoparticles showed a difference in cellular internalization (Boonrungsiman et al., 2017). In a study, the antibacterial activities of spherical silver nanoparticles ranging in size from 15 to 90 nm were compared. The smallest sized spherical nanoparticles displayed the most potent activity (Raza et al., 2016). The effect of larger surface area of the nanomaterial on the antimicrobial activity was also demonstrated in a study where the biocidal efficiency of 17 nm-sized core–shell nanoparticles was found to be threefold higher than that of 28 nm-sized particles (Gupta, Mumtaz, Li, Hussain, & Rotello, 2019).

Surface charge is another characteristic that influences NM–microbe interaction. It plays an important role in governing the initial electrostatic interaction and the cellular uptake by bacteria. Cationic nanomaterials act as effective antimicrobial agents by causing membrane disruption (Hou et al., 2017). A study was conducted to investigate the influence of different surface charges of silver nanoparticles (positive, neutral, and negative) on their antibacterial effectiveness against a range of human pathogens (Abbaszadegan et al., 2015). It was concluded that among different silver nanoparticles tested, the positively charged nanoparticles displayed highest antibacterial activity (Abbaszadegan et al., 2015). Metallic silver and gold nanoparticles have been synthesized with terminal dimethylamine groups that interact with the negatively charged bacterial membranes via strong ionic interactions. Here, in addition to the size, the surface functionalization or surface chemistry influences the antibacterial activity (Daima & Bansal, 2015).

A study investigated the structure–activity relationship of gold nanoparticles with ligands of varying surface hydrophobicity. From the study, it was found that the antimicrobial activity increased with the increase of hydrophobic functional groups on the NM surfaces (X. Li et al., 2014). Many other reports have also confirmed that the surface hydrophobicity influences the interaction between NMs and microbes (Bakirdere et al., 2015; R. Liu, Lee, Chong, & Tomczak, 2016; Mahmoud, Alkilany, Khalil, & Al-Bakri, 2018). The surface roughness of the NMs has also been found to affect their antimicrobial activity. Rough mesoporous hollow silica (MHS) nanoparticles showed enhanced antibacterial effect compared to MHS with a smooth surface (Ahmad Nor et al., 2015). The surface chemistry and the composition of the corona surrounding the NM also influence the NM–microbe interaction. It has been observed that nano $TiO_2$/polyimide composites modified with isocyanatopropyltriethoxysilane show improved corona resistance (H. Lu et al., 2017).

The formation of biomolecule corona around the NM significantly impacts the NM–microbe complex formation. Biomolecules present in physiological/ecological environments have varying affinities to the NM surface (Docter et al., 2015). The biomolecule corona surrounding a NM is what dictates its biological and ecological identity. It is, therefore, important to study the impact of corona on NM–microbe cross talk. The composition of the biomolecule corona varies depending on particle size, material and surface properties, exposure time, and relative concentration of the physiological fluid to the NM dispersion. The composition of the biomolecule corona is highly heterogeneous and includes proteins, lipids, and polysaccharides. Pristine NMs behave differently in physiological environments due to coronation (Treuel et al., 2014). Parameters like pH, temperature, and ionic strength of the physiological environment also determine the composition of the biomolecule corona. The formation of biomolecule corona has been shown to reduce NM–microbe complex formation in a dose-dependent manner (Siemer et al., 2018).

The purity of NMs is an often neglected physicochemical parameter that can interfere with antimicrobial activity of the NM (Barbolina et al., 2016). The impurities present may be responsible for antimicrobial activity rather than the NM itself.

Electron transfer capability (X. Lu et al., 2017), surface defects, crystallinity, oxidizability, and counterion effects are other important physicochemical parameters of NMs that may influence NM–microbe interactions (Daima & Bansal, 2015).

The intense color possessed by the colloidal solution of metal nanoparticles is due to the phenomenon called surface plasmon resonance. This property has been used as an early indicator of the antimicrobial activity of NMs (Mlalila, Swai, Hilonga, & Kadam, 2016). However, the plasmonic behavior of these nanoparticles is dependent on several factors, such as size and shape (H. Kang et al., 2018; Seoudi, Shabaka, El Sayed, & Anis, 2011). Among several parameters, size and polydispersity of the nanoparticles play a significant role in their catalytic, magnetic, and optical property-based applications. Iron oxide nanoparticles (IONPs) in the nanometer size range possess monodomain architecture and hence are superior in their functionality under the external magnetic field (Laurent et al., 2008).

To gain control over the size and morphology of the nanoparticles, stabilizing agents, also known as capping agents, are used. These stabilizing agents also prevent aggregation and oxidation and assist in the controlled assembly of the nanoparticles (H. Kang et al., 2018; Patel et al., 2016). Surface functionalization strategies partially fulfill the role played by stabilizing agents, as discussed earlier. Several materials have been exploited as stabilizing agents to fine-tune the properties of the NMs and vary from organic to inorganic, monomeric to polymeric, and natural to synthetic (Laurent et al., 2008). Biopolymers such as starch and deoxyribonucleic acid (DNA) have greater advantages as these are highly biocompatible and mask the toxicity associated with the NMs (H. Kang et al., 2018; Laurent et al., 2008). The shape of the NMs can also be controlled with the help of stabilizing agents based on their preferential binding to the crystal facets (Al-Saidi, Feng, & Fichthorn, 2012). For example, polyvinylpyrrolidone (PVP) is used as a structure-directing polymeric agent that has higher affinity toward crystal facet (100) of silver than the (111) facet. Hence, silver NMs' growth is longer in the direction of Ag(100) compared to Ag(111) when PVP is used as a stabilizing agent (Al-Saidi et al., 2012). IONPs generate heat when exposed to the alternative magnetic field, which is used for hyperthermic treatment to kill bacteria. It was shown that heat generation by IONPs is dependent on the size and polydispersity (Mohapatra et al., 2018). To control the heat generated, IONPs were coated with oleic acid, which controlled the crystal growth and produced monodispersed nanoparticles (Mohapatra et al., 2018).

## 3.3 Surface Functionalization

The physicochemical properties of the NMs could be altered using various surface functionalization techniques. It is a process by which novel properties are added to material by modifying its surface characteristics. Surface functionalization techniques generally focus on improving material properties and preventing bacterial colonization. A flowchart illustrating various surface functionalization techniques is shown in Figure 3.3a.

### 3.3.1 Techniques Used for Surface Functionalization

The most commonly used techniques of surface functionalization are:

1) Self-assembled monolayers (SAMs)
2) Layer-by-layer (LbL) technique.

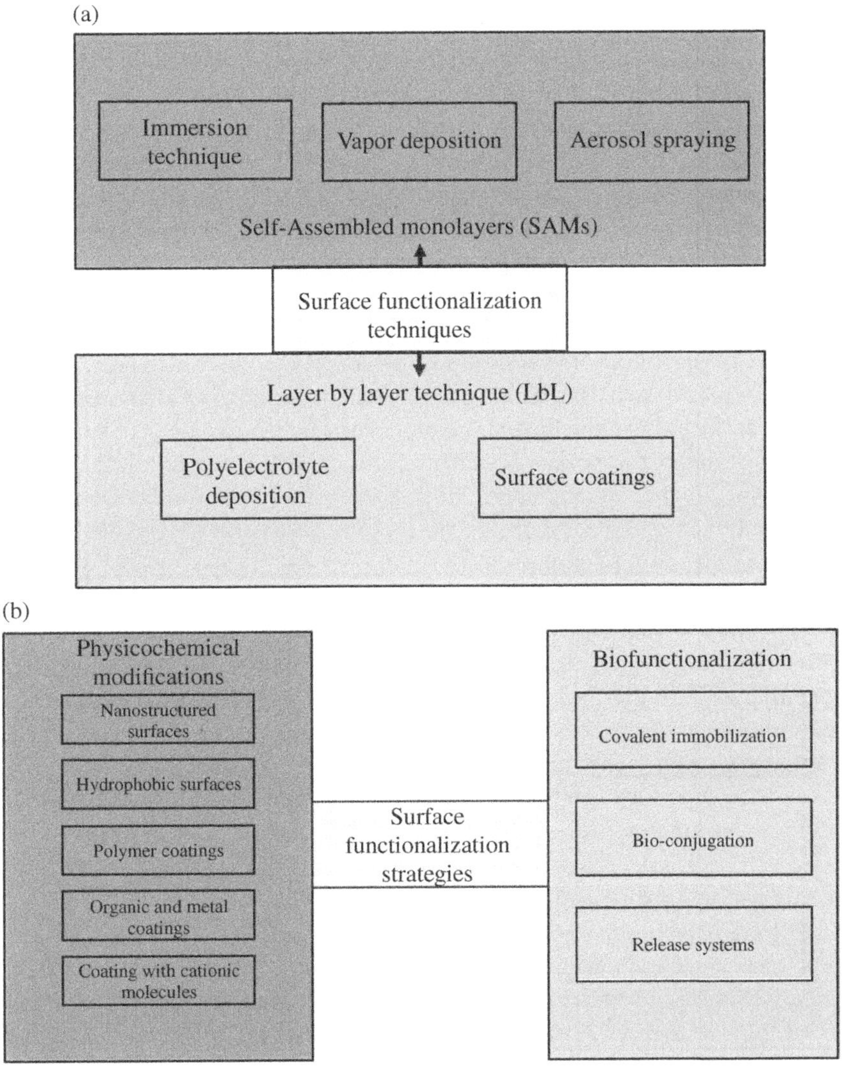


**Figure 3.3** (a) Surface functionalization techniques. (b) Surface functionalization strategies.

#### 3.3.1.1 Self-Assembled Monolayers

Self-assembled monolayers (SAMs) refer to a layer of active surfactant molecules that self-assemble in an organized manner on a solid surface. SAMs of compounds on the desired surface is one of the simpler and promising techniques of surface functionalization (Ulman, 1996). The functionalization of the self-assembled monolayer is achieved by using various reactive groups. The purpose of these reactive groups is to facilitate the conjugation of the desired compound to the surface layer.

The most widely explored method of SAM formation is the immersion technique. This technique is very effective, especially for materials possessing complex surface morphologies (Love, Estroff, Kriebel, Nuzzo, & Whitesides, 2005). In the initial step of the immersion technique, solution deposition is done by immersing a suitable substrate into an active surfactant. A highly purified solvent is used for this procedure, and the whole setup is incubated for an extended period. Following incubation, thorough rinsing is carried out using the same solvent, followed by drying in an inert environment (Love et al., 2005). The surfactants used here comprise of a head group, a linker, and a terminal group. The head group of the surfactant interacts with the substrate's surface. The linker, which is predominantly an alkyl chain, links the head and the terminal groups. The length of the alkyl chain can be maneuvered according to the requirements. The linker contributes to the stabilization of the monolayer as well. The terminal group occupies the monolayer surface above the linker. The monolayer molecules are usually closely packed as a result of the van der Waals interactions. These non-covalent interactions contribute to the self-assembly (Chechik & Stirling, 2009; Kind & Wöll, 2009; Love et al., 2005).

Apart from the immersion technique, other methods such as aerosol spraying and vapor deposition are also used to form SAMs. SAMs are formed on oxide surfaces in an ordered manner by the aerosol spraying technique. Shorter deposition times and lower concentrations are sufficient for monolayer formation using this technique (Quiñones, Raman, & Gawalt, 2008). Chemical immobilization of antimicrobial peptides (AMPs) on polymer surfaces was carried out using chemical vapor deposition technique. It was found that the AMPs immobilized were more stable compared to the other techniques (Xiao et al., 2018). SAMs have been in use in numerous antimicrobial surface functionalization strategies.

#### 3.3.1.2 Layer-by-Layer Technique

Decher, Hong, & Schmitt (1992) introduced the LbL technique, which is a unique form of coating technology, involving the use of charged polyelectrolytes. Here, oppositely charged polyelectrolytes are coated onto a substrate, which is charged in an alternative manner (Izquierdo, Ono, Voegel, Schaaf, & Decher, 2005). A charged surface is the minimum criteria required for any coating material to be applied using the LbL approach.

This procedure is very simple, as well as cost-effective. The first step involves immersing the charged substrate surface into a polyelectrolyte solution. The polyelectrolyte should be chosen in such a way that it carries an opposite charge to that of the substrate. Once the dipping process is complete, the surface is rinsed with deionized water. This step ensures that all the unbound polyelectrolyte molecules are removed from the surface of the substrate. Deposition of an oppositely charged polyelectrolyte results in a reversal of the net charge of the substrate. The subsequent step involves dipping the substrate in a polyelectrolyte of opposite charge so that the charge is brought back to the initial state. At the end of one cycle, a bilayer comprising of two polyelectrolytes of opposite charges is coated on the upper surface of the substrate (Chuang, Smith, & Hammond, 2008). By controlled alteration of the deposition cycles, parameters such as thickness can be precisely controlled. Desired structures can also be formed by maneuvering the dipping and the deposition cycles. The selection of correct charged polyelectrolytes plays a crucial role in obtaining LbL assemblies.

Multifunctional antibacterial LbL films are coatings having both bactericidal and bacterial-adhesion-resistant functions. It should be noted that the major drawback of the nonadhesive bacterial-resistant LbL films is the antibacterial efficacy. To improve the antibacterial properties of these films, the bactericidal components are incorporated, resulting in a fully functional antibacterial LbL system (Chua, Neoh, Kang, & Wang, 2008).

### 3.3.2 Surface Functionalization Strategies

Using the above two techniques, various strategies have been deployed to functionalize the surface of NMs (Figure 3.3b). These can be grouped into:

1) Physicochemical modifications
2) Biofunctionalization.

#### 3.3.2.1 Physicochemical Modifications

Physical methods are successful in imparting structural change on the NM surfaces. This approach is proven to be effective in combating biofilm formation (Fisher et al., 2016). Here, designing the surfaces by increasing its roughness restricts and prevents microbial growth (Langowska, Kowal, Palivan, & Meier, 2014). Creating nanostructured surfaces is the most commonly used technique. These surfaces are prepared by direct addition of nano-objects. These objects when impregnated on the surface kill the bacteria upon contact with the surface (Zabara et al., 2019). When compared to the unstructured surface, the structured alternations are shown to massively reduce bacterial retention (Vasudevan, Kennedy, Merritt, Crocker, & Baney, 2014). Microscale motifs such as hexagonal pillars, pits, and cross pillars can be patterned on to the surface to suppress bacterial growth (Vasudevan et al., 2014).

A deep reactive ion etching technique is used to pattern silicon nanopillars. The etch cycle uses a combination of $SF_6$ and $O_2$. This is followed by a deposition cycle involving $C_4F_6$ gas. It was found that the random interspaces seen in between the silicon nanopillars doubled the contact angle and resulted in significant lysis of bacteria on the surface (Hasan, Raj, Yadav, & Chatterjee, 2015). The functionalization of silicon nanopillars with bacterial binding molecules has been achieved. This strategy entraps the bacteria, eventually killing it (Y.-Q. Li et al., 2014). The antibacterial activity could be further enhanced by the addition of silver-copper nanoparticles to the silicon nanopillars (Palza, Quijada, & Delgado, 2015).

Hydrophobic surfaces decrease bacterial growth on the surface. This could be achieved by generating micro and nanostructured roughness using a laser surface modification technique (Gillett, Waugh, Lawrence, Swainson, & Dixon, 2016). Various biocompatible polymers such as polystyrene, polycarbonate, polydimethylsiloxane, and polyethylene are used for creating superhydrophobic surfaces by simply molding them without any chemical alteration. The appropriate surface roughness and hydrophobicity are maintained to achieve antimicrobial properties (Freschauf, McLane, Sharma, & Khine, 2012). Surface charge of the substrate can also be tuned to make them resistant to bacterial adhesion. Polyelectrolytes, when used to modify the surface charge of the substrate, resulted in significant bacterial repulsion (Xiaoying Zhu et al., 2014).

Natural polymers such as chitosan (Renoud, Toury, Benayoun, Attik, & Grosgogeat, 2012) and synthetic polymers such as PEG, polymethacrylic acid (Tanaka et al., 2010) can be used as a surface coating imparting antibacterial surface properties. An alternative for improving antibacterial effects is to use quaternized polymer brushes. In a study, quaternized polymer brushes were formed by grafting poly[2-(dimethylamino) ethyl methacrylate] from a cyclo-triphosphazene ring. Here, alkyl chain with iodides were used to quaternize the polymers that imparted antibacterial properties (X. Liu, Zhang, Tian, Sen, & Allcock, 2012).

A synthetic strategy for preventing the biofilm formation involves surface modification using polymer brushes where significant arrest in bacterial growth was seen (Roosjen, Norde, Mei, & Busscher, 2006). In addition to this, polymer brushes are used as excellent antifouling agents. Similarly, amphiphilic block copolymers such as polystyrene-block-polyethylene-ran-butylene-block-polyisoprene with PEG-hydrocarbon side chains significantly suppress bacterial density on the monolayer surfaces (Zhou et al., 2014). Nanopatterned polymer brushes having antibacterial properties are coated onto the surface. The bactericidal effects of these polymers could be triggered by altering the temperature (Yu, Ista, & López, 2014). Here the property of temperature-dependent conformational changes of the polymer was explored. Thermoresponsive polymer brushes exposing patterns of various antimicrobial enzymes have also been studied (Yu et al., 2014). After effectively killing bacteria, the bacterial debris is removed using thermoresponsive nanoparticle brushes (Y.-Q. Li, Zhu, et al., 2014; Wei, Yu, Zhan, & Chen, 2016).

Polysaccharide coatings, having anti-adhesive properties, decrease bacterial adhesion onto the surface (Gadenne, Lebrun, Jouenne, & Thebault, 2013). Agarose-crosslinked polymers on a solid support are also shown to have similar effects as other polysaccharide coatings (M. Li, Neoh, Kang, Lau, & Chiong, 2014). The adhesive property of the coating material could also be coupled with the antibacterial effects of the polymer. A mixture of hyaluronic acid and chitosan was shown to not only decrease the bacterial adhesion on the surface but also inhibit their proliferation effectively (Hernandez-Montelongo et al., 2016).

NMs are also incorporated into the LbL films by electrostatic interactions (Shen et al., 2012). Organic molecules or any charged species can also use electrostatic interactions to form LbL films (Chuang et al., 2008). Apart from electrostatic interactions, other interactions such as hydrogen bonding (Stockton & Rubner, 1997) and covalent binding (Such, Anthony Quinn, Tjipto, & Caruso, 2006) are also used in the development of LbL films. The requirements govern the method of choice for the formation of LbL films. This technique has become so advanced that it provides a vast array of choices for users to develop antibacterial surface coatings. A variety of antibacterial surfaces have been developed by LbL approach using different assembly techniques involving many types of substrates. Antibacterial agents are prepared by coating silver nanoparticles and $TiO_2$ layers onto the surface of MNPs (D. Lee, Cohen, & Rubner, 2005). Even though silver and other heavy metals have been used since antiquity, their usage is limited in biological systems owing to toxicity (Mijnendonckx, Leys, Mahillon, Silver, & Van Houdt, 2013). Hence, coating the heavy metal-based NMs can improve their potential in biological applications.

One exciting option is the use of cationic molecules that can target the phospholipid component of the lipid bilayer of the bacterial cell membrane. The negatively charged lipid bilayer interacts with the cationic molecules rapidly, causing protoplast lysis triggered by osmotic stress, eventually disrupting the bacterial cell membrane. Coating silver nanoparticles with cationic galactomannan imparted antibacterial effects (Abdullah, Ghosh, Basu, & Mukherjee, 2015). The versatility of the LbL system is such that it can be used to engineer multilayered systems by incorporating different bactericidal components.

#### 3.3.2.2 Biofunctionalization

The biocidal system is one of the ways to curb microbial adhesion. AMPs and antibiotics are used as surface coatings to prevent microbial adhesion. These systems can be deployed in two ways: (i) as a permanent system imparting a local effect on the microbe, only at the site of action (Bayramov & Neff, 2017), or (ii) as a continuous, gradual-release system sustaining the antimicrobial effect around the NM (Chua et al., 2008; S. M. Kang, Kong, Oh, Choi, & Choi, 2010; Shi, Neoh, Kang, Poh, & Wang, 2009).

#### 3.3.2.2.1 Covalent Immobilization and Bioconjugation-Based Systems

Covalent or affinity binding techniques have been used for modifying SAMs. Glycolipids, when covalently attached to the substrate surface, exhibited antibacterial activity against a wide range of Gram-positive and Gram-negative bacteria (Valotteau et al., 2017). A small peptide, temporin-SHha, and its carboxylic analog were also covalently immobilized onto the surface. This resulted in the rapid lysis of bacterial membranes (Lombana et al., 2014). It was also seen that non-natural D-configuration analog of temporin showed a significant increase in antibacterial activity compared to temporin-SHa (Oger, Piesse, Ladram, & Humblot, 2019).

In a study, con-A conjugated polysaccharide MNPs with the ability to promptly arrest the growth of *Escherichia coli* in blood culture were prepared (Kaittanis, Nath, & Perez, 2008). Chen, Zhou, Chen, Lv, and Yao (2009) immobilized antimicrobial peptide LL-37 on the surface of polycyclic acid-coated nanoparticles and found that these conjugates developed bactericidal activity toward *E. coli* (Chen et al., 2009).

#### 3.3.2.2.2 Release Systems

Releasing active agents from the functionalized surfaces can kill the bacteria. It is also the most effective way to prevent biofilm formation (Ashton et al., 2019). This can be achieved by releasing active agents, such as small biomolecules or silver ions into the environment surrounding the surface (Pasban Ziyarat, Asoodeh, Sharif Barfeh, Pirouzi, & Chamani, 2013). When these active agents are released close to the surface, the bactericidal effects will eliminate biofilm formation. Releasing transition metals is also found to suppress bacterial growth. These molecules use different mechanisms to kill bacteria. Toxic doses of these metals generally act by either disrupting the enzyme activity, or membrane function or by damaging the genetic material (Lemire, Harrison, & Turner, 2013). Entrapping the antibacterial enzymes, such as lysozyme, in biocompatible polymer brushes can also impart antibacterial activity (Wei et al., 2016).

The usage of antibiotics as an active agent is another option. The advantage of using antibiotics is that it targets the bacterial cells alone, causing cell death by inhibiting the synthesis of cell wall components or nucleic acids or proteins (Kohanski, Dwyer, & Collins, 2010). The major limitation of using antibiotics for a prolonged period is the development of resistance by the bacteria. This limitation urged researchers to explore novel molecules having antibacterial properties to be used as LbL films.

## 3.4 Characterization of NM–Microbe Interactions

Several techniques can be used to characterize NM–microbe interactions. In this section, we brief the different parameters that can be exploited for characterization and the different techniques that can be used for the same (Figure 3.4).

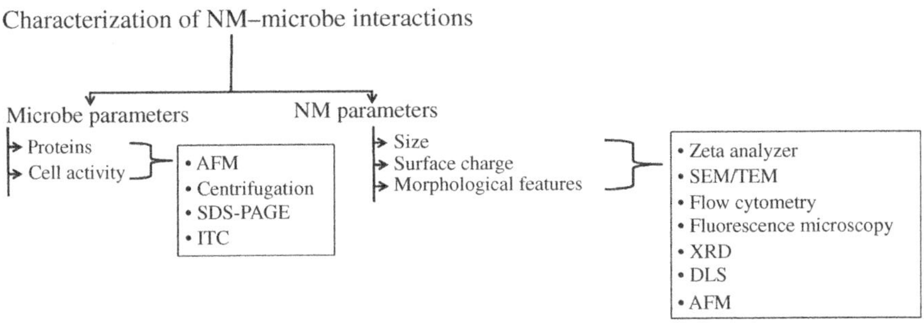


**Figure 3.4** Techniques to characterize NM–microbe interactions.

### 3.4.1 Microbe Parameters

Proteins are the major components of the cell wall that interact with the external environment. Several techniques can be used to detect the interaction of these proteins with the various external moieties, which includes NMs.

Atomic force microscopy (AFM) is a highly sensitive quantitative technique that measures nanoscale changes in the bacterial surface. AFM relies on the nanoscale measurement of the cantilever fluctuations that are due to bacterial activity. AFM has been used to detect the interaction between the bacterial surface and antibiotics in a fraction of time (Longo et al., 2013). They found that fluctuations produced by the cantilever were high when they introduced bacteria into the Luria–Bertani media and it is dependent on bacterial activity. As they introduced ampicillin into the media, the fluctuations started decreasing with respect to ampicillin concentration (Longo et al., 2013). However, attaching microbial cells to the surface for AFM scanning without affecting their viability is one of the major challenges (Louise Meyer et al., 2010). Several approaches have been tried to immobilize cells on to the surface such as physical confinement, covalent binding, physisorption, and immobilization using adhesive proteins. Among these, immobilization using adhesive proteins has proven to be the better choice (Louise Meyer et al., 2010).

Another most commonly used method to detect NM–protein conjugate is centrifugation, followed by gel electrophoresis (Cedervall et al., 2007). *N*-Isopropylacrylamide-*co*-*N*-*tert*-butylacrylamide (NIPAM:BAM) copolymer particles of different sizes were incubated with plasma and allowed to form protein–nanoparticles aggregates. Then, those particles were centrifuged and followed by sodium dodecyl sulfate-polyacrylamide gel electrophoresis (SDS-PAGE) to isolate bound protein from nanoparticles (Cedervall et al., 2007). Several other NMs such as liposomes, polystyrene latex, and carbon nanotubes were also studied using centrifugation followed by gel electrophoresis method (Blunk, Hochstrasser, Sanchez, Müller, & Müller, 1993; Diederichs, 1996; Salvador-Morales et al., 2006).

However, the process is complicated due to sedimentation of larger proteins, protein aggregates, and coprecipitation. The conjunction of other analytical methods, such as size exclusion chromatography, can eliminate most of the contaminants (Cedervall et al., 2007).

Matrix-assisted laser desorption/ionization mass spectrometry (MALDI-MS) is another analytical technique used to detect proteins present on the microorganism. This technique identifies the proteins based on their molecular weight (Lin, Tsai, Weng, & Chen, 2005). Lin et al. (2005) selectively trapped pathogenic bacteria with the help of vancomycin-labeled MNPs. Vancomycin has affinity toward the D-Ala-D-Ala moieties of the tetra-peptide unit present in the cell wall of the bacteria. Once vancomycin bound to these amino acid moieties, external magnetic field was applied to trap pathogenic bacteria selectively. Subsequently, MALDI-MS was carried out to characterize bacteria based on the proteins present in their cell walls (Lin et al., 2005). The quadrupole mass analyzer, another mass spectroscopy technique, can also be used to characterize proteins (Leung et al., 2014).

Isothermal titration calorimetry (ITC) gives information on the NMs adsorbed onto the proteins based on the heat response (Meister et al., 2013; Pasban Ziyarat et al., 2013). NMs interact with macromolecules through weak non-covalent interactions. Through these interactions, there is a change in enthalpy and entropy of the system. Such changes are measured using ITC. The advantage of ITC method is that it provides quantitative information on the binding affinity of the NM–protein complexes (Pasban Ziyarat et al., 2013).

### 3.4.2 NM Parameters

Several parameters of the NMs, such as size, shape, and charge, can be exploited to characterize the NM–microbe interactions. Size variation due to the interactions between the NMs and microbes can be analyzed using various techniques. Dynamic light scattering (DLS) is one of the simplest to use and analyze. It depends on the scattering of light by particles that are suspended in the sample and are in continuous Brownian motion (Hoo, Starostin, West, & Mecartney, 2008). However, particle size given by DLS method is the hydrodynamic radius, rather than the actual size. Moreover, DLS method is not suited for polydispersed samples. It is found that AFM is more accurate in predicting the size of the particles, and also suited for polydispersed samples (Hoo et al., 2008). In addition, electron microscopy and X-ray diffraction (XRD) techniques can also be employed for accurate prediction of the size (Mahmoudi et al., 2011).

Change in surface charge due to interaction between NM and microbe is assessed using a zeta potential analyzer (Westmeier et al., 2018). NMs have surface charge when suspended in water or biological media. Similarly, microbial surfaces also possess surface charge. Once NM comes in contact with microbial surface, there is a change in the surface charge depending on the amount of NM bound to the microbial surface. Such changes in surface charge are measured using zeta potential analyzer (Westmeier et al., 2018).

Labeling NMs and microbes with fluorescent molecules enables the study of NM–microbe interaction using fluorescence microscopy and flow cytometry (Bernabini, Holmes, & Morgan, 2011; Westmeier et al., 2018). Electron microscopy (scanning electron microscopy (SEM) and transmission electron microscopy (TEM)) can also be used to study the interaction between NMs and microbes (Arakha et al., 2015; Treuel et al., 2014). In addition, SEM provides morphological features of the interacting moieties. The energy-dispersive X-ray spectroscope (EDX) attached to the SEM instrument provides elemental composition of the sample (Arakha et al., 2015). However, the number of particles imaged in electron microscopy is not representative of the entire sample, and sample preparation may lead to severe artifacts (Treuel et al., 2014).

## 3.5 Toxicity of the Surface-Modified NMs

Surface modification of NMs is one of the strategies to achieve microbial interaction. In addition, surface modification of NMs improves their colloidal stability, imparts functionality, and improves cytocompatibility (Guerrini, Alvarez-Puebla, & Pazos-Perez, 2018; Westmeier et al., 2018). Stealth liposome wherein liposomal surface is coated with PEG to improve the blood circulation time of liposomes is one such example of surface modification. PEG is well-known for its beneficial roles in pharmacology. Drugs are derivatized with PEG to improve their solubility, reducing toxicity and clearance from the renal system. Conventional liposomes are cleared from the blood system by macrophages. To prevent that, liposomes are coated with PEG to improve their circulation time (Immordino, Dosio, & Cattel, 2006).

On the other hand, one of the major problems associated with surface-modified NMs is the formation of the protein corona. Due to the surface functionalization of the NMs, proteins present in the biological fluid adhere to the NM surface and form a layer around, which is known as the protein corona (Obst et al., 2017). Several other parameters such as size, shape, and charge of the NM, physiological conditions such as pH, ionic strength, and the composition of the biological medium influence the formation of the protein corona (Abdelkhaliq et al., 2018; Westmeier et al., 2018). The formation of protein corona impedes several functionalities of the NMs such as cell adhesion, drug release, and cell uptake (Figure 3.5) (Abdelkhaliq et al., 2018; Obst et al., 2017; Oh et al., 2018). In addition, such NMs pose higher toxicity to the cells (Obst et al., 2017). In a

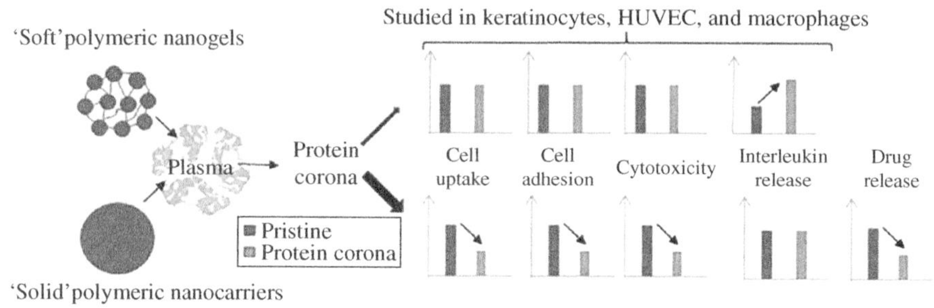


**Figure 3.5** Impact of corona formation on the NM functions. *Source*: Reproduced from Obst et al. (2017).

similar *in vitro* study, it was shown that protein corona formation on zinc oxide nanoparticles is associated with reactive oxygen species (ROS) generation, activation of intrinsic apoptotic and necrotic pathways (Giau, Park, Shim, Son, & An, 2019). Apart from proteins, other biomolecules present in the biological fluids can also induce corona formation and have similar adverse effects. Though surface modification improves cytocompatibility of the NMs, a few reports have shown potential accumulation of nanoparticles in lung and liver, and toxicity to the host tissues (Xie, Sun, Zhong, Shi, & Zhang, 2009).

## 3.6 Challenges and Future Perspectives

NM–microbe cross talk depends on various parameters such as (i) surface functionalization strategy and the chemical moiety used for functionalization; (ii) NM parameters such as size, shape, hydrophobicity/hydrophilicity, charge; and (iii) type of microbe and the physiological environment. Another important aspect of studying NM–microbial cross talk is using a reliable analytical tool to characterize the interaction. Though several techniques have been exploited to characterize the cross talk, every technique is associated with certain disadvantages. Hence, combinatory approaches or improvising the existing technique such as highly sensitive fluctuations (nanomechanical sensor) of AFM cantilevers could be a solution (Longo et al., 2013). One of the major challenges in NM–microbe interaction is the formation of corona. As explained in earlier text, corona formation impedes functionalities of pristine NMs. Strategies to overcome the formation of corona are indeed requiring extensive research. Recently, precoating nanoparticle with protein along with the targeting moiety to act as a shield against serum protein-forming corona has been shown to prevent protein corona formation (Oh et al., 2018). Hence, extensive interdisciplinary research is required to understand and optimize the various aspects of NM–microbe cross talk.

## Questions and Answers

**1** Explain how the physicochemical properties of NMs affect NM–microbe interaction?
Properties of NMs like surface-to-volume ratio, shape, size, surface charge, and chemistry influence NM–microbe interaction, and hence affect their antibacterial activity. These properties also affect the toxicity of the material.

**2** What are the various mechanisms by which NMs act as antibacterial agents?
NMs use any of the following inhibitory mechanisms of antimicrobial action

release of metal ions
production of reactive oxygen species
physical damage of bacterial cell wall
inhibition of polyion of enzyme activity
inhibition of DNA synthesis.

**3** What is the effect of biomolecule corona on NM–microbe complex formation?
The formation of corona can affect the morphology of the NM, which can influence NM–microbe interaction. Also, the biomolecular corona provides a new biological identity to the NM. The presence of the corona may change the surface charge and surface free energy of the materials, altering the possibility of interactions significantly. The biomolecular corona introduces new properties to the NM, which influences its cytotoxicity.

**4** What are the mechanisms by which the active components of the LbL assembly act upon the bacteria?
The active agents act by releasing small molecules or ions into the surrounding environment. This strategy actively kills the bacteria and is considered as an efficient way to prevent biofilm formation.

**5** List out various polymer modifications used for the functionalization of surfaces?
Polysaccharide coatings, agarose crosslinked polymers, carbohydrate-based polymers, nanopatterned polymer brushes, and thermoresponsive polymers.

**6** Name any one parameter of the NM and microbe that can be exploited for the characterization of NM–microbe interaction?

The size of the NM varies after its interaction with the microbe and can be exploited for characterizing NM–microbe interaction. The surface charge of the microbe varies upon interacting with NM that can be exploited for characterization of NM–microbe interaction.

**7** Discuss any two challenges in NM–microbe interaction?

- Formation of corona: Corona formation around the NM is one of the major challenges in NM–microbe interaction. The surface functionalization of NM can induce corona formation around it.
- Characterization of NM–microbe interaction: It is a reliable analytical tool to characterize the NM–microbe interaction.

### Take Home Message

- The study of NM–microbe complex formation is crucial for developing novel and effective antimicrobial agents to fight antibiotic resistance. The interaction of NMs with biofilms is also a crucial aspect to be studied.
- The NM–microbe interactions depend on the difference in surface features of microbes and on the physicochemical properties of the NMs. The formation of biomolecule corona also affects the interaction.
- Self-assembled monolayers (SAMs) and layer-by-layer (LbL) technique are the two most promising techniques that are being widely used for surface functionalization.
- Various physicochemical modifications such as altering of surface hydrophobicity, creating nanostructured surfaces, surface coating using polymers have been performed. Biofunctionalization has been achieved by inhibition of various molecules onto the NM surface or using release-based systems.
- Several techniques can be used to characterize the NM–microbe cross talk. However, a combinatorial approach or improvising on existing techniques can provide more insightful information.

## References

Abbaszadegan, A., Ghahramani, Y., Gholami, A., Hemmateenejad, B., Dorostkar, S., Nabavizadeh, M., & Sharghi, H. (2015). The effect of charge at the surface of silver nanoparticles on antimicrobial activity against Gram-positive and Gram-negative bacteria: A preliminary study. *Journal of Nanomaterials*, 2015, 8. doi:10.1155/2015/720654

Abdelkhaliq, A., van der Zande, M., Punt, A., Helsdingen, R., Boeren, S., Vervoort, J. J. M., . . . Bouwmeester, H. (2018). Impact of nanoparticle surface functionalization on the protein corona and cellular adhesion, uptake and transport. *Journal of Nanobiotechnology*, 16(1). doi:10.1186/s12951-018-0394-6

Abdullah, M. F., Ghosh, S. K., Basu, S., & Mukherjee, A. (2015). Cationic guar gum orchestrated environmental synthesis for silver nano-bio-composite films. *Carbohydrate Polymers*, 134, 30–37. doi:10.1016/j.carbpol.2015.06.029

Achari, G. A., & Kowshik, M. (2018). Recent developments on nanotechnology in agriculture: Plant mineral nutrition, health, and interactions with soil microflora. *Journal of Agricultural and Food Chemistry*, 66(33), 8647–8661. doi:10.1021/acs.jafc.8b00691

Acharya, D., Singha, K. M., Pandey, P., Mohanta, B., Rajkumari, J., & Singha, L. P. (2018). Shape dependent physical mutilation and lethal effects of silver nanoparticles on bacteria. *Scientific Reports*, 8(1), 201. doi:10.1038/s41598-017-18590-6

Ahmad Nor, Y., Niu, Y., Karmakar, S., Zhou, L., Xu, C., Zhang, J., . . . Yu, C. (2015). Shaping manoparticles with hydrophilic compositions and hydrophobic properties as nanocarriers for antibiotic delivery. *ACS Central Science*, 1(6), 328–334. doi:10.1021/acscentsci.5b00199

Akaighe, N., MacCuspie, R. I., Navarro, D. A., Aga, D. S., Banerjee, S., Sohn, M., & Sharma, V. K. (2011). Humic acid-induced silver nanoparticle formation under environmentally relevant conditions. *Environmental Science & Technology*, 45(9), 3895–3901. doi:10.1021/es103946g

Al-Saidi, W. A., Feng, H., & Fichthorn, K. A. (2012). Adsorption of polyvinylpyrrolidone on Ag surfaces: Insight into a structure-directing agent. *Nano Letters*, 12(2), 997–1001. doi:10.1021/nl2041113

Arakha, M., Pal, S., Samantarrai, D., Panigrahi, T. K., Mallick, B. C., Pramanik, K., . . . Jha, S. (2015). Antimicrobial activity of iron oxide nanoparticle upon modulation of nanoparticle–bacteria interface. *Scientific Reports*, 5(1). doi:10.1038/srep14813

Ashton, N. N., Allyn, G., Porter, S. T., Haussener, T. J., Sebahar, P. R., Looper, R. E., & Williams, D. L. (2019). In vitrotesting of a first-in-class tri-alkylnorspermidine-biaryl antibiotic in an anti-biofilm silicone coating. *Acta Biomaterialia*, 93, 25–35. doi:10.1016/j.actbio.2019.02.010

Bakirdere, S., Yilmaz, M. T., Tornuk, F., Keyf, S., Yilmaz, A., Sagdic, O., & Kocabas, B. (2015). Molecular characterization of silver-stearate nanoparticles (AgStNPs): A hydrophobic and antimicrobial material against foodborne pathogens. *Food Research International*, 76, 439–448. doi:10.1016/j.foodres.2015.08.005

Bamrungsap, S., Zhao, Z., Chen, T., Wang, L., Li, C., Fu, T., & Tan, W. (2012). Nanotechnology in therapeutics: A focus on nanoparticles as a drug delivery system. *Nanomedicine*, 7(8), 1253–1271. doi:10.2217/nnm.12.87

Barbolina, I., Woods, C. R., Lozano, N., Kostarelos, K., Novoselov, K. S., & Roberts, I. S. (2016). Purity of graphene oxide determines its antibacterial activity. *2D Materials*, 3(2), 025025.

Bassous, N. J., & Webster, T. J. (2019). The binary effect on methicillin-resistant *Staphylococcus aureus* of polymeric nanovesicles appended by proline-rich amino acid sequences and inorganic nanoparticles. *Small*, 15(18), 1804247. doi:10.1002/smll.201804247

Bayramov, D. F., & Neff, J. A. (2017). Beyond conventional antibiotics—new directions for combination products to combat biofilm. *Advanced Drug Delivery Reviews*, 112, 48–60. doi:10.1016/j.addr.2016.07.010

Bernabini, C., Holmes, D., & Morgan, H. (2011). Micro-impedance cytometry for detection and analysis of micron-sized particles and bacteria. *Lab Chip*, 11(3), 407–412. doi:10.1039/c0lc00099j

Beyth, N., Houri-Haddad, Y., Domb, A., Khan, W., & Hazan, R. (2015). Alternative antimicrobial approach: Nano-antimicrobial materials. *Evidence-Based Complementary and Alternative Medicine*, 2015, 16. doi:10.1155/2015/246012

Blunk, T., Hochstrasser, D. F., Sanchez, J.-C., Müller, B. W., & Müller, R. H. (1993). Colloidal carriers for intravenous drug targeting: Plasma protein adsorption patterns on surface-modified latex particles evaluated by two-dimensional polyacrylamide gel electrophoresis. *Electrophoresis*, 14(1), 1382–1387. doi:10.1002/elps.11501401214

Boonrungsiman, S., Suchaoin, W., Chetprayoon, P., Viriya-empikul, N., Aueviriyavit, S., & Maniratanachote, R. (2017). Shape and surface properties of titanate nanomaterials influence differential cellular uptake behavior and biological responses in THP-1 cells. *Biochemistry and Biophysics Reports*, 9, 203–210. doi:10.1016/j.bbrep.2016.12.014

Bowman, S. M., & Free, S. J. (2006). The structure and synthesis of the fungal cell wall. *BioEssays*, 28(8), 799–808. doi:10.1002/bies.20441

Breus, V. V., Heyes, C. D., Tron, K., & Nienhaus, G. U. (2009). Zwitterionic biocompatible quantum dots for wide pH stability and weak nonspecific binding to cells. *ACS Nano*, 3(9), 2573–2580. doi:10.1021/nn900600w

Cedervall, T., Lynch, I., Foy, M., Berggård, T., Donnelly, S. C., Cagney, G., . . . Dawson, K. A. (2007). Detailed identification of plasma proteins adsorbed on copolymer nanoparticles. *Angewandte Chemie International Edition*, 46(30), 5754–5756. doi:10.1002/anie.200700465

Chechik, V., & Stirling, C. J. M. (2009). Gold-thiol self-assembled monolayers. *PATAI'S Chemistry of Functional Groups*. doi:10.1002/9780470682531.pat0220

Chemla, Y. R., Grossman, H. L., Poon, Y., McDermott, R., Stevens, R., Alper, M. D., & Clarke, J. (2000). Ultrasensitive magnetic biosensor for homogeneous immunoassay. *Proceedings of the National Academy of Sciences*, 97(26), 14268–14272. doi:10.1073/pnas.97.26.14268

Chen, G., Zhou, M., Chen, S., Lv, G., & Yao, J. (2009). Nanolayer biofilm coated on magnetic nanoparticles by using a dielectric barrier discharge glow plasma fluidized bed for immobilizing an antimicrobial peptide. *Nanotechnology*, 20(46), 465706. doi:10.1088/0957-4484/20/46/465706

Cheng, Q., Li, C., Pavlínek, V., Saha, P., & Wang, H. (2006). Surface-modified antibacterial TiO2/Ag+ nanoparticles: Preparation and properties. *Applied Surface Science*, 252, 4154–4160. doi:10.1016/j.apsusc.2005.06.022

Chua, P.-H., Neoh, K.-G., Kang, E.-T., & Wang, W. (2008). Surface functionalization of titanium with hyaluronic acid/chitosan polyelectrolyte multilayers and RGD for promoting osteoblast functions and inhibiting bacterial adhesion. *Biomaterials*, 29(10), 1412–1421. doi:10.1016/j.biomaterials.2007.12.019

Chuang, H. F., Smith, R. C., & Hammond, P. T. (2008). Polyelectrolyte multilayers for tunable release of antibiotics. *Biomacromolecules*, 9(6), 1660–1668. doi:10.1021/bm800185h

Corbierre, M. K., & Lennox, R. B. (2005). Preparation of thiol-capped gold nanoparticles by chemical reduction of soluble Au(I)—thiolates. *Chemistry of Materials*, 17(23), 5691–5696. doi:10.1021/cm051115a

Daima, H., & Bansal, V. (2015). Influence of physicochemical properties of nanomaterials on their antibacterial applications. In *Nanotechnology in diagnosis, treatment, and prophylaxis of infectious disease* (pp. 151–166). Boston: Acadamic Press.

Decher, G., Hong, J. D., & Schmitt, J. (1992). Buildup of ultrathin multilayer films by a self-assembly process: III. Consecutively alternating adsorption of anionic and cationic polyelectrolytes on charged surfaces. *Thin Solid Films*, 210–211(Part 2), 831–835. doi:10.1016/0040-6090(92)90417-A

Della Ventura, B., Iannaccone, M., Funari, R., Pica Ciamarra, M., Altucci, C., Capparelli, R., . . . Velotta, R. (2017). Effective antibodies immobilization and functionalized nanoparticles in a quartz-crystal microbalance-based immunosensor for the detection of parathion. *PLoS One*, 12(2), e0171754. doi:10.1371/journal.pone.0171754

Diederichs, J. E. (1996). Plasma protein adsorption patterns on liposomes: Establishment of analytical procedure. *Electrophoresis*, 17(3), 607–611. doi:10.1002/elps.1150170332

Docter, D., Westmeier, D., Markiewicz, M., Stolte, S., Knauer, S. K., & Stauber, R. H. (2015). The nanoparticle biomolecule corona: lessons learned – challenge accepted? *Chemical Society Reviews*, 44(17), 6094–6121.

Emerson, D., Fleming, E. J., & McBeth, J. M. (2010). Iron-oxidizing bacteria: An environmental and genomic perspective. *Annual Review of Microbiology*, 64(1), 561–583. doi:10.1146/annurev.micro.112408.134208

Fakruddin, M., Hossain, Z., & Afroz, H. (2012). Prospects and applications of nanobiotechnology: A medical perspective. *Journal of Nanobiotechnology*, 10(1), 31. doi:10.1186/1477-3155-10-31

Fisher, L. E., Yang, Y., Yuen, M.-F., Zhang, W., Nobbs, A. H., & Su, B. (2016). Bactericidal activity of biomimetic diamond nanocone surfaces. *Biointerphases*, 11(1), 011014. doi:10.1116/1.4944062

Freschauf, L. R., McLane, J., Sharma, H., & Khine, M. (2012). Shrink-induced superhydrophobic and antibacterial surfaces in consumer plastics. *PLoS One*, 7(8), e40987. doi:10.1371/journal.pone.0040987

Gadenne, V., Lebrun, L., Jouenne, T., & Thebault, P. (2013). Antiadhesive activity of ulvan polysaccharides covalently immobilized onto titanium surface. *Colloids and Surfaces B: Biointerfaces*, 112, 229–236. doi:10.1016/j.colsurfb.2013.07.061

Gao, J., Huang, X., Liu, H., Zan, F., & Ren, J. (2012). Colloidal stability of gold nanoparticles modified with thiol compounds: Bioconjugation and application in cancer cell imaging. *Langmuir*, 28(9), 4464–4471. doi:10.1021/la204289k

García, I., Sánchez-Iglesias, A., Henriksen-Lacey, M., Grzelczak, M., Penadés, S., & Liz-Marzán, L. M. (2015). Glycans as biofunctional ligands for gold nanorods: Stability and targeting in protein-rich media. *Journal of the American Chemical Society*, 137(10), 3686–3692. doi:10.1021/jacs.5b01001

Giau, V.-V., Park, Y.-H., Shim, K.-H., Son, S.-W., & An, S.-S. A. (2019). Dynamic changes of protein corona compositions on the surface of zinc oxide nanoparticle in cell culture media. *Frontiers of Chemical Science and Engineering*, 13(1), 90–97. doi:10.1007/s11705-018-1766-z

Gillett, A., Waugh, D., Lawrence, J., Swainson, M., & Dixon, R. (2016). Laser surface modification for the prevention of biofouling by infection causing *Escherichia coli*. *Journal of Laser Applications*, 28(2), 022503. doi:10.2351/1.4944442

Gref, R., Lück, M., Quellec, P., Marchand, M., Dellacherie, E., Harnisch, S., . . . Müller, R. H. (2000). 'Stealth' corona-core nanoparticles surface modified by polyethylene glycol (PEG): Influences of the corona (PEG chain length and surface density) and of the core composition on phagocytic uptake and plasma protein adsorption. *Colloids and Surfaces B: Biointerfaces*, 18(3), 301–313. doi:10.1016/S0927-7765(99)00156-3

Guerrini, L., Alvarez-Puebla, R., & Pazos-Perez, N. (2018). Surface modifications of nanoparticles for stability in biological fluids. *Materials*, 11(7), 1154. doi:10.3390/ma11071154

Gupta, A., Mumtaz, S., Li, C.-H., Hussain, I., & Rotello, V. M. (2019). Combatting antibiotic-resistant bacteria using nanomaterials. *Chemical Society Reviews*, 48(2), 415–427. doi:10.1039/c7cs00748e

Harper, R. A., Carpenter, G. H., Proctor, G. B., Harvey, R. D., Gambogi, R. J., Geonnotti, A. R., . . . Jones, S. A. (2019). Diminishing biofilm resistance to antimicrobial nanomaterials through electrolyte screening of electrostatic interactions. *Colloids Surface B Biointerfaces*, 173, 392–399.

Hasan, J., Raj, S., Yadav, L., & Chatterjee, K. (2015). Engineering a nanostructured "super surface" with superhydrophobic and superkilling properties. *RSC Advances*, 5(56), 44953–44959. doi:10.1039/c5ra05206h

Hernandez-Montelongo, J., Lucchesi, E. G., Gonzalez, I., Macedo, W. A. A., Nascimento, V. F., Moraes, A. M., . . . Cotta, M. A. (2016). Hyaluronan/chitosan nanofilms assembled layer-by-layer and their antibacterial effect: A study using *Staphylococcus aureus* and *Pseudomonas aeruginosa*. *Colloids and Surfaces B: Biointerfaces*, 141, 499–506. doi:10.1016/j.colsurfb.2016.02.028

Hoo, C. M., Starostin, N., West, P., & Mecartney, M. L. (2008). A comparison of atomic force microscopy (AFM) and dynamic light scattering (DLS) methods to characterize nanoparticle size distributions. *Journal of Nanoparticle Research*, 10(S1), 89–96. doi:10.1007/s11051-008-9435-7

Hortelão, A. C., Patiño, T., Perez-Jiménez, A., Blanco, À., & Sánchez, S. (2018). Enzyme-powered nanobots enhance anticancer drug delivery. *Advanced Functional Materials*, 28(25), 1705086. doi:10.1002/adfm.201705086

Hou, Z., Shankar, Y. V., Liu, Y., Ding, F., Subramanion, J. L., Ravikumar, V., . . . Chan-Park, M. B. (2017). Nanoparticles of short cationic peptidopolysaccharide self-assembled by hydrogen bonding with antibacterial effect against multidrug-resistant bacteria. *ACS Applied Materials & Interfaces*, 9(44), 38288–38303.

Huang, W.-C., Tsai, P.-J., & Chen, Y.-C. (2007). Functional gold nanoparticles as photothermal agents for selective-killing of pathogenic bacteria. *Nanomedicine*, 2(6), 777–787. doi:10.2217/17435889.2.6.777

Immordino, M. L., Dosio, F., & Cattel, L. (2006). Stealth liposomes: Review of the basic science, rationale, and clinical applications, existing and potential. *International Journal of Nanomedicine*, 1(3), 297–315.

Izquierdo, A., Ono, S. S., Voegel, J. C., Schaaf, P., & Decher, G. (2005). Dipping versus spraying: Exploring the deposition conditions for speeding up layer-by-layer assembly. *Langmuir*, 21(16), 7558–7567. doi:10.1021/la047407s

Kaittanis, C., Nath, S., & Perez, J. M. (2008). Rapid nanoparticle-mediated monitoring of bacterial metabolic activity and assessment of antimicrobial susceptibility in blood with magnetic relaxation. *PLoS One*, 3(9), e3253. doi:10.1371/journal.pone.0003253

Kang, H., Buchman, J. T., Rodriguez, R. S., Ring, H. L., He, J., Bantz, K. C., & Haynes, C. L. (2018). Stabilization of silver and gold nanoparticles: Preservation and improvement of plasmonic functionalities. *Chemical Reviews*, 119(1), 664–699. doi:10.1021/acs.chemrev.8b00341

Kang, S. M., Kong, B., Oh, E., Choi, J. S., & Choi, I. S. (2010). Osteoconductive conjugation of bone morphogenetic protein-2 onto titanium/titanium oxide surfaces coated with non-biofouling poly(poly(ethylene glycol) methacrylate). *Colloids and Surfaces B: Biointerfaces*, 75(1), 385–389. doi:10.1016/j.colsurfb.2009.08.039

Kaushal, S., Priyadarshi, N., Pinnaka, A. K., Soni, S., Deep, A., & Singhal, N. K. (2019). Glycoconjugates coated gold nanorods based novel biosensor for optical detection and photothermal ablation of food borne bacteria. *Sensors and Actuators B: Chemical*, 289, 207–215. doi:10.1016/j.snb.2019.03.096

Kind, M., & Wöll, C. (2009). Organic surfaces exposed by self-assembled organothiol monolayers: Preparation, characterization, and application. *Progress in Surface Science*, 84(7), 230–278. doi:10.1016/j.progsurf.2009.06.001

Kohanski, M. A., Dwyer, D. J., & Collins, J. J. (2010). How antibiotics kill bacteria: From targets to networks. *Nature Reviews Microbiology*, 8, –423. doi:10.1038/nrmicro2333 [Review Article]

Koren, E., Apte, A., Jani, A., & Torchilin, V. P. (2012). Multifunctional PEGylated 2C5-immunoliposomes containing pH-sensitive bonds and TAT peptide for enhanced tumor cell internalization and cytotoxicity. *Journal of Controlled Release*, 160(2), 264–273. doi:10.1016/j.jconrel.2011.12.002

Langowska, K., Kowal, J., Palivan, C. G., & Meier, W. (2014). A general strategy for creating self-defending surfaces for controlled drug production for long periods of time. *Journal of Materials Chemistry B*, 2(29), 4684–4693. doi:10.1039/c4tb00277f

Laurent, S., Forge, D., Port, M., Roch, A., Robic, C., Vander Elst, L., & Muller, R. N. (2008). Magnetic iron oxide nanoparticles: Synthesis, stabilization, vectorization, physicochemical characterizations, and biological applications. *Chemical Reviews*, 108(6), 2064–2110. doi:10.1021/cr068445e

Lee, D., Cohen, R. E., & Rubner, M. F. (2005). Antibacterial properties of Ag nanoparticle loaded multilayers and formation of magnetically directed antibacterial microparticles. *Langmuir*, 21(21), 9651–9659. doi:10.1021/la0513306

Lee, S. J., Lee, Y. B., Lim, Y. R., Han, J. K., Jeon, I. S., Bae, G., . . . Lee, S. S. (2019). High energy electron beam stimulated nanowelding of silver nanowire networks encapsulated with graphene for flexible and transparent electrodes. *Scientific Reports*, 9(1), 9376. doi:10.1038/s41598-019-45887-5

Lemire, J. A., Harrison, J. J., & Turner, R. J. (2013). Antimicrobial activity of metals: Mechanisms, molecular targets and applications. *Nature Reviews Microbiology*, 11, 371. doi:10.1038/nrmicro3028 [Review Article]

Leung, Y., Chan, C., Ng, A. M. C., Chan, H. T., Chiang, M., Djurišić, A., . . . Au, D. (2012). Antibacterial activity of ZnO nanoparticles with a modified surface under ambient illumination. *Nanotechnology*, 23, 475703.

Leung, Y. H., Ng, A. M. C., Xu, X., Shen, Z., Gethings, L. A., Wong, M. T., . . . Leung, F. C. C. (2014). Mechanisms of antibacterial activity of MgO: Non-ROS mediated toxicity of MgO nanoparticles towards *Escherichia coli*. *Small*, 10(6), 1171–1183. doi:10.1002/smll.201302434

Li, M., Neoh, K.-G., Kang, E.-T., Lau, T., & Chiong, E. (2014). Surface modification of silicone with covalently immobilized and crosslinked agarose for potential application in the inhibition of infection and omental wrapping. *Advanced Functional Materials*, 24(11), 1631–1643. doi:10.1002/adfm.201302242

Li, X., Robinson, S. M., Gupta, A., Saha, K., Jiang, Z., Moyano, D. F., . . . Rotello, V. M. (2014). Functional gold nanoparticles as potent antimicrobial agents against multi-drug-resistant bacteria. *ACS Nano*, 8(10), 10682–10686. doi:10.1021/nn5042625

Li, Y.-Q., Zhu, B., Li, Y., Leow, W. R., Goh, R., Ma, B., . . . Chen, X. (2014). A synergistic capture strategy for enhanced detection and elimination of bacteria. *Angewandte Chemie International Edition*, 53(23), 5837–5841. doi:10.1002/anie.201310135

Lin, Y.-S., Tsai, P.-J., Weng, M.-F., & Chen, Y.-C. (2005). Affinity capture using vancomycin-bound magnetic nanoparticles for the MALDI-MS analysis of bacteria. *Analytical Chemistry*, 77(6), 1753–1760. doi:10.1021/ac048990k

Liu, R., Lee, Y. Y., Chong, K. S. L., & Tomczak, N. (2016). Transparent hydrophobic antimicrobial coatings by in situ synthesis of Ag nanoparticles in a surface-grafted amphiphilic comb-copolymer. *Advanced Materials Interfaces*, 3(1), 1500448. doi:10.1002/admi.201500448

Liu, X., Zhang, H., Tian, Z., Sen, A., & Allcock, H. R. (2012). Preparation of quaternized organic–inorganic hybrid brush polyphosphazene-co-poly[2-(dimethylamino)ethyl methacrylate] electrospun fibers and their antibacterial properties. *Polymer Chemistry*, 3(8), 2082. doi:10.1039/c2py20170d

Lohan, S., Raza, K., Singla, S., Chhibber, S., Wadhwa, S., Katare, O. P., . . . Singh, B. (2016). Studies on enhancement of anti-microbial activity of pristine MWCNTs against pathogens. *AAPS PharmSciTech*, 17(5), 1042–1048. doi:10.1208/s12249-015-0430-x

Lombana, A., Raja, Z., Casale, S., Pradier, C.-M., Foulon, T., Ladram, A., & Humblot, V. (2014). Temporin-SHa peptides grafted on gold surfaces display antibacterial activity. *Journal of Peptide Science*, 20(7), 563–569. doi:10.1002/psc.2654

Longo, G., Alonso-Sarduy, L., Rio, L. M., Bizzini, A., Trampuz, A., Notz, J., . . . Kasas, S. (2013). Rapid detection of bacterial resistance to antibiotics using AFM cantilevers as nanomechanical sensors. *Nature Nanotechnology*, 8(7), 522–526. doi:10.1038/nnano.2013.120

Louise Meyer, R., Zhou, X., Tang, L., Arpanaei, A., Kingshott, P., & Besenbacher, F. (2010). Immobilisation of living bacteria for AFM imaging under physiological conditions. *Ultramicroscopy*, 110(11), 1349–1357. doi:10.1016/j.ultramic.2010.06.010

Love, J. C., Estroff, L. A., Kriebel, J. K., Nuzzo, R. G., & Whitesides, G. M. (2005). Self-assembled monolayers of thiolates on metals as a form of nanotechnology. *Chemical Reviews*, 105(4), 1103–1170. doi:10.1021/cr0300789

Lu, H., Lin, J., Yang, W., Liu, L., Wang, Y., Chen, G., & Huang, W. (2017). Effect of nano-TiO2 surface modification on polarization characteristics and corona aging performance of polyimide nano-composites. *Journal of Applied Polymer Science*, 134(29), 45101.

Lu, X., Feng, X., Werber, J. R., Chu, C., Zucker, I., Kim, J.-H., . . . Elimelech, M. (2017). Enhanced antibacterial activity through the controlled alignment of graphene oxide nanosheets. *Proceedings of the National Academy of Sciences*, 114(46), E9793–E9801.

Mahmoud, N. N., Alkilany, A. M., Khalil, E. A., & Al-Bakri, A. G. (2018). Nano-photothermal ablation effect of hydrophilic and hydrophobic functionalized gold nanorods on *Staphylococcus aureus* and *Propionibacterium acnes*. *Scientific Reports*, 8(1), 6881. doi:10.1038/s41598-018-24837-7

Mahmoudi, M., Shokrgozar, M. A., Sardari, S., Moghadam, M. K., Vali, H., Laurent, S., & Stroeve, P. (2011). Irreversible changes in protein conformation due to interaction with superparamagnetic iron oxide nanoparticles. *Nanoscale*. doi:10.1039/c0nr00733a

Meister, S., Zlatev, I., Stab, J., Docter, D., Baches, S., Stauber, R. H., . . . Pietrzik, C. U. (2013). Nanoparticulate flurbiprofen reduces amyloid-β42 generation in an in vitroblood–brain barrier model. *Alzheimer's Research & Therapy*, 5(6), 51–51. doi:10.1186/alzrt225

Mijnendonckx, K., Leys, N., Mahillon, J., Silver, S., & Van Houdt, R. (2013). Antimicrobial silver: Uses, toxicity and potential for resistance. *BioMetals*, 26(4), 609–621. doi:10.1007/s10534-013-9645-z

Mlalila, N., Swai, H., Hilonga, A., & Kadam, D. (2016). Antimicrobial dependence of silver nanoparticles on surface plasmon resonance bands against *Escherichia coli*. *Nanotechnology, Science and Applications*, 10, 1–9. doi:10.2147/nsa.s123681

Mohamed, M. M., Fouad, S. A., Elshoky, H. A., Mohammed, G. M., & Salaheldin, T. A. (2017). Antibacterial effect of gold nanoparticles against *Corynebacterium pseudotuberculosis*. *International Journal of Veterinary Science and Medicine*, 5(1), 23–29. doi:10.1016/j.ijvsm.2017.02.003

Mohapatra, J., Zeng, F., Elkins, K., Xing, M., Ghimire, M., Yoon, S., . . . Liu, J. P. (2018). Size-dependent magnetic and inductive heating properties of Fe3O4 nanoparticles: Scaling laws across the superparamagnetic size. *Physical Chemistry Chemical Physics*, 20(18), 12879–12887. doi:10.1039/c7cp08631h

Munawar, A., Ong, Y., Schirhagl, R., Tahir, M. A., Khan, W. S., & Bajwa, S. Z. (2019). Nanosensors for diagnosis with optical, electric and mechanical transducers. *RSC Advances*, 9(12), 6793–6803. doi:10.1039/c8ra10144b

Navarre, W. W., & Schneewind, O. (1999). Surface proteins of gram-positive bacteria and mechanisms of their targeting to the cell wall envelope. *Microbiology and Molecular Biology Reviews*, 63(1), 174–229. [Review]

Nevius, B. A., Chen, Y. P., Ferry, J. L., & Decho, A. W. (2012). Surface-functionalization effects on uptake of fluorescent polystyrene nanoparticles by model biofilms. *Ecotoxicology*, 21(8), 2205–2213.

Nyalosaso, J. L., Rascol, E., Pisani, C., Dorandeu, C., Dumail, X., Maynadier, M., . . . Chopineau, J. (2016). Synthesis, decoration, and cellular effects of magnetic mesoporous silica nanoparticles. *RSC Advances*, 6(62), 57275–57283. doi:10.1039/c6ra09017f

Obst, K., Yealland, G., Balzus, B., Miceli, E., Dimde, M., Weise, C., . . . Hedtrich, S. (2017). Protein corona formation on colloidal polymeric nanoparticles and polymeric nanogels: Impact on cellular uptake, toxicity, immunogenicity, and drug release properties. *Biomacromolecules*, 18(6), 1762–1771. doi:10.1021/acs.biomac.7b00158

Oger, P.-C., Piesse, C., Ladram, A., & Humblot, V. (2019). Engineering of antimicrobial surfaces by using temporin analogs to tune the biocidal/antiadhesive effect. *Molecules*, 24(4), 814. doi:10.3390/molecules24040814

Oh, J. Y., Kim, H. S., Palanikumar, L., Go, E. M., Jana, B., Park, S. A., . . . Ryu, J.-H. (2018). Cloaking nanoparticles with protein corona shield for targeted drug delivery. *Nature Communications*, 9(1). doi:10.1038/s41467-018-06979-4

Palza, H., Quijada, R., & Delgado, K. (2015). Antimicrobial polymer composites with copper micro- and nanoparticles: Effect of particle size and polymer matrix. *Journal of Bioactive and Compatible Polymers*, 30(4), 366–380. doi:10.1177/0883911515578870

Pasban Ziyarat, F., Asoodeh, A., Sharif Barfeh, Z., Pirouzi, M., & Chamani, J. (2013). Probing the interaction of lysozyme with ciprofloxacin in the presence of different-sized Ag nano-particles by multispectroscopic techniques and isothermal titration calorimetry. *Journal of Biomolecular Structure and Dynamics*, 32(4), 613–629. doi:10.1080/07391102.2013.785919

Patel, K., Bharatiya, B., Mukherjee, T., Soni, T., Shukla, A., & Suhagia, B. N. (2016). Role of stabilizing agents in the formation of stable silver nanoparticles in aqueous solution: Characterization and stability study. *Journal of Dispersion Science and Technology*, 38(5), 626–631. doi:10.1080/01932691.2016.1185374

Pelaz, B., del Pino, P., Maffre, P., Hartmann, R., Gallego, M., Rivera-Fernández, S., . . . Parak, W. J. (2015). Surface functionalization of nanoparticles with polyethylene glycol: Effects on protein adsorption and cellular uptake. *ACS Nano*, 9(7), 6996–7008. doi:10.1021/acsnano.5b01326

Qin, D., He, X., Wang, K., & Tan, W. (2008). Using fluorescent nanoparticles and SYBR Green I based two-color flow cytometry to determine *Mycobacterium tuberculosis* avoiding false positives. *Biosensors Bioelectronics*, 24(4), 626–631.

Quiñones, R., Raman, A., & Gawalt, E. S. (2008). Functionalization of nickel oxide using alkylphosphonic acid self-assembled monolayers. *Thin Solid Films*, 516(23), 8774–8781. doi:10.1016/j.tsf.2008.06.055

Rahme, K., Nolan, M. T., Doody, T., McGlacken, G. P., Morris, M. A., O'Driscoll, C., & Holmes, J. D. (2013). Highly stable PEGylated gold nanoparticles in water: Applications in biology and catalysis. *RSC Advances*, 3(43), 21016–21024. doi:10.1039/c3ra41873a

Raj, S., Jose, S., Sumod, U., & Sabitha, M. (2012). Nanotechnology in cosmetics: Opportunities and challenges. *Journal of Pharmacy And Bioallied Sciences*, 4(3), 186–193. doi:10.4103/0975-7406.99016 [Review Article]

Raza, A. M., Kanwal, Z., Rauf, A., Sabri, N. A., Riaz, S., & Naseem, S. (2016). Size- and shape-dependent antibacterial studies of silver nanoparticles synthesized by wet chemical routes. *Nanomaterials*, 6(4). doi:10.3390/nano6040074

Renoud, P., Toury, B., Benayoun, S., Attik, G., & Grosgogeat, B. (2012). Functionalization of titanium with chitosan via silanation: Evaluation of biological and mechanical performances. *PLoS One*, 7(7), e39367. doi:10.1371/journal.pone.0039367

Roosjen, A., Norde, W., Mei, H. C., & Busscher, H. J. (2006). The use of positively charged or low surface free energy coatings versus polymer brushes in controlling biofilm formation. *Characterization of Polymer Surfaces and Thin Films*, 132, 138–144. doi:10.1007/2882_026

Salomoni, R., Léo, P., Montemor, A. F., Rinaldi, B. G., & Rodrigues, M. (2017). Antibacterial effect of silver nanoparticles in *Pseudomonas aeruginosa*. *Nanotechnology, Science and Applications*, 10, 115–121. doi:10.2147/nsa.s133415

Salton, M. R. J. (1953). Studies of the bacterial cell wall: IV. The composition of the cell walls of some gram-positive and gram-negative bacteria. *Biochimica et Biophysica Acta*, 10, 512–523. doi:10.1016/0006-3002(53)90296-0

Salvador-Morales, C., Flahaut, E., Sim, E., Sloan, J., Green, M. L. H., & Sim, R. B. (2006). Complement activation and protein adsorption by carbon nanotubes. *Molecular Immunology*, 43(3), 193–201. doi:10.1016/j.molimm.2005.02.006

Senarathna, U. L. N. H., Fernando, S. S. N., Gunasekara, T. D. C. P., Weerasekera, M. M., Hewageegana, H. G. S. P., Arachchi, N. D. H., . . . Jayaweera, P. M. (2017). Enhanced antibacterial activity of TiO2 nanoparticle surface modified with *Garcinia zeylanica* extract. *Chemistry Central Journal*, 11(1), 7. doi:10.1186/s13065-017-0236-x

Seoudi, R., Shabaka, A., El Sayed, Z. A., & Anis, B. (2011). Effect of stabilizing agent on the morphology and optical properties of silver nanoparticles. *Physica E: Low-dimensional Systems and Nanostructures*, 44(2), 440–447. doi:10.1016/j.physe.2011.09.018

Shen, L., Wang, B., Wang, J., Fu, J., Picart, C., & Ji, J. (2012). Asymmetric free-standing film with multifunctional anti-bacterial and self-cleaning properties. *ACS Applied Materials & Interfaces*, 4(9), 4476–4483. doi:10.1021/am301118f

Shi, Z., Neoh, K. G., Kang, E. T., Poh, C. K., & Wang, W. (2009). Surface functionalization of titanium with carboxymethyl chitosan and immobilized bone morphogenetic protein-2 for enhanced osseointegration. *Biomacromolecules*, 10(6), 1603–1611. doi:10.1021/bm900203w

Siddiqi, K. S., Ur Rahman, A., Tajuddin, & Husen, A. (2018). Properties of zinc oxide nanoparticles and their activity against microbes. *Nanoscale Research Letters*, 13(1), 1–13.

Siemer, S., Westmeier, D., Barz, M., Eckrich, J., Wünsch, D., Seckert, C., . . . Strieth, S. (2018). Biomolecule-corona formation confers resistance of bacteria to nanoparticle-induced killing: Implications for the design of improved nanoantibiotics. *Biomaterials*, 192.

Singh, J., Vishwakarma, K., Ramawat, N., Rai, P., Singh, V. K., Mishra, R. K., . . . Sharma, S. (2019). Nanomaterials and microbes' interactions: A contemporary overview. *Biotech*, 9(3), 68. doi:10.1007/s13205-019-1576-0

Stockton, W. B., & Rubner, M. F. (1997). Molecular-level processing of conjugated polymers. 4. Layer-by-layer manipulation of polyaniline via hydrogen-bonding interactions. *Macromolecules*, 30(9), 2717–2725. doi:10.1021/ma9700486

Such, G. K., Anthony Quinn, J. F. Q., Tjipto, E., & Caruso, F. (2006). Assembly of ultrathin polymer multilayer films by click chemistry. *Journal of American Chemical Society*, 128, 2.

Sweet, M. J., & Singleton, I. (2011). Chapter 5 – Silver nanoparticles: A microbial perspective. In A. I. Laskin, S. Sariaslani, & G. M. Gadd (Eds.), *Advances in Applied Microbiology* (Vol. 77, pp. 115–133). Academic Press.

Tanaka, Y., Matin, K., Gyo, M., Okada, A., Tsutsumi, Y., Doi, H., . . . Hanawa, T. (2010). Effects of electrodeposited poly(ethylene glycol) on biofilm adherence to titanium. *Journal of Biomedical Materials Research Part A*, 95A(4), 1105–1113. doi:10.1002/jbm.a.32932

Tebbe, M., Kuttner, C., Männel, M., Fery, A., & Chanana, M. (2015). Colloidally stable and surfactant-free protein-coated gold nanorods in biological media. *ACS Applied Materials & Interfaces*, 7(10), 5984–5991. doi:10.1021/acsami.5b00335

Treuel, L., Eslahian, K. A., Docter, D., Lang, T., Zellner, R., Nienhaus, K., . . . Maskos, M. (2014). Physicochemical characterization of nanoparticles and their behavior in the biological environment. *Physics Chemistry Chemical Physics*, 16(29), 15053–15067. doi:10.1039/c4cp00058g

Ulman, A. (1996). Formation and structure of self-assembled monolayers. *Chemical Reviews*, 96, 22.

Vallabani, N. V. S., & Singh, S. (2018). Recent advances and future prospects of iron oxide nanoparticles in biomedicine and diagnostics. *Biotech*, 8(6), 279. doi:10.1007/s13205-018-1286-z

Valotteau, C., Banat, I. M., Mitchell, C. A., Lydon, H., Marchant, R., Babonneau, F., . . . Humblot, V. (2017). Antibacterial properties of sophorolipid-modified gold surfaces against Gram positive and Gram negative pathogens. *Colloids and Surfaces B: Biointerfaces*, 157, 325–334. doi:10.1016/j.colsurfb.2017.05.072

Vasudevan, R., Kennedy, A. J., Merritt, M., Crocker, F. H., & Baney, R. H. (2014). Microscale patterned surfaces reduce bacterial fouling-microscopic and theoretical analysis. *Colloids and Surfaces B: Biointerfaces*, 117, 225–232. doi:10.1016/j.colsurfb.2014.02.037

Wang, L., Hu, C., & Shao, L. (2017). The antimicrobial activity of nanoparticles: Present situation and prospects for the future. *International Journal of Nanomedicine*, 12, 1227–1249.

Wang, W., Zeng, Z., Zeng, G., Zhang, C., Xiao, R., Zhou, C., . . . Zhou, Y. (2019). Sulfur doped carbon quantum dots loaded hollow tubular g-C3N4 as novel photocatalyst for destruction of *Escherichia coli* and tetracycline degradation under visible light. *Chemical Engineering Journal*, 378, 122132. doi:10.1016/j.cej.2019.122132

Wei, T., Yu, Q., Zhan, W., & Chen, H. (2016). A smart antibacterial surface for the on-demand killing and releasing of bacteria. *Advanced Healthcare Materials*, 5(4), 449–456. doi:10.1002/adhm.201500700

Weidenmaier, C., & Peschel, A. (2008). Teichoic acids and related cell-wall glycopolymers in Gram-positive physiology and host interactions. *Nature Reviews Microbiology*, 6(4), 276–287. doi:10.1038/nrmicro1861

Westmeier, D., Posselt, G., Hahlbrock, A., Bartfeld, S., Vallet, C., Abfalter, C., . . . Stauber, R. H. (2018). Nanoparticle binding attenuates the pathobiology of gastric cancer-associated *Helicobacter pylori*. *Nanoscale*, 10(3), 1453–1463. doi:10.1039/c7nr06573f

Xiao, M., Jasensky, J., Gerszberg, J., Chen, J., Tian, J., Lin, T., . . . Chen, Z. (2018). Chemically immobilized antimicrobial peptide on polymer and self-assembled monolayer substrates. *Langmuir*, 34(43), 12889–12896. doi:10.1021/acs.langmuir.8b02377

Xie, G., Sun, J., Zhong, G., Shi, L., & Zhang, D. (2009). Biodistribution and toxicity of intravenously administered silica nanoparticles in mice. *Archives of Toxicology*, 84(3), 183–190. doi:10.1007/s00204-009-0488-x

Yu, Q., Ista, L. K., & López, G. P. (2014). Nanopatterned antimicrobial enzymatic surfaces combining biocidal and fouling release properties. *Nanoscale*, 6(9), 4750–4757. doi:10.1039/c3nr06497b

Zabara, M., Senturk, B., Gontsarik, M., Ren, Q., Rottmar, M., Maniura-Weber, K., . . . Salentinig, S. (2019). Multifunctional nano-biointerfaces: Cytocompatible antimicrobial nanocarriers from stabilizer-free cubosomes. *Advanced Functional Materials*, 29(35), 1904007. doi:10.1002/adfm.201904007

Zharov, V. P., Mercer, K. E., Galitovskaya, E. N., & Smeltzer, M. S. (2006). Photothermal nanotherapeutics and nanodiagnostics for selective killing of bacteria targeted with gold nanoparticles. *Biophysical Journal*, 90(2), 619–627. doi:10.1529/biophysj.105.061895

Zhou, Z., Calabrese, D. R., Taylor, W., Finlay, J. A., Callow, M. E., Callow, J. A., . . . Ober, C. K. (2014). Amphiphilic triblock copolymers with PEGylated hydrocarbon structures as environmentally friendly marine antifouling and fouling-release coatings. *Biofouling*, 30(5), 589–604. doi:10.1080/08927014.2014.897335

Zhu, N., Ji, H., Yu, P., Niu, J., Farooq, M., Akram, M. W., . . . Niu, X. (2018). Surface modification of magnetic iron oxide nanoparticles. *Nanomaterials*, 8(10), 810.

Zhu, X., Jańczewski, D., Guo, S., Lee, S. S. C., Parra Velandia, F. J., Teo, S. L.-M., . . . Vancso, G. J. (2014). Polyion multilayers with precise surface charge control for antifouling. *ACS Applied Materials & Interfaces*, 7(1), 852–861. doi:10.1021/am507371a

Zhu, X., Zhu, Y., Murali, S., Stoller, M. D., & Ruoff, R. S. (2011). Nanostructured reduced graphene oxide/$Fe_2O_3$ composite as a high-performance anode material for lithium ion batteries. *ACS Nano*, 5(4), 3333–3338. doi:10.1021/nn200493r

## Glossary

**Gram-positive bacteria:** Large group of bacteria that retain the crystal violet stain and appear purple in the Gram staining test owing to the presence of a thick peptidoglycan layer in the cell wall. The cell wall lacks any outer lipid membrane.

**Gram-negative bacteria:** Large group of bacteria that do not retain the crystal violet stain and appear red/pink in the Gram staining test. They are distinguished from Gram-positive bacteria by the presence of a thin peptidoglycan layer and an outer lipid membrane.

**Structure-activity relationship:** Relationship between the chemical structure (structure related properties) and the biological activity of a compound.

**Surface hydrophobicity:** It is the property of a surface to repel water. It is generally measured by the contact angle between droplets of water with the surface of the material.

**Polydispersity:** A measure of size distribution of nanoparticles.

**Biomolecule corona:** Biomolecules that adsorb onto the surface of a nanoparticle in a biological fluid form the biomolecule corona. It includes proteins and extracellular matrix components.

**SPIONs (superparamagnetic iron oxide nanoparticles):** They are a class of metal oxide nanoparticles with unique magnetic properties and excellent biocompatibility. They consist of an iron oxide core that can be targeted to the required area through external magnets.

**SQUID (superconducting quantum interference device):** SQUID is a magnetometer and typically allows a fully automated measurement of the magnetization of a specimen as a function of magnetic field and/or temperature.

**Zwitterion:** A zwitterion is a molecule with at least one positively charged and one negatively charged functional group. The net charge of the entire molecule is zero.

**Aptamer:** Aptamers are a type of nanoparticles composed of DNA or RNA oligonucleotides that can bind with high affinity to specific antigens, such as peptides and small molecules.

**Extracellular polymeric substances (EPSs):** They are high-molecular weight polymers secreted by microorganisms into their environment. They form a major component of biofilms.

**Mesoporous:** According to IUPAC nomenclature, a mesoporous material is one that consists of pores with diameters between 2 and 50 nm.

**Aerosol spraying:** Aerosol spraying is a form of dispensing system that generates a fine mist of liquid particles of the material to be coated and a suitable volatile component. The droplets of the volatile components evaporate rapidly and the desired material is coated as fine particles onto the surface.

**Vapor deposition:** The vapor deposition technique comes under the category of thin film deposition. A thin film, where the thickness is varied from several nanometers to a few micrometers is deposited onto the surface. The deposition can be carried out by physical and chemical methods.

**Immobilization:** Immobilization is the process by which a biomolecule is physically entrapped or covalently attached to the surface of the substrate. It involves the use of van der Waals forces, electrostatic interactions, affinity recognition, etc.

**Microscale motifs:** Microscale motifs are structured or unstructured patterned surfaces such as hexagonal pillars, cross pillars and pits that are used to impart structural changes on the nanomaterial surface. These surface alterations are used to reduce bacterial retention.

**Polymer brushes:** Polymer brushes are chains of polymer molecules that are tethered or grafted at one or more points onto the solid surface. The polymer chains usually project outward, thereby imparting surface functionalization properties to the substrate.

**Block copolymers:** Block copolymers comprise of alternating fragments of chemically distinct polymers arranged as blocks in a linear manner and covalently linked to each other through their reactive functional groups.

**Biocidal system:** A biocide system denotes any chemical or microorganism that has the capacity to destroy or control the effect of harmful microorganisms through chemical or biological means.

**Biofilm:** Biofilm is a complex cluster of surface-associated microbial community that is enclosed in a self-produced polymer matrix and is closely associated with the surface of the adhered substrate. The mechanisms by which biofilms are formed depend mainly on specific strains and the environmental conditions.

**Bioconjugation:** Bioconjugation is a strategy in which two molecules (at least one of which has to be a biomolecule) are either covalently linked or non-covalently attached to each other. Polymer bioconjugation refers to covalent attachment of synthetic polymers to biomolecules.

**Polyelectrolytes:** Polyelectrolytes are polymers containing dissociating groups in their repeating units. Polyelectrolytes are classified as polyanions and polycations. Since the repeating units have ionizable functional groups, they dissociate in aqueous solutions, thereby making the polymer charged.

**Flow cytometry:** It is a laser-based technique used to measure optical and fluorescence characteristics of the cell.

**Scanning electron microscopy:** It is a technique wherein electron beam is focused onto the surface of the sample, measures various types of electrons emitted from the surface and thereby gives information of the surface topography of the sample.

**Transmission electron microscopy:** It is a technique where electron beam is penetrated through the samples and the two-dimensional image of the samples is created based on how sample interacts with the incident electrons.

**Macrophage:** A type of immune cell that plays a role as an antigen presenting cells.

**Apoptotic pathway:** It is a programmed pathway through which cell undergoes death.

**Necrotic pathway:** It is an un-programmed cell death that occurs due to chemical or physical damage to the cells.

**Enthalpy:** It is a measure of the heat content of the system, measured by using following formula.

$$E = U + PV$$

where, $E$ is the enthalpy of the system; $U$, internal energy of the system; $P$, gas pressure within the system; $V$, volume.

**Entropy:** Entropy is a measure of the randomness in the system. More the entropy, more randomness in the system.

**X-ray diffraction:** It is a technique wherein X-ray is incident on the sample and the scattered X-rays by the atoms present in the samples are collected. These scattered X-rays give information about the sample nature.

**Zeta potential:** Nanoparticles possess certain charge in the colloidal solution system. This attracts counterions around the nanoparticles. This double layer of ions is known as Stern layer. The electric potential around this Stern layer is known as zeta potential.

# 4

# Surface Functionalization of Nanoparticles for Stability in Biological Systems

*Srishti Agarwal and D. Sakthi Kumar*

*Bio-Nano Electronics Research Center, Graduate School of Interdisciplinary New Science, Toyo University, Kawagoe, Saitama, Japan*

## Objectives

- Understanding the key processes affecting nanoparticle stability in biological milieu.
- Knowing diverse measures to augment nanoparticle stability.
- Understanding the importance of nanoparticle design in shaping fruitful interactions at biological interface.
- Understanding the impact of nanomaterial physicochemical properties on microbe–nanomaterial interactions.

## 4.1 Introduction

Nanotechnology has become an attractive field from last decade owing to extraordinary properties displayed by nanomaterials (NMs) like high surface: volume ratio, enhanced biocompatibility, and minimal toxicity occurring due to the nanosize of such materials. The major advantage is the ability to modify the surface properties of the NPs according to the experimental requirements, thus acquiring responsive materials suitable for a specific application. For example, enhanced cellular internalization ability, non-cytotoxicity, selective recognition, stealth properties, and improved delivery can be achieved via specifically designed NP monolayer structures (Elghanian, Storhoff, Mucic, Letsinger, & Mirkin, 1997; Hoffman, 2008; Saha, Bajaj, Duncan, & Rotello, 2011). Despite being tremendously useful, the major drawback hindering the effectiveness of NMs is the interaction of biological fluids with matter (Guerrini, Ramon, & Pazos-Perez, 2018). What happens to NPs when they are transported into biological systems where their components are set to interact with NPs? Will these events be accountable

*Microbial Interactions at Nanobiotechnology Interfaces: Molecular Mechanisms and Applications*, First Edition. Edited by R. Navanietha Krishnaraj and Rajesh K. Sani.

for biological responses of NPs? How can NPs be stabilized in the biological milieu using functionalization strategies? Understanding the physical and chemical reactions of NPs upon contact with biological media is indispensable to disclose answers to the above questions, which will in turn improve the effectiveness of the application of NPs in biological systems.

NPs have been functionalized with various ligands, surfactants, polymers, etc. using different strategies. It is now well-described that the biomaterial surfaces are instantly covered by biomolecules (e.g. enzymes, proteins, detergents, etc.) upon contact with biological medium (known as "protein corona"), which bestows a new "biological identity" to the NP surfaces, hence defining their cellular responses (Mahmoudi et al., 2011). The main characteristics affecting the biological responses to NPs can be classified as electrodynamic, steric, hydrodynamic, solvent characteristics along with intrinsic characteristics like NP dimensions, charge, coatings, surface modifications, crystallinity, electronic states, hydrophobicity, and wettability, which support the growth of the protein corona (Mahmoudi et al., 2011).

In this chapter, we mainly emphasize the consequence of physicochemical properties of NPs (i.e. size, shape, composition, surface chemistry, coatings, etc.) on its interaction with biological milieu, and the functionalization strategies that can be used to enhance NM stability.

## 4.2 Major Processes Affecting NP Stability in Biological Media

NPs interaction with biological media is affected by many processes, which affect its stability (Figure 4.1). Some of the factors are described below.

### 4.2.1 Aggregation

Aggregation is defined as the process wherein a large aggregate is formed on association of dispersed particles. The colloidal stability of NPs is determined by various factors like coating, surface charge as well as the pH and ionic strength of the medium (Curtis, Toghani, Wong, & Nance, 2018). Usually having high ionic strengths, biological media lead to aggregation of electrostatically stabilized NPs followed by their sedimentation due to screening of the surface charge by electrolytes present in the media (Moore et al., 2015). In this case, resulting from shielding of electric field, NPs can come close to each other, thus agglomerating. Colloidal stabilization can be imparted through two major approaches: stabilization via electrostatic repulsion and steric stabilization. However, due to van der Waals forces in solutions of high ionic strengths, the diffuse double layer (DDL) is

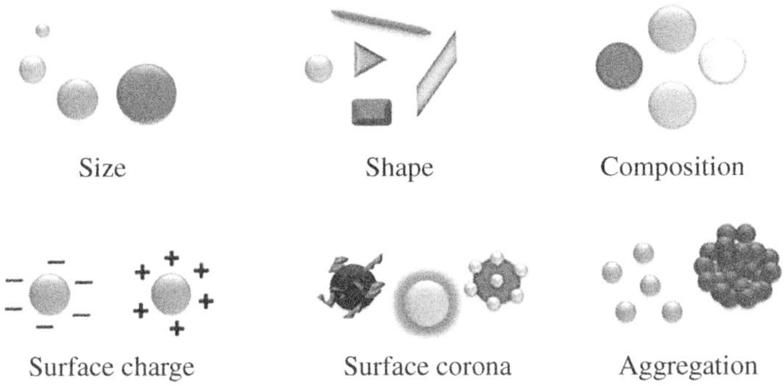


**Figure 4.1** Various physicochemical properties of NMs affecting their stability. *Source*: Reproduced with permission from copyright © Navya and Daima (2016).

neutralized leading to aggregation, thus causing failure of electrostatic stabilization in stipulating colloidal stability in biological media (Guerrini et al., 2018). In case of steric stabilization, a physical barrier at the NP surface like poly(ethylene glycol) (PEG) provides stability to NPs. Short-range repulsive hydration forces occurring due to the hydrophilic nature of the polymers impart an additional layer of stability. Therefore, steric stabilization is more suitable for biological systems. Stabilization of metal-based NMs in the solution can be achieved by addition of shielding agents that are essential to prevent agglomeration. It is also vital to know that in physiological conditions many NMs may aggregate because of their intrinsic reactive nature. Hence, it is likely that NMs will form aggregates rather than prevailing as individual units upon introduction in living cells in physiological environment or biological medium. An example is the size-dependent antibacterial activity of silver nanoparticles since silver nanoparticles can aggregate in media due to excessive electrolyte content hindering bacterial efficacy. To avoid this, silver nanoparticles can be modified with stabilizing agents that can prevent aggregation (Lok et al., 2007). Therefore, NMs need to be premeditated sensibly while exploiting aggregation phenomenon for biomedical applications taking into account their probable stability and toxicity.

### 4.2.2 Nanoparticle Design and Properties

NPs were originally deliberated as benign carriers; however, various studies have established that their contact with serum proteins and cell membrane receptors is determined by the nanoparticle design, which will result in affecting cell uptake, gene expression, and toxicity (Albanese, Tang, & Chan, 2012). The ligand density of a nanomaterial as well as the engineered geometry of a nanomaterial is responsible for interactions between nanoparticle-bound ligands and cellular receptors

(Albanese et al., 2012). The nano-size is an important parameter in governing the kinetics, distribution, metabolism, and excretion of the particles compared to the composition of the bulk material (Longmire, Choyke, & Kobayashi, 2008). A study reported a size-dependent biological profile and biodistribution valuation of three monodisperse drug-silica nanoconjugates of 20, 50, and 200 nm (Tang et al., 2014). It was shown that 50 nm-sized drug-silica nanoconjugate particles had maximum cancer tissue retention over time, thereby causing enhanced uptake and infiltration with slower clearance. While in another study, 2–100 nm-size gold and silver nanoparticles coated with antibodies were described to govern the process of internalization causing down regulation of cellular expression level, which in turn modified the cellular responses vital for basic cell functioning (Navya & Daima, 2016). Besides size of a material, shape/morphology is another important parameter at the nanoscale. Shape of nanoparticle directly influences cellular uptake with rod-shaped particles showing highest uptake trailed by spheres, cylinders, and cubes (Gratton et al., 2008). Silver triangular nanoplates were reported to display superior antimicrobial activity in contrast to rod- and sphere-shaped AgNPs against *E. Coli* (Pal, Tak, & Song, 2007). The strength of the cell–nanoparticle interactions is dictated by the small-radius surface asperities that arise due to result of differences in composition of the corona shell on the surface of NPs (Nel et al., 2009). Simulations of NPs cooperating with synthetic membranes indicate that nanoscale surface roughness significantly diminishes repulsive interactions (for example, electrostatic, hydrophilic), thus supporting adhesion, which may translate into easier uptake by cells. It is essential to know that the number of exterior molecules surges exponentially while NM size drops to below 100 nm; and, NMs' size and number of surface exposed molecules display an inverse relationship (Oberdorster, Oberdorster, & Oberdorster, 2005). Irregular crystal planes can arise from reduction in size of material, which can intensify whole structural defects and might disturb systematic electronic configuration of the material, causing changed electronic properties (Navya & Daima, 2016). All these physical changes can create explicit external groups that might work as reactive sites like hydrophilic or hydrophobic, catalytically active or passive, etc. depending upon the chemical composition of the material (Oberdorster et al., 2005).The ligand that is utilized to coat the NM will also affect downstream biological responses. For example, when the NPs were coated with two dissimilar proteins targeting the same receptor, the uptake and cytotoxicity of NPs were drastically transformed (J. Wang, Tian, Petros, Napier, & Desimone, 2010). The NP surface generates a region of extremely concentrated ligands, which surges avidity and, hypothetically, alters cell signaling. But, the downside is unforeseen changes caused by NMs in cell signaling. For instance, internalization of NPs coated with intercellular adhesion molecule I (ICAM-I) protein is not known to cause endocytosis, which is a rare discovery regarding ICAM-1. Still, placing numerous ICAM-I proteins on a NM's surface led to unexpected uptake (S. Muro et al., 2008). Another study reported activation of the AKT signaling pathway, instigating cell proliferation upon interaction of 14-nm carbon nanoparticles with epidermal growth receptors and $\beta$1-integrins on rat alveolar II epithelial cells (Unfried, Sydlik, Bierhals, Weissenberg, & Abel, 2008). An extra concern with NP–ligand complexes is the likely denaturation of proteins when bound to the engineered surface, thus affecting its binding to the receptor as well as escalating nonspecific interactions. For example, lysozyme bound onto gold nanoparticles denatured as well as produced protein–NP aggregates upon contact with other lysozyme molecules (D. Zhang et al., 2009). Surface charge is also vital in dictating cellular fate. Positively charged NPs are internalized at a faster rate compared to negatively charged or neutral NPs (Slowing, Trewyn, & Lin, 2006; Thorek & Tsourkas, 2008). However, a positive charge can disturb the cell membrane potential instigating $Ca^{2+}$ influx into cells and the inhibition of cell proliferation for small NPs (2 nm) (Arvizo et al., 2010) while for larger NPs (4–20 nm), it causes reconstruction of lipid bilayers (B. Wang, Zhang, Bae, & Granick, 2008). Binding of negatively charged NPs to a lipid bilayer produces local gelation, while binding of positively charged NPs produces fluidity (Albanese et al., 2012). Likewise, it has been stated that NMs with cationic surface charge are more liable to interact with the genetic material generating genotoxicity owing to negative charge on DNA. Functionalization of NPs with negatively charged groups (sulfate, phosphate, carboxylates) leads to their stabilization in physiological media resulting in a negative $\zeta$ potential of about 30–50 mV (Mahmoudi et al., 2011). Despite having a negative charge, NPs are straightaway covered by protein corona leading to drop of $\zeta$ potential to about 5–10 mV, thus, the colloidal stability of these complexes cannot be due to a pure electrostatic effect, but it is intently associated with the complex nature of the protein corona (Ehrenberg, Friedman, Finkelstein, Oberdorster, & McGrath, 2009).

### 4.2.3 Hydrophobicity/Hydrophilicity Effects

A well-defined correlation exists among the affinities of biomolecules for the NPs' exterior and the degree of structural changes (Lei et al., 2015). Having elevated affinities for hydrophobic surfaces, adsorbed proteins can cause severe protein denaturation due to having a less native structure than when adsorbed on hydrophilic surfaces (Mahmoudi et al., 2011). In agreement with literature reports, hydrophobic interactions tend to control the energy balance in most cases to date; yet the consequence of electrostatic interactions cannot be overlooked (Chanteau, Fresnais, & Berret, 2009). Cedervall et al. used isothermal titration colorimetry (ITC) to demonstrate that the number of bound protein molecules increases with the particle hydrophobicity and that there was a clear disparity in the protein affinities for polymeric particles of different hydrophobicities (Cedervall et al., 2007).

### 4.2.4 Effect of Protein Corona

The first tissue encountered by therapeutic NPs upon their administration in the body is the blood as well as the proteins it contains within (X. Q. Zhang et al., 2012). The NPs are covered with a protein corona that might modify their surface charge and properties, which are another major hurdle affecting the stability of colloids in biological media. The corona initially is "soft corona," which is further hardened in few minutes by long-lived or "hard corona" layer, which remains intensely adsorbed to the NP surfaces (Morteza Mahmoudi & HarshalZope, 2016). A number of interactions are involved in protein binding to NP surface : hydrogen bonding (H-bond), van der Walls interactions (VDW), as well as hydrophobic and electrostatic interactions. An early nano-bio interface that undergoes dynamic changes as the particles subsequently move onto or into cells is formed by particles bound to proteins in biological fluids (Nel et al., 2009). The particle's interactions with biological surfaces and receptors and its fate are primarily determined by the kinetics of nanoparticle–protein association and dissociation, and simultaneous exchange with free proteins in the media. Since the protein corona could shape the cellular interactions with NMs, an important query is whether such protein interactions rely on particle composition. The particle's hydrophobicity, size, radius of curvature, charge, and coatings that exercise steric or electrosteric effects will regulate which biomolecules should cooperate with the particles and thus facilitate their admission to cells (Nel et al., 2009). Numerous proteins are recognized to form transitory complexes with NPs; sizeable disparities in their dissociation rates create favored binding interactions amid precise particle types and the biological fluids in which they are disbanded. Also, the binding of the particles will be influenced by the protein concentrations. Like, fibrinogen and human serum albumin could govern the particle surface for short times; however, lesser abundance proteins with greater affinities and slower kinetics may eventually shift them (Cedervall et al., 2007).

#### 4.2.4.1 Effect of Protein Corona on Active Targeting

In order to deliver therapeutic payload to specific sites in the body, surface functionalization of the NP surface is required with ligands/antibodies that are specific for the receptors expressed by the target site. In the absence of serum-free media, the delivery of NPs to target site has been shown to work perfectly in vitro; but the efficacy under in vivo; conditions was comparatively poor (Morteza Mahmoudi & HarshalZope, 2016). Mirshafiee, Mahmoudi, Lou, Cheng, and Kraft (2013) and Anna Salvati et al. (2013) discovered that the efficiency of NP targeting was significantly reduced due to deposition of protein corona on the targeting ligand. Mirshafiee et al. (2013) carried out copper-free click reaction between silicon substrates and NPs that were functionalized with bicyclononyne and azide groups. A variety of techniques like scanning electron microscopy (SEM) and fluorescence microscopy were used in order to probe the targeting efficacy of the NPs toward silicon substrates in the presence and absence of protein corona. The results confirmed the reduced targeting efficiency of the NPs in the presence of protein corona.

### 4.2.5 External Factors

Various external factors also influence NM properties in addition to inherent physicochemical properties. For example, photosensitive materials are efficient in generating superior quantity of reactive oxygen species (ROS) in the presence of light. Three photosensitive NMs, namely titanium dioxide, silicon dioxide, and zinc oxide, have been evaluated to exhibit toxicity with variable degrees toward Gram-positive and Gram-negative bacterial strains in water suspensions in a particle concentration-dependent manner due to increased ROS production because of light stimulus (Adams, Lyon, & Alvarez, 2006). Also, the purity of a NM needs to be considered for its biomedical role. The presence of residual contaminating foreign metals, unreduced metal ions, chemicals or other agents might really be accountable for harmful actions rather than the actual NMs itself (Navya & Daima, 2016). Nowadays, many postproduction nanomaterial processing methods are identified to eradicate these chemical substances from NMs, though there might still be some left. Consequently, the outcomes of such chemical impurities, remaining metals, and occurrence of counterions on both possible biomedical and harmful effects cannot be ignored.

## 4.3 Measures to Enhance NP Stability in Biological Systems

### 4.3.1 Stabilization Against Aggregation

The conjugation of ligands to the NP surface not only controls the growth of the particles during synthesis but also prevents their aggregation (Sperling & Parak, 2010). Stable particles can result from the right choice of ligand depending upon the core material and the solvent in which the particles are dispersed. Some kind of attractive interaction like chemisorption, hydrophobic interaction, or electrostatic interaction must first bind the ligand molecules to the NP surface. Polar ligand molecules deliver solubility in polar solvents, while NPs with apolar ligand molecules like hydrocarbon chains are only soluble in apolar organic solvents, e.g. hexane, toluene, or chloroform. Molecules like PEG are amphiphilic in nature and NP having PEG as ligands can be soluble in many solvents (Sperling & Parak, 2010). The NP surface is covered by hydrophobic ligand molecules in organic solvents that prevent aggregation of the particle cores. But, the bonds among the inorganic particle surface and the electron-donating end group of a ligand molecule, like thiol (Love, Estroff, Kriebel, Nuzzo, & Whitesides, 2005), endure dynamic binding and unbinding processes, which may lead to washing away of the ligand molecules, hence compromising the NP stability leading to aggregation. It has been reported that strongly charged ligand molecules, e.g. sulfonic or carboxy groups, form a dense layer around the particles, hence stabilizing them for longer periods of time even at elevated salt concentrations (Sperling & Parak, 2010). Also, ligand molecules offering steric stabilization are found to be much more resilient to high salt concentrations than electrostatically stabilized NPs (Takeshi Sakura, Kataoka, & Nagasaki, 2005) when they are bound strongly to the NP surface, while poor stability is observed where NPs are not stabilized by ligand molecules.

### 4.3.2 Ligand Exchange

The ligand molecules bound to the NP exterior can be exchanged by other ligands to expand the stability of the given NPs. Mostly, the incoming ligand molecule binds more sturdily to the inorganic NP surface. Gold (Au) NP in an aqueous solution manufactured by citrate reduction is a common example. Negatively charged citrate ions are adsorbed on the NP surface, thereby providing electrostatic repulsion, which may remain stable for years; however, in the presence of salts they can aggregate irreversibly. Ligands like mercaptoacetic acid (MAA), mercaptopropionic acid (MPA), or mercaptoundecanoic acid (MUA) can bind strongly to the nanoparticle surface, replacing the citrate layer (Shu-Yi Lin, Chen, Lin, & Chen, 2004). Other examples for ligand exchange include dendrimers (Y. A. Wang, Li, Chen, & Peng, 2002) and quantum dots wherein proteins can be conjugated by replacing MAA on the surface (Gao, Chan, & Nie, 2002). From these examples, we infer that the new ligand molecules should have strong affinity for the NP surface for ligand exchange to occur.

### 4.3.3 Coating with Additional Layers

The circulation and biodistribution of NPs upon systemic administration are greatly affected by the proteins that adhere to their surface (Owens 3rd & Peppas, 2006). The particles are rapidly cleared from the bloodstream due to complement and immunoglobulin binding, which causes particle opsonization leading to detection by the mononuclear phagocyte system (MPS). Low efficacy, short circulation half-life, and toxicity triggered by buildup of foreign materials in the liver and spleen are the main drawbacks for the systemic administration of

NPs (X. Q. Zhang et al., 2012). These concerns have guided the development of approaches intended at mounting blood residence time. One method is coating the NP surface with an additional molecular layer that adsorbs on the original ligand molecules and alters the external properties (Sperling & Parak, 2010), leading to formation of a ligand bilayer that allows transfer of hydrophilic particles from aqueous phase to organic solvents (Caruso, 2003). The molecules acting as phase-transfer agents have to be amphiphilic and one example is quaternary ammonium salts that transfer from aqueous to organic phase as well as cause transfer of hydrophilic nanoparticles to the organic phase. The ammonium salt is suspended in the organic phase and can adsorb electrostatically onto the negatively charged surface of the NPs. Some examples for quaternary salts are cetyltrimethylammonium chloride (CTAC) (Fendler, 1996) and tetraoctylammonium bromide (TOAB) (Hiroshi, Osamu, Tomoki, & Keisaku, 2000). The ligand shell can be swapped by strongly binding ligands after phase transfer like dodecanethiol in the case of Au nanoparticles. The same approach has also been applied in the case of lipids for coating of hydrophobic NPs (Erogbogbo et al., 2008; Srinivasan et al., 2006). The additional layer preserves the native environment of the inorganic NPs since the original ligand molecules are not substituted. This may be helpful in preventing oxidation of sensitive core materials.

#### 4.3.3.1 Silica Coating

Silica coating is one of the most widely used method for surface coating of NPs as it enhances the water solubility, imparts colloidal stability, biocompatibility as well as minimizes nonspecific interactions (Giaume et al., 2008). The process typically involves exchange of hydrophobic ligands with primer molecules like 3-(trimethoxysilyl) propyl methacrylate (MPS) and 3-aminopropyltriethoxysilane (APTS) by selection of the precursors exclusive for each NP material. Then the growth of amorphous silica shell can be done using tetraethoxysilane in alcohol (Stöber method) or by hydrolysis of sodium silicate (Iler process) (Mulvaney, 2003). To render chemical functionality and hydrophilicity to the NPs, secondary silanes with ionizable hydrophilic groups can be introduced on the initial silica shell. For example, the biocompatibility and stability of silica-coated NPs can be improved by inclusion of PEG chains. Mesoporous silica can also develop on the NPs' surface, just after the creation of a thin layer of amorphous silica using CTAB as a templating material (Gorelikov & Matsuura, 2008). Drugs like doxorubicin have been delivered using mesoporous silica nanoparticles (J. Kim et al., 2006). However, the hydrodynamic diameter of the NPs is increased by the growth of silica shell, which may affect the physical properties of the NPs and also cause NP aggregation by the presence of polymeric silica present in the NP surface (Bagwe, Hilliard, & Tan, 2006).

#### 4.3.3.2 PEG Coating

PEG is a linear polymer having repeated units of $—CH_2—CH_2—O—$ known to be soluble in variety of apolar and polar solvents (Sperling & Parak, 2010). Having nontoxic nature, inertness, simple structure, and chemical stability, PEG has a variety of applications in biomedicine. Due to steric effects, PEG repels other molecules when bound to surfaces and hence the incoming molecule cannot penetrate the hydrated PEG layer (Zareie, Boyer, Bulmus, Nateghi, & Davis, 2008) leading to inert hydrophilic surfaces. Enhanced water solubility, decreased immunogenicity in organisms, and increased half-life have been shown by PEG-modified proteins (Veronese, 2001). PEG facilitates passive targeting of NPs via enhanced permeation and retention (EPR) effect, prevents corona formation, and augments the likelihood of particle permeation into tumor tissues (Niidome et al., 2006; Petros & DeSimone, 2010). PEGylated liposomal doxorubicin (Doxil®) and the PLA-PEG micelle form of paclitaxel (Genexol-PM®, marketed in Korea in 2007) are some examples of "stealth" nanocarriers (F. Wang, Porter, Konstantopoulos, Zhang, & Cui, 2017). Lately, PEGylated AuNPs were encircled with numerous amounts of human transferrin (Tf), producing a series of Tf-coated particles with near-constant size and electrokinetic potentials, and then transported into mice bearing Neuro2A tumors. Enhanced stabilization in salt solution was observed in PEGylated particles compared to unmodified particles (Choi, Alabi, Webster, & Davis, 2010). PEG-modified nanoparticles are more stable at elevated salt concentrations and in biological environments they display less nonspecific binding to proteins and cells (Gref et al., 2000; van Vlerken, Vyas, & Amiji, 2007). PEGylation is the term given to modification of other molecules with PEG. A cumulative variety of mono-, homo-, and heterobifunctional PEG reagents is commercially obtainable, with diverse functional groups, and molecular weights (Dozier & Distefano, 2015). Monofunctional PEG molecules produce more inert and stable particles (Hirsch et al., 2003) while new functional groups can be introduced using bifunctional PEG molecules. PEG-containing ligands can be used for phase transfer of nanoparticles as well as PEG-coated nanoparticles can also be disseminated in polar organic solvents (Peng, Kim, Chen, & Liu, 2005). PEG head groups can be modified enabling tailoring of specific ligands to NP surfaces. For example, functional moieties like amine ($—NH_2$) and carboxylic ($—COOH$) groups are most widely used as they can be incorporated into PEG molecule without compromising the colloidal stability of PEGylated nanoparticles (Carril et al., 2017). However, PEG coating for plasmonic NPs is extremely reliant on NP material and stabilizers used in the production. Diverse methodologies have been established to substitute the remaining surface agents from NP surfaces with PEG after the production.

#### 4.3.3.2.1 PEG Coating of Plasmonic NPs

**Citrate-Stabilized Particles** The use of thiol-terminated derivatives (PEG-SH), which covalently attached onto NP surfaces, is the most common method to coat plasmonic nanoparticles with PEG (Guerrini et al., 2018). For citrate-stabilized NPs it can be accomplished by addition of nanoparticles to a solution of PEG-SH under stirring (Figure 4.2a). The citrate ligands, over time, exchange with PEG-SH resulting in very stable solutions (Kamil Rahme et al., 2013). Rahme et al. (Kamil Rahme et al., 2013) used UV–visible spectroscopy to examine the stability of PEGylated AuNPs in physiological conditions (0.157 M, NaCl). It was inferred that the colloidal stability of PEG-stabilized NPs was merely affected upon addition of very high amount of salts.

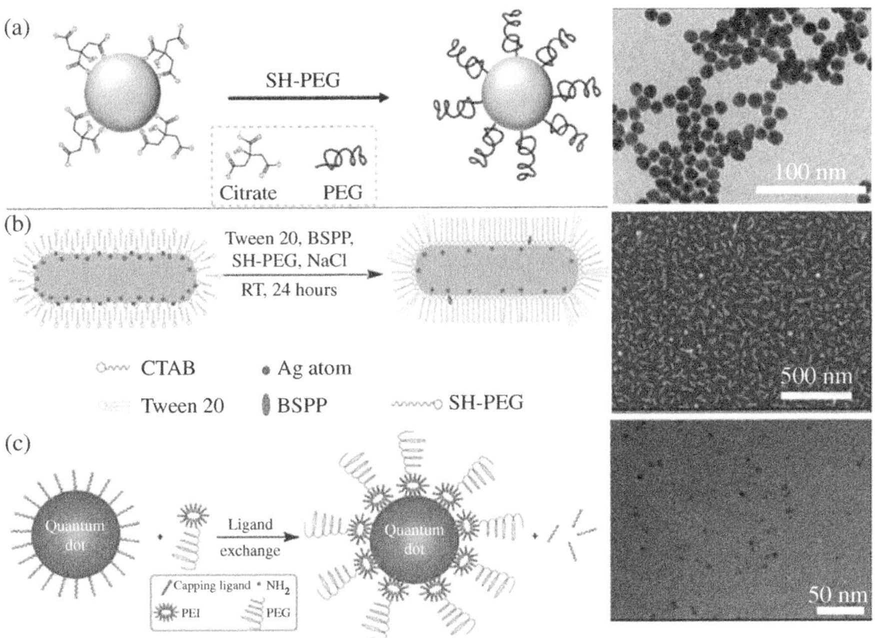


**Figure 4.2** Different strategies for PEG coating on NPs. (a) Thiolated PEG coating of plasmonic NPs. (b) In case of AuNRs, PEGylation is achieved by addition of tween 20, bis(*p*-sulfonatophenyl) phenylphosphine dihydrate dipotassium (BSPP), HS-PEG, and NaCl into a solution of CTAB-capped AuNR. (c) Amines can be utilized for ligand exchange for QDs dispersed in organic phase. Here, polyethylenimine (PEI)-PEG replaced octadecylamine leading to PEG functionalization and dispersal of QDs in water. *Source*: Reproduced with permission from Guerrini et al. (2018), K. Liu et al. (2015) Copyright © 2015 American Chemical Society and Duan and Nie (2007) Copyright © 2007 American Chemical Society.

**CTAB-Stabilized Particles** Using citrate, only spherical nanoparticles can be created and therefore to produce different shapes, diverse surfactants are required. Cetyltrimethylammonium bromide (CTAB) is used as a surfactant for synthesis of Au nanorods (AuNRs) (Figure 4.2b), wherein CTAB molecules are held on the AuNRs as a double layer (K. Liu et al., 2015). PEGylated AuNRs were stated to be tremendously stable in high-ionic strength solutions, like 0.5 M NaCl and 20 mM phosphate buffer (pH 7.5). Hafner and Liao in their study reported a method to substitute CTAB with PEG molecules by a one-step ligand exchange reaction employing PEG-SH (Hafner & Liao, 2005). The reaction kinetics was significantly escalated upon addition of tris-buffer at pH 3 (Lin, 2014). The main drawback of this method was the displacement of CTAB molecules at the tips of nanorods (NRs) leaving a sizeable portion of remaining CTAB at the side surfaces (Yu, Nakshatri, & Irudayaraj, 2007). An alternate method was proposed to promote the elimination of remaining CTAB due to the compatibility of PEGylated Au NRs with a varied set of solvents (Rubio-Garcia et al., 2011). Ethanol was used to desorb surfactant molecules from the NR sides after substituting CTAB from the tips of the NRs with PEG-SH (Kinnear et al., 2013). The complete functionalization of AuNRs with PEG was achieved with this method.

#### 4.3.3.2.2 PEG Coating of Magnetic NPs

PEG was successfully attached to $Fe_3O_4$ NPs through a covalent bond as reported by Xie, X, Kohler, Hou, and Sun (2007). Dopamine was first connected to one carboxylic group of a diacid PEG using EDC/NHS chemistry. The so-derivatized PEG could then covalently anchor the surface of the particles because of high affinity of dopamine to $Fe_3O_4$. Dopamine-modified PEG in a CHCl3/DMF solution displaced the oleate and oleylamine residues used during the synthesis. Upon introducing these particles in normal cell culture conditions, no agglomeration was detected. Silane-terminated PEG is another approach to bind PEG onto oxide-based NPs (Kohler, Fryxell, & Zhang, 2004).

#### 4.3.3.2.3 PEG Coating on Quantum Dots (QDs)

PEGylation of QDs is usually achieved via two main approaches depending on the specific synthetic approach (in organic vs. aqueous solvent). A ligand exchange is required to phase transfer the particle into water when quantum dots are created in organic solvents. This step can be executed instantaneously to the PEGylation step. For example, in the presence of PEG-polyethlenimine, CdSe/CdS/ZnS QDs attained in chloroform using octadecylamine (ODA) as a stabilizing agent can be transported into buffer solutions (Duan & Nie, 2007). The polyethylenimine moieties work as anchoring groups replacing ODA from the QD surfaces (Figure 4.2c). Another approach makes use of thiol-terminated molecules that tightly attach to particle surface via sulfur bonds to solubilize hydrophobic QDs (Tyrakowski & Snee, 2014).

#### 4.3.3.3 Lipid Bilayer Coating

Lipid bilayers have been used as coating agents to impart stability and generate a stealth effect on NPs in biological media. It is basically done to imitate the cell membrane bilayer for achieving biostability. Homogeneous shells of around 5 nm thickness are created by lipid bilayers around the NPs, thereby not causing any drastic changes in hydrodynamic diameter of the NPs (Mashaghi, Jadidi, Koenderink, & Mashaghi, 2013). It has been described that a lipid bilayer coating of 1,2-dimyristoyl-sn-glycero-3-phosphocholine (DMPC) on AuNPs is very consistent, consequently delivering enormous stability in a suspension of HBS 150 mM NaCl, pH 7.4 and in cell culture medium (RPMI) (Rascol et al., 2017). A lipid bilayer constituted of 1-palmitoyl-2-oleoyl-sn-glycero-3-phosphocholine (POPC), (1-palmitoyl-2-oleoyl-sn-glycero-3-phosphoglycerol (POPG), and cholesterol was reported by Bhowmik et al. (2015) to imitate the cell membrane. High stability was depicted by the coated NP suspension in a high salt including solution.

##### *4.3.3.3.1 Lipid Bilayer Coating of Plasmonic NPs*

Formation of a lipid film is required for the coating of plasmonic NPs with lipid bilayers in aqueous suspensions. Lipids like dioleoylphosphatidylcholine (DOPC), egg sphingomyelin (ESM), and ovine cholesterol (Chol) are suspended together in a mixture of chloroform/methanol and are then dried under vacuum to remove any remaining solvent (Mahmoudi et al., 2011). The films are hydrated with water first for NP encapsulation and the plasmonic particles are added and sonicated to create unilamellar vesicles (Ip, MacLaughlin, Gunari, & Walker, 2011). A single Au NP covered with a lipid bilayer and the consequent UV–vis spectra of the colloidal suspension earlier and after coating can be seen in Figure 4.3a and b.

##### *4.3.3.3.2 Lipid Bilayer Coating of Magnetic NPs*

By mixing a suspension of DMPC with the NPs, lipid bilayer coating of iron oxide is achieved. Small unilamellar vesicles can be obtained through sonication. High stability has been shown by such DPMC-coated magnetic NPs in high-ionic strength medium (Nyalosaso et al., 2016).

##### *4.3.3.3.3 Lipid Bilayer Coating of Fluorescent NPs*

For coating of fluorescent NPs, generally chloroform solutions of QDs and lipids are mixed and dried. Hydration of the film is done using water that leads to formation of multilamellar vesicles. These multilamellar vesicles can be converted into small unilamellar vesicles by sonicating the dispersion (Zheng et al., 2014).

#### 4.3.3.4 Zwitterionic Coating

Zwitterionic ligands have two or more functional groups and are amphoteric in nature with a net charge of zero, which makes them less perceptive to elevated

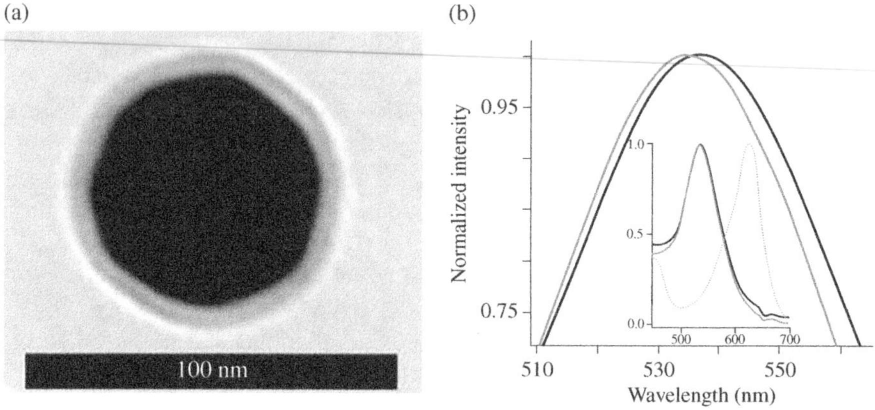


**Figure 4.3** Plasmonic NPs coated with lipid bilayers. (a) TEM image of AuNP having a lipid bilayer (dioleoylphosphatidylcholine (DOPC), egg sphingomyelin (ESM), and ovine cholesterol (Chol)). (b) Normalized UV–vis absorption spectra of lipid-coated (black) and citrate (grey) AuNPs. Shift of LSPR absorption peaks after lipid coating from 534 to 538 nm. *Source*: Reproduced with permission from Ip et al. (2011) Copyright © 2011 American Chemical Society.

ionic strength solvents (Pelaz et al., 2017). They have been anticipated as outstanding ligands for NPs' stability in biofluids. NPs having small hydrodynamic radii, lower degrees of opsonization, and better stealth properties are known to result from the use of zwitterionic ligands compared to PEG (Nguyen & Thanh, 2010; Pelaz et al., 2017). Carboxybetaines, sulfobetaines, and other zwitterionic ligands have been used efficaciously as antifouling coatings on NP surfaces (Cheng, Zhang, Chen, Bryers, & Jiang, 2007) and they have been proved to increase the specific uptake and blood circulation times as well as reduce the low adsorption of proteins (Yang, Zhang, Wang, White, & Jiang, 2009). Yang et al. (2009) validated that gold nanoparticles functionalized with poly(carboxybetaine) were more steady in serum and blood plasma compared to PEG-coated NPs. Agasti et al. (2009) stated intracellular delivery of therapeutics and negligible cytotoxicity from the nanomaterials. Arvizo et al. (2011) further added that zwitterionic gold NPs exhibit superior blood circulation times and heightened tumor accumulation, whereas positively and negatively charged NPs were speedily removed. But the uptake and biodistribution of zwitterionic NPs can be influenced by the surface charge distribution. Zwitterionic NPs having positive charges in outermost layer display non-specific adsorption in vitro; and in vivo;, but zwitterionic NPs showing negative charges are far less vulnerable to interactions with proteins (Han et al., 2013).

#### *4.3.3.4.1 Zwitterionic Coating on Plasmonic NPs*

Citrate-stabilized NPs and thiolated zwitterionic ligands are most commonly used to fabricate zwitterionic plasmonic NPs, wherein NPs are added drop wise to a solution comprising these ligands under constant mixing (Gupta et al., 2016). Another method involves ligand exchange to obtain aqueous stable dispersions in organic solvents. C. K. Kim et al. (2009) first used the Brust–Schiffrin two-phase method to produce AuNPs in presence of pentanethiol. Using this, the tetrachloroaurate ions were transferred into toluene with tetraoctylammonium bromide for its reduction with sodium borohydride. Ligand exchange was then used to perform the zwitterionic functionalization by mixing the AuNPs with thiolated zwitterionic ligands in dichloromethane.

##### *4.3.3.4.2 Zwitterionic Coating on QDs*

Like for plasmonic NPs, thiolated ligands are used for zwitterionic coating of QDs (W. Liu et al., 2007). A phase step is typically required as most of the synthesis of QD occurs in organic solvents. A biphasic exchange of QDs disseminated in chloroform into an aqueous NaCl solution was reported by E. Muro et al. (2010) while Breus, Heyes, Tron, and Nienhaus (2009) used zwitterionic D-penicillamine (DPA) to transfer the oleylamine surfactant from the QDs surfaces by refluxing the nanoparticles in 2-propanol and dioxane/methanol. However, an exchange process is not required when QDs are produced in aqueous solutions using thiolated ligands.

##### *4.3.3.4.3 Zwitterionic Coating on Magnetic NPs*

The generation of water-soluble zwitterionic particles is typically attained via ligands with functional moieties that deliver strong coordination to oxide surfaces (Mahmoudi et al., 2011). Since, organometallic synthesis is used for producing homogeneous magnetic NPs, a phase transfer step is necessary. Iron oxide NPs were produced from thermal decomposition of $Fe(CO)_5$ in dioctyl ether using oleic acid as ligand and trimethylamine N-oxide as additional oxidizing reagent following which water-soluble zwitterionic NPs were acquired by a two-step ligand exchange process (Wei et al., 2012).

#### 4.3.3.5 Protein Coating

Protein corona usually triggers NP destabilization and passivation of surface functional elements (Corbo et al., 2016). However this process can be utilized to be advantageous. It is recognized that proteins are macromolecules having decent stability under physiological conditions and with plentiful charged groups that can offer NPs steric and electrostatic stabilization. Furthermore, the capability of proteins to create a hydration shell via hydrogen bonds with water molecules averts the adhesion of other proteins (Jan Karolin, Strandberg, & Ny, 1994). Pre-protein coating on a NP surface before its introduction in the biological media can impart colloidal stability as well as prevent the surface attachment of undesired proteins. The most commonly used is human serum albumin (HSA), having cysteine groups, which are extremely reactive toward metallic surfaces (Casals, Pfaller, Duschl, Oostingh, & Puntes, 2010).

#### 4.3.3.5.1 Protein Coating on Plasmonic NPs

Ligand exchange is the process for coating NPs with proteins. For citrate-coated gold NPs, a pH 7–8 protein/citrate solution is added dropwise to colloidal suspension under dynamic mixing. The process of protein coating differs for different surface agents. For CTAB-capped AuNRs, the surfactant concentration has to be first dropped to critical micelle concentration (CMC), following which NR dispersion is added dropwise under sonication to a BSA/citrate solution at pH 12 (Tebbe, Kuttner, Mannel, Fery, & Chanana, 2015). Human serum albumin (HSA) and apoliprotein A (apoE), which are often found in the protein corona, were used to coat AuNPs by Schaffler et al. (2014). On the other hand, Chanana, Rivera Gil, Correa-Duarte, Liz-Marzan, and Parak (2013) coated citrate AuNPs with proteins of different molecular weights like ovalbumin (45 kDa), insulin (5808 Da), and $\beta$-lactoglobulin (18.3 kDa). Extremely high colloidal stability in physiological and other culture media was shown by the protein-encapsulated NPs.

#### 4.3.3.5.2 Protein Coating on QDs

One-step synthesis of protein-functionalized QDs was reported by Xuewen et al. (He, Gao, & Ma, 2013). $Hg(ClO_4)_2$ and $Zn(OAc)_2$ were mixed in PBS with a protein solution (hemoglobin, trypsin, lysozyme, BSA) in the presence of mercaptopropionic acid and $NH_4HCO_3$. Injecting a solution of NaHSe induced $ZnxHg_{1-x}Se$ QDs formation.

#### 4.3.3.5.3 Protein Coating on Magnetic NPs

Hiremath, Kariduraganavar, and Hiremath (2018) reported iron oxide nanoparticles coated with human serum albumin for synergistic delivery of curcumin and 5-fluorouracil. Folic acid was also decorated on human serum albumin using EDC-NHS coupling (5FU-CUR-C-MNP-HAS-FA) to confer targetability (Figure 4.4). The nanoparticles displayed good aqueous dispersibility and zeta potential of 49.1 mV. The effect of protein adsorption on the colloidal stability and the dynamical magnetic response of functionalized magnetic nanoparticles by using BSA were observed by Antonio Aires et al. (2015). The results concluded that the colloidal properties of MNP were improved by the protein coating under physiological conditions, thereby reducing aggregation and enhancing their stability over time.

### 4.3.3.6 Aptamer Coating

Although colloidal stability is imparted by many functionalization strategies in biological media, they principally fail to convey any chemical specificity, thereby making way for active targeting elements (Keefe, Pai, & Ellington, 2010). Numerous efforts have been taken to tackle this without compromising on the particle stability and biorecognition functions. Aptamers surfaced as valued options compared to antibodies because of high stability, small size, ease of preparation, immunogenicity, etc.

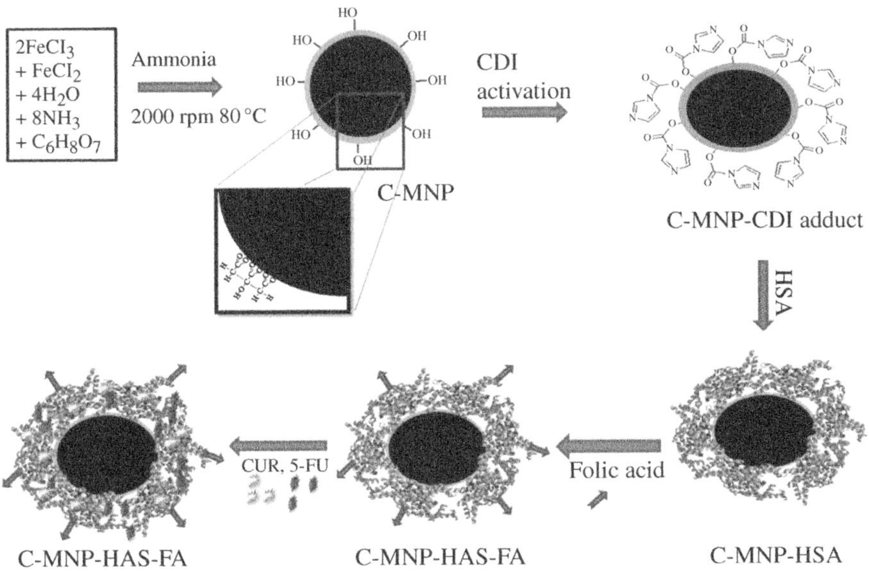


**Figure 4.4** Synthesis of 5FU-CUR-C-MNP-HAS-FA nanoparticles. Curcumin and 5-fluorouracil were entrapped inside human serum albumin-coated citrate-modified iron oxide nanoparticles further decorated with folic acid. *Source*: Reproduced with permission from Hiremath et al. (2018).

#### 4.3.3.6.1 Aptamer Coating on Plasmonic NPs

Thiol bonding is the method to decorate thiol-modified aptamers on plasmonic NPs surface. For example, a 5′-alkyl-thiol-modified *Staphylococcus aureus* aptamer was self-assembled on the exterior of silver nanoparticles (AgNPs) by directly adding it to a colloidal suspension in the presence of sodium dodecyl sulfate (SDS), tris-borate-EDTA (TBE) buffer, and NaCl (Carme Catala et al., 2016).The colloidal integrity of the NPs was maintained by the aptamer conjugation leading to their stability in several real human fluids – blood, urine, etc.

#### 4.3.3.6.2 Aptamer Coating on QDs

QD functionalization can also be done using thiol-modified aptamers. Cocaine and thrombin were detected by thiolated aptamers that were covalently conjugated to the surface of CdSe/ZnS QDs (C. Y. Zhang & Johnson, 2009; Zhou et al., 2005). QDs stabilized in organic solvents are usually first shifted to water before the aptamer conjugation. Zhou et al. (2005) first used 3-mercaptopropionic acid (MPA) in a chloroform/methanol mixture to substitute the QDs ligand (TOP) with a thiol at basic pH to deprotonate the carboxylic groups of MPA and achieve water-soluble QDs. The QDs were then disseminated in a solution comprising thiol-ended aptamer after being transferred to water.

#### *4.3.3.6.3 Aptamer Coating on Magnetic NPs*

To functionalize aptamers on magnetic nanoparticles, the most common approach is to couple the terminal amino group of ligands to carboxyl or hydroxyl functionalized NPs using EDC chemistry. Aptamer-functionalized magnetic NPs have been indicated to be steady in PBS solutions (Najmeh Delaviz, Ajamia, & Aarabi, 2015). Najmeh Delaviz et al. (2015) stimulated hydroxyl groups of starch-functionalized magnetic NPs to pair them with the amino groups of aptamers.

### 4.3.4 Subsiding the Nonspecific Protein Interaction

The full potential of ex vivo; nano-based diagnostics and arrays is impeded by the nonspecific interaction of nanomaterials with proteins from convoluted biological samples (Spisak, Tulassay, Molnar, & Guttman, 2007). New diagnostic nanomaterials are developing for the recognition and quantification of fewer biomarkers in biological samples and they hold many exclusive and beneficial physical properties when applied as ultrasensitive signal transducers and protein biosensors in the fields of molecular diagnostics and proteomics (Johnson, Zhukovsky, Cass, & Nagy, 2008; X. Q. Zhang et al., 2012). Some of the examples are nanotube arrays, nanowires, nanocantilevers, and oligonucleotide-modified gold NP-based bio-barcode assays that permit multibiomarker recognition (Ferrari, 2005). Nevertheless, deconvolution of noise from the signal particularly in respect to biofouling forms a principal block in expansion of these vastly sensitive and selective approaches. It is important to block probable sites for nonspecific binding or treat the NMs with surface coatings that pool an ultralow fouling background with copious biorecognition elements for applications like affinity biosensors or nanoarrays since nonspecific binding may lead to generation of false positives (X. Q. Zhang et al., 2012).

### 4.3.5 Nanoparticle Design

Nano-bio interactions in the past decade have prominently manipulated NP design. Three generations of NPs have been engineered for biomaterial applications by far. Novel NMs functionalized with basic surface chemistries to evaluate biocompatibility and toxicity form the first generation. The second generation comprises of NMs with improved surface chemistries that enhance stability and targeting in biological systems while the third generation consists of environment-responsive systems expanding targeted delivery (Albanese et al., 2012). First-generation NPs were tailored with non-stealth surface chemistries and used in experiments to evaluate cell uptake and toxicity (Kirchner et al., 2005) as well as define the effects of surface charge (Cho, Xie, Wurm, & Xia, 2009). However, the major drawback was the conductance of experiments in serum-free media, thereby neglecting the interaction of nanomaterials with proteins under physiological conditions in the body. The in vivo; data reported rapid clearance of the nanoparticles due to absence of PEG (De Jong et al., 2008). These initial studies were significant in underlining the biocompatibility of these innovative materials and marked a key step in the shift from chemical synthesis to biological usage.

Upon standardizing nanoparticle synthesis, research moved to surface chemistry optimization for cancer diagnosis and treatment. Second-generation NP studies primarily concentrated on tumor delivery as a proof of concept (Albanese et al., 2012). These findings exploited EPR effect for passive accumulation of drugs at tumor site and incorporated two significant design concepts: stealth and active targeting. Augmented half-life of NMs was reported by many studies simply by PEG addition (Owens 3rd & Peppas, 2006). Active-targeting of NPs also relies on the EPR effect to gain entry to the intratumoral space but it uses targeting moieties to hypothetically upsurge total accrual by anchoring the NPs onto the cells. However, chances of clearance by the MPS also upsurged due to the addition of excess targeting ligands, which became visible on the surface (Shmeeda, Tzemach, Mak, & Gabizon, 2009). Second-generation NMs also raised several concerns like: overdependence on the EPR effect, no single nanoparticle size could enter all areas of the tumor and gather in substantial quantities, and restriction of active-targeted nanoparticles to the first few layers of cells since they stick to their targeted receptors (Albanese et al., 2012).

In order to counter these concerns, researchers established a new generation of NMs that do not depend on EPR effect or on endogenous tumor ligands. The third generation of NMs had "stimulus-responsive" properties and became activated via any physical, biological, or chemical cues in their target environment (Albanese et al., 2012). Use of low pH, low $O_2$, and matrix metalloproteinase enzymatic activity formed the first approach for nanotherapeutics delivery while use of near-infrared light inside the target tissue comprised the second approach. Some examples are: pH-triggered deshielding of the PEG surface layer to expose a positively charged surface that instigated nonspecific uptake of drug-loaded NPs (Poon, Chang, Zhao, & Hammond, 2011); use of enzymatic activity for fluorescence or drug release (Sarkar et al., 2008; Watkins et al., 2009); and the use of NIR for heat generation inside the tumor via excitation of nanorods or nanoshells, thus triggering the payload release (Agarwal, Mackey, El-Sayed, & Bellamkonda, 2011). All these methods smartly evaded the need for universal tumor antigens and did not depend on the EPR effect. By taking advantage of local signals inside the tumor to elicit drug release, the NPs inside the liver and spleen did not cause toxicity.

#### 4.3.5.1 Particle Functionalization

Covalent coupling delivers the stable conjugation of biomolecules with NPs. Carboxylic group, thiol and amide group are the functional groups present on the NP surface that can cross-link with biomolecules via different strategies.

#### 4.3.5.1.1 Chemical Functional Groups (Covalent Coupling)

The ligand particles imparting stability to NPs can either comprise of an inert molecular chain (PEG or hydrocarbon chain) or have functional groups that are terminating linear molecules (Sperling & Parak, 2010). In the case of water-soluble NPs, these functional groups are frequently carboxylic acids stabilizing the NPs by electrostatic repulsion, and can be manipulated for the conjugation of other molecules to the particles. AuNPs or quantum dots stabilized with mercaptocarboxylic acids are some common examples. Additional functional groups can be introduced to the NP by their ligand molecules in a similar way. For example, MAA was used to stabilize quantum dots to which BSA (bovine serum albumin), a protein, was adsorbed. The amino groups of the protein could then be used for further conjugation chemistry (Gao et al., 2002).

#### 4.3.5.1.2 Amide Group Involved Coupling

Using bifunctional molecules, the functional groups on the NP exterior can be changed to other functional groups (Sperling & Parak, 2010). Particularly in the case of NPs dispersed in an aqueous solution, the reaction conditions may hinder NP stability, hence mild reactions like bi-conjugation chemistry needs to be chosen. Water-soluble carbodiimide (e.g. EDC) is normally used to carry out a condensation reaction between carboxylic groups and primary amines to produce amide bonds. *N*-Hydroxy-succinimide (NHS) can be used in the case of primary amines existing on the particle surface to form zero-length amide bonds; for example succinimidyl-4-(*N*-maleimidomethyl)cyclohexane-1-carboxylate (SMCC) containing an NHS group reacts with primary amines, converting them to maleimides that are reactive toward thiols (Sam, Finch, Hallahan, & Giorgio, 2007). The majority of ligand molecules or supplementary amphiphilic coatings stabilize the nanoparticles by negatively charged carboxylic groups. These can be transformed, e.g. to hydroxyl groups (Kairdolf, Mancini, Smith, & Nie, 2008) or tertiary amines (Yezhelyev, Qi, O'Regan, Nie, & Gao, 2008).

#### 4.3.5.1.3 Thiol Group Coupling

Under mediation of reagents like iodoacetamides and maleimides, the thiol group can form specific conjugates with primary amine groups (Brinkley, 1992). Sulfosuccinimidyl-4-(maleimidomethyl) cyclohexane-1-carboxylate (sulfo-SMCC) is the most common coupling reagent. Conjugation of amine-functionalized NPs can be done with various thiolated molecules like proteins and peptides with cysteine, thiol-modified peptides taking advantage of this reaction (Cai et al., 2006). Reversible coupling of NPs can also be done using disulfide bridges. Tian et al. used glutathione (GSH) disulfide and dithiothreitol to depict reversible oxidation and reduction of disulfide bridges between $Fe_3O_4$ NPs and silica NPs (Tian, Shi, & Zhu, 2007).

#### 4.3.5.1.4 *Click Chemistry*

Click chemistry has gained much attention over the last couple of years and has also been used for nanoparticle derivatization in both aqueous and organic solutions (Brennan et al., 2006), as well with carbon nanotubes and polymeric nanoparticles (Rakesh Voggu, Chandrasekaran, & Rao, 2007). Click chemistry leads to high yield and selectivity due to spontaneously occurring chemical reactions between stable functional groups under mild conditions (Kolb, Finn, & Sharpless, 2001). In the reaction, coupling of alkyne to an azide gives a 1,2,3-triazole ring under catalyst of Cu(I). The only limitation is that it demands particular preparation of alkyne or azide functional bioactive species, which is quite labor intensive (Sun, Stabler, Cazalis, & Chaikof, 2006).

### 4.3.6 Influence of NM Physicochemical Properties on Microbe–NM Interaction

A vast amount of data is available on interactions of human cells with NMs; however, very little is known about the microbe–nanomaterial cross talk (Stauber et al., 2018). The behavior of NMs at nano-bio interfaces as well as their potential applications is determined by their characteristics like charge, size, roughness, and surface modifications (Docter et al., 2015; Stauber et al., 2018). However, it is important to elucidate which physicochemical character of NM is most pertinent and for which type of microbe as a variety of shapes are available for both microbes and NMs. Bacterial shape is an important determinant of its interaction with NMs. Greater effect of NMs has been stated on Gram-positive versus Gram-negative bacteria (Gupta et al., 2016). Most microbes usually exhibit a negative charge and hence it was suggested that positively charged NMs bind more proficiently to microbes compared to negatively charged NMs (Gupta et al., 2016). However, latest studies have confirmed successful binding of various negatively charged NMs to microbes like algae, fungi, and bacteria (Westmeier et al., 2018). Selective reduction of the surface-negative charge of NMs by chemical modification keeping other variables constant did not enhance binding efficiency. Smaller NMs were depicted to bind more efficiently to microbes compared to larger ones (Figure 4.5) with stealth modification also contributing to reduced binding (Westmeier et al., 2018). As previously discussed in the chapter, protein corona is an important factor affecting nanoparticle stability in biological systems. Similarly, the corona effect has also been considered on microbe–nanomaterial cross talk. In a study by Westmeier et al. (2018), effect of environmental factors comprising corona-forming biomolecules was studied on nanomaterial–microbe interactions. The researchers condensed NM–microbe complex formation by the gathering of diverse biomolecule coronas. Due to preincubation of bacteria with biomolecules, they were still able to bind to NMs implying that adsorption of biomolecules to

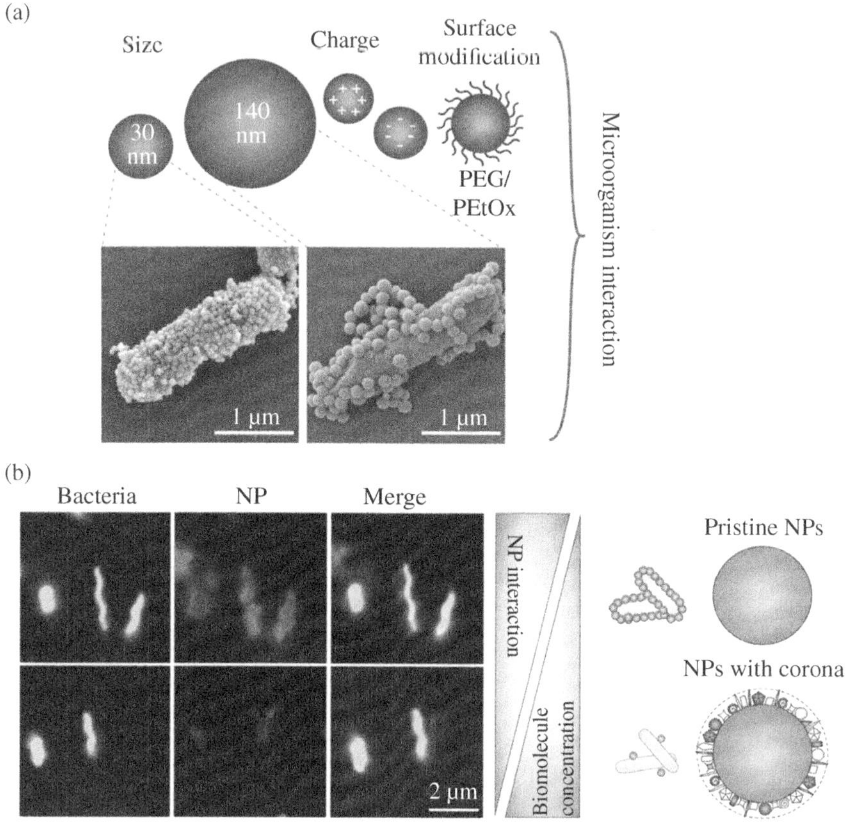


**Figure 4.5** Effect on NMs' physicochemical properties and biomolecule coronas on NM–microbe interactions. (a) Size of nanoparticle plays a critical role as smaller silica nanoparticles (30 nm) bind more efficiently to *E. coli* bacteria compared to larger NPs (140 nm). (b) Formation of biomolecule corona reduces the NM–bacteria complex formation. In situ complex formation of autofluorescent green bacteria with red fluorescent silica NPs is demonstrated by fluorescence microscopy. *Source*: Reproduced with permission from Stauber et al. (2018) Copyright © 2018 American Chemical Society.

NMs and not to microbes hampered with complex formation. NM binding to bacteria was prevented at high biomolecule concentrations (Figure 4.5). This study strongly proposes that the extent of microbe–nanomaterial complex formation is influenced by biomolecule corona in a concentration-dependent manner at first contact of microbes and NMs.

Another type of interaction that occurs is between NPs and biofilms, which is usually an assembly of surface-attached microbial cells bounded by a matrix of extracellular polymers (Ikuma, Decho, & Lau, 2015). Surface modification of nanoparticles via ligand capping and corona formation plays an important role in biofilm–NP interactions. For example, cadmium selenide quantum dots conjugated with PEG were found to easily infiltrate *Pseudomonas aeruginosa* biofilms compared to QDs having carboxyl groups (Morrow, Arango, & Holbrook, 2010). The NP–biofilm interactions are also affected by size and charge of the NPs. For example, reduction rates of iron oxide NPs by the dissimilatory iron-reducing bacteria, Shewanella oneidensis MR-1, were induced by the size of the NPs (SaumyadityaBose et al., 2009). Smaller NPs have been reported to exhibit higher antibacterial activity and some of the reasons can be: the penetration of the smaller NPs inside the bacterial cells and also generation of free radicals by smaller NPs (Aruguete & Hochella, 2010). Surface sulfate ($SO^{4-}$) groups on fluorescent polystyrene NPs led to enhanced sorption on *Alteromonas macleodii* biofilms compared to NPs functionalized with amine or carboxy groups (Nevius, Chen, Ferry, & Decho, 2012). The core composition is one of the most critical features to govern the behavior of a nanoparticle (Aruguete & Hochella, 2010). AgNPs are known to possess antimicrobial properties and thus are of great interest. A key role is played by $Ag^+$ ions released by nano-Ag in conferring bacterial toxicity. Lok et al. showed a proteomic analysis to inspect the communication between *Escherichia coli*, Ag+, and nano-Ag. Comparable protein expression profiles were discovered and both forms of silver dissolved the plasma membrane potential, instigated ATP depletion, and undermined the outer membrane (Lok et al., 2007). There are numerous examples of influence of surface coatings on microbe–NM interaction; however, the results can be puzzling. For example, in one study nano-Ag coated with sodium dodecyl sulfate (Ag-SDS) had no antibacterial activity (Kyung-Hwan Cho, Osaka, & Park, 2005) while in another study, Ag-SDS was stated to have the most antibacterial activity (Libor Kvítek et al., 2008). Hence, NP properties play an important role in affecting microbe–NM cross talk.

## 4.4 Conclusion and Future Perspectives

A dynamic series of interactions between nanomaterial surfaces and biological surfaces occur at the interface between NMs and biological systems. Whether the NM is bioavailable and may participate in bio-adverse reactions is determined via these interactions, which are shaped, by a large number of forces. The interactions regulate processes like the protein corona formation, cellular contact, particle wrapping at cell surfaces, endocytosis, and intracellular biocatalysis. Surface functionalization of NPs is contemplated as a necessity for biomedical applications. Having good chemical functionality and water stability; preserving NPs' physical and chemical properties; and biocompatibility are some of the characteristics of an ideal surface. However, the nanoparticle properties can largely affect its interaction with biological milieu. Various types of coatings are available to enhance functionality of biomaterials like PEG incorporation, which can enhance the biocompatibility; however, it leads to enlargement of particle size. Hence, a challenge remains to satisfy all the criteria. Covalent bioconjugation approach perhaps is the best used bioconjugation method as it leads to formation of stable linkages. However, it is hard to control the stoichiometry of the NP/biomolecule, which may lead to self-assembly, and aggregation. Hence, challenges persist in regard to the expansion of proper approaches for surface functionalization and bioconjugation of NPs with heightened chemical and physical properties for preferred applications in biomedicine. In order to have highly effectual nanoparticle-based delivery systems, nano-bio interactions must be deeply studied. The outcomes will help in understanding how biology works, as most biological components are nanometer-sized. Engineered nanoparticles suggest an extraordinary chance to discover the morphology and chemistry of nanoscale objects in mediating biological responses.

## 4.5 Summary

- In this chapter, we accentuated the significance of NP physicochemical properties (i.e. size, shape, composition, surface chemistry, coatings, etc.) on its interaction with biological environment.
- NP interaction is mainly affected by the design of the NPs, aggregation of the media, hydrophobic/hydrophilic effects, and various other external factors.
- Protein corona is an important determinant of NP stability in biological media as adsorption of different biomolecules on the NP surface can amend its properties, thereby modifying its function.
- Ligand exchange processes, coating the NPs with different types of coatings, and covalent conjugation of different chemical groups may enhance NP stability and upsurge blood residence time of the NPs.
- Microbe–NM cross talk is also largely affected by NP properties like size, shape, surface charge as well as surface corona and hence it is important to analyze these interactions carefully to minimize any undesired effects.

## Questions and Answers

1 Why it is important to take the NP–biological media interactions into consideration?

Along with NP physicochemical properties, the NP function is also influenced by the properties of the biological media like protein corona as well processes like aggregation, which all confer a new biological identity to cells, thereby altering their cellular processes. Hence, it is important to consider all the parameters that may affect NP stability.

**2** What properties of the nanoparticles are important in determining biomolecule interaction with NPs, thereby facilitating their uptake inside cells?
Particle's hydrophobicity, size, radius of curvature, charge, coatings that exercise steric or electrosteric effects will govern which biomolecules should interact with the particles.

**3** How can the process of ligand exchange enhance nanoparticle stability?
The ligand molecules bound to the nanoparticle surface can be swapped by other ligands to magnify the nanoparticles' stability. For example, MAA can strongly bind to nanoparticle surface and replace the citrate layer formed in synthesis of Au nanoparticles in order to prevent their aggregation in presence of salts.

**4** What are the different coatings that can be used to cover the core of the nanoparticles to enhance stability as well as prevent surface oxidation?
There are various types of coatings that can be used to cover NP core like: silica coatings, PEG coatings, protein coatings, zwitterionic coatings, coating with ligands and aptamers.

**5** How does covalent coupling enhance nanoparticle stability?
Covalent coupling produces the stable conjugation of biomolecules with NPs. Carboxylic group, thiol and amide groups are the functional groups existing on the NP surface that can cross-link with biomolecules via diverse strategies imparting stability.

## References

Adams, L. K., Lyon, D. Y., & Alvarez, P. J. (2006). Comparative eco-toxicity of nanoscale $TiO_2$, $SiO_2$, and ZnO water suspensions. *Water Research*, 40(19), 3527–3532. doi:10.1016/j.watres.2006.08.004

Agarwal, A., Mackey, M. A., El-Sayed, M. A., & Bellamkonda, R. V. (2011). Remote triggered release of doxorubicin in tumors by synergistic application of thermosensitive liposomes and gold nanorods. *ACS Nano*, 5(6), 4919–4926. doi:10.1021/nn201010q

Agasti, S. S., Chompoosor, A., You, C. C., Ghosh, P., Kim, C. K., & Rotello, V. M. (2009). Photoregulated release of caged anticancer drugs from gold nanoparticles. *Journal of the American Chemical Society*, 131(16), 5728–5729. doi:10.1021/ja900591t

Albanese, A., Tang, P. S., & Chan, W. C. (2012). The effect of nanoparticle size, shape, and surface chemistry on biological systems. *Annual Review of Biomedical Engineering*, 14, 1–16. doi:10.1146/annurev-bioeng-071811-150124

Anna Salvati, A. S. P., Monopoli, M. P., Prapainop, K., Bombelli, F. B., Hristov, D. R., Kelly, P. M., . . . Dawson, K. A. (2013). Transferrin-functionalized nanoparticles lose their targeting capabilities when a biomolecule corona adsorbs on the surface. *Nature Nanotechnology*, 8, 137–143.

Antonio Aires, S. M. O., Cabrera, D., de la Cueva, L., Salas, G., Teran, F. J., & Cortajarena, A. L. (2015). BSA-coated magnetic nanoparticles for improved therapeutic properties. *Journal of Materials Chemistry B*, 3(30), 6239–6247.

Aruguete, D. M., & Hochella, M. F. (2010). Bacteria–nanoparticle interactions and their environmental implications. *Environment and Chemistry*, 7(1), 3–9.

Arvizo, R. R., Miranda, O. R., Moyano, D. F., Walden, C. A., Giri, K., Bhattacharya, R., . . . Mukherjee, P. (2011). Modulating pharmacokinetics, tumor uptake and biodistribution by engineered nanoparticles. *Plos One*, 6(9), e24374. doi:10.1371/journal.pone.0024374

Arvizo, R. R., Miranda, O. R., Thompson, M. A., Pabelick, C. M., Bhattacharya, R., Robertson, J. D., . . . Mukherjee, P. (2010). Effect of nanoparticle surface charge at the plasma membrane and beyond. *Nano Letters*, 10(7), 2543–2548. doi:10.1021/nl101140t

Bagwe, R. P., Hilliard, L. R., & Tan, W. (2006). Surface modification of silica nanoparticles to reduce aggregation and nonspecific binding. *Langmuir*, 22(9), 4357–4362. doi:10.1021/la052797j

Bhowmik, D., Mote, K. R., MacLaughlin, C. M., Biswas, N., Chandra, B., Basu, J. K., . . . Maiti, S. (2015). Cell-membrane-mimicking lipid-coated nanoparticles confer Raman enhancement to membrane proteins and reveal membrane-attached amyloid-beta conformation. *ACS Nano*, 9(9), 9070–9077. doi:10.1021/acsnano.5b03175

Brennan, J. L., Hatzakis, N. S., Tshikhudo, T. R., Dirvianskyte, N., Razumas, V., Patkar, S., . . . Brust, M. (2006). Bionanoconjugation via click chemistry: The creation of functional hybrids of lipases and gold nanoparticles. *Bioconjugate Chemistry*, 17(6), 1373–1375. doi:10.1021/bc0601018

Breus, V. V., Heyes, C. D., Tron, K., & Nienhaus, G. U. (2009). Zwitterionic biocompatible quantum dots for wide pH stability and weak nonspecific binding to cells. *ACS Nano*, 3(9), 2573–2580. doi:10.1021/nn900600w

Brinkley, M. (1992). A brief survey of methods for preparing protein conjugates with dyes, haptens, and cross-linking reagents. *Bioconjugate Chemistry*, 3(1), 2–13. doi:10.1021/bc00013a001

Cai, W., Shin, D. W., Chen, K., Gheysens, O., Cao, Q., Wang, S. X., . . . Chen, X. (2006). Peptide-labeled near-infrared quantum dots for imaging tumor vasculature in living subjects. *Nano Letters*, 6(4), 669–676. doi:10.1021/nl052405t

Carme Catala, B. M. S., Feng, X., Cardozo, C., Pazos-Perez, N., Pazos, E., Pedro, S. G.-d., . . . Alvarez-Puebla, R. A. (2016). Online SERS quantification of *Staphylococcus aureus* and the application to diagnostics in human fluids. *Advanced Materials Technologies*, 1(8), 1600163.

Carril, M., Padro, D., Del Pino, P., Carrillo-Carrion, C., Gallego, M., & Parak, W. J. (2017). In situ detection of the protein corona in complex environments. *Nature Communications*, 8(1), 1542. doi:10.1038/s41467-017-01826-4

Caruso, K. S. M. a. F. (2003). Phase transfer of surface-modified gold nanoparticles by hydrophobization with alkylamines. *Langmuir*, 19(17), 6987–6993.

Casals, E., Pfaller, T., Duschl, A., Oostingh, G. J., & Puntes, V. (2010). Time evolution of the nanoparticle protein corona. *ACS Nano*, 4(7), 3623–3632. doi:10.1021/nn901372t

Cedervall, T., Lynch, I., Lindman, S., Berggard, T., Thulin, E., Nilsson, H., . . . Linse, S. (2007). Understanding the nanoparticle-protein corona using methods to quantify exchange rates and affinities of proteins for nanoparticles. *Proceedings of the National Academy of Sciences of the United States of America*, 104(7), 2050–2055. doi:10.1073/pnas.0608582104

Chanana, M., Rivera Gil, P., Correa-Duarte, M. A., Liz-Marzan, L. M., & Parak, W. J. (2013). Physicochemical properties of protein-coated gold nanoparticles in biological fluids and cells before and after proteolytic digestion. *Angewandte Chemie International Edition In English*, 52(15), 4179–4183. doi:10.1002/anie.201208019

Chanteau, B., Fresnais, J., & Berret, J.-F. (2009). Electrosteric enhanced stability of functional sub-10 nm cerium and iron oxide particles in cell culture medium. *Langmuir*, 25(16), 9064–9070.

Cheng, G., Zhang, Z., Chen, S., Bryers, J. D., & Jiang, S. (2007). Inhibition of bacterial adhesion and biofilm formation on zwitterionic surfaces. *Biomaterials*, 28(29), 4192–4199. doi:10.1016/j.biomaterials.2007.05.041

Cho, E. C., Xie, J., Wurm, P. A., & Xia, Y. (2009). Understanding the role of surface charges in cellular adsorption versus internalization by selectively removing gold nanoparticles on the cell surface with a I2/KI etchant. *Nano Letters*, 9(3), 1080–1084. doi:10.1021/nl803487r

Choi, C. H., Alabi, C. A., Webster, P., & Davis, M. E. (2010). Mechanism of active targeting in solid tumors with transferrin-containing gold nanoparticles. *Proceedings of the National Academy of Sciences of the United States of America*, 107(3), 1235–1240. doi:10.1073/pnas.0914140107

Corbo, C., Molinaro, R., Parodi, A., Toledano Furman, N. E., Salvatore, F., & Tasciotti, E. (2016). The impact of nanoparticle protein corona on cytotoxicity, immunotoxicity and target drug delivery. *Nanomedicine (London)*, 11(1), 81–100. doi:10.2217/nnm.15.188

Curtis, C., Toghani, D., Wong, B., & Nance, E. (2018). Colloidal stability as a determinant of nanoparticle behavior in the brain. *Colloids and Surfaces B: Biointerfaces*, 170, 673–682. doi:10.1016/j.colsurfb.2018.06.050

De Jong, W. H., Hagens, W. I., Krystek, P., Burger, M. C., Sips, A. J., & Geertsma, R. E. (2008). Particle size-dependent organ distribution of gold nanoparticles after intravenous administration. *Biomaterials*, 29(12), 1912–1919. doi:10.1016/j.biomaterials.2007.12.037

Docter, D., Westmeier, D., Markiewicz, M., Stolte, S., Knauer, S. K., & Stauber, R. H. (2015). The nanoparticle biomolecule corona: Lessons learned – Challenge accepted? *Chemical Society Reviews*, 44(17), 6094–6121. doi:10.1039/c5cs00217f

Dozier, J. K., & Distefano, M. D. (2015). Site-specific pegylation of therapeutic proteins. *International Journal of Molecular Sciences*, 16(10), 25831–25864. doi:10.3390/ijms161025831

Duan, H., & Nie, S. (2007). Cell-penetrating quantum dots based on multivalent and endosome-disrupting surface coatings. *Journal of the American Chemical Society*, 129(11), 3333–3338. doi:10.1021/ja068158s

Ehrenberg, M. S., Friedman, A. E., Finkelstein, J. N., Oberdorster, G., & McGrath, J. L. (2009). The influence of protein adsorption on nanoparticle association with cultured endothelial cells. *Biomaterials*, 30(4), 603–610. doi:10.1016/j.biomaterials.2008.09.050

Elghanian, R., Storhoff, J. J., Mucic, R. C., Letsinger, R. L., & Mirkin, C. A. (1997). Selective colorimetric detection of polynucleotides based on the distance-dependent optical properties of gold nanoparticles. *Science*, 277(5329), 1078–1081. doi:10.1126/science.277.5329.1078

Erogbogbo, F., Yong, K. T., Roy, I., Xu, G., Prasad, P. N., & Swihart, M. T. (2008). Biocompatible luminescent silicon quantum dots for imaging of cancer cells. *ACS Nano*, 2(5), 873–878. doi:10.1021/nn700319z

Fendler, Y. T. H. (1996). Langmuir—Blodgett film formation from fluorescence-activated, surfactant-capped, size-selected CdS nanoparticles spread on water surfaces. *Chemistry of Materials*, 8(4), 969–974.

Ferrari, M. (2005). Cancer nanotechnology: Opportunities and challenges. *Nature Reviews Cancer*, 5(3), 161–171. doi:10.1038/nrc1566

Gao, X., Chan, W. C., & Nie, S. (2002). Quantum-dot nanocrystals for ultrasensitive biological labeling and multicolor optical encoding. *Journal of Biomedical Optics*, 7(4), 532–537. doi:10.1117/1.1506706

Giaume, D., Poggi, M., Casanova, D., Mialon, G., Lahlil, K., Alexandrou, A., . . . Boilot, J. P. (2008). Organic functionalization of luminescent oxide nanoparticles toward their application as biological probes. *Langmuir*, 24(19), 11018–11026. doi:10.1021/la8015468

Gorelikov, I., & Matsuura, N. (2008). Single-step coating of mesoporous silica on cetyltrimethyl ammonium bromide-capped nanoparticles. *Nano Letters*, 8(1), 369–373. doi:10.1021/nl0727415

Gratton, S. E., Ropp, P. A., Pohlhaus, P. D., Luft, J. C., Madden, V. J., Napier, M. E., & DeSimone, J. M. (2008). The effect of particle design on cellular internalization pathways. *Proceedings of the National Academy of Sciences of the United States of America*, 105(33), 11613–11618. doi:10.1073/pnas.0801763105

Gref, R., Luck, M., Quellec, P., Marchand, M., Dellacherie, E., Harnisch, S., . . . Muller, R. H. (2000). 'Stealth' corona-core nanoparticles surface modified by polyethylene glycol (PEG): Influences of the corona (PEG chain length and surface density) and of the core composition on phagocytic uptake and plasma protein adsorption. *Colloids and Surfaces B: Biointerfaces*, 18(3-4), 301–313.

Guerrini, L., Ramon, A.-P. A., & Pazos-Perez, N. (2018). Surface modifications of nanoparticles for stability in biological fluids. *Materials*, 11(7), 1154.

Gupta, A., Moyano, D. F., Parnsubsakul, A., Papadopoulos, A., Wang, L. S., Landis, R. F., . . . Rotello, V. M. (2016). Ultrastable and biofunctionalizable gold nanoparticles. *ACS Applied Materials & Interfaces*, 8(22), 14096–14101. doi:10.1021/acsami.6b02548

Hafner, J. H., & Liao, H. (2005). Gold nanorod bioconjugates. *Chemistry of Materials*, 17(18), 4636–4641.

Han, H. S., Martin, J. D., Lee, J., Harris, D. K., Fukumura, D., Jain, R. K., & Bawendi, M. (2013). Spatial charge configuration regulates nanoparticle transport and binding behavior in vivo;. *Angewandte Chemie International Edition In English*, 52(5), 1414–1419. doi:10.1002/anie.201208331

He, X., Gao, L., & Ma, N. (2013). One-step instant synthesis of protein-conjugated quantum dots at room temperature. *Scientific Reports*, 3, 2825. doi:10.1038/srep02825

Hiremath, C. G., Kariduraganavar, M. Y., & Hiremath, M. B. (2018). Synergistic delivery of 5-fluorouracil and curcumin using human serum albumin-coated iron oxide nanoparticles by folic acid targeting. *Progress in Biomaterials*, 7(4), 297–306. doi:10.1007/s40204-018-0104-3

Hiroshi, Y., Osamu, M., Tomoki, H., & Keisaku, K. (2000). Phase transfer of gold nanoparticles across a water/oil interface by stoichiometric ion-pair formation on particle surfaces. *Bulletin of the Chemical Society of Japan*, 73(12), 2675–2678.

Hirsch, L. R., Stafford, R. J., Bankson, J. A., Sershen, S. R., Rivera, B., Price, R. E., . . . West, J. L. (2003). Nanoshell-mediated near-infrared thermal therapy of tumors under magnetic resonance guidance. *Proceedings of the National Academy of Sciences of the United States of America*, 100(23), 13549–13554. doi:10.1073/pnas.2232479100

Hoffman, A. S. (2008). The origins and evolution of "controlled" drug delivery systems. *Journal of Controlled Release*, 132(3), 153–163. doi:10.1016/j.jconrel.2008.08.012

Ikuma, K., Decho, A. W., & Lau, B. L. (2015). When nanoparticles meet biofilms-interactions guiding the environmental fate and accumulation of nanoparticles. *Frontiers in Microbiology*, 6, 591. doi:10.3389/fmicb.2015.00591

Ip, S., MacLaughlin, C. M., Gunari, N., & Walker, G. C. (2011). Phospholipid membrane encapsulation of nanoparticles for surface-enhanced Raman scattering. *Langmuir*, 27(11), 7024–7033. doi:10.1021/la200212c

Jan Karolin, L. B.-A. J., Strandberg, L., & Ny, T. (1994). Fluorescence and absorption spectroscopic properties of dipyrrometheneboron difluoride (BODIPY) derivatives in liquids, lipid membranes, and protein. *Journal of the American Chemical Society*, 116(17), 7801–7806.

Johnson, C. J., Zhukovsky, N., Cass, A. E., & Nagy, J. M. (2008). Proteomics, nanotechnology and molecular diagnostics. *Proteomics*, 8(4), 715–730. doi:10.1002/pmic.200700665

Kairdolf, B. A., Mancini, M. C., Smith, A. M., & Nie, S. (2008). Minimizing nonspecific cellular binding of quantum dots with hydroxyl-derivatized surface coatings. *Analytical Chemistry*, 80(8), 3029–3034. doi:10.1021/ac800068q

Kamil Rahme, M. T. N., Doody, T., McGlacken, G. P., Morris, M. A., O'Driscoll, C., & Holmes, J. D. (2013). Highly stable PEGylated gold nanoparticles in water: Applications in biology and catalysis. *RSC Advances*, 3(43), 21016–21024.

Keefe, A. D., Pai, S., & Ellington, A. (2010). Aptamers as therapeutics. *Nature Reviews Drug Discovery*, 9(7), 537–550. doi:10.1038/nrd3141

Kim, C. K., Ghosh, P., Pagliuca, C., Zhu, Z. J., Menichetti, S., & Rotello, V. M. (2009). Entrapment of hydrophobic drugs in nanoparticle monolayers with efficient release into cancer cells. *Journal of the American Chemical Society*, 131(4), 1360–1361. doi:10.1021/ja808137c

Kim, J., Lee, J. E., Lee, J., Yu, J. H., Kim, B. C., An, K., . . . Hyeon, T. (2006). Magnetic fluorescent delivery vehicle using uniform mesoporous silica spheres embedded with monodisperse magnetic and semiconductor nanocrystals. *Journal of the American Chemical Society*, 128(3), 688–689. doi:10.1021/ja0565875

Kinnear, C., Dietsch, H., Clift, M. J., Endes, C., Rothen-Rutishauser, B., & Petri-Fink, A. (2013). Gold nanorods: Controlling their surface chemistry and complete detoxification by a two-step place exchange. *Angewandte Chemie International Edition In English*, 52(7), 1934–1938. doi:10.1002/anie.201208568

Kirchner, C., Liedl, T., Kudera, S., Pellegrino, T., Munoz Javier, A., Gaub, H. E., . . . Parak, W. J. (2005). Cytotoxicity of colloidal CdSe and CdSe/ZnS nanoparticles. *Nano Letters*, 5(2), 331–338. doi:10.1021/nl047996m

Kohler, N., Fryxell, G. E., & Zhang, M. (2004). A bifunctional poly(ethylene glycol) silane immobilized on metallic oxide-based nanoparticles for conjugation with cell targeting agents. *Journal of the American Chemical Society*, 126(23), 7206–7211.

Kolb, H. C., Finn, M. G., & Sharpless, K. B. (2001). Click chemistry: Diverse chemical function from a few good reactions. *Angewandte Chemie International Edition In English*, 40(11), 2004–2021. doi:10.1002/1521-3773(20010601)40:11<2004::aid-anie2004>3.3.co;2-x

Kyung-Hwan Cho, J.-E. P., Osaka, T., & Park, S.-G. (2005). The study of antimicrobial activity and preservative effects of nanosilver ingredient. *Electrochimica Acta*, 51(5), 956–960.

Lei, C., Xu, C., Noonan, O., Meka, A. K., Zhang, L., Nouwens, A., & Yu, C. (2015). Mesoporous materials modified by aptamers and hydrophobic groups assist ultra-sensitive insulin detection in serum. *Chemical Communications (Cambridge)*, 51(71), 13642–13645. doi:10.1039/c5cc4458h

Libor Kvítek, A. P., Soukupová, J., Kolář, M., Večeřová, R., Prucek, R., Holecová, M., & Zbořil, R. (2008). Effect of surfactants and polymers on stability and antibacterial activity of silver nanoparticles (NPs). *Journal of Physical Chemistry C*, 112(15), 5825–5834.

Lin, Z. Z. a. M. (2014). Fast loading of PEG-SH on CTAB-protected gold nanorods. *RSC Advances*, 4(34), 17760–17767.

Liu, K., Zheng, Y., Lu, X., Thai, T., Lee, N. A., Bach, U., & Gooding, J. J. (2015). Biocompatible gold nanorods: One-step surface functionalization, highly colloidal stability, and low cytotoxicity. *Langmuir*, 31(17), 4973–4980. doi:10.1021/acs.langmuir.5b00666

Liu, W., Choi, H. S., Zimmer, J. P., Tanaka, E., Frangioni, J. V., & Bawendi, M. (2007). Compact cysteine-coated CdSe(ZnCdS) quantum dots for in vivo; applications. *Journal of the American Chemical Society*, 129(47), 14530–14531. doi:10.1021/ja073790m

Lok, C. N., Ho, C. M., Chen, R., He, Q. Y., Yu, W. Y., Sun, H., . . . Che, C. M. (2007). Silver nanoparticles: Partial oxidation and antibacterial activities. *Journal of Biological Inorganic Chemistry*, 12(4), 527–534. doi:10.1007/s00775-007-0208-z

Longmire, M., Choyke, P. L., & Kobayashi, H. (2008). Clearance properties of nano-sized particles and molecules as imaging agents: Considerations and caveats. *Nanomedicine (London)*, 3(5), 703–717. doi:10.2217/17435889.3.5.703

Love, J. C., Estroff, L. A., Kriebel, J. K., Nuzzo, R. G., & Whitesides, G. M. (2005). Self-assembled monolayers of thiolates on metals as a form of nanotechnology. *Chemical Reviews*, 105(4), 1103–1169. doi:10.1021/cr0300789

Mahmoudi, M., Lynch, I., Ejtehadi, M. R., Monopoli, M. P., Bombelli, F. B., & Laurent, S. (2011). Protein-nanoparticle interactions: Opportunities and challenges. *Chemical Reviews*, 111(9), 5610–5637. doi:10.1021/cr100440g

Mashaghi, S., Jadidi, T., Koenderink, G., & Mashaghi, A. (2013). Lipid nanotechnology. *International Journal of Molecular Sciences*, 14(2), 4242–4282. doi:10.3390/ijms14024242

Mirshafiee, V., Mahmoudi, M., Lou, K., Cheng, J., & Kraft, M. L. (2013). Protein corona significantly reduces active targeting yield. *Chemical Communications (Cambridge)*, 49(25), 2557–2559. doi:10.1039/c3cc37307j

Moore, T. L., Rodriguez-Lorenzo, L., Hirsch, V., Balog, S., Urban, D., Jud, C., . . . Petri-Fink, A. (2015). Nanoparticle colloidal stability in cell culture media and impact on cellular interactions. *Chemical Society Reviews*, 44(17), 6287–6305. doi:10.1039/c4cs00487f

Morrow, J. B., Arango, C. P., & Holbrook, R. D. (2010). Association of quantum dot nanoparticles with *Pseudomonas aeruginosa* biofilm. *Journal of Environmental Quality*, 39(6), 1934–1941. doi:10.2134/jeq2009.0455

Morteza Mahmoudi, N. B., & HarshalZope, O. C. F. (2016). Emerging understanding of the protein corona at the nano-bio interfaces. *Nano Today*, 11(6), 817–832.

Mulvaney, L. M. L.-M. A. P. (2003). The assembly of coated nanocrystals. *The Journal of Physical Chemistry B*, 107(30), 7312–7326.

Muro, E., Pons, T., Lequeux, N., Fragola, A., Sanson, N., Lenkei, Z., & Dubertret, B. (2010). Small and stable sulfobetaine zwitterionic quantum dots for functional live-cell imaging. *Journal of the American Chemical Society*, 132(13), 4556–4557. doi:10.1021/ja1005493

Muro, S., Garnacho, C., Champion, J. A., Leferovich, J., Gajewski, C., Schuchman, E. H., . . . Muzykantov, V. R. (2008). Control of endothelial targeting and intracellular delivery of therapeutic enzymes by modulating the size and shape of ICAM-1-targeted carriers. *Molecular Therapy*, 16(8), 1450–1458. doi:10.1038/mt.2008.127

Najmeh Delaviz, P. G., Ajamia, A., & Aarabi, M. (2015). Aptamer-conjugated magnetic nanoparticles for the efficient removal of HCV particles from human plasma samples. *RSC Advances*, 5(97), 79433–79439.

Navya, P. N., & Daima, H. K. (2016). Rational engineering of physicochemical properties of nanomaterials for biomedical applications with nanotoxicological perspectives. *Nano Convergence*, 3(1), 1. doi:10.1186/s40580-016-0064-z

Nel, A. E., Madler, L., Velegol, D., Xia, T., Hoek, E. M., Somasundaran, P., . . . Thompson, M. (2009). Understanding biophysicochemical interactions at the nano-bio interface. *Nature Materials*, 8(7), 543–557. doi:10.1038/nmat2442

Nevius, B. A., Chen, Y. P., Ferry, J. L., & Decho, A. W. (2012). Surface-functionalization effects on uptake of fluorescent polystyrene nanoparticles by model biofilms. *Ecotoxicology*, 21(8), 2205–2213. doi:10.1007/s10646-012-0975-3

Nguyen, T. K., & Thanh, L. A. W. G. (2010). Functionalisation of nanoparticles for biomedical applications. *Nano Today*, 5(3), 213–230.

Niidome, T., Yamagata, M., Okamoto, Y., Akiyama, Y., Takahashi, H., Kawano, T., . . . Niidome, Y. (2006). PEG-modified gold nanorods with a stealth character for in vivo; applications. *Journal of Controlled Release*, 114(3), 343–347. doi:10.1016/j.jconrel.2006.06.017

Nyalosaso, J. L., Rascol, E., Pisani, C., Dorandeu, C., Dumail, X., Maynadier, M., . . . Chopineau, J. (2016). Synthesis, decoration, and cellular effects of magnetic mesoporous silica nanoparticles. *RSC Advances*, 6(62), 57275–57283.

Oberdorster, G., Oberdorster, E., & Oberdorster, J. (2005). Nanotoxicology: An emerging discipline evolving from studies of ultrafine particles. *Environmental Health Perspectives*, 113(7), 823–839. doi:10.1289/ehp.7339

Owens, D. E., 3rd, & Peppas, N. A. (2006). Opsonization, biodistribution, and pharmacokinetics of polymeric nanoparticles. *International Journal of Pharmaceutics*, 307(1), 93–102. doi:10.1016/j.ijpharm.2005.10.010

Pal, S., Tak, Y. K., & Song, J. M. (2007). Does the antibacterial activity of silver nanoparticles depend on the shape of the nanoparticle? A study of the Gram-negative bacterium *Escherichia coli*. *Applied and Environmental Microbiology*, 73(6), 1712–1720. doi:10.1128/AEM.02218-06

Pelaz, B., Alexiou, C., Alvarez-Puebla, R. A., Alves, F., Andrews, A. M., Ashraf, S., . . . Parak, W. J. (2017). Diverse applications of nanomedicine. *ACS Nano*, 11(3), 2313–2381. doi:10.1021/acsnano.6b06040

Peng, X., Kim, M., Chen, Y., & Liu, Y. (2005). Super-stable, high-quality $Fe_3O_4$ dendron–nanocrystals dispersible in both organic and aqueous solutions. *Advanced Materials*, 17(11), 1429–1432.

Petros, R. A., & DeSimone, J. M. (2010). Strategies in the design of nanoparticles for therapeutic applications. *Nature Reviews Drug Discovery*, 9(8), 615–627. doi:10.1038/nrd2591

Poon, Z., Chang, D., Zhao, X., & Hammond, P. T. (2011). Layer-by-layer nanoparticles with a pH-sheddable layer for in vivo; targeting of tumor hypoxia. *ACS Nano*, 5(6), 4284–4292. doi:10.1021/nn200876f

Rakesh Voggu, P. S., Chandrasekaran, S., & Rao, C. N. R. (2007). Assembling covalently linked nanocrystals and nanotubes through click chemistry. *Chemical Physics Letters*, 443(1–3), 118–121.

Rascol, E., Daurat, M., Da Silva, A., Maynadier, M., Dorandeu, C., Charnay, C., . . . Chopineau, J. (2017). Biological fate of Fe(3)O(4) core-shell mesoporous silica nanoparticles depending on particle surface chemistry. *Nanomaterials (Basel)*, 7(7). doi:10.3390/nano7070162

Rubio-Garcia, J., Coppel, Y., Lecante, P., Mingotaud, C., Chaudret, B., Gauffre, F., & Kahn, M. L. (2011). One-step synthesis of metallic and metal oxide nanoparticles using amino-PEG oligomers as multi-purpose ligands: Size and shape control, and quasi-universal solvent dispersibility. *Chemical Communications (Cambridge)*, 47(3), 988–990. doi:10.1039/c0cc02615h

Saha, K., Bajaj, A., Duncan, B., & Rotello, V. M. (2011). Beauty is skin deep: A surface monolayer perspective on nanoparticle interactions with cells and bio-macromolecules. *Small*, 7(14), 1903–1918. doi:10.1002/smll.201100478

Sam, J. K., Finch, S. K., Hallahan, D. E., & Giorgio, T. D. (2007). Facile production of multivalent enzyme-nanoparticle conjugates. *Journal of Magnetism and Magnetic Materials*, 311(1), 68–72.

Sarkar, N., Banerjee, J., Hanson, A. J., Elegbede, A. I., Rosendahl, T., Krueger, A. B., . . . Srivastava, D. K. (2008). Matrix metalloproteinase-assisted triggered release of liposomal contents. *Bioconjugate Chemistry*, 19(1), 57–64. doi:10.1021/bc070081p

SaumyadityaBose, M. F. H. J., Yuri, A. G., David, W. K., McCready, D. E., Andrew, S. M., & Brian, H. L. (2009). Bioreduction of hematite nanoparticles by the dissimilatory iron reducing bacterium *Shewanella oneidensis* MR-1. *Geochimica et Cosmochimica Acta*, 73(4), 962–976.

Schaffler, M., Sousa, F., Wenk, A., Sitia, L., Hirn, S., Schleh, C., . . . Krol, S. (2014). Blood protein coating of gold nanoparticles as potential tool for organ targeting. *Biomaterials*, 35(10), 3455–3466. doi:10.1016/j.biomaterials.2013.12.100

Shmeeda, H., Tzemach, D., Mak, L., & Gabizon, A. (2009). Her2-targeted PEGylated liposomal doxorubicin: Retention of target-specific binding and cytotoxicity after in vivo; passage. *Journal of Controlled Release*, 136(2), 155–160. doi:10.1016/j.jconrel.2009.02.002

Shu-Yi Lin, Y.-T. T., Chen, C.-C., Lin, C.-M., & Chen, C.-h. (2004). Two-step functionalization of neutral and positively charged thiols onto citrate-stabilized Au nanoparticles. *The Journal of Physical Chemistry. B*, 108(7), 2134–2139.

Slowing, I., Trewyn, B. G., & Lin, V. S. (2006). Effect of surface functionalization of MCM-41-type mesoporous silica nanoparticles on the endocytosis by human cancer cells. *Journal of the American Chemical Society*, 128(46), 14792–14793. doi:10.1021/ja0645943

Sperling, R. A., & Parak, W. J. (2010). Surface modification, functionalization and bioconjugation of colloidal inorganic nanoparticles. *Philosophical Transactions. Series A, Mathematical, Physical, and Engineering Sciences*, 368(1915), 1333–1383. doi:10.1098/rsta.2009.0273

Spisak, S., Tulassay, Z., Molnar, B., & Guttman, A. (2007). Protein microchips in biomedicine and biomarker discovery. *Electrophoresis*, 28(23), 4261–4273. doi:10.1002/elps.200700539

Srinivasan, C., Lee, J., Papadimitrakopoulos, F., Silbart, L. K., Zhao, M., & Burgess, D. J. (2006). Labeling and intracellular tracking of functionally active plasmid DNA with semiconductor quantum dots. *Molecular Therapy*, 14(2), 192–201. doi:10.1016/j.ymthe.2006.03.010

Stauber, R. H., Siemer, S., Becker, S., Ding, G. B., Strieth, S., & Knauer, S. K. (2018). Small meets smaller: Effects of nanomaterials on microbial biology, pathology, and ecology. *ACS Nano*, 12(7), 6351–6359. doi:10.1021/acsnano.8b03241

Sun, X. L., Stabler, C. L., Cazalis, C. S., & Chaikof, E. L. (2006). Carbohydrate and protein immobilization onto solid surfaces by sequential Diels–Alder and azide-alkyne cycloadditions. *Bioconjugate Chemistry*, 17(1), 52–57. doi:10.1021/bc0502311

Takeshi Sakura, T. T., Kataoka, K., & Nagasaki, Y. (2005). One-pot preparation of mono-dispersed and physiologically stabilized gold colloid. *Colloid & Polymer Science*, 284(1), 97–101.

Tang, L., Yang, X., Yin, Q., Cai, K., Wang, H., Chaudhury, I., . . . Cheng, J. (2014). Investigating the optimal size of anticancer nanomedicine. *Proceedings of the National Academy of Sciences of the United States of America*, 111(43), 15344–15349. doi:10.1073/pnas.1411499111

Tebbe, M., Kuttner, C., Mannel, M., Fery, A., & Chanana, M. (2015). Colloidally stable and surfactant-free protein-coated gold nanorods in biological media. *ACS Applied Materials & Interfaces*, 7(10), 5984–5991. doi:10.1021/acsami.5b00335

Thorek, D. L., & Tsourkas, A. (2008). Size, charge and concentration dependent uptake of iron oxide particles by non-phagocytic cells. *Biomaterials*, 29(26), 3583–3590. doi:10.1016/j.biomaterials.2008.05.015

Tian, L., Shi, C., & Zhu, J. (2007). Reversible formation of hybrid nanostructures via an organic linkage. *Chemical Communications (Cambridge)*, (37), 3850–3852. doi:10.1039/b706889a

Tyrakowski, C. M., & Snee, P. T. (2014). A primer on the synthesis, water-solubilization, and functionalization of quantum dots, their use as biological sensing agents, and present status. *Physical Chemistry Chemical Physics*, 16(3), 837–855. doi:10.1039/c3cp53502a

Unfried, K., Sydlik, U., Bierhals, K., Weissenberg, A., & Abel, J. (2008). Carbon nanoparticle-induced lung epithelial cell proliferation is mediated by receptor-dependent Akt activation. *American Journal of Physiology—Lung Cellular and Molecular Physiology*, 294(2), L358–L367. doi:10.1152/ajplung.00323.2007

van Vlerken, L. E., Vyas, T. K., & Amiji, M. M. (2007). Poly(ethylene glycol)-modified nanocarriers for tumor-targeted and intracellular delivery. *Pharmaceutical Research*, 24(8), 1405–1414. doi:10.1007/s11095-007-9284-6

Veronese, F. M. (2001). Peptide and protein PEGylation: A review of problems and solutions. *Biomaterials*, 22(5), 405–417. doi:10.1016/s0142-9612(00)00193-9

Wang, B., Zhang, L., Bae, S. C., & Granick, S. (2008). Nanoparticle-induced surface reconstruction of phospholipid membranes. *Proceedings of the National Academy of Sciences of the United States of America*, 105(47), 18171–18175. doi:10.1073/pnas.0807296105

Wang, F., Porter, M., Konstantopoulos, A., Zhang, P., & Cui, H. (2017). Preclinical development of drug delivery systems for paclitaxel-based cancer chemotherapy. *Journal of Controlled Release*, 267, 100–118. doi:10.1016/j.jconrel.2017.09.026

Wang, J., Tian, S., Petros, R. A., Napier, M. E., & Desimone, J. M. (2010). The complex role of multivalency in nanoparticles targeting the transferrin receptor for cancer therapies. *Journal of the American Chemical Society*, 132(32), 11306–11313. doi:10.1021/ja1043177

Wang, Y. A., Li, J. J., Chen, H., & Peng, X. (2002). Stabilization of inorganic nanocrystals by organic dendrons. *Journal of the American Chemical Society*, 124(10), 2293–2298. doi:10.1021/ja016711u

Watkins, G. A., Jones, E. F., Scott Shell, M., VanBrocklin, H. F., Pan, M. H., Hanrahan, S. M., . . . Franc, B. L. (2009). Development of an optimized activatable MMP-14 targeted SPECT imaging probe. *Bioorganic & Medicinal Chemistry*, 17(2), 653–659. doi:10.1016/j.bmc.2008.11.078

Wei, H., Insin, N., Lee, J., Han, H. S., Cordero, J. M., Liu, W., & Bawendi, M. G. (2012). Compact zwitterion-coated iron oxide nanoparticles for biological applications. *Nano Letters*, 12(1), 22–25. doi:10.1021/nl202721q

Westmeier, D., Posselt, G., Hahlbrock, A., Bartfeld, S., Vallet, C., Abfalter, C., . . . Stauber, R. H. (2018). Nanoparticle binding attenuates the pathobiology of gastric cancer-associated *Helicobacter pylori*. *Nanoscale*, 10(3), 1453–1463. doi:10.1039/c7nr06573f

Xie, J., X, C., Kohler, N., Hou, Y., & Sun, S. (2007). Controlled pegylation of monodisperse $Fe_3O_4$ nanoparticles for reduced non-specific uptake by macrophage cells. *Advanced Materials*, 19(20), 3163–3166.

Yang, W., Zhang, L., Wang, S., White, A. D., & Jiang, S. (2009). Functionalizable and ultra stable nanoparticles coated with zwitterionic poly(carboxybetaine) in undiluted blood serum. *Biomaterials*, 30(29), 5617–5621. doi:10.1016/j.biomaterials.2009.06.036

Yezhelyev, M. V., Qi, L., O'Regan, R. M., Nie, S., & Gao, X. (2008). Proton-sponge coated quantum dots for siRNA delivery and intracellular imaging. *Journal of the American Chemical Society*, 130(28), 9006–9012. doi:10.1021/ja800086u

Yu, C., Nakshatri, H., & Irudayaraj, J. (2007). Identity profiling of cell surface markers by multiplex gold nanorod probes. *Nano Letters*, 7(8), 2300–2306. doi:10.1021/nl070894m

Zareie, H. M., Boyer, C., Bulmus, V., Nateghi, E., & Davis, T. P. (2008). Temperature-responsive self-assembled monolayers of oligo(ethylene glycol): Control of biomolecular recognition. *ACS Nano*, 2(4), 757–765. doi:10.1021/nn800076h

Zhang, C. Y., & Johnson, L. W. (2009). Single quantum-dot-based aptameric nanosensor for cocaine. *Analytical Chemistry*, 81(8), 3051–3055. doi:10.1021/ac802737b

Zhang, D., Neumann, O., Wang, H., Yuwono, V. M., Barhoumi, A., Perham, M., . . . Halas, N. J. (2009). Gold nanoparticles can induce the formation of protein-based aggregates at physiological pH. *Nano Letters*, 9(2), 666–671. doi:10.1021/nl803054h

Zhang, X. Q., Xu, X., Bertrand, N., Pridgen, E., Swami, A., & Farokhzad, O. C. (2012). Interactions of nanomaterials and biological systems: Implications to personalized nanomedicine. *Advanced Drug Delivery Reviews*, 64(13), 1363–1384. doi:10.1016/j.addr.2012.08.005

Zheng, W., Liu, Y., West, A., Schuler, E. E., Yehl, K., Dyer, R. B., . . . Salaita, K. (2014). Quantum dots encapsulated within phospholipid membranes: Phase-dependent structure, photostability, and site-selective functionalization. *Journal of the American Chemical Society*, 136(5), 1992–1999. doi:10.1021/ja411339f

Zhou, D., Piper, J. D., Abell, C., Klenerman, D., Kang, D. J., & Ying, L. (2005). Fluorescence resonance energy transfer between a quantum dot donor and a dye acceptor attached to DNA. *Chemical Communications (Cambridge)*, (38), 4807–4809. doi:10.1039/b508911e

# Glossary

**NMs:** Nanomaterials
**NPs:** Nanoparticles
**EPR:** Enhanced permeation and retention effect
**PEG:** Poly(ethylene glycol)
**DDL:** Diffuse double layer
**ICAM:** Intercellular adhesion molecule I
**E. coli:** Escherichia coli
**AKT:** Protein kinase B
**SEM:** Scanning electron microscopy
**DNA:** Deoxyribonucleic acid
**ITC:** Isothermal titration colorimetry
**VDW:** Van der Walls interactions
**ROS:** Reactive oxygen species
**MAA:** Mercaptoacetic acid
**MPA:** Mercaptopropionic acid
**MUA:** Mercaptoundecanoic acid
**MPS:** Mononuclear phagocyte system
**CTAC:** Cetyltrimethylammonium chloride
**TOAB:** Tetraoctylammonium bromide
**MPS:** 3-(Trimethoxysilyl) propyl methacrylate
**APTS:** 3-Aminopropyltriethoxysilane
**AuNRs:** Gold nanorods
**QDs:** Quantum dots
**DOPC:** Dioleoylphosphatidylcholine
**ESM:** Egg sphingomyelin
**Chol:** Cholestrol
**AuNPs:** Gold nanoparticles
**DPA:** d-Penicillamine
**CMC:** Critical micelle concentration
**ApoE:** Apolipoprotein A
**Tf:** Transferrin
**BSA:** Bovine serum albumin
**NHS:** N-Hydroxy-succinimide
**EDC:** Carbodiimide
**UV–vis:** Ultraviolet–visible
**SDS:** Sodium dodecyl sulfate
**AgNPs:** Silver nanoparticles
**POPC:** Palmitoyl-2-oleoyl-sn-glycero-3-phosphocholine
**POPG:** 1-Palmitoyl-2-oleoyl-sn-glycero-3 phosphoglycerol

**CTAB:** Cetyltrimethylammonium bromide
**ODA:** Octadecylamine
**MPA:** 3-Mercaptopropionic acid
**TBE:** Tris-borate-EDTA
**GSH:** Glutathione
**SMCC:** Succinimidyl-4-(N-maleimidomethyl)cyclohexane-1- carboxylate

# 5

# Molecular Mechanisms Behind Nano-Cancer Therapeutics

*Surya Prakash Singh and Aravind Kumar Rengan*

*Department of Biomedical Engineering, Indian Institute of Technology Hyderabad, Kandi, Sangareddy, Hyderabad, India*

## Objectives

- Understand the role of nanotechnology at nano–bio interfaces.
- Understand the role of size, surface charge, topography, and chemical composition of nanoparticles at nano–bio interfaces.
- Know how to modulate the nanoparticle edge for tumor targeting and enhanced permeability and retention effect (EPR).
- Understand the role of protein interaction, circulation time, and triggered drug release of nanoparticles for cancer therapeutics.
- Know the role of intracellular uptake and nanoparticle trafficking.
- Understand the role of multidrug resistance (MDR) for cancer treatment.
- Nanotechnology: challenges and opportunities for clinical applications.

## 5.1 Nanotechnology at Nano–Bio Interfaces

Nanotechnology has made significant contributions to cancer theranostics. Nanoparticles (NPs) with sizes ranging from 1 to 100 nm can interact with biomolecules giving rise to the nano–bio interface (Kah, Yeo, Koh, Poinard, & Neo, 2014). The term "nanotechnology" was coined in the 1970s by Norio Taniguchi. Eric Drexler and Robert Freitas Jr. coined the term "nanomedicine" in the 1990s denotes the role of nanoscience in medicine. The design of a nano–bio interface formed by a nanoparticle and its biological surroundings is decisive to its biological application. The therapeutic outcomes of the NPs are dependent on their properties including surface texture (Niu et al., 2013), size (X. Liu et al., 2016), geometry (Herd et al., 2013), surface charge (Fröhlich, 2012), elasticity (Guo et al., 2018), and targeting agents (Q. Wei et al., 2018). NPs can enhance the pharmaceutical

*Microbial Interactions at Nanobiotechnology Interfaces: Molecular Mechanisms and Applications*, First Edition. Edited by R. Navanietha Krishnaraj and Rajesh K. Sani.

properties of the drugs in terms of its solubility, stability, tumor uptake, and blood circulation time (Ali, Hanafy, & Alqurshi, 2019; Ji et al., 2016; Saroj & Rajput, 2018; Wang, Ma, Chen, Zhao, & Mou, 2019). NPs can focus on the tumor area by selective uptake and sustained or stimulus-responsive triggered drug release (Zhang, Sun, Adu-Frimpong, Yu, & Xu, 2019; W. Xu et al., 2018). The study of nano–bio interface having a crucial role in targeting specific tumor markers present over the cell surface through active targeting to enhance cellular uptake and prevents in vivo nonspecific biodistribution (Zhong et al., 2019). Drug-loaded NPs can also be accumulated at the tumor site through passive targeting via enhanced permeability and retention (EPR) effect. EPR is the result of poor lymphatic drainage system due to leaky tumor vasculature in the tumor microenvironment (Maeda, 2015). Cancer stem cell (CSC) population present in tumor cells has distinct features as compared to other cancer cells. CSCs can show drug resistance and initiate rapid tumor proliferation (L. Li & Neaves, 2006). Nanoparticle-mediated drug delivery can overcome the drug resistance due to its distinct endocytic mechanism (Yuan et al., 2016). Therefore, CSC targeted drug delivery through NPs can enhance the therapeutic index by reducing systemic toxicity to healthy cells.

A biological membrane contains membrane proteins, glycocalyx, several lipid types, and cholesterol. Therefore it forms a complex system and it is consequently complicated to set up a model that can envisage the NPs' interaction with a biological membrane system. The foremost interaction of NPs happens with biological cells via the biological membrane, therefore it is important to identify the mechanism behind the interaction between NPs and biological membrane at nano–bio interfaces.

## 5.2 Armory of Nanomedicine at Nano–Bio Interfaces

The interactions between NPs and cells can vary depending upon the optimal sizes of the NPs. Cellular uptake may vary according to the distinct sizes of the NPs. When the nanoparticle enters ectogenically and touches cells, the cell membranes deform, allowing the drug-loaded NPs to phagocytize within the cell. Xu et al. reported a possible mechanism for NPs' uptake based on the bending and stretching energy. The aims of the studies are based on to improve the oral bioavailability of drugs by nanoparticle modifications. The cellular uptake of didodecyl dimethylammonium bromide (DMAB)-modified NPs was investigated in Caco-2 and HT-29 cells (colorectal adenocarcinoma). Here, Deserno's model has been developed to study the kinetic rates of cellular uptake mechanisms. The results indicated that active targeting via receptor-mediated endocytosis is not only the main driving force for the cellular uptake. The different sizes of the NPs and their electrostatic interaction with the cellular membrane could control the uptake process. The cellular uptake mechanism can be influenced by various driving forces, such as bending and stretching forces, electrostatic forces, and hydrophobic forces. The enhanced cellular uptake of the NPs based on competition between the bending energy and the stretching energy at the nano–bio interfaces. The competition between the bending energy and the stretching energy could be the non-competitive to each other due to the change in membrane conformation after NP binding. This model is consistent with the experimental explanation. For small NPs, the bending energy of the membrane is larger, which further induces higher uptake for small NPs. When the particle size increases, higher stretching energies in the membrane are required for uptake. So there is a competition between stretching and bending energies due to the difference in sizes of the NPs (A. Xu et al., 2012). The anisotropic properties of the NPs are also one of the aspects influencing cellular uptake. Rod-like NPs showed enhanced internalization efficiency compared to the spherical NPs due to their inherent optical, chemical, magnetic, and electrical anisotropy (Daum, Tscheka, Neumeyer, & Schneider, 2012). Tobacco mosaic virus (TMV) is a characteristic rod-like bionanoparticle that is developed as a model to clarify the consequence of aspect ratio of rod-like NPs on cellular internalization (Adams, Dogic, Keller, & Fraden, 1998). The shape effects of NPs have a chief role in NPs' flow in circulation and drug delivery. When comparing between filaments versus spherical particles, the NPs with aspect ratio showed up to ten times longer blood circulation as compared to its spherical counterparts studied in rodents (Geng et al., 2007). Zeta potential measurement represents the surface charge of NPs. It is recommended as one of the key factors deciding NPs' internalization in cells and their biological impact (Yue et al., 2011). In addition to size, surface properties such as surface charge and hydrophobicity are generally necessary for cellular uptake. Positively charged NPs have a higher chance of cytotoxicity due to their higher uptake. Cationic NPs cause more prominent damage by disruption of plasma-membrane integrity compared to anionic NPs (Fröhlich, 2012). Positively charged NPs based on gold functionalized with cationic side chains, organic monolayer-coated silicon, silica-titania hollow NPs act more cytotoxically than the negative forms of NPs of comparable size in nonphagocytic cells (Bhattacharjee et al., 2010; Goodman, McCusker, Yilmaz, & Rotello, 2004; Oh et al., 2010). Phagocytic cells show more pronounced interaction with negatively charged NPs in contrast to nonphagocytic cells. There are significant evidence such as ingestion of bacteria having a net negative charge at their surface can interact well with phagocytic cells (Cooper, 2011), e.g. poly (lacticco-glycolic acid) (PLGA) NPs coated with different targeting moieties. For some NPs, e.g. poly (lactic-co-glycolic acid) (PLGA) NPs coated with different targeting moieties. Cationic surface coatings over PLGA NPs for tumor targeting showed more effective accumulation in tumor tissue than bare Pluronic-coated ones (Chung et al., 2010). Besides this, the charge polarity and charge density also play a very important role in cytotoxicity. In one of the studies under in vivo conditions, dendrimers-based NPs with high numbers of primary amine groups over their surface were found to be related with increased cytotoxicity of the NPs (Chauhan, Diwan, Jain, & Tomalia, 2009). There are different modes of cytotoxic action depending on the surface charge of NPs. Positively charged NPs cause membrane damage while negatively charged NPs cause intracellular damage. It has been investigated that there is a relationship between cytotoxicity and lysosomal localization as negatively charged nanoceria NPs are localized more in the lysosome (Asati, Santra, Kaittanis, & Perez, 2010). Although cationic NPs can interact with serum protein with a higher extent and showed nonspecific uptake as compared to anionic or neutrally charged NPs (Yamamoto, Nagasaki, Kato, Sugiyama, & Kataoka, 2001). Nanoparticle elasticity also plays an important role in tumor uptake. In one of the studies, nanolipogels (NLGs) encapsulating an alginate core were found to impart tunable elasticity in NPs. Results demonstrated that the elastic NLGs better accumulated in the liver while soft NLGs accumulated considerably more in the tumor region (Guo et al., 2018). The shape of the NPs could alter the phagocytic potential whereas curvature and flexibility could determine the ability for internalization of NPs (Champion & Mitragotri, 2009). Some groups have shown toll-like receptor 2 upregulation and rough edges spiky NPs could enhance the uptake of non-spherical NPs (Albanese, Sykes, & Chan, 2010; Doshi & Mitragotri, 2010). In another study, discoidal carriers showed more enhanced efficacy as compared to spherical carriers with superior targeting efficacy because of active targeting via receptor binding. This strategy is used effectively to design nanovehicles for photodynamic therapy. The trends for the better efficacies under in vitro and in vivo condition is in following order i.e. folate nanodiscs > folate vesicles > nonfolate nanodiscs > nonfolate vesicles (Tahmasbi Rad et al., 2019)

## 5.3 Nanoparticle Edge in Modulating Biological Process

Serum plays a very important role in the cytotoxicity of the NPs. Coating of NPs with bovine serum albumin (BSA) or incubation of NPs in growth medium containing serum could reduce its cytotoxicity. Cytotoxicity of the NPs based on a polysaccharide, iron oxide, PLGA gets reduced on nonphagocytic cells in the presence of serum (Nafee, Schneider, Schaefer, & Lehr, 2009; Petri-Fink, Steitz, Finka, Salaklang, & Hofmann, 2008). Both types of NPs either positively or negatively charged can bind to serum or albumin protein. The interaction of serum with carboxylated polystyrene NPs was higher as compared to positively charged NPs (Ehrenberg, Friedman, Finkelstein, Oberdörster, & McGrath, 2009). Studies shows that the positively charged CeO NPs could interact better with BSA as compared to negatively charged NPs (Patil, Sandberg, Heckert, Self, & Seal, 2007). The BSA-coated NPs showed less cellular uptake as compared to uncoated NPs (Baier et al., 2011). After NPs' internalization in tissue fluids, the NPs get to interact with biological fluid molecules such as protein. Protein adsorbed modified NPs formed protein corona, which is easily recognized by biological cells. The blood plasma-derived protein corona over NPs help in persisting long-lived NPs without affecting NPs' structure (Walczyk, Bombelli, Monopoli, Lynch, & Dawson, 2010).

Stimulus-responsive smart NPs are the better alternative for triggered drug release at the specific tumor sites. Mesoporous silica nanoparticles (MSNs) are more efficient for drug loading but drawbacks associated with them are burst release and nonspecific leakage of loaded drugs. To resolve this problem the surface of the MSN can be grafted via thermoresponsive poly(*N*-isopropylacrylamide) (PNIPAM) or pH-sensitive polymers, for example, methacrylic-type ionic liquid terpolymer (Parhi, Mohanty, & Sahoo, 2012). These polymers act like switchable "gatekeepers" for triggered drug release. Doxorubicin (DOX)-loaded MSN coated with pH-responsive polymers such as poly (styrene sulfonate) and poly (allylamine hydrochloride) reacted to low pH and released the drug at acidic pH 5.0; however, they did not release the drug at physiological pH 7.4 (Feng et al., 2013). Charge-convertible polymer-based NPs are another alternative for stimulus-responsive triggered drug release. Charge-convertible polymeric NPs have the ability for charge conversion from negative/neutral to positive at the specific site. Li and coworkers synthesized core–shell NPs including paclitaxel (PTX) loaded in hydrophobic core and shell made up of charge-convertible polyethylenimine (PEI). The zeta potential of NP showed that a negative charge due to coated polymer shell prevents its cellular uptake at physiological pH 7.4 into normal cells. However, the surface charge gets changed a slightly acidic pH (6.8–7.0), became positive, and showed better tumor uptake (Tang et al., 2017). For in situ drug release, the magnetic thermosensitive micelles were used for cancer therapy. Hydrophobic $Fe_3O_4$ magnetic NPs loaded with DOX and coated with an amphiphilic polymer, poly (AAm-*co*-AN)-g-PEG (PAAP), (DOX-$Fe_3O_4$@PAAP) micelles showed near infrared (NIR) triggered DOX release after laser irradiation (Wu et al., 2019). In another report, capsule nanosystem was proposed based on matrix metalloproteinase-2 (MMP-2)-cleavable shell with oil-core. The 3D tumor spheroid model was used to examine the efficacy of nanocapsule because of its high-level expression of MMP-2 enzymes. MMP-2 enzymes are able to distort nanocapsule integrity preferentially at tumor microenvironment (Iaccarino, Profeta, Vecchione, & Netti, 2019). The conformational transformation of cyclic RGD sequence due to stimulus could substitute for enhancing the cellular uptake into tumor cells. It can act as on–off gatekeeper for on-demand drug release stimulated by glutathione levels of the cancer cells (Lee et al., 2017).

Studies on biodistribution suggest that the enhanced accumulation of NPs at the tumor site is due to the absence of proper vasculature support, leaky tumor vasculature with pore size ranging from 100 nm to 2 μm diameter. This process is known as the enhanced permeability and retention (EPR) effect and can be useful for the development of novel cancer therapeutics (Kalyane et al., 2019). The tumor vasculatures are hyperpermeable, possess tortuosity, lack basement membrane, anomalous bulges, and irregular shape. So for appropriate growth and development tumor tissues build up new vessels for the oxygen delivery, ingestion of nutrition, and waste excretion. The progression of the new vasculature is called angiogenesis (Byrne, Betancourt, & Brannon-Peppas, 2008; Siemann, 2011). EPR effects were shown by Evans blue dye after intravenous injection due to selective uptake at the tumor region. The concentration of dye was found ~10-fold higher compared to its concentration in the blood after 145 hours (Kalyane et al., 2019).

Some of the factors that can affect the EPR-based accumulation include the reticuloendothelial system (RES) present in spleen and liver, nanoparticle aggregation, and protein binding. So the surface modification of NPs is very important to avoid such processes (Blanco, Shen, & Ferrari, 2015). The accumulation of NPs at tumor site depends on both tumor size as well as the size of the nanoparticle, gold NPs <45 nm can penetrate and accumulate effectively (Kalyane et al., 2019). The tumor uptake of varying sizes of NPs depends upon the interaction between perivascular extracellular matrixes with NPs. The larger NPs follow Brownian motion to enter in the tumor microenvironment (Sykes et al., 2016). In one of the reports, the pretreatment of angiotensin-II (AT-II) caused normal blood vessel constriction because of hypertension but tumor vessels did not show constriction due to lack of smooth muscle layer. An example of such a study is the specific accumulation of $^{51}Cr$ labeled bovine serum albumin (BSA) in Walker carcinoma (Suzuki, Hori, Abe, Sato, & Saito, 1981). This strategy may enhance the NPs uptake specifically at the tumor regions. The other method includes disrupting the tumor vasculature by photodynamic therapy (PDT). PDT is one of the alternative therapies that can be used for cancer treatment. It consists of three components such as photosensitizers (PS), tissue oxygen, and a specific wavelength of light. PS gets activated upon the exposure of a fixed wavelength of light and transfers its energy at excited state to tissue oxygen molecule. The resulting oxygen molecule gets converted to cytotoxic oxygen species. These oxygen species are very aggressive and can damage tumor vasculature (Lepor, 2008). In a study, it was observed that the irradiation of PS along with RGD-modified ferritin exclusively conveyed $^1O_2$ to the tumor vasculature. This process steered the NPs' accumulation around 20-fold in the tumor region as compared to normal tissues (Zhen et al., 2014).

## 5.4 Intracellular Uptake and Trafficking of Nanoparticle

Multidrug resistance (MDR) is one of the hurdles for cancer treatment and diagnosis. Tumor cells overexpress the transporters for drug efflux to be found at the plasma membrane. This has resulted in a reduction in cytotoxicity due to reduced drug concentration at the cellular level. Some of the channel regulators are linked with potassium ABCC8/sulfonylurea receptor 1 (SUR1), and ABCC9/SUR2 and chloride ABCC7/cystic fibrosis transmembrane conductance regulator (CFTR) are extensively studied for MDR. Another MDR protein is MRP1 (190 kDa) having P-gp-like core with 17 transmembrane domains and extra N-terminal membrane-spanning domain (TMD0) (Sodani, Patel, Kathawala, & Chen, 2012; Yuan et al., 2016). P-glycoprotein (P-gp) is one of the important causes of drug resistance in cancer cells. This is a 170 kDa protein expressed over the plasma membrane encoded via MDR1 gene (Gottesman, Fojo, & Bates, 2002). The multidrug resistance protein (MRP) family includes 13 types of proteins, most of these proteins are involved in the transport of xenobiotics as well as endogenous substances. MDR is also associated with dysfunctional apoptosis, elevated DNA repair mechanism, and commencement of prosurvival regulatory pathways (Parhi et al., 2012). As MDR develops, the tumor sensitivity to drugs progressively decreases. Nanoparticle system encapsulated with doxorubicin hydrochloride (DOX) and quercetin (QUR) combinedly called it QD. QD is now loaded in hollow mesoporous silica nanoparticles (HMSNs) further coated with methoxypolyethylene glycol amine (mPEG-NH2) and functionalized by polydopamine (PDA) (HMSNs-PDA-PEG@QD). The proposed nanosystem enhanced the anticancer effects on taxol (TAX) and DOX double resistant human colorectal cancer cell line HCT-8 (HCT-8/TAX cells) and further reverse the MDR. HMSNs-PDA-PEG@QD nanosystem showed the better efficiency of drug delivery and overcome MDR by reducing the expression of P-gp (Figure 5.1) (Shao et al., 2019).

Nanoparticle-mediated drug delivery can overcome the drawbacks associated with MDR. NPs possess size-exclusion effect due to which they are resistant to MDR-mediated drug efflux. Thus the drug gets accumulated at their effective concentration at the intracellular level (J. Liu et al., 2012). Some of the amphiphilic copolymers such as methoxy poly(ethylene glycol)–poly(lactide) copolymer (mPEGPLA), Pluronic P123/F127 can inhibit the expression of P-gp via ATP depletion (W. Li et al., 2014; z. Wei, Yuan, Hao, & Fang, 2013). Despite of P-gp suppression, NPs may enhance drug activity by increasing the intracellular uptake. Liu and coworkers used a cMLV encapsulated with PTX and DOX tested into the drug-resistant 4T1 cells. The combination of both the drugs inhibited the P-gp expression of 4T1 cells (W. Li et al., 2014). The therapeutic efficacy and fate of NPs depend on their cellular uptake mechanism. The cellular uptake mechanism may follow two distinct routes: pinocytosis and phagocytosis. Uptake of

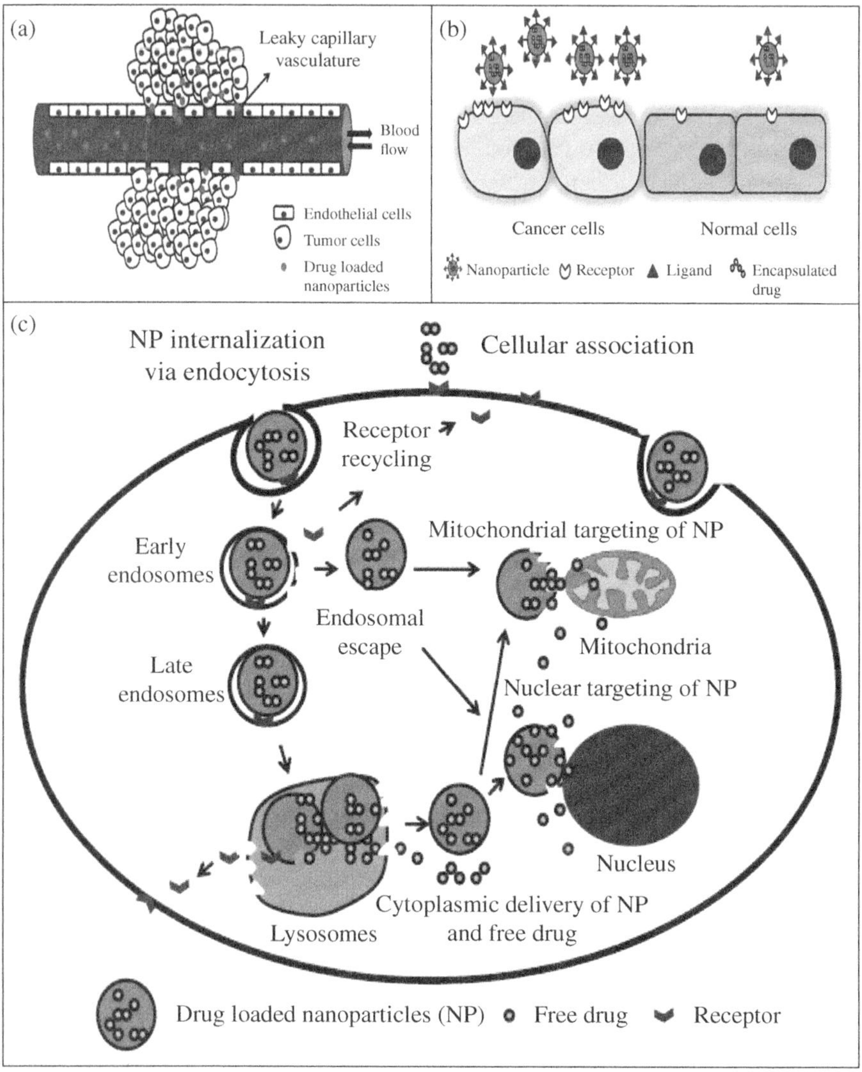


**Figure 5.1** (a) Passive targeting of nanoparticles through enhanced permeability and retention effect (EPR), (b) receptor-mediated active targeting of nanoparticles to cancer cells, (c) cytosolic delivery and organelle-specific drug targeting through receptor-mediated endocytosis. Nanoparticles are engulfed by lysosomes followed by early and late endosomes. Degradation of nanoparticles takes place in lysosomes. A fraction of nanoparticles can be organelle targeted through endosomal escape. *Source*: Reprinted with permission from Elsevier article (Biswas & Torchilin, 2014.

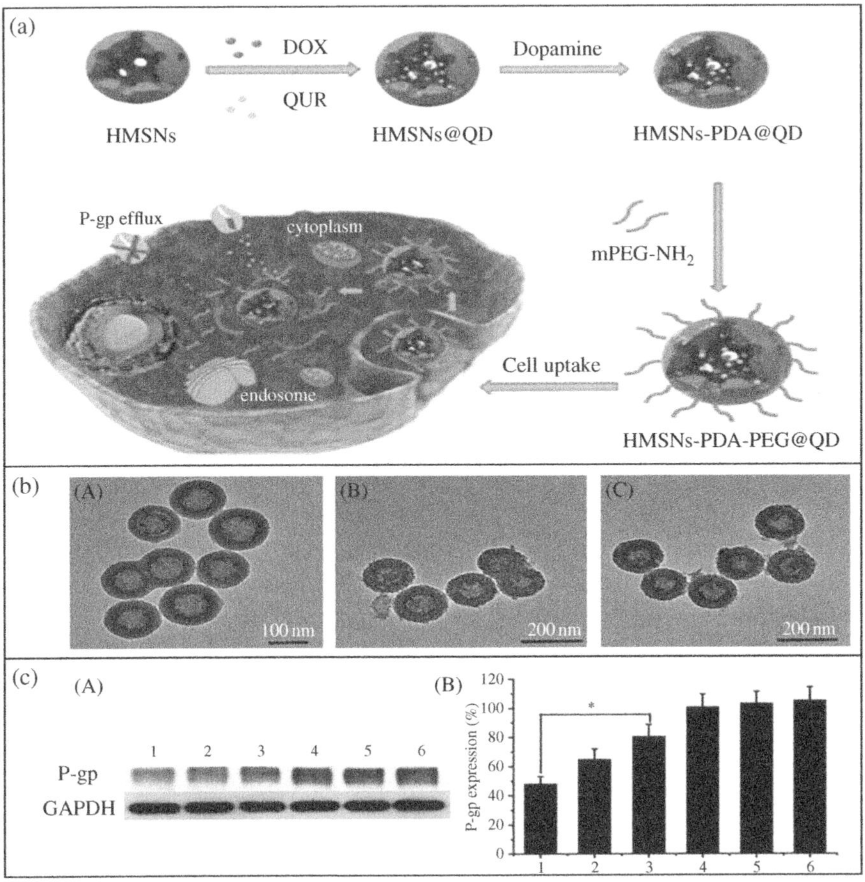


**Figure 5.2** (a) Schematic diagram of nanoparticle system encapsulated with doxorubicin hydrochloride (DOX) and quercetin (QUR) together called QD. QD-loaded hollow mesoporous silica nanoparticles (HMSNs) coated with methoxypolyethylene glycol amine (mPEG-NH2) functionalized by polydopamine (PDA) (HMSNs-PDA-PEG@QD). (b) TEM images of (A) HMSNs, (B) HMSNs-PDA, and (C) HMSNs-PDA-PEG. (c) Western blot analysis for P-gp expression in HCT-8/TAX cells: (A) Lane 1, HMSNs-PDA-PEG@QD (10 μg/ml); 2, HMSNs-PDA-PEG@QD (5 μg/ml); 3, free DOX + QUR; 4, HMSNs-PDA-PEG@DOX; 5, free DOX; 6, untreated control. (B) Relative P-gp expression (quantitative analysis). *Source*: Reprinted with permission from Elsevier article (Shao et al., 2019).

fluid through small vesicles is called pinocytosis while engulfment of microbes as well as cell debris and large particles is called phagocytosis (Kuhn et al., 2014). Cellular uptake through pinocytosis includes clathrin-mediated endocytosis, clathrin-independent endocytosis, caveolin-mediated endocytosis, caveolin-independent endocytosis, and macropinocytosis. NPs based on microvesicles and exosomes are also called extracellular vesicles (EVs). EVs are important for the transport of RNA, protein, and biological molecules endogenously among cells. EVs-based NPs' uptake via macropinocytosis depends upon phosphoinositide 3-kinase activity and $Na^+/H^+$ exchange while clathrin-independent endocytosis depends on tyrosine kinase activity as well as cholesterol. (Behzadi et al., 2014; X. Wei et al., 2019; Costa Verdera, Gitz-Francois, Schiffelers, & Vader, 2017). NPs' uptake can show enhanced accumulation of drugs through the EPR effect (passive targeting) or by receptor-mediated cellular uptake (active targeting). The organelle-specific drug targeting could be possible through endosomal escape from early endosomes to deliver their therapeutic payload. Cytosolic delivery of NPs could degrade in lysosomes. The NPs get engulfed by the early endosome vesicles and make their way to lysosomes via late endosome vesicles (Figure 5.2) (Biswas & Torchilin, 2014).

## 5.5 Challenges in Clinical Applications

The optimization of NPs' behavior under physiological condition is essential for successful clinical translation. in vitro assays in multiwell cell culture plates lack complexity associated with the physiological environment in tumor tissues. Most of the studies on NPs' performance have been done under in vivo conditions based on animal experimental data. However, the translational efficacy of NPs in humans remains uncharted. So the prediction of NPs' efficacy and biocompatibility based on the preclinical studies over animal models may be unusual with human trials. Several studies were performed at the preclinical and clinical level for understanding NPs' biodistribution and pharmacokinetics (Eliasof et al., 2013; Zuckerman et al., 2014). NPs' behavior can be modified in the complex condition under in vivo environment due to its surface charge, composition, and aggregation, and NPs' protein corona formation can persuade cellular uptake efficacy (Yue et al., 2011). Current studies based on the varied animal models include genetically engineered mouse models (GEMMs), orthotopic xenografts, subcutaneous models, and xenografts models derived from the patient. These models cannot fully reproduce all facets required for cancer treatment (Bertrand, Wu, Xu, Kamaly, & Farokhzad, 2014). Current research based on the 3D tumor spheroids, organ model, and tumor on chip-based model can mimic the physiological environment existing inside tumors (Albanese, Lam, Sykes, Rocheleau, & Chan, 2013; Kankala, Wang, & Chen, 2019; Lu et al., 2018; Toh et al., 2009). Another challenge for clinical translation is the scale-up process of NPs. Complex NPs require more complicated technologies for reproducible synthesis. Translation of the nanomedicine from preclinical to clinical level entails good manufacturing practice (GMP) as well as chemistry, manufacturing and controls (CMC) for making the materialistic/commercialized market product. There is also the possibility of batch-to-batch variation during nanoparticle synthesis. The newer technology based on particle replication in non-wetting template (PRINT) is helpful in the production of monodisperse NPs with ideal properties for their clinical action (J. Xu et al., 2013; Kim et al., 2012; Rhee et al., 2011; Beletskii, Galloway, Rele, Stone, & Malinoski, 2014).

## 5.6 Conclusion

NPs have shown immense potential for the construction of multimodal nanovehicles for cancer theranostics. This book chapter has explored the use of nanotherapeutics for cancer biology and its thriving translation at the clinical level. The physicochemical properties of NPs include size, surface charge, topography, chemical composition, and targeting moieties having a very important role in nano-bio interactions. NPs can be modified to modulate its biological impacts such as circulation time, protein interaction, biodistribution, EPR, and drug release at a specific tumor site. MDR in cancer cells is still one of the hurdles for successful clinical translation of anticancer drugs. To avoid the MDR-related pathways nanoparticle-mediated drug delivery holds excellent potential in specific tumor targeting and bypassing the drug efflux through MDR cells. Nanoformulated drugs can avoid drug transporters by following the different uptake mechanism for entering the target cells. Nanoparticle-mediated drug delivery can enhance drug accumulation by reversing the MDR process. Biocompatibility of NPs can be achieved by the conjugation/coating of different ligand molecules or post-surface modification. Although NPs bear versatile properties, another scientific advancement is also required for their clinical translation that includes identifying specific tumor markers, understanding of tumor microenvironment, and better tumor model. Successful translation of nanomedicine at the clinical level requires GMP and CMC protocol for making the commercialized product. Opportunities offered by cancer nanomedicines shortly can shift the standard of cancer theranostics that could be more prominent to enhance the patient survival.

### Take Home Message

- Nanosized vehicles could offer a better opportunity to interact with diseased target proteins by transporting drug molecules as well as biomacromolecules at the specific site.
- The in vitro 3D tumor model testing could be very helpful in filling the gaps in-between 2D culture and in vivo studies. It may avoid unnecessary animal killing.

- The precise testing of actively targeted nanoparticles is required for its possible consequences at the clinical level.
- Nanomaterials should be biocompatible. Some parameters should be analyzed before the nanomaterials' application such as oxidative stress, toxicity to human tissue, and inflammatory cytokine production.
- For the successful translation of nanoparticles at the clinical level, a good manufacturing practice (GMP), as well as chemistry, manufacturing, and controls (CMC) protocol, is compulsory.

## Questions and Answers

1 Explain the role of enhanced permeability and retention (EPR) effect for nanoparticles?
Enhanced permeability and retention (EPR) effect is a mechanism that happens due to pathophysiological processes. The abnormal growth of tumor cells resulting from leaky tumor vasculature and poor lymphatic drainage system leads to the higher accumulation of nanoparticles in tumor tissues as compared to normal tissues.

2 What are nano–bio interactions?
The fate of nanoparticle's interaction with the biological system such as lipid, protein, extracellular matrix (ECM), as well as in intracellular organelles is known as nano–bio interaction. This will describe the tumor localization, internalization routes, circulation half-life, retention time, and biodistribution of the nanoparticle.

3 What do you understand by the payload of nanoparticles?
The payload is the diagnostic or therapeutic agents loaded with nanoparticles.

4 What is the polydispersity index (PDI) of nanoparticles?
The nanoparticle size distribution and size range characterization can be defined by the polydispersity index (PDI). PDI value ranges from 0 to 1. Larger PDI values correspond with larger size distribution while PDI value close to 0.1 is considered monodisperse.

5 What are the roles of opsonins?
An opsonin is a plasma component that can enhance phagocytosis by generating an immune response. Opsonins include fibrinogen, immunoglobulins, and several complement proteins that can coat nanoparticles and cause their destruction.

**6** What is the circulation half-life time for nanoparticles?
The circulation half-life time for the nanoparticles is the period of time required to reduce one-half of its initial amounts in the blood.

**7** What is the role of the reticuloendothelial system (RES) in nanoparticle clearance?
The reticuloendothelial system (RES) is associated with the mononuclear phagocyte system (MPS). It is part of our immune system and includes macrophages and monocytes. Nanoparticles are usually taken up by the RES depending on its physiological properties such as size and surface charge. Nanoparticles mostly taken up by the macrophages reside in the liver (Kupffer cells) and spleen. Serum-coated nanoparticles get opsonized and phagocytize for destruction. More hydrophobic nanoparticles are preferentially cleared by the liver macrophages, followed by the spleen and the lungs.

## Acknowledgements

Author SPS would like to thank DBT-RA for post-doctoral fellowship. Financial support from the DBT-RA Program in Biotechnology and Life Sciences is gratefully acknowledged.

## References

Adams, M., Dogic, Z., Keller, S. L., & Fraden, S. (1998). Entropically driven microphase transitions in of colloidal rods and spheres. *Nature*. doi:10.1038/30700

Albanese, A., Lam, A. K., Sykes, E. A., Rocheleau, J. V., & Chan, W. C. W. (2013). Tumour-on-a-chip provides an optical window into nanoparticle tissue transport. *Nature Communications*. doi:10.1038/ncomms3718

Albanese, A., Sykes, E. A., & Chan, W. C. W. (2010). Rough around the edges: The inflammatory response of microglial cells to spiky nanoparticles. *ACS Nano*. doi:10.1021/nn100776z

Ali, H. S., Hanafy, A., & Alqurshi, A. (2019). Engineering of solidified glyburide nanocrystals for tablet formulation via loading of carriers: Downstream processing, characterization, and bioavailability. *International Journal of Nanomedicine*. doi:10.2147/ijn.s194734

Asati, A., Santra, S., Kaittanis, C., & Perez, J. M. (2010). Surface-charge-dependent cell localization and cytotoxicity of cerium oxide nanoparticles. *ACS Nano*. doi:10.1021/nn100816s

Baier, G., Costa, C., Zeller, A., Baumann, D., Sayer, C., Araujo, P. H. H., . . . Landfester, K. (2011). BSA adsorption on differently charged polystyrene nanoparticles using isothermal titration calorimetry and the influence on cellular uptake. *Macromolecular Bioscience*. doi:10.1002/mabi.201000395

Behzadi, S., Serpooshan, V., Sakhtianchi, R., Müller, B., Landfester, K., Crespy, D., & Mahmoudi, M. (2014). Protein corona change the drug release profile of nanocarriers: The "overlooked" factor at the nanobio interface. *Colloids and Surfaces B: Biointerfaces*. doi:10.1016/j.colsurfb.2014.09.009

Beletskii, A., Galloway, A., Rele, S., Stone, M., & Malinoski, F. (2014). Engineered PRINT® nanoparticles for controlled delivery of antigens and immunostimulants. *Human Vaccines and Immunotherapeutics*. doi:10.4161/hv.28817

Bertrand, N., Wu, J., Xu, X., Kamaly, N., & Farokhzad, O. C. (2014). Cancer nanotechnology: The impact of passive and active targeting in the era of modern cancer biology. *Advanced Drug Delivery Reviews*. doi:10.1016/j.addr.2013.11.009

Bhattacharjee, S., de Haan, L. H. J., Evers, N. M., Jiang, X., Marcelis, A. T. M., Zuilhof, H., . . . Alink, G. M. (2010). Role of surface charge and oxidative stress in cytotoxicity of organic monolayer-coated silicon nanoparticles towards macrophage NR8383 cells. *Particle and Fibre Toxicology*. doi:10.1186/1743-8977-7-25

Biswas, S., & Torchilin, V. P. (2014). Nanopreparations for organelle-specific delivery in cancer. *Advanced Drug Delivery Reviews*. doi:10.1016/j.addr.2013.11.004

Blanco, E., Shen, H., & Ferrari, M. (2015). Principles of nanoparticle design for overcoming biological barriers to drug delivery. *Nature Biotechnology*. doi:10.1038/nbt.3330

Byrne, J. D., Betancourt, T., & Brannon-Peppas, L. (2008). Active targeting schemes for nanoparticle systems in cancer therapeutics. *Advanced Drug Delivery Reviews*. doi:10.1016/j.addr.2008.08.005

Champion, J. A., & Mitragotri, S. (2009). Shape induced inhibition of phagocytosis of polymer particles. *Pharmaceutical Research*. doi:10.1007/s11095-008-9626-z

Chauhan, A. S., Diwan, P. V., Jain, N. K., & Tomalia, D. A. (2009). Unexpected in vivo anti-inflammatory activity observed for simple, surface functionalized poly(amidoamine) dendrimers. *Biomacromolecules*. doi:10.1021/bm9000298

Chung, Y. I., Kim, J. C., Kim, Y. H., Tae, G., Lee, S. Y., Kim, K., & Kwon, I. C. (2010). The effect of surface functionalization of PLGA nanoparticles by heparin- or chitosan-conjugated Pluronic on tumor targeting. *Journal of Controlled Release*. doi:10.1016/j.jconrel.2010.01.017

Cooper, S. P. (2011). *Nanocomposite strategies for limiting medical and marine biofouling*. ProQuest Dissertations and Theses.

Costa Verdera, H., Gitz-Francois, J. J., Schiffelers, R. M., & Vader, P. (2017). Cellular uptake of extracellular vesicles is mediated by clathrin-independent endocytosis and macropinocytosis. *Journal of Controlled Release*. doi:10.1016/j.jconrel.2017.09.019

Daum, N., Tscheka, C., Neumeyer, A., & Schneider, M. (2012). Novel approaches for drug delivery systems in nanomedicine: Effects of particle design and shape. *Wiley Interdisciplinary Reviews: Nanomedicine and Nanobiotechnology*. doi:10.1002/wnan.165

Doshi, N., & Mitragotri, S. (2010). Macrophages recognize size and shape of their targets. *PLoS ONE*. doi:10.1371/journal.pone.0010051

Ehrenberg, M. S., Friedman, A. E., Finkelstein, J. N., Oberdörster, G., & McGrath, J. L. (2009). The influence of protein adsorption on nanoparticle association with cultured endothelial cells. *Biomaterials*. doi:10.1016/j.biomaterials.2008.09.050

Eliasof, S., Lazarus, D., Peters, C. G., Case, R. I., Cole, R. O., Hwang, J., . . . Davis, M. E. (2013). Correlating preclinical animal studies and human clinical trials of a multifunctional, polymeric nanoparticle. *Proceedings of the National Academy of Sciences of the United States of America*. doi:10.1073/pnas.1309566110

Feng, W., Zhou, X., He, C., Qiu, K., Nie, W., Chen, L., . . . Zhang, Y. (2013). Polyelectrolyte multilayer functionalized mesoporous silica nanoparticles for pH-responsive drug delivery: Layer thickness-dependent release profiles and biocompatibility. *Journal of Materials Chemistry B*. doi:10.1039/c3tb21193b

Fröhlich, E. (2012). The role of surface charge in cellular uptake and cytotoxicity of medical nanoparticles. *International Journal of Nanomedicine*. doi:10.2147/IJN.S36111

Geng, Y., Dalhaimer, P., Cai, S., Tsai, R., Tewari, M., Minko, T., & Discher, D. E. (2007). Shape effects of filaments versus spherical particles in flow and drug delivery. *Nature Nanotechnology*. doi:10.1038/nnano.2007.70

Goodman, C. M., McCusker, C. D., Yilmaz, T., & Rotello, V. M. (2004). Toxicity of gold nanoparticles functionalized with cationic and anionic side chains. *Bioconjugate Chemistry*. doi:10.1021/bc049951i

Gottesman, M. M., Fojo, T., & Bates, S. E. (2002). Multidrug resistance in cancer: Role of ATP–dependent transporters. *Nature Reviews Cancer*. doi:10.1038/nrc706

Guo, P., Liu, D., Subramanyam, K., Wang, B., Yang, J., Huang, J., . . . Moses, M. A. (2018). Nanoparticle elasticity directs tumor uptake. *Nature Communications*. doi:10.1038/s41467-017-02588-9

Herd, H., Daum, N., Jones, A. T., Huwer, H., Ghandehari, H., & Lehr, C. M. (2013). Nanoparticle geometry and surface orientation influence mode of cellular uptake. *ACS Nano*. doi:10.1021/nn304439f

Iaccarino, G., Profeta, M., Vecchione, R., & Netti, P. A. (2019). Matrix metalloproteinase-cleavable nanocapsules for tumor-activated drug release. *Acta Biomaterialia*. doi:10.1016/j.actbio.2019.02.043

Ji, P., Yu, T., Liu, Y., Jiang, J., Xu, J., Zhao, Y., . . . Wu, C. (2016). Naringenin-loaded solid lipid nanoparticles: Preparation, controlled delivery, cellular uptake, and pulmonary pharmacokinetics. *Drug Design, Development and Therapy*. doi:10.2147/DDDT.S97738

Kah, J. C. Y., Yeo, E. L. L., Koh, W. L., Poinard, B. E. A., & Neo, D. J. H. (2014). Nanoparticle interface to biology: Applications in probing and modulating biological processes. *Critical Reviews in Biomedical Engineering*. doi:10.1615/critrevbiomedeng.2014010490

Kalyane, D., Raval, N., Maheshwari, R., Tambe, V., Kalia, K., & Tekade, R. K. (2019). Employment of enhanced permeability and retention effect (EPR): Nanoparticle-based precision tools for targeting of therapeutic and diagnostic agent in cancer. *Materials Science and Engineering C*. doi:10.1016/j.msec.2019.01.066

Kankala, R. K., Wang, S.-B., & Chen, A.-Z. (2019). Microengineered organ-on-a-chip platforms towards personalized medicine. *Current Pharmaceutical Design*. doi:10.2174/1381612825666190222143542

Kim, Y., Lee Chung, B., Ma, M., Mulder, W. J. M., Fayad, Z. A., Farokhzad, O. C., & Langer, R. (2012). Mass production and size control of lipid–polymer hybrid nanoparticles through controlled microvortices. *Nano Letters*. doi:10.1021/nl301253v

Kuhn, D. A., Vanhecke, D., Michen, B., Blank, F., Gehr, P., Petri-Fink, A., & Rothen-Rutishauser, B. (2014). Different endocytotic uptake mechanisms for nanoparticles in epithelial cells and macrophages. *Beilstein Journal of Nanotechnology*. doi:10.3762/bjnano.5.174

Lee, J., Oh, E. T., Song, J., Kim, H. G., Park, H. J., & Kim, C. (2017). Stimulus-induced conformational transformation of a cyclic peptide for selective cell-targeting on–off gatekeeper for mesoporous nanocarriers. *Chemistry – An Asian Journal*. doi:10.1002/asia.201701050

Lepor, H. (2008). Vascular targeted photodynamic therapy for localized prostate cancer. *Reviews in Urology*, 10(4), 254–261.

Li, L., & Neaves, W. B. (2006). Normal stem cells and cancer stem cells: The niche matters. *Cancer Research*. doi:10.1158/0008-5472.CAN-05-3986

Li, W., Li, X., Gao, Y., Zhou, Y., Ma, S., Zhao, Y., . . . Yin, D. (2014). Inhibition mechanism of *P*-glycoprotein mediated efflux by mPEG-PLA and influence of PLA chain length on *P*-glycoprotein inhibition activity. *Molecular Pharmaceutics*. doi:10.1021/mp4004223

Liu, J., Zhao, Y., Guo, Q., Wang, Z., Wang, H., Yang, Y., & Huang, Y. (2012). TAT-modified nanosilver for combating multidrug-resistant cancer. *Biomaterials*. doi:10.1016/j.biomaterials.2012.05.035

Liu, X., Wu, F., Tian, Y., Wu, M., Zhou, Q., Jiang, S., & Niu, Z. (2016). Size dependent cellular uptake of rod-like bionanoparticles with different aspect ratios. *Scientific Reports*. doi:10.1038/srep24567

Lu, S., Cuzzucoli, F., Jiang, J., Liang, L. G., Wang, Y., Kong, M., . . . Wang, S. Q. (2018). Development of a biomimetic liver tumor-on-a-chip model based on decellularized liver matrix for toxicity testing. *Lab on a Chip*. doi:10.1039/C8LC00852C

Maeda, H. (2015). Toward a full understanding of the EPR effect in primary and metastatic tumors as well as issues related to its heterogeneity. *Advanced Drug Delivery Reviews*. doi:10.1016/j.addr.2015.01.002

Nafee, N., Schneider, M., Schaefer, U. F., & Lehr, C. M. (2009). Relevance of the colloidal stability of chitosan/PLGA nanoparticles on their cytotoxicity profile. *International Journal of Pharmaceutics*. doi:10.1016/j.ijpharm.2009.04.049

Niu, Y., Yu, M., Hartono, S. B., Yang, J., Xu, H., Zhang, H., . . . Yu, C. (2013). Nanoparticles mimicking viral surface topography for enhanced cellular delivery. *Advanced Materials*. doi:10.1002/adma.201302737

Oh, W. K., Kim, S., Choi, M., Kim, C., Jeong, Y. S., Cho, B. R., . . . Jang, J. (2010). Cellular uptake, cytotoxicity, and innate immune response of silica-Titania hollow nanoparticles based on size and surface functionality. *ACS Nano*. doi:10.1021/nn100561e

Parhi, P., Mohanty, C., & Sahoo, S. K. (2012). Nanotechnology-based combinational drug delivery: An emerging approach for cancer therapy. *Drug Discovery Today*. doi:10.1016/j.drudis.2012.05.010

Patil, S., Sandberg, A., Heckert, E., Self, W., & Seal, S. (2007). Protein adsorption and cellular uptake of cerium oxide nanoparticles as a function of zeta potential. *Biomaterials*. doi:10.1016/j.biomaterials.2007.07.029

Petri-Fink, A., Steitz, B., Finka, A., Salaklang, J., & Hofmann, H. (2008). Effect of cell media on polymer coated superparamagnetic iron oxide nanoparticles (SPIONs): Colloidal stability, cytotoxicity, and cellular uptake studies. *European Journal of Pharmaceutics and Biopharmaceutics*. doi:10.1016/j.ejpb.2007.02.024

Rhee, M., Valencia, P. M., Rodriguez, M. I., Langer, R., Farokhzad, O. C., & Karnik, R. (2011). Synthesis of size-tunable polymeric nanoparticles enabled by 3D hydrodynamic flow focusing in single-layer microchannels. *Advanced Materials*. doi:10.1002/adma.201004333

Saroj, S., & Rajput, S. J. (2018). Etoposide encased folic acid adorned mesoporous silica nanoparticles as potent nanovehicles for enhanced prostate cancer therapy: Synthesis, characterization, cellular uptake and biodistribution. *Artificial Cells, Nanomedicine and Biotechnology*. doi:10.1080/21691401.2018.1533843

Shao, M., Chang, C., Liu, Z., Chen, K., Zhou, Y., Zheng, G., . . . Lu, B. (2019). Polydopamine coated hollow mesoporous silica nanoparticles as pH-sensitive nanocarriers for overcoming multidrug resistance. *Colloids and Surfaces B: Biointerfaces*. doi:10.1016/j.colsurfb.2019.110427

Siemann, D. W. (2011). The unique characteristics of tumor vasculature and preclinical evidence for its selective disruption by tumor-vascular disrupting agents. *Cancer Treatment Reviews*. doi:10.1016/j.ctrv.2010.05.001

Sodani, K., Patel, A., Kathawala, R. J., & Chen, Z. S. (2012). Multidrug resistance associated proteins in multindrung resistance. *Chinese Journal of Cancer*. doi:10.5732/cjc.011.10329

Suzuki, M., Hori, K., Abe, I., Sato, H., & Saito, S. (1981). A new approach to cancer chemotherapy: Selective enhancement of tumor blood flow with angiotensin ii. *Journal of the National Cancer Institute*. doi:10.1093/jnci/67.3.663

Sykes, E. A., Dai, Q., Sarsons, C. D., Chen, J., Rocheleau, J. V., Hwang, D. M., . . . Chan, W. C. W. (2016). Tailoring nanoparticle designs to target cancer based on tumor pathophysiology. *Proceedings of the National Academy of Sciences*. doi:10.1073/pnas.1521265113

Tahmasbi Rad, A., Chen, C. W., Aresh, W., Xia, Y., Lai, P. S., & Nieh, M. P. (2019). Combinational effects of active targeting, shape, and enhanced permeability and retention for cancer theranostic nanocarriers. *ACS Applied Materials and Interfaces*. doi:10.1021/acsami.8b21609

Tang, S., Meng, Q., Sun, H., Su, J., Yin, Q., Zhang, Z., . . . Li, Y. (2017). Dual pH-sensitive micelles with charge-switch for controlling cellular uptake and drug release to treat metastatic breast cancer. *Biomaterials*. doi:10.1016/j.biomaterials.2016.06.005

Toh, Y. C., Lim, T. C., Tai, D., Xiao, G., Van Noort, D., & Yu, H. (2009). A microfluidic 3D hepatocyte chip for drug toxicity testing. *Lab on a Chip*. doi:10.1039/b900912d

Walczyk, D., Bombelli, F. B., Monopoli, M. P., Lynch, I., & Dawson, K. A. (2010). What the cell "sees" in bionanoscience. *Journal of the American Chemical Society*. doi:10.1021/ja910675v

Wang, S.-B., Ma, Y.-Y., Chen, X.-Y., Zhao, Y.-Y., & Mou, X.-Z. (2019). Ceramide-graphene oxide nanoparticles enhance cytotoxicity and decrease HCC xenograft development: A novel approach for targeted cancer therapy. *Frontiers in Pharmacology*. doi:10.3389/fphar.2019.00069

Wei, Q., Huang, C., Zhang, Y., Zhao, T., Zhao, P., Butler, P., & Zhang, S. (2018). Mechanotargeting: Mechanics-dependent cellular uptake of nanoparticles. *Advanced Materials*. doi:10.1002/adma.201707464

Wei, X., Wei, R., Jiang, G., Jia, Y., Lou, H., Yang, Z., . . . Gao, Q. (2019). Mechanical cues modulate cellular uptake of nanoparticles in cancer via clathrin-mediated and caveolae-mediated endocytosis pathways. *Nanomedicine*. doi:10.2217/nnm-2018-0334

Wei, Z., Yuan, S., Hao, J., & Fang, X. (2013). Mechanism of inhibition of P-glycoprotein mediated efflux by Pluronic P123/F127 block copolymers: Relationship between copolymer concentration and inhibitory activity. *European Journal of Pharmaceutics and Biopharmaceutics*. doi:10.1016/j.ejpb.2012.09.014

Wu, L., Zong, L., Ni, H., Liu, X., Wen, W., Feng, L., . . . Shen, S. (2019). Magnetic thermosensitive micelles with upper critical solution temperature for NIR triggered drug release. *Biomaterials Science*. doi:10.1039/c8bm01672k

Xu, A., Yao, M., Xu, G., Ying, J., Ma, W., Li, B., & Jin, Y. (2012). A physical model for the size-dependent cellular uptake of nanoparticles modified with cationic surfactants. *International Journal of Nanomedicine*. doi:10.2147/IJN.S32188

Xu, J., Wong, D. H. C., Byrne, J. D., Chen, K., Bowerman, C., & Desimone, J. M. (2013). Future of the particle replication in nonwetting templates (PRINT) technology. *Angewandte Chemie – International Edition*. doi:10.1002/anie.201209145

Xu, W., Qian, J., Hou, G., Wang, Y., Wang, J., Sun, T., . . . Yao, Y. (2018). PEGylated hydrazided gold nanorods for pH-triggered chemo/photodynamic/photothermal triple therapy of breast cancer. *Acta Biomaterialia*. doi:10.1016/j.actbio.2018.10.019

Yamamoto, Y., Nagasaki, Y., Kato, Y., Sugiyama, Y., & Kataoka, K. (2001). Long-circulating poly(ethylene glycol)-poly(d,l-lactide) block copolymer micelles with modulated surface charge. *Journal of Controlled Release*. doi:10.1016/S0168-3659(01)00451-5

Yuan, Y., Cai, T., Xia, X., Zhang, R., Chiba, P., & Cai, Y. (2016). Nanoparticle delivery of anticancer drugs overcomes multidrug resistance in breast cancer. *Drug Delivery*. doi:10.1080/10717544.2016.1178825

Yue, Z. G., Wei, W., Lv, P. P., Yue, H., Wang, L. Y., Su, Z. G., & Ma, G. H. (2011). Surface charge affects cellular uptake and intracellular trafficking of chitosan-based nanoparticles. *Biomacromolecules*. doi:10.1021/bm101482r

Zhang, H. Y., Sun, C. Y., Adu-Frimpong, M., Yu, J., & Xu, X. (2019). Glutathione-sensitive PEGylated curcumin prodrug nanomicelles: Preparation, characterization, cellular uptake and bioavailability evaluation. *International Journal of Pharmaceutics*. doi:10.1016/j.ijpharm.2018.11.049

Zhen, Z., Tang, W., Chuang, Y. J., Todd, T., Zhang, W., Lin, X., . . . Xie, J. (2014). Tumor vasculature targeted photodynamic therapy for enhanced delivery of nanoparticles. *ACS Nano*. doi:10.1021/nn501134q

Zhong, L., Xu, L., Liu, Y., Li, Q., Zhao, D., Li, Z., . . . He, Z. (2019). Transformative hyaluronic acid-based active targeting supramolecular nanoplatform improves long circulation and enhances cellular uptake in cancer therapy. *Acta Pharmaceutica Sinica B*. doi:10.1016/j.apsb.2018.11.006

Zuckerman, J. E., Gritli, I., Tolcher, A., Heidel, J. D., Lim, D., Morgan, R., . . . Yen, Y. (2014). Correlating animal and human phase Ia/Ib clinical data with CALAA-01, a targeted, polymer-based nanoparticle containing siRNA. *Proceedings of the National Academy of Sciences*. doi:10.1073/pnas.1411393111

## Glossary

**Nanoparticle:** Particle with size ranging from 1 and 100 nm.

**Nanotechnology:** Technology dealing with matter on an atomic or molecular level.

**Nanomedicine:** Nanotechnology employed for the treatment and prevention of diseases.

**Bioimaging:** The methods that help to envisage the biological processes in real time.

**Biosensors:** Biosensors comprise the three different components a reader device, transducer, and an analyte recognizing component. It can detect the presence of biological analytes.

**Theranostics:** Theranostics is a specific targeted therapy based on a specific targeted treatment and diagnostic.

**Nano–bio interfaces:** Nanomaterials provide an ideal interface to react with biological systems.

**Enhanced permeability and retention (EPR) effect:** Leaky tumor vasculature and poor lymphatic drainage system consequential from the abnormal tumor growth leads to the higher accumulation of nanoparticles in tumor tissues as compared to normal tissues.

**Multidrug resistance:** Resistance shown by cancerous cells against multiple anticancer drugs.

**Triggered drug release:** The drug release in the presence of external stimuli.

**Targeting agents:** Targeting agents are ligand molecule used for the specific uptake (i.e. folic acid specifically binds with the folate receptors overexpressed on cancer cells).

**Biodistribution:** This is the method of tracking drug molecules/nanoparticles of interest in different organs of the experimental animal.

**Tumor markers:** These are the substances made by the cells in response to cancer.

**Tumor microenvironment:** The environment in the region of a tumor such as extracellular matrix, blood vessels, and fibroblasts cells.

**Cancer stem cells:** Cancer stem cells are totipotent cells that can form all cell types found in the cancer sample.

**Endocytosis:** The intracellular uptake of nanomaterials in living cells by the invagination of the cell membrane.

**Proliferation:** The rapid cell division of a cell, tissue, or organism.

**Therapeutic index:** This is a comparative measurement of the virtual toxicity and safety of a drug.

**Topography:** Features present on the surface.

**Systemic toxicity:** The toxicity associated with healthy cells due to the failure of an array of biological pathways.

**Cytotoxicity:** A quantity of cellular toxicity in the presence of any substance.

**Hydrophobicity:** The physical property of any molecule that enables it to repel a mass of water.

**Phagocytosis:** This is the phenomenon associated with the engulfment of any particle resulting in the formation of the phagosome.

**Pinocytosis:** The cell drinking process due to the ingestion of liquid by the cells.

**Thermosensitive:** Temperature-sensitive nanomaterials for control drug delivery.

**Xenografts:** Organ transplantation from a donor of a dissimilar species.

**Good manufacturing practice:** The production and control of the products according to quality standards.

# 6

# Protein Nanoparticle Interactions and Factors Influencing These Interactions

*R. Mala and R. Keerthana*

*Department of Biotechnology, MepcoSchlenk Engineering College, Sivakasi, Tamil Nadu, India*

## Objectives

- Understand the structure and composition of protein corona.
- Understand the influence of various factors on protein corona formation.
- Understand the cellular uptake of protein corona.
- Summarize the techniques used to study protein corona.

## 6.1 Introduction

Nanomaterial is "a material containing particles either natural or anthropogenic, present bound or in free form where more than 50% of particles are less than 100 nm in any one dimension" (European Union Law, 2011).When the size of materials is confined to the nanoscale at least in one dimension, their energy, optical properties, magnetic properties, and chemical reactivity are radically different and unique. Miniaturization of materials to the nanoscale offers scientists and researchers innumerable possibilities to apply them in all fields of science and technology. Especially in the biomedical field, nanomaterials are used in therapy and diagnostics. The physicochemical properties of materials are manipulated and controlled to form sensitive, specific, and selective candidate for bio assays. These are used in the early detection of heritable and pathogenic diseases at low cost (Michalet et al., 2005; Rosi & Mirkin, 2005). Nanopatterns created by different types of lithography are used in the diagnosis at the molecular level in high throughput assay (Ngunjiri, Daniels, Li, Serem, & Garno, 2008). Furthermore

*Microbial Interactions at Nanobiotechnology Interfaces: Molecular Mechanisms and Applications*, First Edition. Edited by R. Navanietha Krishnaraj and Rajesh K. Sani.

the compatibility of nanomaterials is tailored by functionalization with various biomolecules such as DNA (Shete, Muniswamy, Pandit, & Khandelwal, 2015), proteins (Liao et al., 2015, and polysaccharides (Akbarzadeh et al., 2012). Engineered nanomaterials are used in all fields of technology including aerospace, civil, aviation, energy, batteries, environment, medicine, and healthcare.

## 6.2 Types and Biomedical Application of Nanoparticles

According to their chemical nature, nanomaterials are classified into organic (dendrimers, liposomes etc.) and inorganic materials (metal nanoparticles and carbon-based nanoparticles) as represented in Figure 6.1.

Dynamic properties of nanomaterials like reactivity, catalytic power, and thermal and electrical conductivity are exploited for drug delivery and therapy. Nanomaterials are preferably used as contrast agents in MRI due to their biocompatibility, lack of toxicity, and minimum time required to image and discriminate diseased and normal tissues. Ferrite, cobalt, and manganese-doped ferrites are used as contrast

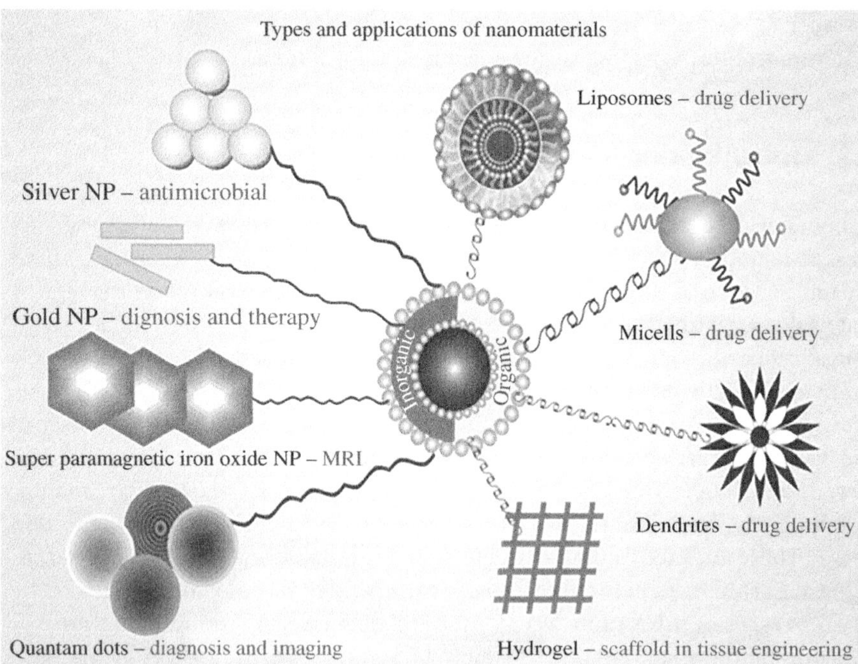


**Figure 6.1** Schematic showing nanomaterials' classification and their biomedical application.

agents in MRI. The hyperthermic property of ferrites is exploited in cancer therapy (Amiri & Shokrollahi, 2013; Sabale et al., 2015). Gold nanoparticles are also used in imaging (Meir et al., 2015), photothermal therapy (Rengan et al., 2015), and in biosensors (G. Wang, Papasani, Cheguru, Hrdlicka, & Hill, 2012). Bionanoscience exploits the high aspect ratio of nanoparticles and uses it to access highly restricted sites like brain and other intracellular machinery. Nanoparticles are used as a carrier in gene therapy, tissue engineering, and stem cell therapy.

## 6.3 Methods and Mechanisms of Nanomaterials Synthesis

Nanomaterials are synthesized in many different forms like clusters, powders, and films. Physical, chemical, and biological methods are adopted for the biosynthesis of nanomaterials. Physical methods include high-energy ball milling, melt mixing, physical vapor deposition, sputtering, laser ablation, etc. Chemical methods include sol–gel synthesis, colloidal synthesis, inverse micelle formation, etc. (Shah et al., 2019). Biological methods include the use of bacteria, fungi, yeast, plant (Anantha Suganiya & Mala, 2014) blue green algae, foley catheter (Mala et al., 2017), enzymes, and biomolecules. Few methods like electrochemical methods and chemical vapor deposition are hybrid methods.

Among the various methods available for the synthesis of nanomaterials, green methods using biological materials are preferable due to their biocompatibility, ease of preparation and lack of hazardous chemicals. The applications of noble metal nanoparticles are diverse and wide. They find their application in nanomedicine, diagnostic, medical devices, electronics, photonics, sensors, imaging, catalysis fuels, etc. (Daraee et al., 2016; Rana et al., 2018). Biological methods are operated under normal temperature, pH, and pressure. The materials produced by the methods are highly reactive and biocompatible to cellular mechanisms (Bhattacharya & Mukherjee, 2008).

Microbe-mediated synthesis of nanoparticles proceeds either by intracellular or extracellular modes. In the intracellular synthesis, bulkmaterials are uptaken through the cell wall and then reduced to nanomaterials. In extracellular mode, the enzymes are released into the medium. *Corynebacterium glutamicum* (Sneha, Sathishkumar, Mao, Kwak, & Yun, 2010), *Rhodococcus* sp. *Escherichia coli, Pseudomonas aeruginosa, Brevibacterium casei* (Kalishwaralal et al., 2010), *Bacillus licheniformi* (Kalimuthu, Suresh Babu, Venkataraman, Bilal, & Gurunathan, 2008), and *Bacillus cereus* are some of the bacteria used to synthesize metal nanoparticles. Among fungi, *Aspergillus fumigatus* (Bhainsa & D'Souza, 2006), *Aspergillus flavus* (Vigneshwaran et al., 2007), *Trichoderma viride,* and *Fusarium oxysporum* (Senapati, Ahmad, Khan, Sastry, & Kumar, 2005) are were reported in the synthesis of nanoparticles.

Many different mechanisms were proposed to explain the microbe-mediated biosynthesis of metal nanoparticles.Yin et al. (2016) proposed that microbial cells either entrap or bind at the cell surface with the metals and reduce them by enzymes. Mukherjee et al. (2002) suggested that metals are attracted electrostatically to the negatively charged carboxylate ions on the fungal cell walls, which are further reduced to form a nucleus. Further accumulation and reduction increase the synthesis of metal nanoparticles. The key role of reducing agents like NADH and the enzyme nitrate reductase in *B. licheniformis* in the reduction and biosynthesis of silver nanoparticles was demonstrated by Duran et al. (2011). Another hypothesis based on pH was proposed by Sintubin et al. (2009) that proposes the competition between protons and metal ions for binding to enzymes. Few studies reported that the reduction of metals is catalyzed by quinine-based shuttle systems. The role of peptides in the reduction of metal nanoparticles was documented. Peptides with sulfur-containing amino acids like methionine and cystein and basic amino acids like lysine and arginine were reported by Selvakannan et al. (2004). Similarly, few studies recorded the role of aromatic amino acids like tyrosine (Shankar et al., 2015) and tryptophane in the biosynthesis of nanoparticles. Fungi-mediated synthesis of nanoparticles produces higher amount of nanoparticles. They are stabilized by proteins and enzymes secreted outside the fungi cells (Shah et al., 2019). Nitrate reductase secretion and synthesis of nanoparticles by *F. oxysporum* was reported (Senapati et al., 2005). Devi and Joshi (2012) reported the intracellular synthesis of gold and silver nanoparticles by *Verticillium*. Exopolysaccharides (EPSs) secreted by microbes are used in the biosynthesis of metal nanoparticles. The carbohydrates (xanthun, dextran, rhamnose, etc.) present in the EPS serve as the reducing, capping, and stabilizing agents in the biosynthesis of nanoparticles (Yan et al., 2018). Nanoparticles synthesized by green methods are highly stable. The colloidal stability of the nanoparticles can be tailored based on the requirements by complexation or surface functionalization techniques (Sharma, Siskova, Zboril, & Gardea-Torresdey, 2014).

## 6.4 Routes of Entry of Nanoparticles into Biological System

Nanoparticles enter into body through different routes like inhalation, injection, and dermal exposure. Once the nanoparticle is exposed to the biological fluids, the surface energy is quenched by the interaction with proteins. The concept of protein corona was first proposed by Cedervall et al. (2007). The study reported that proteins do not cover the nanoparticles as a layer but as a shell forming corona. The structure and composition of protein corona vary based on the route of entry into the body. For example, the structure and composition of protein corona on nanoparticles entering the body through the respiratory system interact with the proteins, lipids, and surfactants of lungs (Kapralov et al., 2012).

Nanoparticles entering through blood stream will be covered by clotting proteins, apolipoproteins, immunoglobulins, complements, etc. to form a corona on their surface (Tenzer et al., 2011). It creates a bio-nano surface where the binding of proteins to the surface of the nanoparticles is influenced by the chemical composition, size, and surface characteristics of nanoparticles (Maiolo, Bergese, Mahon, Dawson, & Monopoli, 2014; Tenzer et al., 2013) and the confirmation, abundance, and binding affinity of protein in the microenvironment. Shielding the surface of the nanoparticles by proteins masks the original identity of nanoparticles; changes their size, surface charge, and reactivity; and endows them with a new biological identity (Ahsan, Rao, & Ahmad, 2018). Cells perceive only the protein corona and not the pristine nanoparticles (Ahsan et al., 2018).

## 6.5 Rationale for Studying Nanoparticles–Protein Interactions

Nanoparticles are widely studied for their application in diagnosis and therapy. They are engineered to fit the application. The protein corona changes continuously as it traverses through the biological system. They determine the properties of nanomaterials (Tenzer et al., 2013). The behavior of pristine nanoparticles is different from nanoparticles within the protein shell. The metabolic fate and elimination of nanoparticles covered by corona differ significantly from pristine nanoparticles. As nanoparticles are used in theranostics as a carrier, understanding the fundamental principle behind the mechanism and rate of formation of corona is of paramount significance. Nanomaterials intended for therapeutic approach enter the body via injection and they are encountered by plasma proteins, which determine their uptake, distribution, and clearance (Aggarwal, Hall, McLeland, Dobrovolskaia, & McNeil, 2009). The corona is termed as adsorbome and it is dynamic (Walkey & Chan, 2012). Major proteins that dominate the adsorbome are haptoglobin, anti-trypsin, plasminogen, etc. (Tenzer et al., 2011). Protein corona determines the distribution and internalization of nanoparticles inside the biological system (Treuel, Jiang, & Nienhaus, 2013). Fundamental knowledge about the structure, composition, and stability of corona is essential to study the pharmacokinetics of nanoparticles.

## 6.6 Formation of Protein Corona

### 6.6.1 Structure and Composition of Corona

Protein corona on nanoparticle is distinguished as soft or hard corona as illustrated in Figure 6.2. The structure of protein corona consists of top-most shell of soft layer formed of weakly associated proteins termed soft corona. Internal to the

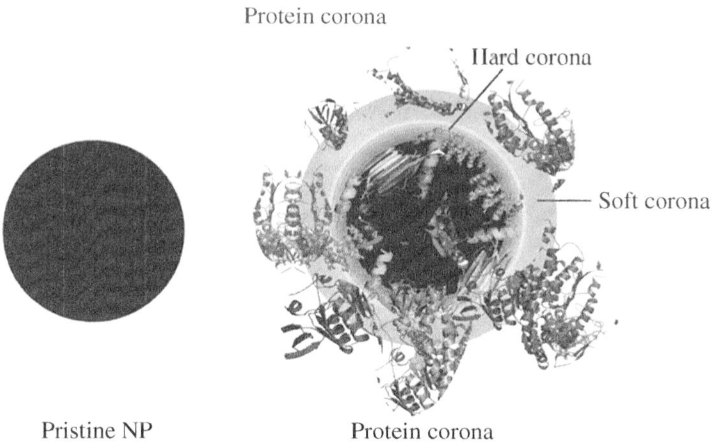


**Figure 6.2** Types of protein corona.

soft corona is a strongly bound monolayer of proteins, which are in direct contact with the nanoparticles. Protein corona is soft or hard depending on the affinity of the proteins. Proteins with low affinity make soft corona and those with high binding affinity form hard corona (Ahsan et al., 2018). Soft corona is comparatively less stable than hard corona. As hard corona is stable, the interaction of nanomaterials with the microenvironment and cells is governed by the hard corona (Maiolo et al., 2014).

The adsorption and desorption of proteins from corona are determined by thermodynamic principles. Proteins with less free energy of adsorption are rapidly desorbed and vice versa (Walkey & Chan, 2012). The composition of protein corona changes continuously and it is termed as Vroman effect (Vroman, Adams, Fischer, & Munoz, 1980). Low-affinity and high-mobility proteins (albumin, fibrinogen, and immunoglobulin) dominate the protein corona at the initial stage of corona formation, which vary from seconds to minutes. Weakly bound proteins are replaced by high-affinity, low-mobility proteins (clotting factors and apolipoprotein) at later stages of corona formation, which vary from hours to days (Casals, Pfaller, Duschl, Oostingh, & Puntes, 2010; Tenzer et al., 2011). Nanoparticles covered by corona made up of IgG and C3 are rapidly cleared from the cells (Nagayama, Ogawara, Fukuoka, Higaki, & Kimura, 2007). Anionic, cationic, and hydrophobic nanomaterials interact strongly with proteins than neutral and hydrophilic nanomaterials. Some proteins assist nanomaterials in their transport and others in their metabolic elimination (Owens 3rd & Peppas, 2006).

Protein corona formation is driven by entropy/enthalpy. Entropy-driven formation occurs via the elimination of water from the surface. Albumin binding to nanomaterials is driven by enthalpy. Binding of lysozyme, fibrinogen, etc. is driven by entropy (Ahsan et al., 2018). Albumin binding to nanomaterials is driven by enthalpy. Proteins undergo conformational change upon binding with the nanoparticle, which exposes the hidden core of the proteins (Nel et al., 2009). Proteins in the corona are always in dynamic state of adsorption and desorption (Donahue, Acar, & Wilhelm, 2019). During the initial stages of corona formation, competition between proteins and nanoparticles is a common phenomenon. In the advanced stages of corona formation, proteins inside the corona compete with the proteins in the surrounding medium. This is the driving force for the formation of late-stage corona layer that is dominated by low abundance proteins. Polystyrene nanoparticles functionalized with sulfonated groups and carboxyl groups were covered by multilayer of transferin (Milani, Bombelli, Pitek, Dawson, & Radler, 2012). The type of protein on the corona is governed not only by charge and electrostatic attraction but also by the chemical nature of nanoparticle. For example, the type of corona was similar between titanium and silica nanoparticles. Titanium and silica nanoparticles were enriched with clusterin, apolipoprotein D, and alpha-2-acid glycoprotein. The corona on zinc oxide nanoparticles was totally distinct and made up of transferrin, Ig heavy chain, and haptoglobin (Deng, Liang, Toth, Monteiro, & Minchin, 2012). Corona composition differs with the crystalline and amorphous nature of nanoparticles. Hydrogel nanoparticles have less affinity to proteins and the density of proteins is less compared to solid nanoparticles. The phenomenon was confirmed in poly(*N*-isopropylacrylamide) hydrogel and carboxylated polystyrene nanoparticles (O'Brien & Shea, 2016). Monopoli, Aberg, Salvati, and Dawson (2012) studied the protein corona on 63 different nanoparticles and identified the binding of 125 unique proteins in the protein corona. The most dominant proteins independent of physical and chemical properties of nanoparticle are those involved in foreign body recognition, clotting cascade, and lipoprotein metabolism. Early stage proteins like albumin and immunoglobulin are progressively replaced by fibrinogen, which is also replaced by clotting proteins. The study demonstrated that these proteins are substituted by trypsin inhibitor and apolipoproteins.

### 6.6.2 Kinetics of Formation of Nanoparticles–Corona

Kinetics of corona formation vary from minutes to days. Initiation of protein corona formation is very fast and takes place within few minutes. Pisani et al. (2017) observed instantaneous coating of nanoparticles by proteins on exposure to serum. The study recorded the adsorption of 134 proteins on to core shell nanoparticle made up of magnetic core and silica shell. Variation in protein composition was observed till seven days. In the first few seconds, highly abundant proteins of serum like albumin predominantly bind on to the surface of nanoparticles followed by transferrin, $\alpha$-1 antitrypsin, and hemopexin. IgG constant chain from heavy chain and light chain (kappa and lambda) were the dominant immunoglobulin chains adsorbed within 30 seconds. The proteins that bind after 30 seconds constitute hard corona composed of apolipoproteins, complements, and coagulation proteins. The adsorption followed Vroman effect. Dynamics of soft corona vary significantly with time of exposure to the serum, the abundance and affinity of the protein to the nanoparticle. Least abundant proteins form corona on nanoparticles within few minutes to hours. Peak abundance of proteins in PEGylated liposome was recorded at one hour. A2 macroglobulin showed its peak at 10 minutes, ApoE at 30 minutes, and hemoglobin at 3 hours. Vroman et al. (1980) proposed that the as time of interaction between nanoparticles and serum proteins increases, high-abundant and low-affinity proteins are substituted by low-abundant, high-affinity proteins. The effect is described as Vroman effect (Vroman et al., 1980).

Pisani et al. (2017) has used computational tools, protein–protein interaction database, and kinetic behavior of protein corona to explain the corona interactome. The study evaluated the interaction between the corona proteins independent of the characteristics of nanoparticles. Two different kinetic models are available for explaining the formation of protein corona. Coarse grained atomic simulation model is specific for the study of dynamic soft corona (Sakulkhu, Mahmoudi, Maurizi, Salaklang, & Hofmann, 2014) and the mean field model for the study of hard corona (Angioletti-Uberti, Ballauff, & Dzubiella, 2014). Zhdanov and Cho (2016) described the kinetic mean model for the description of hard corona. The study considered two major factors like diffusion limitation of proteins and the competitive interplay between different proteins and their reconfiguration into their native structures. Diffusion plays a vital role in first few minutes, whereas competitive interaction plays a key role in later hours. The study highlighted the significance of surrounding pH and ionic strength in determining the charge of the proteins, diffusion, interaction, and kinetics of protein adsorption onto the nanoparticles.

Computational model developed by Vilanova et al. (2016) proposed that the initially adsorbing proteins have faster diffusion rates with weak interaction and binding. Immediately after the exposure of nanoparticles to the biological fluid containing a complex mixture of proteins, proteins with higher diffusion rate form the corona. As time of exposure to biological fluid increases, the protein corona evolves differently by exchanging the weak interacting proteins by strongly bound proteins with less diffusion.

Vilanova et al. (2016) used two technical tools to study the kinetics of protein corona formation on 100 nm-sized silica nanoparticles. One is differential centrifugal sedimentation, used to separate nanoparticles, and the other is microscale thermophoresis used to study the interaction between proteins (Krpetic et al., 2013; Salvati et al., 2013). Vilanova et al. (2016) applied molecular simulation studies in ternary protein solution of human serum albumin, transferrin, and fibrinogen for 10 seconds and extrapolated to hours. The theoretical prediction was tested by fluorescence correlation spectroscopy. The study demonstrated that the binding of fibrinogen to nanoparticles occurred within few seconds. When the nanoparticle was precoated with human serum albumin, then binding of fibrinogen slowed down significantly. The same was true if the nanoparticles were precoated with transferrin. The study proved that the affinity of human serum albumin and transferrin is greater than that of fibrinogen to nanoparticles. In addition, the study confirmed the memory effect wherein the concentration of the first incubated protein was high in the protein corona.

## 6.7 Nanoparticles-Induced Structural Changes in Proteins

### 6.7.1 Reversible

According to Vroman effect, the binding of proteins to nanoparticle is reversible. The adsorbed proteins desorb from the coronal soft layer if the concentration of the protein in the surrounding medium is less. The time-dependent evolution of protein corona was studied at different time scales. Some studies mapped the dynamics of protein corona from few seconds to long hours. Nagayama et al. (2007) studied the evolution of protein corona for 360 minutes. Most of the studies reported quantitative increase in corona with time. Qualitatively albumin was constant. Defense proteins like complement and clotting factors are high over time.

### 6.7.2 Irreversible

Monopoli, Bombelli, and Dawson (2011) proposed that the time taken for the internalization of nanoparticle can be used as the basic time for differentiating reversible and irreversible binding. Nanoparticles that reside in the corona or on the surface of the nanoparticle more than few minutes are considered to be irreversible. Interaction of proteins with nanoparticles induces changes in proteins. Folded confirmations are induced to unfold during their interaction with nanoparticles. The binding of proteins to nanoparticles in unfolded conformation denatures the protein and the changes are irreversible with the loss of protein function. The unfolded conformation is thermodynamically unfavorable for the protein. Denaturation exposes the shielded residues and favors protein–protein and protein–nanoparticle interactions. The denaturation is dependent on the nature of protein and surface of nanoparticles. For example, the transferrin is denatured from compact to open conformation on binding with super paramagnetic iron oxide nanoparticles. The magnitude of denaturation is more in pristine nanoparticles than nanoparticles functionalized with PEG. The secondary and tertiary structure of albumin was altered by its binding with gold nanoparticles at different pH.

Structural change in the protein exposes the shielded hydrophobic residues. The exposure of the hidden hydrophobic portion will be stabilized by the unusual interaction with other proteins within the corona and with the surrounding proteins. These unfolded or partially unfolded proteins bind strongly to the nanoparticle than their native folded state. These proteins may stabilize or destabilize the surrounding proteins bound to the nanoparticle. This will attract other proteins and facilitate in acquiring new proteins during the evolution of protein corona (H. Wang, Lin, Nienhaus, & Nienhaus, 2018). In hard corona, the protein adsorption is irreversible and they do not change with time (Zhao et al., 2011). The biological fate of nanoparticle is determined by the hard corona. The proteins of corona interact with the host cells and tissues, which determine the ADME properties of nanoparticles (Jia et al., 2012). Hard corona is formed by protein–nanoparticle interaction and soft corona is formed by protein–protein interaction.

## 6.8 Factors Influencing Corona Formation

The structure and composition of protein corona is determined by the characteristics of nanomaterials, proteins, and the microenvironment as illustrated in Figure 6.3. Properties of nanomaterials like size, shape, charge, and the surface properties play a key role in the formation of corona (Dobrovolskaia et al., 2014).

### 6.8.1 Properties of Nanoparticles

Adsorption of proteins onto nanoparticles is determined by physicochemical properties of nanoparticles (Walczyk, Bombelli, Monopoli, Lynch, & Dawson, 2010).

#### 6.8.1.1 Size

Particle size plays a key role in determining the composition of protein corona (Walczyk et al., 2010). There are controversial reports on the influence of particle size on qualitative and quantitative composition of protein. Cedervall et al. (2007) reported that qualitatively the protein in corona remained the same on polymeric nanoparticles. But quantitatively, the amount of protein adsorbed varied significantly. In contrast, metal nanoparticles like gold exhibited both quantitative and qualitative difference in the protein corona with respect to size (Aggarwal et al., 2009). Size-dependent corona composition was studied by Lacerda et al. (2010). The smaller the size of the nanoparticle, the greater the thickness and density of protein corona. This is due to the increased curvature available for the interaction with proteins (Shang, Nienhaus, & Nienhaus, 2014). Lynch reported that the particle size determines the composition of corona by an unknown

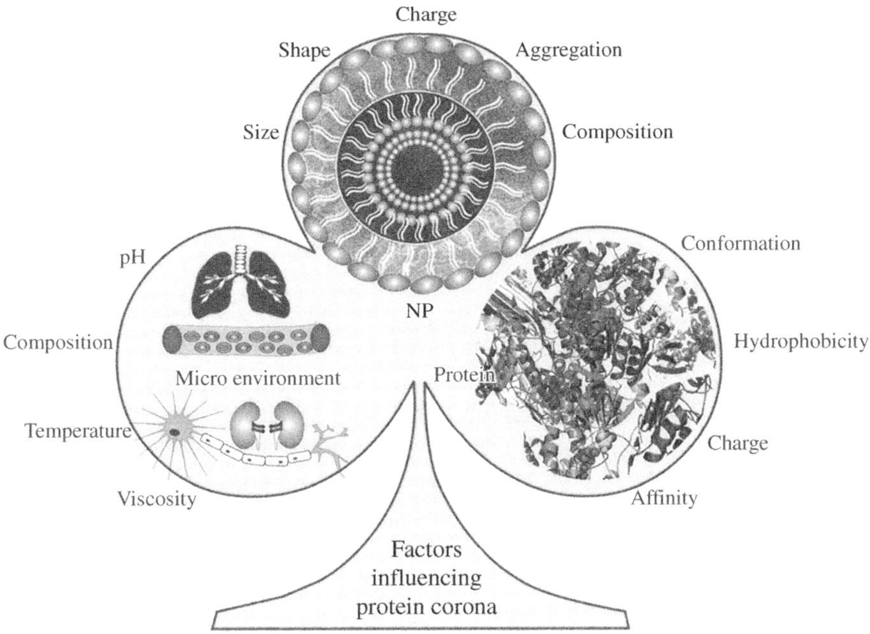


**Figure 6.3** Factors influencing protein corona formation.

mechanism. The quantity and quality of protein corona are governed by the characteristics of nanoparticles and the biological environment it is exposed to like intestinal fluid, blood, etc. (Brun & Sicard-Roselli, 2014; Saptarshi, Duschl, & Lopata, 2013). Nanomaterials with minimum size have increased curvature on the surface, which restricts the adsorption of proteins (Lynch, Salvati, & Dawson, 2009).The protein corona was formed as single- or multiple layers on the surface of citrate-stabilized gold nanoparticles with increase in size of the particle (Piella, Bastus, & Puntes, 2017; Tenzer et al., 2011). In the absence of surface capping or functionalization, the nanoparticles have the tendency to aggregate. The aggregates have different protein corona. Hu, Zhang, Zhang, Wu, and Zou (2014) reported that the protein corona was quantitatively different with 50, 200, and 400 nm iron oxide nanoparticles. Albumin, apolipoprotein, and platelets were common in all iron oxide nanoparticles irrespective of the size. Ceruloplasmin and amyloid P component were unique to 30 nm, fibronectin and serotransferin were exclusive to 200 nm, and coagulation factor was unique to 400 nm. The type of proteins in the protein corona determines the cellular uptake and distribution of nanoparticles. Those nanoparticles covered by lipoproteins and serum albumin increase their circulation in blood. Those nanoparticles covered by IGg increased their clearance from blood (Karmali & Simberg, 2011).

#### 6.8.1.2 Shape

Spherically shaped nanoparticles have rich protein corona compared to other shapes like tubes and rods. This was demonstrated in gold nanoparticles and titanium nanoparticles (Deng et al., 2012). The surface of the nanoparticle provides the necessary driving force like van der Waals, hydrophobic, hydrophilic, or electrostatic forces that foster protein corona formation and their interaction with cells and uptake (H. Wang et al., 2018).

#### 6.8.1.3 Charge

Charge constitutes the major surface property that determines the structure and composition of protein corona. Increased charge density in nanoparticles increases the adsorption of proteins. Electrostatic attraction and isoelectric point of protein determine their adsorption onto nanomaterials. Most of the plasma proteins are negatively charged at physiological pH and they exhibit negative zeta potential between −10 and −20 Mv (Walkey & Chan, 2012). Proteins bound to charged nanoparticles denature (Lynch et al., 2009). Ahsan et al. (2018) reported that proteins with $pI > 5.5$ bind with high affinity to cationic nanoparticles. Cationic nanoparticles are readily recognized by opsonins and they are cleared easily by reticuloendothelial system. Binding of proteins with negatively charged nanoparticles reduced their zeta potential (Ehrenberg, Friedman, Finkelstein, Oberdorster, & McGrath, 2009). Plasma protein density was high in negatively charged polymeric nanoparticles (Gessner et al., 2000). Negatively charged nanoparticles are rich in proteins with PI higher than 5.5 like IgG. Proteins with PI less than 5.5 are the rich components of positively charged proteins. Proteins with $pI < 5.5$ bind with affinity to anionic nanoparticles. Neutrally charged nanoparticles have very less dense corona than anionic or cationic nanoparticles.

#### 6.8.1.4 Surface Functionalization

To extend the life span of nanoparticles in circulation and to evade the immune system recognition and response, these coatings are barriers to the binding of proteins to nanoparticles. Generally, the nanoparticles are functionalized with PEG, amino group, carboxyl group, sulfur groups, etc. In all cases, the protein corona was less than unfunctionalized pristine nanoparticles (Ehrenberg et al., 2009). Coating polystyrene nanoparticles with polaxime reduced the adsorption of fibronectin (Moghimi, Muir, Illum, Davis, & Kolb-Bachofen, 1993). Coating of carbon nanotubes and silica nanoparticles by Pluronic F127 reduces the adsorption of serum proteins. In addition, the type of proteins adsorbed depends on the functional group or surface charge. For example, apoproteins adsorb rapidly and freely than other proteins to nanoparticles functionalized with $NH_2$ and $SO_3$. Functionalization of proteins with PEG increases the hydrophilicity of proteins and hinders the attachment of serum proteins (Walkey et al., 2014). Reports by

Lundqvist et al. (2008) reported that 80% of the protein adsorbed was similar between unfunctionalized polystyrene nanoparticles of size between 50 nm and 100 nm. On the contrary if the nanoparticles were functionalized with carboxyl groups, a significant difference in the protein corona was recorded (Lundqvist et al., 2008). Nanoparticle surface is engineered to possess long circulation time by PEGylation, which avoids the formation protein corona. At the same time, functionalization of a nanoparticle with BSA decreases nonspecific binding of protein, increases its stability in circulation, and minimizes the elimination by mononuclear phagocytes. This was demonstrated by Chow, Brown, and Merad (2011) wherein coating of silicon nanoparticle increased its half-life in vivo.

#### 6.8.1.5 Surface Reactivity

Increased surface area-to-volume ratio increases the surface reactivity of nanomaterials. Increased reactivity of nanomaterials reduces the adsorption energy of proteins and corona formation is favored thermodynamically (Caracciolo, 2015). Competition of protein binding is dependent on the respective proteins' binding energy, abundance, association, and dissociation kinetics. Docter et al. (2014) and Walkey et al. (2014) suggested that the core nature of the nanoparticle plays an equally influential role in the binding of proteins. For example, silver and gold nanoparticles share only 36.9% of proteins bound to the surface. Rest of the proteins is unique to the type of nanomaterials.

#### 6.8.1.6 Solubility

Solubility of the nanoparticles determines the type and amount of proteins in the corona. Hydrophobic surfaces of nanoparticles are bound by plasma proteins like apolipoprotein that increase their clearance from blood (Owens 3rd & Peppas, 2006). The proteins lose their native conformation and bind strongly to nanoparticles. Albumin, IgG, and fibrinogen were dominant on hydrophilic nanoparticles (Gessner et al., 2000). Proteins preferably bind to hydrophobic nanoparticles than hydrophilic proteins (Roach, Farrar, & Perry, 2005).

### 6.8.2 Properties of Protein

Conformation of protein is altered by binding with nanoparticles (Liu et al., 2013). Protein corona favors or prevents the aggregation of nanoparticles. Fibrinogen and bovine serum albumin (BSA) were observed to induce aggregation of nanoparticles (Deng et al., 2012). Dynamics of nanoparticle corona is evaluated in in vivo conditions. The composition of corona is dependent on the constituents present in the medium. The concentration of proteins bound to nanoparticle increased with the concentration of plasma used. The type and concentration of proteins in the corona change with the surrounding microenvironment.

The predominant proteins in the corona of silica nanoparticles were related to clotting and defense proteins like complements if incubated in serum instead of plasma (Mirshafiee, Kim, Mahmoudi, & Kraft, 2016). The media used for the in vitro culturing of cells like RPMI media or DMEM media also contribute to the difference in the composition of protein corona.

Difference in the protein corona composition was recorded with different dispersion media like water, saline, and buffer. Silicon nanoparticles and cobalt ferrite nanoparticle tagged with polyacrylic acid dispersed in different solutions like water, saline, and buffer incubated in RPMI medium showed significant difference in corona composition (Strojan et al., 2017). Exploring the composition and equilibrium of protein corona is essential for understanding the cellular uptake distribution and half-life of nanoparticles (Dobrovolskaia et al., 2009). Many studies reported the abundance of serum albumin on nanoparticles. Contrast to the general hypothesis of direct relationship between the abundance of protein in the medium and the corona formation on nanoparticles was recorded by Chen, Ganesh, Wang, and Amiji (2017). The study observed silica nanoparticles encased with apolipoprotein A1, which is less abundant than $\alpha$ 2 macroglobulin. Binding may be governed by charge in addition to abundance of proteins (Mahmoudi et al., 2011). There are cases where those proteins present in plasma below the lower most detection limit enriched the protein corona. For example prothrombin was not detected in plasma but it was increased 35-fold in silica nanoparticles exposed to human plasma. A sum of 125 different proteins was recorded in silica nanoparticles exposed to human plasma. But the total number of proteins recorded in plasma was 85. Disparity in the types of proteins between plasma and the nanoparticle incubated in the plasma was also reported earlier (Mahmoudi et al., 2011). Based on the functional classification, most of the proteins that form the protein corona are those related to immune response, lipid metabolism, clotting, and proteins related to disease like acute phase proteins. Proteins not a component of blood but a product of tissue leakage were also found in corona. Approximately 37% of proteins adsorbed to nanoparticle is size dependent. A slightest change of 10 nm difference is enough to vary the composition of protein corona. Protein coronas on differently sized silica nanoparticles have a cocktail of proteins with different binding affinities to nanoparticle of different size. Lipoproteins bind preferentially to smaller sized silica nanoparticle, whereas proteins related to blood clotting bind to large-sized silica nanoparticles

Protein corona on silica, metal, polystyrene, and magnetic nanoparticles was studied extensively (Hu et al., 2014). Hu et al. (2014) reported that some proteins like albumin, apolipoprotein A1, and platelet activating factor were common in nanoparticles irrespective of size. Ceruloplasmin was found only in small-sized nanoparticle (30 nm). Coagulation protein XIII was found only in 400 nm particle. In 30 nm-sized nanoparticles, 33% of proteins were related to metabolic process.

In particles with size of 200 nm, the proteins related to metabolic process were 9.5 and 16.7% in 200 and 400 nm, particle size respectively. Proteins related to cell cycle were found exclusively in 10 nm-sized particles and not found in 200 and 400 nm-sized particles.

### 6.8.3 Effect of Surrounding Environment

#### 6.8.3.1 Effect of Media Composition on Corona Formation

At *in vitro* levels, protein corona formation and cellular uptake are studied in culturing the cells in RMPI or DMEM medium supplemented with fetal bovine serum albumin. As the cell grows at the expense of cell culture medium, the concentration of the protein always varies distinctly based on the growth rate of cells. Therefore, the abundance of proteins differs significantly. This is one of the variables that affect the protein corona formation (Albanese et al., 2014). Maiorano et al. (2010) studied the impact of media composition on the evolution of protein corona on gold nanoparticles capped with citrate. The studies confirmed that the kinetics of corona evolution was low in RPMI medium than with DMEM medium. Cellular uptake of gold nanoparticles by HeLa cells (human cervical cancer cell lines) and U937 (human leukemic monocyte lymphoma cells) cultured in RPMI medium was high compared to RPMI medium. The concentration of plasma or serum used in the culture directly influences the hydrodynamic size of the protein corona. This is confirmed with silica nanoparticles and carboxylated polystyrene nanoparticles. Binding of complement and immunoglobulins onto charged nanoparticles increases their elimination through opsonization (Roser, Fischer, & Kissel, 1998).

#### 6.8.3.2 Effect of Temperature

Increase in temperature reduces the thickness of protein corona (Z. Wang et al., 2017). Temperature not only influences the protein corona thickness and composition but also the cellular uptake. Cellular uptake of nanoparticle corona was greater at 37 compared to 39 °C (Mahmoudi et al., 2012).

#### 6.8.3.3 Static In Vitro Model Vs. Dynamic In Vivo System

All in vitro studies for the formation and characterization of nanoparticle corona were evaluated in static experimental conditions. But the original in vivo system is absolutely dynamic where the nanoparticles move at the rate of 60 cm/second. Pozzi et al. (2015) reported that in dynamic system, the corona contains less concentration of complements and it influences the metabolic clearance of nanoparticles. Ye et al. (2016) compared the protein corona formed on plain PEGYlated and antibody functionalized PEGylated liposomes in CD1 mice and static in vitro conditions. The study reported the complexity of corona formed on the nanoparticle in dynamic in vivo system.

## 6.9 Interaction of Nanoparticles with Cells and Their Uptake

Protein corona determines the fate of nanoparticles and their distribution inside the cell. Proteins like IgG and complements are recognized by macrophages for the elimination of foreign substances by enhanced phagocytosis termed opsonization (Kotagiri & Kim, 2014). Nanoparticle corona formed by these proteins enhances their clearance. Proteins likeCD47 and serum albumin in corona increase the circulation of nanoparticles and residence time in blood stream and decrease their clearance from blood.

The type of response of a living system to nanoparticle depends on the type of nanoparticle, its surrounding microenvironment and proteins. Some proteins in corona increase the cellular uptake of nanoparticles. For example, coating of silica nanoparticles with BSA facilitates its uptake by endothelial system. Protein corona formed by ApoE and Apo A-I also facilitates the early cellular uptake of silica nanoparticles (Barran-Berdon et al., 2013). Nanoparticles are admirable for their application in targeted therapy in cancer. Transferrin is tagged to nanoparticles for selective accumulation in cancer cells. But it was discovered by Salvati et al. (2013) that formation of protein corona masks both cell surface and free receptors of transferrin ruining the selective targeting and accumulation of nanoparticles at target. Moreover, the protein corona drives the nanoparticles to take alternative cellular uptake and non-target accumulation.

Increased cellular uptake of nanoparticles after formation of protein corona was established using gold nanorods (Qiu et al., 2010). Proteins bound to the nanoparticle surface are well-recognized by the cell surface receptors, which increases their internalization (Zhao et al., 2011). The dynamics of hard corona govern the metabolic fate and distribution of nanoparticles. For example, those nanoparticles coated by apolipoproteins can cross the blood–brain barrier. Whereas those coated with complement proteins are recognized by the immune system and eliminated from the system. But PEGylation minimizes the adsorption of proteins, especially complements, and reduces their elimination and increases their survival and circulation. Once the PEG is oxidized within the biological system, it cannot prevent the adsorption of proteins (Walkey & Chan, 2012).

Cellular uptake of smaller nanoparticles with protein corona was more than that of larger nanoparticles (Brun & Sicard-Roselli, 2014). Smith et al. (2012) demonstrated the reduction in the cellular uptake of carboxylated polystyrene nanoparticles by the addition of serum to medium. Addition of higher volume of serum increases the fluidity and decreases the interaction between nanoparticle and cell membrane (Smith et al., 2012). Reduction in cellular uptake is supported in other studies in silica (Docter et al., 2014), silver nanoparticle (Lesniak et al., 2010), and

FePt nanoparticles (Saptarshi et al., 2013). Increased cellular uptake of nanoparticles after formation of protein corona was established using gold nanorods (Qiu et al., 2010). Proteins bound to the nanoparticle surface are well-recognized by the cell surface receptors, which increases their internalization (Zhao et al., 2011). Cellular uptake is dependent on many factors like the quantity and quality of proteins in the corona, surrounding pH, composition of the medium, type of cells, and the unique receptors on the cell surface. There are different results on the effect of charge on cellular uptake. Positively charged nanoparticle uptake was generally hypothesized to be more than that of other charges due to the negative charge of the cell membrane. The uptake of nanoparticles by umbilical cord vein endothelial cells is independent of charge; but human breast cancer cells preferentially uptake positively charged proteins at a higher rate (Osaka, Nakanishi, Shanmugam, Takahama, & Zhang, 2009).

The protein corona formed on the nanoparticles masks their original identity. The cellular uptake is dependent on the specific protein adsorbed and its recognition by the cell surface receptor. Nanoparticles with protein corona are taken by specific process of internalization mediated by receptors. Whereas pristine nanoparticles are internalized by their interaction with cells directly by nonspecific interaction without being mediated by ligand receptor interaction (Allouni, Gjerdet, Cimpan, & Hol, 2015).

The discrepancies in cellular uptake suggest the possibility of uptake of nanoparticles by different mechanism (Albanese & Chan, 2011). Uptake of aggregated nanoparticles is very less due to their larger size. But exposure of nanoparticle to in vivo system brings proteins closer to nanoparticle surface and enhances corona formation. Corona formation reduces or prevents the formation of nanoparticle aggregates and stabilizes them from nanoparticle–nanoparticle interaction (Sabuncu et al., 2012). Nanoparticles with protein corona are distributed exclusively in lysosome (Lesniak et al., 2012). The type of cells used in the in vitro culture partially determines the cellular uptake and toxicity. The surface characteristics of each cell have its own fingerprint of unique proteins. The effect was demonstrated with super paramagnetic iron oxide nanoparticles in different cells like human neuroblastoma (BE(2)-C), glioblastoma (A172), lung adenocarcinoma (A549), cardiac myocytes (HCM), epithelial carcinoma (A-431), hepatocellular carcinoma (Hep G2), colon adenocarcinoma (SW480), cervical adenocarcinoma (HeLa), pancreatic adenocarcinoma (Capan-2), pancreatic carcinoma (Panc-2), Tell lymphoblast like Jurkat and mouse connective tissue fibroblast (L929). Neuronal cells, glial cells, and lung cells were most sensitive to toxicity by nanoparticles. The toxicity was moderate in all other types of cells. Therefore the protein corona formation on the nanoparticles determines their cellular uptake and toxicity (Laurent, Burtea, Thirifays, Hafeli, & Mahmoudi, 2012).

Therapeutic effect of drugs is determined by their absorption, distribution, metabolism, and elimination in the living system. Similarly, the therapeutic or diagnostic value of nanoparticles is also governed by the absorption, distribution, metabolism, and excretion (ADME) properties. The rate of distribution, cellular uptake, and elimination depends on the protein corona that encircles or covers the nanoparticles. Cellular uptake is determined by a triad of nanoparticle, cell, and protein. In mouse myogenic cells, uptake of 20 nm-sized gold nanoparticle conjugated with cystein-terminated peptides was studied (G. Wang et al., 2012). Protein corona decreased the hepatocellular uptake of 40–80 nm-sized gold nanoparticle functionalized with polyethylene amine (Choi, Riviere, & Monteiro-Riviere, 2017). The charge-bearing group of nanoparticles governs the cellular uptake. Nanoparticles functionalized with an amino group coated with BSA are up taken via BSA receptor whereas cationic nanoparticle BSA binds through scavenger receptor (Fleischer & Payne, 2014).

## 6.10 Pleiotrophic Effect of Nanoparticles

Nanoparticles are taken inside the cell by different endocytic pathways in different cells and it has different cellular responses and effect (Mahmoudi et al., 2012). The same nanoparticle exhibits different actions in different sites (Mahmoudi et al., 2012). The effect of ApoH on nanoparticles is different in phagocytic and non-phagocytic cells (Ritz et al., 2015). Macrophages rank first in the cellular uptake of nanoparticles followed by dendritic cells, monocytes, and neutrophils. All these are first line of nonspecific defense cells. Least uptake was reported in non-dividing parenchymal cells. These are the warriors in cell-mediated immunity. IgM and complement C3 are the effectors of humoral immunity, which induce opsonization. Polyacrylic acid-conjugated sliver nanoparticles were bound by fibrinogen that stimulates the secretion of inflammation-mediating cytokines. The possible reason for the release of inflammatory mediators is the unfolding of fibrinogen induced by the binding of nanoparticles.

## 6.11 Analytical Methods to Study Nanoparticles–Protein Interaction

### 6.11.1 Spectral Properties

#### 6.11.1.1 UV–Vis Spectroscopy

Shannahan, Podila, and Brown (2015) demonstrated hyper spectral properties of silver nanoparticle corona due to interparticle association and plasmon hybridization of PVP and citrate-coated silver nanoparticle corona in DMEM medium supplemented with 10% FBS. Hybridization of individual nanoparticle plasmons due to Coulombic interaction of charge density leads to change in plasmon energy causing red shift (Shannahan et al., 2015). The formation of protein corona changes the interparticle interaction, which is reflected as hyper spectral property. The study proposed that increase in aggregation and spectral properties are an indicator of macrophage uptake. Cellular uptake causes either red shift by aggregation or blue shift by repulsion due to corona formation by Vroman effect. The study observed a red shift of PVP-coated silver nanoparticle of 110 nm within four hours, indicating cellular uptake. No further change in spectral property was observed after two hours. This indicates that the diffusion of proteins is restricted, leading to the formation of stable corona. Similar hyperspectral property was reported in gold nanoparticle clusters (Chen et al., 2017).

#### 6.11.1.2 FTIR

FTIR of each molecule is unique based on the nature of the molecule and the functional groups present in the molecule. In electromagnetic spectrum, infrared region ranges from 100 to 1 μm. Molecules absorb IR radiations at a specific wavelength corresponding to the vibrational motion of the molecules. Proteins have hierarchical structural organization from primary to quaternary structure. Proteins have three-dimensional structures, which is obtained by the folding of secondary structures like helical and $\beta$-sheets. Each secondary structure has its own IR absorbance. The change in IR absorbance occurs due to structural changes in protein. The intensity may be reduced or lost after interaction with nanoparticles. So, nanoparticles–protein corona formation will be reflected in the FTIR spectrum. The peak change or disappearance of peaks in protein corona is due to unfolding or partial denaturation or instability of helix or $\beta$-sheets. Sengupta et al. (2015) evaluated the denaturation of BSA and fibrinogen on binding with single-walled carbon nanotube, graphene oxide nanoribbon. BSA loses the stability on binding with multiwalled carbon nanotube, which is translated as significant reduction in the absorbance of $\alpha$-helical sheet indicating the complete loss of helix. Comparatively, the loss of secondary structure of BSA in the corona of graphene was less. In nanotubes, the BSA completely stretches by unfolding and binds strongly. The opposite is true with fibrinogen, where it binds strongly by denaturation on to graphene oxide nanoribbons mediated by hydrogen bonds.

#### 6.11.1.3 Raman Spectroscopy

Protein corona formation by BSA and fibrinogen on defective carbon nanotubes, and graphene oxide nanoribbons was studied using FTIR and Raman spectroscopy (Sengupta et al., 2015). Raman Spectroscopy measures the inelastic scattering of light by molecules due to their vibrational and rotational movement.

Surface-enhanced Raman spectroscopic measurements (SERS) using the plasmon resonance of metal oxide nanoparticles measures the characteristics of protein (Bonifacio et al., 2014; Fazio et al., 2016; Szekeres & Kneipp, 2019).They also studied the SERS with citrate-coated gold nanoparticle with BSA. Raman spectral intensity was greatly enhanced in SERS compared to Raman spectrum. So, increase in spectral intensity is a measure of nanoparticle protein interaction. Increase in the intensity indicates the presence of individual nanoparticles without aggregation. The study proposes that the aromatic amino acids in BSA are near gold nanoparticles. The aromatic amino acid tryptophan increased significantly indicating the denaturation of helical structure of helix (Bujacz, Zielinski, & Sekula, 2014), but the peaks due to sulfur containing amino acid cysteine remained intact indicating the compactness of certain regions of protein. In native proteins, cysteine, and methionine the sulfur containing amino acids are shielded inside the hydrophobic core. When proteins interact with nanoparticles, the residues are exposed due to unfolding of secondary structures indicating denaturation (Bujacz et al., 2014).

#### 6.11.1.4 Fluorescence Spectroscopy

The fluorescence property of proteins is modified by their complex formation with nanomaterials that is measured by real-time fluorescence spectroscopy (Fischer, Verma, Goodman, Simard, & Rotello, 2003; Lakowicz, 1988). In situ quantitative measurement of nanoparticle–protein interaction is studied by DLS and fluorescence correlation spectroscopy (FCS). FCS is a high-resolution technique to study the extremely low concentration of biologically important molecules. It measures the very slightest fluctuation in fluorescence intensity caused by spontaneous changes of the molecules. Adsorption and desorption of proteins onto the nanoparticles exert deviations in the physicochemical properties of corona. All types of changes result in the deviation of fluorescence intensity which is measured. The kinetics of adsorption and desorption occur within few minutes. Further the protein corona is constantly under dynamic change of proteins. All these changes cause change in the fluorescence intensity that is measured by FCS. The duration and the strength of the fluorescent intensity signal were measured by a mathematical procedure termed as autocorrelation.

In 1972 it was used to study the chemical dynamics and diffusion attributes of DNA-drug. Later the technique was used to study the mobility and concentration of proteins in corona. Few femtolitres of sample are sufficient to give very good signal-to-noise ratio. A small change in the diffusion of proteins from and into the corona causes large deviations in signal that is recorded. Signals vary with the evolution of protein corona. Extremely low concentration of proteins is determined precisely by FCS. Autocorrelation amplitude G(0) is equal to the reciprocal number of molecules in the volume of sample under study. FCS is a highly sensitive tool to study the intracellular diffusion and mobility of biomolecules at sub-micrometer scale on the surface of nanoparticles. The temporal resolution of FCS is about microseconds during which proteins in the corona diffuse and change the composition of protein corona. The protein corona is dynamic over days. FCS records signals for a very long time. Diffusion and concentration dynamics of a protein are studied by FCS. The diffusion coefficient of folded and unfolded protein is different, which is exploited to develop nanocarriers suitable for a specific purpose. Hydrodynamic radius of nanoparticle varies based on the size and concentration of nanoparticles and confirmation of proteins. Hydrodynamic radius Rh of protein corona changes the mobility of proteins, which induce signal changes. The signals are recorded. The diffusion coefficient of proteins is evaluated by Stokes–Einstein relationship as given below

$$D = \frac{k \cdot T}{6\pi \cdot \eta_{\mathrm{v}} \cdot R_{\mathrm{h}}}$$

The hydrodynamic radius of the particle increases with the increase in the protein adsorption and decrease in desorption. Hence, minute changes in the protein binding will be reflected as changes which are monitored for long time. In addition, the mobility of the protein varies with the folding and unfolding of proteins.

H. Wang et al. (2018) studied the adsorption of bovine serum albumin on the surface of fluorescently labeled FePt NPs and CdSe/ZnS core–shell quantum dots. The study explored the concentration-dependent adsorption of proteins onto nanoparticles as monolayers. The formation of protein corona increased the hydrodynamic size of nanoparticles by 3 nm. Similarly, many studies emerged with the different types of proteins like apo transferrin, different types of apolipoprotein and complement proteins (Klapper et al., 2015; Treuel et al., 2013). The study demonstrated that the binding affinity of the protein is dominantly determined by electrostatic interaction on the surface of the nanoparticle. The affinity is not dependent on the net charge but on the surface charge, which is in direct contact with the nanoparticle.

The protein corona formed on incubation with whole human serum is different from that studied with the single protein. In the case of serum, vast array of proteins are present. The competition between proteins for binding with nanoparticle is determined by the relative abundance, charge, affinity of the protein, its interaction with other proteins, etc.

H. Wang et al. (2018) documented that the protein binding was reversible in single protein solution and irreversible in serum. The hydrodynamic particle size differed in the initial hours and it was static after two hours confirming the hypothesis.

Fluorescence correlation spectroscopy provides information about the hydrodynamic radius of protein corona via Stokes–Einstein equation through diffusion coefficient.

The binding of BSA to gold nanoparticles provides information about the number of binding sites and binding constant. The aromatic amino acids, phenylalanine, tyrosine, and tryptophan are the three natural fluorophores present in BSA. Among these fluorophores, tryptophan reflects the local environment better than other amino acids. Fluorescence emission spectrum of tryptophan is used to identify the interaction of BSA with nanoparticles. Tryptophan exhibited a strong intrinsic fluorescence at 336 nm, which is quenched by its interaction with nanoparticles. A red shift of fluorescence by 4 nm indicates the change of polarity in the microenvironment near tryptophan. Decrease in the fluorescence intensity of the protein is an indication of protein corona formation. Quenching of fluorescence by protein with compounds was reported earlier (Toprak, 2016).

The protein corona is dissected to study its structure and composition either inside the biological system (in situ) or outside the biological system (ex situ). The composition, diffusion coefficient, binding affinity of proteins to nanoparticles, their dynamic mobility and exchange are studied by many techniques.

### 6.11.2 Surface Plasmon Resonance

Kinetics of nanomaterials–protein interaction cause changes in the surface plasmon resonance. The technique to investigate surface plasmon resonance involves the immobilization of either the nanoparticle or the protein onto the metal surface and flowing the counterpart through the immobilized material. Adsorption of the flow material onto the immobilized material causes resonance shift, which is a measure of binding kinetics.

### 6.11.3 Cellular Uptake of Nanoparticles–Protein

#### 6.11.3.1 Flow Cytometry

Cellular uptake of nanoparticle protein corona increases the side scatter intensity in flow cyotometer. Silver nanoparticle protein corona mimics pseudo granules and increases the intensity. The magnitude of intensity increase was proportional to the size of the particle. In their study 110 nm particles showed higher intensity than 20 nm particle. So increase in the side scatter intensity is a sensitive tool to confirm the cellular uptake (Zucker et al., 2013).

#### 6.11.3.2 Confocal Microscopy

Cell morphology is studied by confocal microscopy, which permits live cell imaging. Addition of fluorescent dyes distinguishes the nucleus. The impact of pristine nanoparticle and nanoparticle corona formation on cell structure and intracellular organelles and the viability of the cells are studied by confocal fluorescence microscopy. Induction of human lung fibroblast proliferation by airborne nanoparticulate matter was studied by Li et al. (2018). The study proved the significant proliferation of fibroblast in 3D culture exposed to nanoparticle protein corona.

### 6.11.4 Binding Affinity

#### 6.11.4.1 Differential Scanning Calorimetry and Isothermal Calorimetry

Differential scanning calorimetry is used to study the transition temperature of biomolecules. The thermal behavior of long unilamellar vesicles protein corona formed by BSA was studied by Sangra, Estelrich, Sabate, Espargaro, and Busquets (2017). Thermal behavior of PEGylated liposomes is different in the presence of any surface attached molecules. Therefore, it is used as a tool to study the adsorption of protein corona to the lipid vesicle. Change in the transition temperature was recorded due to BSA adsorption to the PEGylated liposomes.

#### 6.11.4.2 Isothermal Titration Calorimetry

Sangra et al. (2017) compared the thermal behavior of large unilamellar vesicles and liposomes with super paramagnetic iron oxide nanoparticles on corona formation with BSA. Both the liposomes were functionalized with cyclic arginine peptides to target integrin receptors in cancer cells. The expression of integrin receptors is high in cancer cells. ITC of large unilamellar vesicles and magnetic liposomes exhibited no significant endothermic heat generation upon binding with BSA.

## Questions and Answers

1. Why protein corona is dynamic?
   The structure and composition of protein corona are dynamic due to the adsorption and desorption of proteins onto the surface of nanoparticles. The proteins in the corona differ based on the route of entry of nanoparticles and the abundance of proteins and their interaction with nanoparticles.

2. What are soft corona and hard corona?
   The part of proteins present in the corona that varies widely due to reversible binding is called soft corona. In contrast, the other part of proteins is stable and does not change over time. So, they are called hard corona.

3. How does functionalization of nanomaterials with PEG influence the formation of protein corona?
   Functionalization of nanoparticles by PEGylation minimizes the adsorption of proteins on to the pristine nanoparticles.

**4** How is fluorescence correlated with protein corona?
The fluorescence property of proteins decreases on interaction.

**5** How is the fate of nanoparticles modified by the adsorption of complements?
Complements facilitate the immune recognition of protein corona and influence the rapid clearance of nanoparticles.

## References

Aggarwal, P., Hall, J. B., McLeland, C. B., Dobrovolskaia, M. A., & McNeil, S. E. (2009). Nanoparticle interaction with plasma proteins as it relates to particle biodistribution biocompatibility and therapeutic efficacy. *Advanced Drug Delivery Reviews*, 61(6), 428–437. Retrieved from https://www.ncbi.nlm.nih.gov/pubmed/19376175. doi:10.1016/j.addr.2009.03.009

Ahsan, S. M., Rao, C. M., & Ahmad, M. F. (2018). Nanoparticle–protein interaction: The significance and role of protein corona. *Advance in Experimental Medicine and Biology*, 1048, 175–198. Retrieved from https://www.ncbi.nlm.nih.gov/pubmed/29453539. doi:10.1007/978-3-319-72041-8_11

Akbarzadeh, A., Samiei, M., Joo, S. W., Anzaby, M., Hanifehpour, Y., Nasrabadi, H. T., & Davaran, S. (2012). Synthesis, characterization and in vitro studies of doxorubicin-loaded magnetic nanoparticles grafted to smart copolymers on A549 lung cancer cell line. *Journal of Nanobiotechnology*, 10, 46. Retrieved from https://www.ncbi.nlm.nih.gov/pubmed/23244711. doi:10.1186/1477-3155-10-46

Albanese, A., & Chan, W. C. (2011). Effect of gold nanoparticle aggregation on cell uptake and toxicity. *ACS Nano*, 5(7), 5478–5489. Retrieved from https://www.ncbi.nlm.nih.gov/pubmed/21692495. doi:10.1021/nn2007496

Albanese, A., Walkey, C. D., Olsen, J. B., Guo, H., Emili, A., & Chan, W. C. (2014). Secreted biomolecules alter the biological identity and cellular interactions of nanoparticles. *ACS Nano*, 8(6), 5515–5526. Retrieved from https://www.ncbi.nlm.nih.gov/pubmed/24797313. doi:10.1021/nn4061012

Allouni, Z. E., Gjerdet, N. R., Cimpan, M. R., & Hol, P. J. (2015). The effect of blood protein adsorption on cellular uptake of anatase $TiO_2$ nanoparticles. *International Journal of Nanomedicine*, 10, 687–695. Retrieved from https://www.ncbi.nlm.nih.gov/pubmed/25632230. doi:10.2147/IJN.S72726

Amiri, S., & Shokrollahi, H. (2013). The role of cobalt ferrite magnetic nanoparticles in medical science. *Materials Science & Engineering. C, Materials for Biological Applications*, 33(1), 1–8. Retrieved from https://www.ncbi.nlm.nih.gov/pubmed/25428034. doi:10.1016/j.msec.2012.09.003

Anantha Suganiya, S., & Mala, R. (2014). Susceptibility profile of bacteria isolated from diabetic wound to silver nanoparticles synthesized using *Murraya koenigii*.

*Asian Journal of Microbiology and Biotechnology Environmental Science*, 16(4), 985–992.

Angioletti-Uberti, S., Ballauff, M., & Dzubiella, J. (2014). Dynamic density functional theory of protein adsorption on polymer-coated nanoparticles. *Soft Matter*, 10(40), 7932–7945. Retrieved from https://www.ncbi.nlm.nih.gov/pubmed/25052205. doi:10.1039/c4sm01170h

Barran-Berdon, A. L., Pozzi, D., Caracciolo, G., Capriotti, A. L., Caruso, G., Cavaliere, C., . . . Lagana, A. (2013). Time evolution of nanoparticle–protein corona in human plasma: Relevance for targeted drug delivery. *Langmuir*, 29(21), 6485–6494. Retrieved from https://www.ncbi.nlm.nih.gov/pubmed/23631648. doi:10.1021/la401192x

Bhainsa, K. C., & D'Souza, S. F. (2006). Extracellular biosynthesis of silver nanoparticles using the fungus *Aspergillus fumigatus*. *Colloids and Surfaces, B: Biointerfaces*, 47(2), 160–164. Retrieved from https://www.ncbi.nlm.nih.gov/pubmed/16420977. doi:10.1016/j.colsurfb.2005.11.026

Bhattacharya, R., & Mukherjee, P. (2008). Biological properties of "naked" metal nanoparticles. *Advanced Drug Delivery Reviews*, 60(11), 1289–1306. Retrieved from https://www.ncbi.nlm.nih.gov/pubmed/18501989. doi:10.1016/j.addr.2008.03.013

Bonifacio, A., Dalla Marta, S., Spizzo, R., Cervo, S., Steffan, A., Colombatti, A., & Sergo, V. (2014). Surface-enhanced Raman spectroscopy of blood plasma and serum using Ag and Au nanoparticles: A systematic study. *Analytical and Bioanalytical Chemistry*, 406(9–10), 2355–2365. Retrieved from https://www.ncbi.nlm.nih.gov/pubmed/24493335. doi:10.1007/s00216-014-7622-1

Brun, E., & Sicard-Roselli, C. (2014). Could nanoparticle corona characterization help for biological consequence prediction? *Cancer Nanotechnology*, 5(1), 7. Retrieved from https://www.ncbi.nlm.nih.gov/pubmed/25309635. doi:10.1186/s12645-014-0007-5

Bujacz, A., Zielinski, K., & Sekula, B. (2014). Structural studies of bovine, equine, and leporine serum albumin complexes with naproxen. *Proteins*, 82(9), 2199–2208. Retrieved from https://www.ncbi.nlm.nih.gov/pubmed/24753230. doi:10.1002/prot.24583

Caracciolo, G. (2015). Liposome-protein corona in a physiological environment: Challenges and opportunities for targeted delivery of nanomedicines. *Nanomedicine*, 11(3), 543–557. Retrieved from https://www.ncbi.nlm.nih.gov/pubmed/25555353. doi:10.1016/j.nano.2014.11.003

Casals, E., Pfaller, T., Duschl, A., Oostingh, G. J., & Puntes, V. (2010). Time evolution of the nanoparticle protein corona. *ACS Nano*, 4(7), 3623–3632. Retrieved from https://www.ncbi.nlm.nih.gov/pubmed/20553005. doi:10.1021/nn901372t

Cedervall, T., Lynch, I., Lindman, S., Berggard, T., Thulin, E., Nilsson, H., . . . Linse, S. (2007). Understanding the nanoparticle–protein corona using methods to quantify exchange rates and affinities of proteins for nanoparticles. *Proceedings of the National Academy of Sciences of the United States of America*, 104(7), 2050–2055. Retrieved from https://www.ncbi.nlm.nih.gov/pubmed/17267609. doi:10.1073/pnas.0608582104

Chen, D., Ganesh, S., Wang, W., & Amiji, M. (2017). Plasma protein adsorption and biological identity of systemically administered nanoparticles. *Nanomedicine (London, England)*, 12(17), 2113–2135. Retrieved from https://www.ncbi.nlm.nih.gov/pubmed/28805542. doi:10.2217/nnm-2017-0178

Choi, K., Riviere, J. E., & Monteiro-Riviere, N. A. (2017). Protein corona modulation of hepatocyte uptake and molecular mechanisms of gold nanoparticle toxicity. *Nanotoxicology*, 11(1), 64–75. Retrieved from https://www.ncbi.nlm.nih.gov/pubmed/27885867. doi:10.1080/17435390.2016.1264638

Chow, A., Brown, B. D., & Merad, M. (2011). Studying the mononuclear phagocyte system in the molecular age. *Nature Reviews. Immunology*, 11(11), 788–798. Retrieved from https://www.ncbi.nlm.nih.gov/pubmed/22025056. doi:10.1038/nri3087

Daraee, H., Eatemadi, A., Abbasi, E., Fekri Aval, S., Kouhi, M., & Akbarzadeh, A. (2016). Application of gold nanoparticles in biomedical and drug delivery. *Artificial Cells, Nanomedicine, and Biotechnology*, 44(1), 410–422. Retrieved from https://www.ncbi.nlm.nih.gov/pubmed/25229833. doi:10.3109/21691401.2014.955107

Deng, Z. J., Liang, M., Toth, I., Monteiro, M. J., & Minchin, R. F. (2012). Molecular interaction of poly(acrylic acid) gold nanoparticles with human fibrinogen. *ACS Nano*, 6(10), 8962–8969. Retrieved from https://www.ncbi.nlm.nih.gov/pubmed/22998416. doi:10.1021/nn3029953

Devi, L. S., & Joshi, S. R. (2012). Antimicrobial and synergistic effects of silver nanoparticles synthesized using soil fungi of high altitudes of eastern himalaya. *Mycobiology*, 40(1), 27–34. Retrieved from https://www.ncbi.nlm.nih.gov/pubmed/22783131. doi:10.5941/MYCO.2012.40.1.027

Dobrovolskaia, M. A., Neun, B. W., Man, S., Ye, X., Hansen, M., Patri, A. K., . . . McNeil, S. E. (2014). Protein corona composition does not accurately predict hematocompatibility of colloidal gold nanoparticles. *Nanomedicine*, 10(7), 1453–1463. Retrieved from https://www.ncbi.nlm.nih.gov/pubmed/24512761. doi:10.1016/j.nano.2014.01.009

Dobrovolskaia, M. A., Patri, A. K., Zheng, J., Clogston, J. D., Ayub, N., Aggarwal, P., . . . McNeil, S. E. (2009). Interaction of colloidal gold nanoparticles with human blood: Effects on particle size and analysis of plasma protein binding profiles. *Nanomedicine*, 5(2), 106–117. Retrieved from https://www.ncbi.nlm.nih.gov/pubmed/19071065. doi:10.1016/j.nano.2008.08.001

Docter, D., Bantz, C., Westmeier, D., Galla, H. J., Wang, Q., Kirkpatrick, J. C., . . . Stauber, R. H. (2014). The protein corona protects against size- and dose-dependent toxicity of amorphous silica nanoparticles. *Beilstein Journal of Nanotechnology*, 5, 1380–1392. Retrieved from https://www.ncbi.nlm.nih.gov/pubmed/25247121. doi:10.3762/bjnano.5.151

Donahue, N. D., Acar, H., & Wilhelm, S. (2019). Concepts of nanoparticle cellular uptake, intracellular trafficking and kinetics in nanomedicine. *Advanced Drug Delivery Reviews*, 143, 68–96. Retrieved from https://www.ncbi.nlm.nih.gov/pubmed/31022434. doi:10.1016/j.addr.2019.04.008

Duran, N., Marcato, P. D., Duran, M., Yadav, A., Gade, A., & Rai, M. (2011). Mechanistic aspects in the biogenic synthesis of extracellular metal nanoparticles by peptides, bacteria, fungi and plants. *Applied Microbiology and Biotechnology*, 90(5), 1609–1624. Retrieved from https://www.ncbi.nlm.nih.gov/pubmed/21484205. doi:10.1007/s00253-011-3249-8

Ehrenberg, M. S., Friedman, A. E., Finkelstein, J. N., Oberdorster, G., & McGrath, J. L. (2009). The influence of protein adsorption on nanoparticle association with cultured endothelial cells. *Biomaterials*, 30(4), 603–610. Retrieved from https://www.ncbi.nlm.nih.gov/pubmed/19012960. doi:10.1016/j.biomaterials.2008.09.050

European Union Law. (2011). *Official Journal of European Union.*

Fazio, B., D'Andrea, C., Foti, A., Messina, E., Irrera, A., Donato, M. G., . . . Gucciardi, P. G. (2016). SERS detection of biomolecules at physiological pH via aggregation of gold nanorods mediated by optical forces and plasmonic heating. *Scientific Reports*, 6, 26952. Retrieved from https://www.ncbi.nlm.nih.gov/pubmed/27246267. doi:10.1038/srep26952

Fischer, N. O., Verma, A., Goodman, C. M., Simard, J. M., & Rotello, V. M. (2003). Reversible "irreversible" inhibition of chymotrypsin using nanoparticle receptors. *Journal of the American Chemical Society*, 125(44), 13387–13391. Retrieved from https://www.ncbi.nlm.nih.gov/pubmed/14583034. doi:10.1021/ja0352505

Fleischer, C. C., & Payne, C. K. (2014). Nanoparticle-cell interactions: Molecular structure of the protein corona and cellular outcomes. *Accounts of Chemical Research*, 47(8), 2651–2659. Retrieved from https://www.ncbi.nlm.nih.gov/pubmed/25014679. doi:10.1021/ar500190q

Gessner, A., Waicz, R., Lieske, A., Paulke, B., Mader, K., & Muller, R. H. (2000). Nanoparticles with decreasing surface hydrophobicities: Influence on plasma protein adsorption. *International Journal of Pharmaceutics*, 196(2), 245–249. Retrieved from https://www.ncbi.nlm.nih.gov/pubmed/10699728. doi:10.1016/s0378-5173(99)00432-9

Hu, Z., Zhang, H., Zhang, Y., Wu, R., & Zou, H. (2014). Nanoparticle size matters in the formation of plasma protein coronas on Fe3O4 nanoparticles. *Colloids and Surfaces, B: Biointerfaces*, 121, 354–361. Retrieved from https://www.ncbi.nlm.nih.gov/pubmed/24974013. doi:10.1016/j.colsurfb.2014.06.016

Jia, Y., Yuan, M., Yuan, H., Huang, X., Sui, X., Cui, X., . . . Guo, Q. (2012). Co-encapsulation of magnetic $Fe_3O_4$ nanoparticles and doxorubicin into biodegradable PLGA nanocarriers for intratumoral drug delivery. *International Journal of Nanomedicine*, 7, 1697–1708. Retrieved from https://www.ncbi.nlm.nih.gov/pubmed/22619520. doi:10.2147/IJN.S28629

Kalimuthu, K., Suresh Babu, R., Venkataraman, D., Bilal, M., & Gurunathan, S. (2008). Biosynthesis of silver nanocrystals by *Bacillus licheniformis*. *Colloids and Surfaces, B: Biointerfaces*, 65(1), 150–153. Retrieved from https://www.ncbi.nlm.nih.gov/pubmed/18406112. doi:10.1016/j.colsurfb.2008.02.018

Kalishwaralal, K., Deepak, V., Ram Kumar Pandian, S., Kottaisamy, M., BarathmaniKanth, S., Kartikeyan, B., & Gurunathan, S. (2010). Biosynthesis of silver and gold nanoparticles using *Brevibacterium casei*. *Colloids and Surfaces, B: Biointerfaces*, 77(2), 257–262. Retrieved from https://www.ncbi.nlm.nih.gov/pubmed/20197229. doi:10.1016/j.colsurfb.2010.02.007

Kapralov, A. A., Feng, W. H., Amoscato, A. A., Yanamala, N., Balasubramanian, K., Winnica, D. E., . . . Kagan, V. E. (2012). Adsorption of surfactant lipids by single-walled carbon nanotubes in mouse lung upon pharyngeal aspiration. *ACS Nano*, 6(5), 4147–4156. Retrieved from https://www.ncbi.nlm.nih.gov/pubmed/22463369. doi:10.1021/nn300626q

Karmali, P. P., & Simberg, D. (2011). Interactions of nanoparticles with plasma proteins: Implication on clearance and toxicity of drug delivery systems. *Expert Opinion on Drug Delivery*, 8(3), 343–357. Retrieved from https://www.ncbi.nlm.nih.gov/pubmed/21291354. doi:10.1517/17425247.2011.554818

Klapper, Y., Maffre, P., Shang, L., Ekdahl, K. N., Nilsson, B., Hettler, S., . . . Nienhaus, G. U. (2015). Low affinity binding of plasma proteins to lipid-coated quantum dots as observed by in situ fluorescence correlation spectroscopy. *Nanoscale*, 7(22), 9980–9984. Retrieved from https://www.ncbi.nlm.nih.gov/pubmed/25975280. doi:10.1039/c5nr01694k

Kotagiri, N., & Kim, J. W. (2014). Stealth nanotubes: Strategies of shielding carbon nanotubes to evade opsonization and improve biodistribution. *International Journal of Nanomedicine*, 9(Suppl 1), 85–105. Retrieved from https://www.ncbi.nlm.nih.gov/pubmed/24872705. doi:10.2147/IJN.S51854

Krpetic, Z., Davidson, A. M., Volk, M., Levy, R., Brust, M., & Cooper, D. L. (2013). High-resolution sizing of monolayer-protected gold clusters by differential centrifugal sedimentation. *ACS Nano*, 7(10), 8881–8890. Retrieved from https://www.ncbi.nlm.nih.gov/pubmed/24063653. doi:10.1021/nn403350v

Lacerda, S. H., Park, J. J., Meuse, C., Pristinski, D., Becker, M. L., Karim, A., & Douglas, J. F. (2010). Interaction of gold nanoparticles with common human blood proteins. *ACS Nano*, 4(1), 365–379. Retrieved from https://www.ncbi.nlm.nih.gov/pubmed/20020753. doi:10.1021/nn9011187

Lakowicz, J. R. (1988). Principles of frequency-domain fluorescence spectroscopy and applications to cell membranes. *Sub-Cellular Biochemistry*, 13, 89–126. Retrieved from https://www.ncbi.nlm.nih.gov/pubmed/2577864

Laurent, S., Burtea, C., Thirifays, C., Hafeli, U. O., & Mahmoudi, M. (2012). Crucial ignored parameters on nanotoxicology: The importance of toxicity assay modifications and "cell vision". *PLoS One*, 7(1), e29997. Retrieved from https://www.ncbi.nlm.nih.gov/pubmed/22253854. doi:10.1371/journal.pone.0029997

Lesniak, A., Campbell, A., Monopoli, M. P., Lynch, I., Salvati, A., & Dawson, K. A. (2010). Serum heat inactivation affects protein corona composition and nanoparticle uptake. *Biomaterials*, 31(36), 9511–9518. Retrieved from https://www.ncbi.nlm.nih.gov/pubmed/21059466. doi:10.1016/j.biomaterials.2010.09.049

Lesniak, A., Fenaroli, F., Monopoli, M. P., Aberg, C., Dawson, K. A., & Salvati, A. (2012). Effects of the presence or absence of a protein corona on silica nanoparticle uptake and impact on cells. *ACS Nano*, 6(7), 5845–5857. Retrieved from https://www.ncbi.nlm.nih.gov/pubmed/22721453. doi:10.1021/nn300223w

Li, Y., Wang, P., Hu, C., Wang, K., Chang, Q., Liu, L., . . . Wu, Y. (2018). Protein corona of airborne nanoscale PM2.5 induces aberrant proliferation of human lung fibroblasts based on a 3D organotypic culture. *Scientific Reports*, 8(1), 1939. Retrieved from https://www.ncbi.nlm.nih.gov/pubmed/29386584. doi:10.1038/s41598-018-20445-7

Liao, S. H., Liu, C. H., Bastakoti, B. P., Suzuki, N., Chang, Y., Yamauchi, Y., . . . Wu, K. C. (2015). Functionalized magnetic iron oxide/alginate core-shell nanoparticles for targeting hyperthermia. *International Journal of Nanomedicine*, 10, 3315–3327. Retrieved from https://www.ncbi.nlm.nih.gov/pubmed/26005343. doi:10.2147/IJN.S68719

Liu, X., Chen, Y., Li, H., Huang, N., Jin, Q., Ren, K., & Ji, J. (2013). Enhanced retention and cellular uptake of nanoparticles in tumors by controlling their aggregation behavior. *ACS Nano*, 7(7), 6244–6257. Retrieved from https://www.ncbi.nlm.nih.gov/pubmed/23799860. doi:10.1021/nn402201w

Lundqvist, M., Stigler, J., Elia, G., Lynch, I., Cedervall, T., & Dawson, K. A. (2008). Nanoparticle size and surface properties determine the protein corona with possible implications for biological impacts. *Proceedings of the National Academy of Sciences of the United States of America*, 105(38), 14265–14270. Retrieved from https://www.ncbi.nlm.nih.gov/pubmed/18809927. doi:10.1073/pnas.0805135105

Lynch, I., Salvati, A., & Dawson, K. A. (2009). Protein-nanoparticle interactions: What does the cell see? *Nature Nanotechnology*, 4(9), 546–547. Retrieved from https://www.ncbi.nlm.nih.gov/pubmed/19734922. doi:10.1038/nnano.2009.248

Mahmoudi, M., Lynch, I., Ejtehadi, M. R., Monopoli, M. P., Bombelli, F. B., & Laurent, S. (2011). Protein-nanoparticle interactions: Opportunities and challenges. *Chemical Reviews*, 111(9), 5610–5637. Retrieved from https://www.ncbi.nlm.nih.gov/pubmed/21688848. doi:10.1021/cr100440g

Mahmoudi, M., Saeedi-Eslami, S. N., Shokrgozar, M. A., Azadmanesh, K., Hassanlou, M., Kalhor, H. R., . . . Vali, H. (2012). Cell "vision": Complementary factor of protein corona in nanotoxicology. *Nanoscale*, 4(17), 5461–5468. Retrieved from https://www.ncbi.nlm.nih.gov/pubmed/22842341. doi:10.1039/c2nr31185b

Maiolo, D., Bergese, P., Mahon, E., Dawson, K. A., & Monopoli, M. P. (2014). Surfactant titration of nanoparticle-protein corona. *Analytical Chemistry*, 86(24), 12055–12063. Retrieved from https://www.ncbi.nlm.nih.gov/pubmed/25350777. doi:10.1021/ac5027176

Maiorano, G., Sabella, S., Sorce, B., Brunetti, V., Malvindi, M. A., Cingolani, R., & Pompa, P. P. (2010). Effects of cell culture media on the dynamic formation of protein–nanoparticle complexes and influence on the cellular response. *ACS Nano*, 4(12), 7481–7491. Retrieved from https://www.ncbi.nlm.nih.gov/pubmed/21082814. doi:10.1021/nn101557e

Mala, R., Annie Aglin, A., Ruby Celsia, A. S., Geerthika, S., Kiruthika, N., VazagaPriya, C., & Srinivasa Kumar, K. (2017). Foley catheters functionalised with a synergistic combination of antibiotics and silver nanoparticles resist biofilm formation. *IET Nanobiotechnology*, 11(5), 612–620. Retrieved from https://www.ncbi.nlm.nih.gov/pubmed/28745297. doi:10.1049/iet-nbt.2016.0148

Meir, R., Shamalov, K., Betzer, O., Motiei, M., Horovitz-Fried, M., Yehuda, R., . . . Cohen, C. J. (2015). Nanomedicine for cancer immunotherapy: Tracking cancer-specific T-cells in vivo with gold nanoparticles and CT imaging. *ACS Nano*, 9(6), 6363–6372. Retrieved from https://www.ncbi.nlm.nih.gov/pubmed/26039633. doi:10.1021/acsnano.5b01939

Michalet, X., Pinaud, F. F., Bentolila, L. A., Tsay, J. M., Doose, S., Li, J. J., . . . Weiss, S. (2005). Quantum dots for live cells, in vivo imaging and diagnostics. *Science*, 307(5709), 538–544. Retrieved from https://www.ncbi.nlm.nih.gov/pubmed/15681376. doi:10.1126/science.1104274

Milani, S., Bombelli, F. B., Pitek, A. S., Dawson, K. A., & Radler, J. (2012). Reversible versus irreversible binding of transferrin to polystyrene nanoparticles: Soft and hard corona. *ACS Nano*, 6(3), 2532–2541. Retrieved from https://www.ncbi.nlm.nih.gov/pubmed/22356488. doi:10.1021/nn204951s

Mirshafiee, V., Kim, R., Mahmoudi, M., & Kraft, M. L. (2016). The importance of selecting a proper biological milieu for protein corona analysis in vitro: Human plasma versus human serum. *The International Journal of Biochemistry & Cell Biology*, 75, 188–195. Retrieved from https://www.ncbi.nlm.nih.gov/pubmed/26643610. doi:10.1016/j.biocel.2015.11.019

Moghimi, S. M., Muir, I. S., Illum, L., Davis, S. S., & Kolb-Bachofen, V. (1993). Coating particles with a block co-polymer (poloxamine-908) suppresses opsonization but permits the activity of dysopsonins in the serum. *Biochimica et Biophysica Acta*, 1179(2), 157–165. Retrieved from https://www.ncbi.nlm.nih.gov/pubmed/8218358. doi:10.1016/0167-4889(93)90137-e

Monopoli, M. P., Aberg, C., Salvati, A., & Dawson, K. A. (2012). Biomolecular coronas provide the biological identity of nanosized materials. *Nature Nanotechnology*, 7(12), 779–786. Retrieved from https://www.ncbi.nlm.nih.gov/pubmed/23212421. doi:10.1038/nnano.2012.207

Monopoli, M. P., Bombelli, F. B., & Dawson, K. A. (2011). Nanobiotechnology: Nanoparticle coronas take shape. *Nature Nanotechnology*, 6(1), 11–12. Retrieved from https://www.ncbi.nlm.nih.gov/pubmed/21179098. doi:10.1038/nnano.2011.267

Mukherjee, P., Senapati, S., Mandal, D., Ahmad, A., Khan, M. I., Kumar, R., & Sastry, M. (2002). Extracellular synthesis of gold nanoparticles by the fungus *Fusarium oxysporum*. *Chembiochem*, 3(5), 461–463. Retrieved from https://www.ncbi.nlm.nih.gov/pubmed/12007181. doi:10.1002/1439-7633(20020503)3:5<461::AID-CBIC461>3.0.CO;2-X

Nagayama, S., Ogawara, K., Fukuoka, Y., Higaki, K., & Kimura, T. (2007). Time-dependent changes in opsonin amount associated on nanoparticles alter their hepatic uptake characteristics. *International Journal of Pharmaceutics*, 342(1–2), 215–221. Retrieved from https://www.ncbi.nlm.nih.gov/pubmed/17566676. doi:10.1016/j.ijpharm.2007.04.036

Nel, A. E., Madler, L., Velegol, D., Xia, T., Hoek, E. M., Somasundaran, P., . . . Thompson, M. (2009). Understanding biophysicochemical interactions at the nano-bio interface. *Nature Materials*, 8(7), 543–557. Retrieved from https://www.ncbi.nlm.nih.gov/pubmed/19525947. doi:10.1038/nmat2442

Ngunjiri, J. N., Daniels, S. L., Li, J. R., Serem, W. K., & Garno, J. C. (2008). Controlling the surface coverage and arrangement of proteins using particle lithography. *Nanomedicine (London, England)*, 3(4), 529–541. Retrieved from https://www.ncbi.nlm.nih.gov/pubmed/18694315. doi:10.2217/17435889.3.4.529

O'Brien, J., & Shea, K. J. (2016). Tuning the protein corona of hydrogel nanoparticles: The synthesis of abiotic protein and peptide affinity reagents. *Accounts of Chemical Research*, 49(6), 1200–1210. Retrieved from https://www.ncbi.nlm.nih.gov/pubmed/27254382. doi:10.1021/acs.accounts.6b00125

Osaka, T., Nakanishi, T., Shanmugam, S., Takahama, S., & Zhang, H. (2009). Effect of surface charge of magnetite nanoparticles on their internalization into breast cancer and umbilical vein endothelial cells. *Colloids and Surfaces, B: Biointerfaces*, 71(2), 325–330. Retrieved from https://www.ncbi.nlm.nih.gov/pubmed/19361963. doi:10.1016/j.colsurfb.2009.03.004

Owens, D. E., 3rd, & Peppas, N. A. (2006). Opsonization, biodistribution and pharmacokinetics of polymeric nanoparticles. *International Journal of Pharmaceutics*, 307(1), 93–102. Retrieved from https://www.ncbi.nlm.nih.gov/pubmed/16303268. doi:10.1016/j.ijpharm.2005.10.010

Piella, J., Bastus, N. G., & Puntes, V. (2017). Size-dependent protein–nanoparticle interactions in citrate-stabilized gold nanoparticles: The emergence of the protein corona. *Bioconjugate Chemistry*, 28(1), 88–97. Retrieved from https://www.ncbi.nlm.nih.gov/pubmed/27997136. doi:10.1021/acs.bioconjchem.6b00575

Pisani, C., Gaillard, J. C., Odorico, M., Nyalosaso, J. L., Charnay, C., Guari, Y., . . . Prat, O. (2017). The timeline of corona formation around silica nanocarriers highlights the role of the protein interactome. *Nanoscale*, 9(5), 1840–1851. Retrieved from https://www.ncbi.nlm.nih.gov/pubmed/27858044. doi:10.1039/c6nr04765c

Pozzi, D., Caracciolo, G., Digiacomo, L., Colapicchioni, V., Palchetti, S., Capriotti, A. L., . . . Lagana, A. (2015). The biomolecular corona of nanoparticles in circulating biological media. *Nanoscale*, 7(33), 13958–13966. Retrieved from https://www.ncbi.nlm.nih.gov/pubmed/26222625. doi:10.1039/c5nr03701h

Qiu, Y., Liu, Y., Wang, L., Xu, L., Bai, R., Ji, Y., . . . Chen, C. (2010). Surface chemistry and aspect ratio mediated cellular uptake of Au nanorods. *Biomaterials*, 31(30), 7606–7619. Retrieved from https://www.ncbi.nlm.nih.gov/pubmed/20656344. doi:10.1016/j.biomaterials.2010.06.051

Rana, S., Nazar, U., Ali, J., Ali, Q. U. A., Ahmad, N. M., Sarwar, F., . . . Jamil, S. U. U. (2018). Improved antifouling potential of polyether sulfone polymeric membrane containing silver nanoparticles: Self-cleaning membranes. *Environmental Technology*, 39(11), 1413–1421. Retrieved from https://www.ncbi.nlm.nih.gov/pubmed/28504053. doi:10.1080/09593330.2017.1330900

Rengan, A. K., Bukhari, A. B., Pradhan, A., Malhotra, R., Banerjee, R., Srivastava, R., & De, A. (2015). In vivo analysis of biodegradable liposome gold nanoparticles as efficient agents for photothermal therapy of cancer. *Nano Letters*, 15(2), 842–848. Retrieved from https://www.ncbi.nlm.nih.gov/pubmed/25554860. doi:10.1021/nl5045378

Ritz, S., Schottler, S., Kotman, N., Baier, G., Musyanovych, A., Kuharev, J., . . . Mailander, V. (2015). Protein corona of nanoparticles: Distinct proteins regulate the cellular uptake. *Biomacromolecules*, 16(4), 1311–1321. Retrieved from https://www.ncbi.nlm.nih.gov/pubmed/25794196. doi:10.1021/acs.biomac.5b00108

Roach, P., Farrar, D., & Perry, C. C. (2005). Interpretation of protein adsorption: Surface-induced conformational changes. *Journal of the American Chemical Society*, 127(22), 8168–8173. Retrieved from https://www.ncbi.nlm.nih.gov/pubmed/15926845. doi:10.1021/ja042898o

Roser, M., Fischer, D., & Kissel, T. (1998). Surface-modified biodegradable albumin nano- and microspheres. II: Effect of surface charges on in vitro phagocytosis and biodistribution in rats. *European Journal of Pharmaceutics and Biopharmaceutics*, 46(3), 255–263. Retrieved from https://www.ncbi.nlm.nih.gov/pubmed/9885296

Rosi, N. L., & Mirkin, C. A. (2005). Nanostructures in biodiagnostics. *Chemical Reviews*, 105(4), 1547–1562. Retrieved from https://www.ncbi.nlm.nih.gov/pubmed/15826019. doi:10.1021/cr030067f

Sabale, S., Jadhav, V., Khot, V., Zhu, X., Xin, M., & Chen, H. (2015). Superparamagnetic MFe2O 4 (M = Ni, Co, Zn, Mn) nanoparticles: Synthesis, characterization, induction heating and cell viability studies for cancer hyperthermia applications. *Journal of Materials Science. Materials in Medicine*, 26(3), 127. Retrieved from https://www.ncbi.nlm.nih.gov/pubmed/25690622. doi:10.1007/s10856-015-5466-7

Sabuncu, A. C., Grubbs, J., Qian, S., Abdel-Fattah, T. M., Stacey, M. W., & Beskok, A. (2012). Probing nanoparticle interactions in cell culture media. *Colloids and Surfaces, B: Biointerfaces*, 95, 96–102. Retrieved from https://www.ncbi.nlm.nih.gov/pubmed/22421416. doi:10.1016/j.colsurfb.2012.02.022

Sakulkhu, U., Mahmoudi, M., Maurizi, L., Salaklang, J., & Hofmann, H. (2014). Protein corona composition of superparamagnetic iron oxide nanoparticles with various physico-chemical properties and coatings. *Scientific Reports*, 4, 5020.

Retrieved from https://www.ncbi.nlm.nih.gov/pubmed/24846348. doi:10.1038/srep05020

Salvati, A., Pitek, A. S., Monopoli, M. P., Prapainop, K., Bombelli, F. B., Hristov, D. R., . . . Dawson, K. A. (2013). Transferrin-functionalized nanoparticles lose their targeting capabilities when a biomolecule corona adsorbs on the surface. *Nature Nanotechnology*, 8(2), 137–143. Retrieved from https://www.ncbi.nlm.nih.gov/pubmed/23334168. doi:10.1038/nnano.2012.237

Sangra, M., Estelrich, J., Sabate, R., Espargaro, A., & Busquets, M. A. (2017). Evidence of protein adsorption in pegylated liposomes: Influence of liposomal decoration. *Nanomaterials (Basel)*, 7(2), 37–53. Retrieved from https://www.ncbi.nlm.nih.gov/pubmed/28336870. doi:10.3390/nano7020037

Saptarshi, S. R., Duschl, A., & Lopata, A. L. (2013). Interaction of nanoparticles with proteins: Relation to bio-reactivity of the nanoparticle. *Journal of Nanobiotechnology*, 11, 26. Retrieved from https://www.ncbi.nlm.nih.gov/pubmed/23870291. doi:10.1186/1477-3155-11-26

Selvakannan, P. R., Swami, A., Srisathiyanarayanan, D., Shirude, P. S., Pasricha, R., Mandale, A. B., & Sastry, M. (2004). Synthesis of aqueous Au core–Ag shell nanoparticles using tyrosine as a pH-dependent reducing agent and assembling phase-transferred silver nanoparticles at the air–water interface. *Langmuir*, 20(18), 7825–7836. Retrieved from https://www.ncbi.nlm.nih.gov/pubmed/15323537. doi:10.1021/la049258j

Senapati, S., Ahmad, A., Khan, M. I., Sastry, M., & Kumar, R. (2005). Extracellular biosynthesis of bimetallic Au–Ag alloy nanoparticles. *Small*, 1(5), 517–520. Retrieved from https://www.ncbi.nlm.nih.gov/pubmed/17193479. doi:10.1002/smll.200400053

Sengupta, B., Gregory, W. E., Zhu, J., Dasetty, S., Karakaya, M., Brown, J. M., . . . Podila, R. (2015). Influence of carbon nanomaterial defects on the formation of protein corona. *RSC Advances*, 5(100), 82395–82402. Retrieved from https://www.ncbi.nlm.nih.gov/pubmed/26877870. doi:10.1039/C5RA15007H

Shah, J. H., Fiaz, M., Athar, M., Ali, J., Rubab, M., Mehmood, R., . . . Djellabi, R. (2019). Facile synthesis of N/B-double-doped $Mn_2O_3$ and $WO_3$ nanoparticles for dye degradation under visible light. *Environmental Technology*, 1–10. Retrieved from https://www.ncbi.nlm.nih.gov/pubmed/30623749. doi:10.1080/09593330.2019.1567604

Shang, L., Nienhaus, K., & Nienhaus, G. U. (2014). Engineered nanoparticles interacting with cells: Size matters. *Journal of Nanobiotechnology*, 12, 5. Retrieved from https://www.ncbi.nlm.nih.gov/pubmed/24491160. doi:10.1186/1477-3155-12-5

Shankar, S., Soni, S. K., Daima, H. K., Selvakannan, P. R., Khire, J. M., Bhargava, S. K., & Bansal, V. (2015). Charge-switchable gold nanoparticles for enhanced enzymatic thermostability. *Physical Chemistry Chemical Physics*, 17, 21517–21524.

Shannahan, J. H., Podila, R., & Brown, J. M. (2015). A hyperspectral and toxicological analysis of protein corona impact on silver nanoparticle properties, intracellular modifications and macrophage activation. *International Journal of Nanomedicine*, 10, 6509–6521. Retrieved from https://www.ncbi.nlm.nih.gov/pubmed/26508856. doi:10.2147/IJN.S92570

Sharma, V. K., Siskova, K. M., Zboril, R., & Gardea-Torresdey, J. L. (2014). Organic-coated silver nanoparticles in biological and environmental conditions: Fate stability and toxicity. *Advances in Colloid and Interface Science*, 204, 15–34. Retrieved from https://www.ncbi.nlm.nih.gov/pubmed/24406050. doi:10.1016/j.cis.2013.12.002

Shete, R. B., Muniswamy, V. J., Pandit, A. P., & Khandelwal, K. R. (2015). Formulation of eco-friendly medicated chewing gum to prevent motion sickness. *AAPS PharmSciTech*, 16(5), 1041–1050. Retrieved from https://www.ncbi.nlm.nih.gov/pubmed/25652732. doi:10.1208/s12249-015-0296-y

Sintubin, L., De Windt, W., Dick, J., Mast, J., van der Ha, D., Verstraete, W., & Boon, N. (2009). Lactic acid bacteria as reducing and capping agent for the fast and efficient production of silver nanoparticles. *Applied Microbiology and Biotechnology*, 84(4), 741–749. Retrieved from https://www.ncbi.nlm.nih.gov/pubmed/19488750. doi:10.1007/s00253-009-2032-6

Smith, P. J., Giroud, M., Wiggins, H. L., Gower, F., Thorley, J. A., Stolpe, B., . . . Rappoport, J. Z. (2012). Cellular entry of nanoparticles via serum sensitive clathrin-mediated endocytosis and plasma membrane permeabilization. *International Journal of Nanomedicine*, 7, 2045–2055. Retrieved from https://www.ncbi.nlm.nih.gov/pubmed/22619541. doi:10.2147/IJN.S29334

Sneha, K., Sathishkumar, M., Mao, J., Kwak, I. S., & Yun, Y. S. (2010). Corynebacterium glutamicum-mediated crystallization of silver ions through sorption and reduction processes. *Chemical Engineering Journal*, 162(3), 989–996.

Strojan, K., Leonardi, A., Bregar, V. B., Krizaj, I., Svete, J., & Pavlin, M. (2017). Dispersion of nanoparticles in different media importantly determines the composition of their protein corona. *PLoS One*, 12(1), e0169552. Retrieved from https://www.ncbi.nlm.nih.gov/pubmed/28052135. doi:10.1371/journal.pone.0169552

Szekeres, G. P., & Kneipp, J. (2019). SERS probing of proteins in gold nanoparticle agglomerates. *Frontiers in Chemistry*, 7, 30. Retrieved from https://www.ncbi.nlm.nih.gov/pubmed/30766868. doi:10.3389/fchem.2019.00030

Tenzer, S., Docter, D., Kuharev, J., Musyanovych, A., Fetz, V., Hecht, R., . . . Stauber, R. H. (2013). Rapid formation of plasma protein corona critically affects nanoparticle pathophysiology. *Nature Nanotechnology*, 8(10), 772–781. Retrieved from https://www.ncbi.nlm.nih.gov/pubmed/24056901. doi:10.1038/nnano.2013.181

Tenzer, S., Docter, D., Rosfa, S., Wlodarski, A., Kuharev, J., Rekik, A., . . . Stauber, R. H. (2011). Nanoparticle size is a critical physicochemical determinant of the human blood plasma corona: A comprehensive quantitative proteomic analysis. *ACS Nano*, 5(9), 7155–7167. Retrieved from https://www.ncbi.nlm.nih.gov/pubmed/21866933. doi:10.1021/nn201950e

Toprak, M. (2016). Fluorescence study on the interaction of human serum albumin with Butein in liposomes. *Spectrochimica Acta. Part A, Molecular and Biomolecular Spectroscopy*, 154, 108–113. Retrieved from https://www.ncbi.nlm.nih.gov/pubmed/26519918. doi:10.1016/j.saa.2015.10.023

Treuel, L., Jiang, X., & Nienhaus, G. U. (2013). New views on cellular uptake and trafficking of manufactured nanoparticles. *Journal of Royal Society of Interface*, 10(82), 20120939. Retrieved from https://www.ncbi.nlm.nih.gov/pubmed/23427093. doi:10.1098/rsif.2012.0939

Vigneshwaran, N., Ashtaputre, N. M., Varadarajan, P. V., Nachane, R. P., Paralikar, K. M., & Balasubramanya, R. H. (2007). Biological synthesis of silver nanoparticles using the fungus *Aspergillus flavus*. *Materials Letters*, 61(6), 1413–1418.

Vilanova, O., Mittag, J. J., Kelly, P. M., Milani, S., Dawson, K. A., Radler, J. O., & Franzese, G. (2016). Understanding the kinetics of protein–nanoparticle corona formation. *ACS Nano*, 10(12), 10842–10850. Retrieved from https://www.ncbi.nlm.nih.gov/pubmed/28024351. doi:10.1021/acsnano.6b04858

Vroman, L., Adams, A. L., Fischer, G. C., & Munoz, P. C. (1980). Interaction of high molecular weight kininogen, factor XII, and fibrinogen in plasma at interfaces. *Blood*, 55(1), 156–159. Retrieved from https://www.ncbi.nlm.nih.gov/pubmed/7350935

Walczyk, D., Bombelli, F. B., Monopoli, M. P., Lynch, I., & Dawson, K. A. (2010). What the cell "sees" in bionanoscience. *Journal of the American Chemical Society*, 132(16), 5761–5768. Retrieved from https://www.ncbi.nlm.nih.gov/pubmed/20356039. doi:10.1021/ja910675v

Walkey, C. D., & Chan, W. C. (2012). Understanding and controlling the interaction of nanomaterials with proteins in a physiological environment. *Chemical Society Reviews*, 41(7), 2780–2799. Retrieved from https://www.ncbi.nlm.nih.gov/pubmed/22086677. doi:10.1039/c1cs15233e

Walkey, C. D., Olsen, J. B., Song, F., Liu, R., Guo, H., Olsen, D. W., . . . Chan, W. C. (2014). Protein corona fingerprinting predicts the cellular interaction of gold and silver nanoparticles. *ACS Nano*, 8(3), 2439–2455. Retrieved from https://www.ncbi.nlm.nih.gov/pubmed/24517450. doi:10.1021/nn406018q

Wang, G., Papasani, M. R., Cheguru, P., Hrdlicka, P. J., & Hill, R. A. (2012). Gold-peptide nanoconjugate cellular uptake is modulated by serum proteins. *Nanomedicine*, 8(6), 822–832. Retrieved from https://www.ncbi.nlm.nih.gov/pubmed/22051699. doi:10.1016/j.nano.2011.10.007

Wang, H., Lin, Y., Nienhaus, K., & Nienhaus, G. U. (2018). The protein corona on nanoparticles as viewed from a nanoparticle-sizing perspective. Wiley interdisciplinary reviews. *Nanomedicine and Nanobiotechnology*, 10(4), e1500. Retrieved from https://www.ncbi.nlm.nih.gov/pubmed/29071798. doi:10.1002/wnan.1500

Wang, Z., Zhao, Q., Cui, M., Pang, S., Wang, J., Liu, Y., & Xie, L. (2017). Probing temperature- and pH-dependent binding between quantum dots and bovine serum albumin by fluorescence correlation spectroscopy. *Nanomaterials (Basel)*, 7(5), 93–106. Retrieved from https://www.ncbi.nlm.nih.gov/pubmed/28441351. doi:10.3390/nano7050093

Yan, J. K., Qiu, W. Y., Wang, Y. Y., Wang, W. H., Yang, Y., & Zhang, H. N. (2018). Fabrication and stabilization of biocompatible selenium nanoparticles by carboxylic curdlans with various molecular properties. *Carbohydrate Polymers*, 179, 19–27. Retrieved from https://www.ncbi.nlm.nih.gov/pubmed/29111042. doi:10.1016/j.carbpol.2017.09.063

Ye, L., Zhang, Y., Wang, Q., Zhou, X., Yang, B., Ji, F., . . . Yao, F. (2016). Physical cross-linking starch-based zwitterionic hydrogel exhibiting excellent biocompatibility, protein resistance and biodegradability. *ACS Applied Materials & Interfaces*, 8(24), 15710–15723. Retrieved from https://www.ncbi.nlm.nih.gov/pubmed/27249052. doi:10.1021/acsami.6b03098

Yin, Y., Yang, X., Hu, L., Tan, Z., Zhao, L., Zhang, Z., . . . Jiang, G. (2016). Superoxide-mediated extracellular biosynthesis of silver nanoparticles by the fungus *Fusarium oxysporum*. *Environmental Science & Technology Letters*, 3(4).

Zhao, F., Zhao, Y., Liu, Y., Chang, X., Chen, C., & Zhao, Y. (2011). Cellular uptake, intracellular trafficking and cytotoxicity of nanomaterials. *Small*, 7(10), 1322–1337. Retrieved from https://www.ncbi.nlm.nih.gov/pubmed/21520409. doi:10.1002/smll.201100001

Zhdanov, V. P., & Cho, N. J. (2016). Kinetics of the formation of a protein corona around nanoparticles. *Mathematical Biosciences*, 282, 82–90. Retrieved from https://www.ncbi.nlm.nih.gov/pubmed/27720880. doi:10.1016/j.mbs.2016.09.018

Zucker, R. M., Daniel, K. M., Massaro, E. J., Karafas, S. J., Degn, L. L., & Boyes, W. K. (2013). Detection of silver nanoparticles in cells by flow cytometry using light scatter and far-red fluorescence. *Cytometry. Part A*, 83(10), 962–972. Retrieved from https://www.ncbi.nlm.nih.gov/pubmed/23943267. doi:10.1002/cyto.a.22342

## Glossary

**Adsorbome:** Group of proteins present in blood plasma that bind to the nanoparticles once it is exposed to blood or serum in in vitro conditions.

**Apolipoprotein:** Proteins that bind to lipids for their transport.

**Complements:** Group of blood proteins that enhances phagocytosis and induces inflammation.

**Conformation:** Arrangement of atoms in space in a molecule.

**Enthalpy:** Measure of energy of in a thermodynamic system.

**Extrinsic:** Originating from outside or extraneous.

**Fibrinogen:** Precursor of fibrin that is essential for blood clotting.

**Haptoglobin:** Protein synthesized by liver to clear free hemoglobin.

**Hydrogel:** Three-dimensional network of polymers that absorbs large quantities of water.

**Hydrophilic:** Water soluble.

**Hydrophobic:** Water insoluble.

**Hyperthermia:** Increased temperature or heat energy.

**Immunoglobulins:** Glycoproteins produced by plasma cells.

**Interactome:** It is a network of protein–protein and protein–DNA interaction. It encompasses transcription factors and regulatory proteins and their target genes. It represents a whole set of interactions within a cell.

**Intrinsic:** Basic property.

**Lithography:** Greek meaning: Lithos, stone; Garphene, to write. In technical aspect it is defined as a "transfer of a pattern to a photosensitive material by selective exposure to a radiation source."

**Plasminogen:** Precursor of plasmin.

**Plasma:** Clear straw yellow-colored liquid of blood devoid of cellular components.

**Pristine:** Original and pure.

**Serum:** Fluid containing solute components of blood after coagulated blood.

**Theranostics:** Concept of combining diagnostics and therapy.

**Transferrin:** Iron binding and transporting glycoprotein.

**Trypsin:** It is a serine protease of digestive system.

**Vroman effect:** Postulates about the adsorption of plasma proteins onto a biomaterial surface. It states that the identity of the protein adsorbed to a biomaterial or a nanomaterial is dynamic and changes over time.

# 7

# Interaction Effects of Nanoparticles with Microorganisms Employed in the Remediation of Nitrogen-Rich Wastewater

*Parmita Chawley and Sheeja Jagadevan*

*Department of Environmental Science and Engineering, Indian Institute of Technology (Indian School of Mines), Dhanbad, Jharkhand, India*

## Objectives

- To understand biological nitrogen removal process, particularly nitrification and denitrification.
- To learn about various genes and enzymes involved in nitrification and denitrification processes.
- To understand the effect of nanoparticles on microorganisms used for treatment of nitrogen-rich wastewater.
- To understand the changes in functional abundance, gene expression, and enzymatic response of nitrifying and denitrifying microorganisms upon nanoparticle exposure.

## 7.1 Introduction

Nanoparticles (NPs) found in the environment either occur naturally or exist as engineered nanoparticles. Ash and smoke from forest fires, dust storms, humic substances, oxides and oxyhydroxides of iron, manganese, aluminum, and aluminosilicates generated during chemical weathering of rocks are the main sources of naturally occurring nanoparticles (Hartland, Lead, Slaveykova, O'Carroll, &

*Microbial Interactions at Nanobiotechnology Interfaces: Molecular Mechanisms and Applications*, First Edition. Edited by R. Navanietha Krishnaraj and Rajesh K. Sani.

Valsami-Jones, 2013). These nanoparticles bind to trace metals and greatly influence the bioavailability of nutrients and toxic metals in microorganisms, thereby producing a buffering effect on the environment (Hartland et al., 2013; Tipping, 2001). NPs in the size range of 1–100 nm impart unique properties to materials, thus furthering the development of nanotechnology for implementation in various spheres at a rapid pace (Nazarenko, Han, Lioy, & Mainelis, 2011). Nanoparticles are engineered for unique characteristics such as shape, size, and surface properties imparting special functionalities including catalytic behavior, improved strength, etc. However, the widespread use of engineered NPs such as ZnO, $TiO_2$, CuO, and $SiO_2$ leads to the release of these NPs into the environment as nanopollutants.

Engineered NPs reach the environment from cosmetics, personal care products, paints and surface coatings, nanopesticides, nanofertilizers, medical nanoformulations, etc. (Domercq, Praetorius, & Boxall, 2018). In addition, improper management in dumping of construction wastes such as concrete, tiles, and plasterboard contributes to the release of several metal and metalloid NPs into the environment, posing serious health risks to the population (Oliviera et al., 2019). In addition, engineered NPs contribute hugely to sewage sludge during waste water treatment (Bundschuh et al., 2018). NPs' concentration may reach as high as dozens of mg/l following swarming of industrial wastewater containing engineered NPs into wastewater treatment (WWT) plants.

Nitrogen removal from wastewater is found to be particularly vulnerable to NP toxicity (Zheng, Chen, & Wu, 2011a; Zheng, Wu, & Chen, 2011b). Biological treatment of nitrogen-rich wastewater involves nitrifying, denitrifying, and anammox microorganisms. Use of membrane bioreactor (MBR) is one of the technologies employed in WWT, which uses these microorganisms. Over time, an MBR system usually faces membrane fouling, which hinders the WWT process. In such scenarios, NPs are used to coat membrane surfaces in MBR to achieve an antifouling effect in MBR (Cheng & Hong, 2020). NPs from such coated membranes leach into the system and as a result microorganisms present in WWT systems constantly come in contact with engineered NPs while being exposed to NPs coming from the discharge of manufacturing industries, municipal sewage, agricultural wastewater, etc.

Engineered NPs are unique in their specific structure, to which microorganisms may not be capable of adapting and thus produce toxicity based on dosage. Engineered metal oxide NPs are reported to have more enhanced toxic effects than carbon containing NPs (Slavin, Asnis, Häfeli, & Bach, 2017). In a bacterial system, metal NP generates reactive oxygen species (ROS), which consequently damages DNA, protein, and lipids within cells and has a high affinity to bind to —SH groups in proteins and disrupt enzymatic activities (Slavin et al., 2017). Thus, it becomes imperative to understand the interaction effect of NPs on microbial response employed in WWT. This chapter discusses gene transcriptional response and enzymatic responses of nitrifying and denitrifying bacteria to NPs. In addition, morphological changes in the nitrifying bacteria due to NP exposure have also been dealt with.

## 7.2 Bacterial Nitrification Process

Biological nitrification is a two-step oxidation process in which oxidation of ammonia is carried out by two-step sequential biochemical reactions. The first step is the oxidation of ammonia into nitrite by ammonia oxidizing bacteria (AOB) (Eq. 7.1) and the second step involves oxidation of nitrite into nitrate by nitrite oxidizing bacteria (NOB) (Eq. 7.2). Nitrifying bacteria therefore involve two groups: AOB and NOB.

$$NH_3^+ + 1.5O_2 \xrightarrow{\text{AOB}} NO_2^- + H_2O + H^+ \tag{7.1}$$

$$NO_2^- + 0.5O_2 \xrightarrow{\text{NOB}} NO_3^- \tag{7.2}$$

The above set of biochemical oxidation reactions are performed by ammonia monooxygenase (*amo*) and hydroxylamine oxidoreductase (*hao*) gene clusters present in AOB. The final product of the full nitrification process is achieved by oxidation of nitrite into nitrate by nitrite oxido-reductase (*nxrA*), which is performed by NOB.

$$NH_3 + O_2 + 2H^+ + 2e^- \xrightarrow{amo} NH_2OH + H_2O \tag{7.3}$$

$$NH_2OH + H_2O \xrightarrow{hao} NO_2^- + 5H^+ + 4e^- \tag{7.4}$$

### 7.2.1 Effect of NPs on Functional Gene Abundance and Transcriptional Response

The effect of NPs on transcriptional and other microbial responses occurs in a dose-dependent manner. In *Nitrosomonas europaea*, ZnO NP at 10 mg/l was found to enhance *amoA* and *hao* gene expression after six hours of incubation, which decreased substantially after a period of three days, probably due to loss of cell membrane integrity (J. Wu et al., 2018). The initial upregulation of functional gene transcription might be due to the cell's intrinsic ability to cope with NP-induced toxicity. Higher levels of ZnO NP (50 mg/l) were found to decrease the expression of *amoA* and *hao*, which in turn resulted in ceasing ammonia oxidation activity (J. Wu et al., 2018). Toxicity at higher dosage is due to intracellular accumulation of ZnO NP, thereby inhibiting the cell's metabolic activities. In an enriched nitrifying community, *amoA* expression has been found to increase at lower dose (0.1–10 mg/l) of ZnO NP and declined on increasing its dosage. On the contrary, the reverse trend was observed for *hao* and *nxrA*, thereby suggesting that the expression patterns of *amoA* and *hao* follow different regulatory mechanisms upon ZnO NP exposure, although harbored inside the same bacteria (Q. Wu et al., 2018). Moreover, the activity of NOB was found to be higher than that of AOB at higher ZnO NP dose.

Long-term exposure of 1 mg/l CuO NP to nitrification sludge has demonstrated an increase in the abundance of AOB (X. Zhang et al., 2019, 2020). Moreover, the metal resistance gene, *copA* expression increased twofold (X. Zhang et al., 2019). The *copA* gene encodes for an ATP-ase pump that extrudes copper ions from the cytoplasm to the periplasmic space. Enhanced expression of *copA* gene in response to 1 mg/l CuO NP is possible because of the cell's ability for self-protection against toxic levels of $Cu^{2+}$ ions.

In another recent study conducted on compost soil, the copy numbers of *amoA* declined on addition of 10 mg/kg AgNPs and increased on addition of 20 mg/kg AgNPs. The abundance of *nxrA* was detrimentally affected at all concentrations of AgNPs, suggesting vulnerability of NOB (Zeng et al., 2018). Higher concentration of AgNPs (50 mg/l) demonstrated inhibitory effect on nitrogen removal in constructed wet land due to a net accumulation of nitrite and nitrate (X. Liu et al., 2019). In contrast, long-term dose of AgNP at lower concentration (50–200 μg/l) has shown to induce stimulatory effect on *amoA* and *nxrA* gene expression by mixed nitrifying bacteria in a constructed wetland (Huang, Cao, Liu, Yan, & Xiao, 2019). Furthermore, 0.1 mg/l AgNP produced no significant effect on the copy numbers of *amoA* gene on a long-term exposure of 60 days in a membrane bioreactor (C. Zhang, Liang, & Hu, 2014). This might have been due to the result of microbe-induced nanoparticle aggregation, which reduced bioavailability and toxicity to microbes in MBR (Choi, Yu, Esteban Fernández, & Hu, 2010). Interestingly, the gene copy number of silver resistance gene (*silE*) increased upon exposure to long-term AgNP stress, which indicates sensitivity of *silE* gene to $Ag^{+}$ exposure (C. Zhang et al., 2014).

The effects of binary mixtures of NPs on *N. europaea* have also been studied because mixtures of multiple NPs can exert synergistic or antagonistic response (Choi & Hu, 2009; Tong et al., 2014, 2015). ZnO NPs showed a higher cytotoxic impact on *N. europaea* than $CeO_2$ and $TiO_2$ NP at the same dosage (Yu, Wu, Liu, Zhu et al., 2016). Binary mixtures of $CeO_2$ and $TiO_2$ resulted in upregulation of *amoA* gene expression with increasing dose of both the NPs, whereas NP stress by mixture of ZnO and $CeO_2$ lead to downregulation of *amoA* expression (Yu, Wu, Liu, Zhu et al., 2016). This clearly indicated that $CeO_2$ NP has a synergistic effect with $TiO_2$ NP whereas it produced antagonistic effect along with ZnO NP on nitrification activity by *N. europaea*.

### 7.2.2 Effect of NPs on Enzyme Activity

The enzymatic activity reflects functionality of a living system and is dependent on availability of metal ion cofactor. Any physiological or enzymatic response of microorganism due to any interaction/s with NP is largely dependent upon its functional behavior. The functional behavior of NPs is largely influenced by the dissolution of metal ions from metallic NP into the aqueous media (Turan, Erkan, Engin, & Bilgili, 2019). Several studies have reported that ionic form of metal ions is more toxic than nanoparticle (Batley, Kirby, & McLaughlin, 2013; Bielmyer, Grosell, & Brixti, 2006). Metal ions thus released are either adsorbed onto the microbial surface or taken inside the microbial cell.

In a constructed wetland that consisted of mixed nitrifying consortium, low dosage of AgNP (50–200 μg/l) for long-term exposure (450 days) resulted in enhanced AMO enzyme activity (Huang et al., 2019). Another recent study reported decline in AMO activity at higher dose of AgNPs (30 mg/l) (Zheng et al., 2017). Toxic effects exerted by NPs on key enzyme activities are due to the generation of reactive oxygen species (ROS), which causes oxidative stress in the bacterial cells via lipid peroxidation, destruction of proteins/enzymes and nucleic acids (Yazdanbakhsh, Rafiee, Daraei, & Amoozegar, 2019). In contrast, Alito and Gunsch (2014) suggested that replacement of $Cu^{2+}$ ion (cofactor for AMO enzyme) from the surface of AMO enzyme by $Ag^{+}$ ion might have caused reduction in its activity. Although $Cu^{2+}$ ion acts as a cofactor for AMO enzyme, higher concentration of CuO NP (60 mg/l) has been found to cause pronounced effect in reduction of AMO and nitrite oxidoreductase in activated sludge (Wang, Li, Gao, She, Guo, et al., 2017).

Other NPs such as NiO and $CeO_2$ have also demonstrated toxic effects on the nitrifying bacterial population by causing a reduction in enzymatic activity when used at concentration upto 60 mg/l (Wang et al., 2016, Wang, Li, Gao, She, Guo, et al., 2017). The toxic effect of $CeO_2$ NP on the membrane integrity was checked by the release of lactate dehydrogenase (LDH). Long-term exposure to $CeO_2$ NP had shown increase in LDH release with rise in $CeO_2$ dosage, leading to the rupture of membrane (Wang et al., 2016). Similar responses to LDH have been obtained for stress to Ag NP (Z. Zhang et al., 2016). Loss of membrane integrity depends on fatty acid contents of the cell membrane, physical permeation of small-sized NPs into the cell membrane, and membrane degradation by reactive oxygen species (Ma, Quan, Si, & Wu, 2013). After penetration inside the microbial cell, NPs disturb the oxidation–reduction process in the cell, bringing about reduction in enzyme activity. Similar responses have been obtained in activated sludge upon long-term treatment with 5–60 mg/l $TiO_2$ NP in a sequential batch reactor. The activity of AMO and nitrite oxidoreductase enzymes reduced to 30.80 and 31.18%, respectively at 60 mg/l $TiO_2$ (Z. Li et al., 2017), thereby suggesting that enzymatic activities of AOB and NOB are almost equal in response to $TiO_2$ NP exposure.

The effect of bimetallic NPs on nitrifying microorganisms has also been studied. Ag NP and nano zero valent iron are widely used in nanotechnology and thus their combined effect was studied on activated sludge. Ag–Fe NP at 0–100 mg/l concentration had shown reduction in AMO and nitrite reductase (NIR) activity in activated sludge (Yazdanbakhsh et al., 2019). The detrimental effect of Ag–Fe NP was more pronounced on NIR activity than upon AMO enzyme activity (Yazdanbakhsh et al., 2019). The viability of activated sludge upon NP stress was also studied by measuring catalase activity, dehydrogenase activity, and LDH release. Catalase acts as an antioxidant and protects the cell from the toxic effects of ROS and dehydrogenase takes part in the bacterial aerobic respiratory activities. Upon introduction of Ag–Fe NP into activated sludge, low catalase and dehydrogenase activities were found, indicating loss in cell viability. Concurrent marked increase in LDH release indicates loss of membrane integrity and cell death due to exposure to NPs.

### 7.2.3 Effect on Cellular Morphology

NPs have been shown to produce deformative changes in the *N. europaea* cell morphology. Control cells without exposure to NPs show an ellipsoidal shape and have been found to contain empty cavities within their cytosolic compartment. Upon exposure to NPs, cell membrane was found to be folded, concaved, or disorganized and intracellular cavities were found to release its contents (Yu, Wu, Liu, Chen et al. 2016; Yu, Wu, Liu, Zhu et al. 2016). When exposed to 24-hour Ag NP stress, nitrifying biofilm resulted in shrinkage and detachment of plasma membrane from the outer membrane (Sheng, Van Nostrand, Zhou, & Liu, 2015). This ultimately led to an apoptosis-like response in bacteria (Pandian, Deepak, Kalishwaralal, Viswanathan, & Gurunathan, 2010). Ag NPs were found to be adsorbed onto the surface of extra-polymeric substances (EPS) layer on 45 minutes of exposure and transmission electron microscope (TEM) images (Figure 7.1) have shown that Ag NPs entered the cells after 24-hour exposure (Sheng et al., 2015). Microorganisms protect themselves against NP toxicity by developing a protective layer of EPS, which helps to impart a protective mechanism against the intrusion of NPs by adsorbing them onto its surface. This is facilitated by electrostatic interactions between ionic functional groups such as hydroxyl, carboxyl, phosphate, and amine groups on EPS with the charged ionic NPs (Tang, Wu, Esquivel-Elizondo, Sørensen, & Rittmann, 2018). *N. europaea* when treated with 25 and 200 nm anatase $TiO_2$ NP caused distorted morphology when compared to control (Fang, Yu, Li, Somasundaran, & Chandran, 2010). However, when *N. europaea* cells were treated with same concentration of both 25 and 200 nm anatase

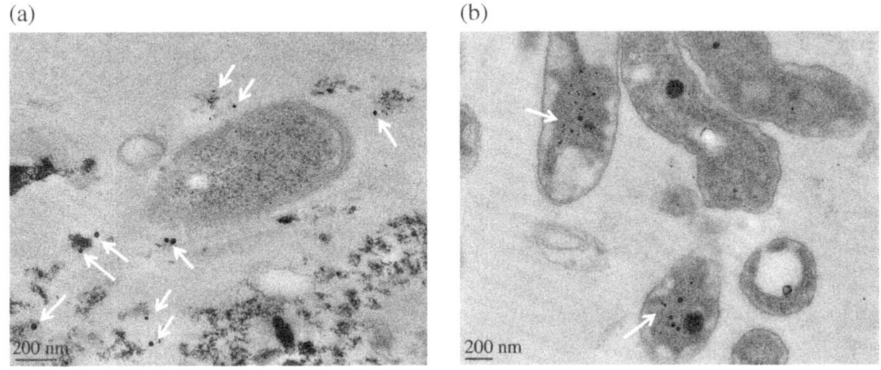


**Figure 7.1** Transmission electron microscopy (TEM) images. (a) Wastewater biofilms incubated with Ag NPs for 45 minutes. (b) wastewater biofilms incubated with Ag NPs for 24 hours. Ag NPs are indicated by white arrows (Sheng et al., 2015).

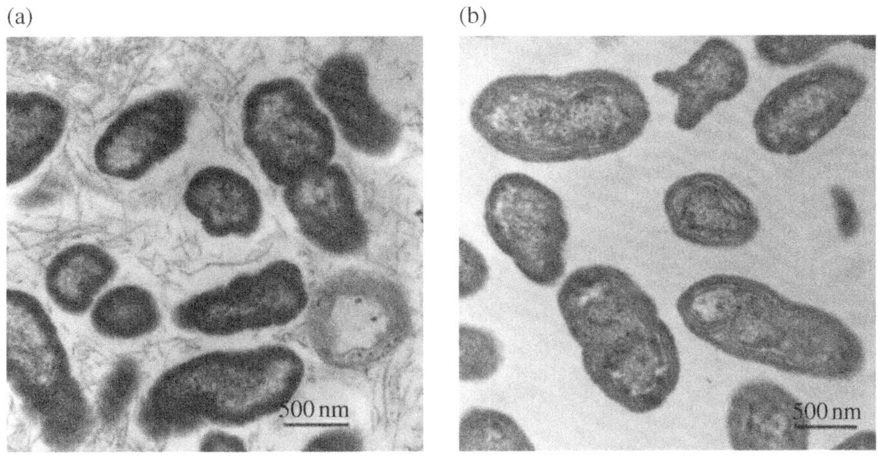


**Figure 7.2** *N. europaea* exposed to (a) 200 ppm of 25 nm anatase $TiO_2$ NPs. (b) 200 ppm of 200 nm anatase $TiO_2$ NPs exposed for four hours (Fang et al., 2010).

$TiO_2$ NP, smaller sized NPs produced more pronounced morphological damage to cells than 200 nm anatase $TiO_2$ NP (Figure 7.2).

## 7.3 Effect of NPs on Denitrifying Bacteria

Denitrification is an effective biological process for removal of nitrate from polluted water. Denitrifying bacteria convert $NO_3^-$ to $N_2$ via NO, and $N_2O$ under hypoxic/anoxic conditions and thus contribute to $NO_3^-$–N removal from wastewater.

This entire reaction pathway is carried out by nitrate reductase (*nar*), nitrite reductase (*nir*), nitric oxide reductase (*nor*), nitrous oxide reductase (*nos*) gene clusters (Lycus et al., 2018). The denitrifying gene structure is highly organized into operon systems.

$$NO_3^- \xrightarrow{NAR} NO_2^- \xrightarrow{NIR} NO \xrightarrow{NOR} N_2O \xrightarrow{NOS} N_2$$

In the cytoplasm, nitrate is reduced to nitrite (Eq. 7.5) by a membrane-bound respiratory nitrate reductase, which is encoded by *NarGHJI* operon (Bueno, Mesa, Bedmar, Richardson, & Delgado, 2012). This reaction is also catalyzed by a periplasmic nitrate reductase (NapA), which is a part of the *NapABC* operon.

$$NO_3^- + 2e^- + 2H^+ \rightarrow NO_2^- + H_2O \tag{7.5}$$

The nitrite produced is then reduced by the periplasmic nitrite reductase (NIR) to nitric oxide (Eq. 7.6). Two kinds of *nir* genes are found in denitrifying organisms: (i) heme requiring *nirS* gene and (ii) copper requiring *nirK* gene.

$$2NO_2^- + 2e^- + 4H^+ \rightarrow 2NO + 2H_2O \tag{7.6}$$

Further reduction of nitric oxide into nitrous oxide gas is catalyzed by NOR enzyme (Eq. 7.7). Three types of NOR enzymes have been characterized: cNOR, qNOR, and $qCu_ANOR$. Out of these, product of cNOR, encoded by *norCBQD* operon, has been most widely studied.

$$2NO + 2e^- + 2H^+ \rightarrow N_2O + H_2O \tag{7.7}$$

The final step of denitrification is catalyzed by nitrous oxide (NOS) enzyme (Eq. 7.8), which consists of copper requiring homo-dimeric subunits.

$$N_2O + 2e^- + 2H^+ \rightarrow N_2 + H_2O \tag{7.8}$$

These denitrifying microorganisms come in contact with NPs in the polluted water and, hence, it becomes important to summarize the latter's effect during the conversion of $NO_3^-$ to $N_2$.

### 7.3.1 Effect on Functional Gene Abundance and Transcriptional Response

In a recent study conducted in a granule-based denitrification system, expression of *napA*, *narG*, and *nosZ* genes increased at 1 mg/l CuNP, eventually decreasing at higher doses (5 mg/l) (Cheng et al., 2019). The denitrification activity was found to increase at lower doses of CuNP treatment by increasing the alkalinity of the system, which is conducive to the dissolution of NPs. Thus, increasing dissolution of $Cu^{2+}$ ions from CuNP causes toxicity, which was evident in the upregulation of *copB* gene at 5 mg/l CuNP (Cheng et al., 2019). *copB* gene encodes for the Cu(II)-exporting ATPase, which extrudes excess $Cu^{2+}$ ions outside the cell. Overexpression of *copB* is henceforth a self-protective mechanism of the cell.

The type of coating on the surface of NP and different chemical species of same metal ion NP exhibit different physiological responses on the denitrifying community (S. Liu et al., 2018). Although some studies report that smaller NPs exhibit higher toxic response than larger sized NPs, polyvinylpyrrolidone coated Ag NP, i.e. (PVP)-Ag NP, smaller in size than Ag NP produced less toxic effect than Ag NP when both were used at the same concentration (S. Liu et al., 2018). The coating material hinders direct contact of NP to microbial surfaces, which leads to lesser toxicity than uncoated NP (Huynh, McCaffery, & Chen, 2014; Kwok et al., 2016; Sheng & Liu, 2017).

In an anaerobic environment, Ag NP readily forms $Ag_2S$ NP, as a natural detoxification mechanism of Ag NP (Reinsch et al., 2012). At the same concentration of Ag NP and $Ag_2S$ NP, the former exerted more toxicity than $Ag_2S$ NP (S. Liu et al., 2018). These findings are consistent with the functional gene abundance of the denitrifying community. The sensitivities of the functional genes to PVP-Ag, $Ag_2S$ NPs and $Ag^+$ in sediments have been found to be in the order of *nirK* > *nosZ* > *nirS* (S. Liu et al., 2018). In a recent study conducted in an agricultural soil, CuO NP treatment resulted in significant reduction in copy abundance of *narG*, *nirS*, and *norB* genes, whereas abundance of *nosZ* gene increased marginally (S. Zhao et al., 2019). Such a pattern of functional gene abundance leads to accumulation of nitrite in the soil sediments.

It has been reported that $TiO_2$ NP is most abundantly found in WWT plants and bioreactors. Denitrifying community in a laboratory-scale WWT system when stressed with $TiO_2$ NP at concentration of 50 mg/l led to decline in the abundance of functional genes (*napB*, *nirK*, *norB*, and *norC*) with respect to the control. In contrast, negative effects were not found on *narI*, *napA*, *nirS*, and *nosZ* gene (X. Yang et al., 2019). The abundance of nitrifying functional genes (*amo* and *hao*) showed much sharper decline at the same concentration of $TiO_2$ NP, thus implying that nitrification is more susceptible to $TiO_2$ NP than denitrification (X. Yang et al., 2019). Conversely, ZnO NP at similar concentration of 50 mg/l has shown downregulation of all the denitrifying genes (Zheng et al., 2014). Such contradictory responses of $TiO_2$ and ZnO NP might be because of the low solubility of titanium ions.

Denitrification is an anaerobic process, although some aerobic denitrifiers such as *Thiosphaera pantotropha*, *Pseudomonas* spp., and *Alcaligenes faecalis* have been reported (J. Zhao et al., 2017). Denitrifying enzyme system in denitrifiers such as *Paracoccus denitrificans* follows diauxic enzymatic activity in response to oxic and anoxic environment (Lycus et al., 2018). Denitrifying bacteria (*Pseudomonas stutzeri* CFY1) under aerobic conditions demonstrated positive response on the expression of *norB*, *napB*, and *nirH* genes when treated with 50 mg/l $TiO_2$ NP (D. Li et al., 2016). The *nor* gene (responsible for oxidation of NO into $N_2O$) encodes for a membrane associated protein and nap gene encodes for a periplasmic protein. The membrane associated proteins are more susceptible to damage as bacterial cell membrane ruptures and comes in direct contact with the nanoparticle. The upregulation of these genes is essentially a defense mechanism, where bacteria overexpress the genes for synthesis of these membrane associated proteins to restore normal activity within the cell (Y. Yang et al., 2011).

In aerobic *P. stutzeri* CFY1, higher concentration of $TiO_2$ NP increased the generation of ROS, which induces the expression of metal ion resistance genes such as *czcB*, *czcC*, and *sodB* gene (encodes for superoxide dismutase) (D. Li et al., 2016). Again, this finding supports the self-protective mechanisms of *P. stutzeri* CFY1 under aerobic condition upon exposure to $TiO_2$ NP. Studies on the effect of ZnO NP on aerobic denitrification by *P. stutzeri* PCN-1 reported downregulated expression of *nirS*, *napA*, *norB*, and *nosZ* gene with increasing concentration of ZnO NP (Chen et al., 2017). From the above discussions, it is clearly evident that gene expression pattern varies for aerobic and anoxic denitrifying microbial culture and different NPs exhibit different responses on aerobic denitrifying bacteria. Table 7.1 represents the effect of various NPs to the expression of nitrifying and denitrifying genes.

### 7.3.2 Enzymatic Response

Agricultural soil is oftentimes exposed to copper NPs via application of nanopesticides, nanofertilizers. Nitrate reductase (NAR) and nitric oxide reductase (NOR) activity in agricultural soil has been found to decrease at CuO NP concentration above 10 mg/kg of soil (S. Zhao et al., 2019). Reduction in NOR activity due to the inhibitory effect of CuO NP finally results in reduction in emission of $N_2O$ gas. This reduced NAR and NOR activity are consistent with lower gene abundance of *nar* and *norB*, as mentioned previously. In contrast, the nitrite reductase (NIR) activity did not exhibit any inhibitory effect due to the presence of CuO NP in agricultural soil (S. Zhao et al., 2019). This is probably due to lesser metal ions being bioavailable from soil matrix as compared to aqueous system. The bioavailable fraction of $Cu^{2+}$ released by NP might have been used up by the $Cu^{2+}$ requiring NIR enzyme as its cofactor, whereas Fe–S center requiring enzymes (NAR and NOR) demonstrated inhibitory response.

In a WWT system consisting of mature denitrifying granule, 5 mg/l Cu NP was found to inhibit cell viability, which is indicated by reduced dehydrogenase activity (Cheng et al., 2019). This reduction in enzymatic activity correlates with reduced functional gene abundance in the WWT system (Cheng et al., 2019).

**Table 7.1** The effect of various NPs to the expression of nitrifying and denitrifying genes.

| S. no. | Nanoparticles with concentration | Genes | Response | References |
|---|---|---|---|---|
| Nitrifying genes | | | | |
| 1 | AgNP; 50–200 μg/l | *amoA, nxrA* | Upregulation | Huang et al. (2019) |
| 2 | ZnO NP; 1 mg/l | *amoA, hao* | Upregulation | Q. Wu et al. (2018) |
| 3 | $TiO_2$ NP anatase; 10 mg/l | *amoA* | Downregulation | Li et al. (2019) |
| Denitrifying genes | | | | |
| 4 | AgNP; 50–200 μg/l | *nirK, nirS, nosZ* | Upregulation | Huang et al. (2019) |
| 5 | CuNP; 5 mg/l | *nirK, napA, narG,* and *nosZ* | Downregulation | Cheng et al. (2019) |
| 6 | AgNP, $Ag_2S$ NP, PVP-Ag NP; 10 mg/l | *nirK* | Downregulation | S. Liu et al. (2018) |
| | | *nirS, nosZ* | Not much affected | S. Liu et al. (2018) |
| 7 | ZnO NP; 50 mg/l | *narG, narH, narH, narI, napA, norB, norC, norQ, norD, norE, nozZ* | Downregulation | Zheng et al. (2014) |
| 8 | AgNP; 135, and 1350 $mg\,l^{-1}$ | *narG and nirS* | Downregulation | Bai et al. (2018) |

Similar results are reported when activated sludge containing denitrifying bacteria was stressed with $TiO_2$ NP. The activity of denitrifying enzymes, i.e. nitrite reductase (NIR) and nitrate reductase (NAR), was found to be reduced with increasing concentration of $TiO_2$ NP in a sequencing batch bioreactor (Z. Li et al., 2017). Increasing concentration of NiO NPs in activated sludge caused reduction in the activity of AMO, NAR, NIR, NOR enzymes, thereby demonstrating that NiO NP causes detrimental effect to both nitrifying and denitrifying microorganisms (Wang, Li, Gao, She, Ma, et al., 2017). The NiO NPs led to a rise in ROS level, resulting in an imbalance between the oxidation and anti-oxidation processes. This eventually leads to a loss in cell viability and LDH release. In aerobic denitrifying bacteria *P. stutzeri* PCN-1, ZnO NPs' treatment led to delayed nitrate and nitrite reduction capability owing to lower activity of NIR and NAR (Chen et al., 2017).

Another recent study reveals that stress due to $TiO_2$ NPs leads to decline in the activity of glycolytic enzymes – 6-phosphofructose kinase PFK and pyruvate kinase PK, which play key role in generation of ATP and NADH (X. Yang et al., 2019). In this way, $TiO_2$ NP creates deficit in net energy production and reduces the availability of electron donor within the cell, ultimately affecting nitrifying and denitrifying microorganisms, which require energy as well as reducing equivalent to complete their physiological processes.

## 7.4 Impact of Nanoparticles on Nitrogen Removal

Accumulation of NPs in an aquatic system results in its attachment to the surface of microbial cells. The higher the concentration of NPs in the aquatic system more is the attachment to the microbial cell surface. This ultimately leads to cell membrane disruption, which serves as the main impact of NP toxicity. Such impacts greatly affect the efficiency of WWT plants. Sludge from these wastewater treatment plants are often used as fertilizers, and once contaminated with nanoparticles, these particles can accumulate in soil, disrupting the natural nitrogen cycle bacteria. Exposure to increasing concentration of NPs is likely to enhance in future, so an understanding of the effect of NPs on biological nitrogen removal will greatly help to optimize treatment efficiency. This would also provide theoretical support for development of emergency control strategies for biological nitrogen removal system in WWT plants.

## 7.5 Conclusion

Although nanotechnology finds application in diverse areas such as human consumer products, agricultural sector, medicines, and various other industries, its accidental or deliberate discharge into the environment poses great risk to microorganisms, which maintain the delicate ecosystem health. This study summarizes the effect of metal NPs on microorganisms involved in nitrogen removal from wastewater. The release of NPs from various sources remains unquantified, which causes detrimental effects in the functionalities of microorganisms in WWT. The effect of NPs on the response of nitrifying and denitrifying bacteria highlights the need for safe disposal of various NPs.

## Take Home Message

- Excessive use and improper disposal of consumer products containing nanoparticles lead to exposure of nanoparticles to the environment.
- Nanoparticles on coming in contact with soil sediments and aquatic ecosystem react undesirably with native microorganisms.

- Nanoparticles affect biological nitrogen removal process, primarily nitrification and denitrification, thus hampering efficiency of various treatment plants.
- Morphological damage, altered gene expression, and changed enzymatic activity of nitrifying and denitrifying bacteria upon nanoparticle exposure have been reported.
- Proper use and discharge of nanoparticle-containing products are the need of the moment in order to protect the environment. Effluent from industries that manufacture such products should be treated before discharge to the environment.

## Questions and Answers

**1** What is the potential environmental hazard from biological nitrogen removal?
Emission of greenhouse gas ($N_2O$) due to incomplete denitrification.

**2** How can the microbial toxicity posed by nanoparticles be evaluated?
By testing the effect of nanoparticle on pure culture strains, next-generation sequencing and metagenomics.

**3** Are there any protection/defense mechanism available in microorganisms to combat the toxic effects of nanoparticles?
Forming microbial aggregates as a result of extra-polymeric substances, which adsorb toxicants on its surface.

**4** How can we utilize our understanding of microbial interaction with nanoparticles in enhancing wastewater treatment efficiency?
Nanoparticles that are more readily retained by extra-polymeric substances exert lesser antimicrobial effects, thus producing the desired outcome in wastewater treatment and aquatic ecosystem.

**5** How can we stop discharge of industrial effluent containing nanoparticles into the environment?
By strictly implementing and following EPA guidelines for safe disposal of wastes.

## Acknowledgements

The authors are thankful to University Grants Commission (UGC), New Delhi for providing research fellowship to P.C. The authors thank DST-FIST for the support to the Department of Environmental Science and Engineering, Indian Institute of Technology (Indian School of Mines) Dhanbad.

## References


Alito, C. L., & Gunsch, C. K. (2014). Assessing the effects of silver nanoparticles on biological nutrient removal in bench-scale activated sludge sequencing batch reactors. *Environmental Science & Technology*, 48(2), 970–976.

Bai, J., Tian, Y. Z., Sun, P. F., Bai, X. Y., Li, K. R., & Zhao, Y. G. (2018). Effect of silver nanoparticles on denitrification and functional gene abundances of sediment in Dagu river estuary and northwest of Jiaozhou Bay. *Huan Jing Ke Xue*, 39(11), 4956–4963.

Batley, G. E., Kirby, J. K., & McLaughlin, M. J. (2013). Fate and risks of nanomaterials in aquatic and terrestrial environments. *Accounts of Chemical Research*, 46(3), 854–862.

Bielmyer, G. K., Grosell, M., & Brixti, K. V. (2006). Toxicity of silver, zinc, copper, and nickel to the copepod *Acartia tonsa* exposed via a phytoplankton diet. *Environmental Science & Technology*, 40(6), 2063–2068.

Bueno, E., Mesa, S., Bedmar, E. J., Richardson, D. J., & Delgado, M. J. (2012). Bacterial adaptation of respiration from oxic to microoxic and anoxic conditions: Redox control. *Antioxidants & Redox Signaling*, 16(8), 819–852.

Bundschuh, M., Filser, J., Lüderwald, S., McKee, M. S., Metreveli, G., Schaumann, G. E., . . . Wagner, S. (2018). Nanoparticles in the environment: Where do we come from, where do we go to? *Environmental Sciences Europe*, 30(1), 6.

Chen, Q., Li, T., Gui, M., Liu, S., Zheng, M., & Ni, J. (2017). Effects of ZnO nanoparticles on aerobic denitrification by strain *Pseudomonas stutzeri* PCN-1. *Bioresource Technology*, 239, 21–27.

Hong Cheng and Pei-ying Hong (2020). Nanoparticles applied in membrane bioreactors: Potential impact on reactor performance and microbial communities. Published in *Contaminants of emerging concern in water and wastewater: Advanced treatment processes*, Edited by Hernández-Maldonado A.J. and Blaney L., Amsterdam, The Netherlands: Elsevier, 207–236.

Cheng, Y. F., Zhang, Q., Li, G. F., Xue, Y., Zheng, X. P., Cai, S., . . . Jin, R. C. (2019). Long-term effects of copper nanoparticles on granule-based denitrification systems: Performance, microbial communities, functional genes and sludge properties. *Bioresource Technology*, 289, 121707.

Choi, O., & Hu, Z. (2009). Role of reactive oxygen species in determining nitrification inhibition by metallic/oxide nanoparticles. *Journal of Environmental Engineering*, 135(12), 1365–1370.

Choi, O., Yu, C. P., Esteban Fernández, G., & Hu, Z. (2010). Interactions of nanosilver with *Escherichia coli* cells in planktonic and biofilm cultures. *Water Research*, 44(20), 6095–6103.

Domercq, P., Praetorius, A., & Boxall, A. B. A. (2018). Emission and fate modelling framework for engineered nanoparticles in urban aquatic systems at high spatial and temporal resolution. *Environmental Science: Nano*, 5, 533.

Fang, X., Yu, R., Li, B., Somasundaran, P., & Chandran, K. (2010). Stresses exerted by ZnO, $CeO_2$ and anatase $TiO_2$ nanoparticles on the *Nitrosomonas europaea*. *Journal of Colloid and Interface Science*, 348(2), 329–334.

Hartland, A., Lead, J. R., Slaveykova, V. I., O'Carroll, D., & Valsami-Jones, E. (2013). The environmental significance of natural nanoparticles. *Nature Education Knowledge*, 4(8), 7.

Huang, J., Cao, C., Liu, J., Yan, C., & Xiao, J. (2019). The response of nitrogen removal and related bacteria within constructed wetlands after long-term treating wastewater containing environmental concentrations of silver nanoparticles. *Science Total Environmental*, 667, 522–531.

Huynh, K. A., McCaffery, J. M., & Chen, K. L. (2014). Heteroaggregation reduces antimicrobial activity of silver nanoparticles: Evidence for nanoparticle cell proximity effects. *Environmental Science & Technology Letters*, 1(9), 361e366.

Kwok, K. W., Dong, W., Marinakos, S. M., Liu, J., Chilkoti, A., Wiesner, M. R., . . . Hinton, D. E. (2016). Silver nanoparticle toxicity is related to coating materials and disruption of sodium concentration regulation. *Nanotoxicology*, 10(9), 1306–1317.

Li, D., Li, B., Wang, Q., Hou, N., Li, C., & Cheng, X. (2016). Toxicity of $TiO_2$ nanoparticle to denitrifying strain CFY1 and the impact on microbial community structures in activated sludge. *Chemosphere*, 144, 1334–1341.

Li, K., Qian, J., Wang, P., Wang, C., Lu, B., Tian, X., ... & Liu, Y. (2020). Differential responses of encoding-amoA nitrifiers and nir denitrifiers in activated sludge to anatase and rutile TiO2 nanoparticles: What is active functional guild in rate limiting step of nitrogen cycle?. *Journal of Hazardous Materials*, 384, 121388.

Li, Z., Wang, X., Ma, B., Wang, S., Zheng, D., She, Z., . . . Gao, M. (2017). Long term impacts of titanium dioxide nanoparticles ($TiO_2$ NPs) on performance and microbial community of activated sludge. *Bioresource Technology*, 238, 361–368.

Liu, S., Wang, C., Hou, J., Wang, P., Miao, L., Fan, X., . . . Xu, Y. (2018). Effects of Ag and $Ag_2S$ nanoparticles on denitrification in sediments. *Water Research*, 137, 28–36.

Liu, X., Yang, X., Hu, X., He, Q., Zhai, J., Chen, Y., . . . Vymazal, J. (2019). Comprehensive metagenomic analysis reveals the effects of silver nanoparticles on nitrogen transformation in constructed wetlands. *Chemical Engineering Journal*, 358, 1552–1560.

Lycus, P., Soriano-Laguna, M. J., Kjos, M., Richardson, D. J., Gates, A. J., Milligan, D. A., . . . Bakken, L. R. (2018). A bet-hedging strategy for denitrifying bacteria curtails their release of $N_2O$. *Proceeeding of the National Academy Science of the United States of America*, 115(46), 11820–11825.

Ma, J., Quan, X., Si, X., & Wu, Y. (2013). Responses of anaerobic granule and flocculent sludge to ceria nanoparticles and toxic mechanisms. *Bioresource Technology*, 149, 346–352.

Nazarenko, Y., Han, T. W., Lioy, P. J., & Mainelis, G. (2011). Potential for exposure to engineered nanoparticles from nanotechnology-based consumer spray products. *Journal of Exposure Science & Environmental Epidemiology*, 5, 515–528.

Oliviera, M. L. S., Izquierdo, M., Querol, X., Lieberman, R. N., Saikia, B. K., & Silva, S. F. O. (2019). Nanoparticles from construction wastes: A problem to health and the environment. *Journal of Cleaner Production*, 219, 236–243.

Pandian, S. R. K., Deepak, V., Kalishwaralal, K., Viswanathan, P., & Gurunathan, S. (2010). Mechanism of bactericidal activity of silver nitrate: A concentration dependent bi-functional molecule. *Brazilian Journal of Microbiology*, 41, 805–809.

Reinsch, B. C., Levard, C., Li, Z., Ma, R., Wise, A., Gregory, K. B., . . . Lowry, G. V. (2012). Sulfidation of silver nanoparticles decreases *Escherichia coli* growth inhibition. *Environmental Science & Technology*, 46(13), 6992–7000.

Sheng, Z., & Liu, Y. (2017). Potential impacts of silver nanoparticles on bacteria in the aquatic environment. *Journal of Environmental Management*, 191, 290–296.

Sheng, Z., Van Nostrand, J. D., Zhou, J., & Liu, Y. (2015). The effects of silver nanoparticles on intact wastewater biofilms. *Frontiers in Microbiology*, 6, 680.

Slavin, Y. N., Asnis, J., Häfeli, U. O., & Bach, H. (2017). Metal nanoparticles: Understanding the mechanisms behind antibacterial activity. *Journal of Nanobiotechnology*, 15(1), 65.

Tang, J., Wu, Y., Esquivel-Elizondo, S., Sørensen, S. J., & Rittmann, B. E. (2018). How microbial aggregates protect against nanoparticle toxicity. *Trends in Biotechnology*, 36(11), 1171–1182.

Tipping, E. (2001). *Cation Binding by Humic Substances*. Cambridge, UK: Cambridge University Press.

Tong, T., Fang, K., Thomas, S. A., Kelly, J. J., Gray, K. A., & Gaillard, J. F. (2014). Chemical interactions between Nano-ZnO and Nano-$TiO_2$ in a natural aqueous medium. *Environmental Science & Technology*, 48(14), 7924–7932.

Tong, T., Wilke, C. M., Wu, J., Binh, C. T., Kelly, J. J., Gaillard, J. F., & Gray, K. A. (2015). Combined toxicity of nano-ZnO and nano-$TiO_2$: From single-to multinanomaterial systems. *Environmental Science & Technology*, 49(13), 8113–8123.

Turan, N. B., Erkan, H. S., Engin, G. O., & Bilgili, G. S. (2019). Nanoparticles in the aquatic environment: Usage, properties, transformation and toxicity—A review. *Process Safty and Environmental Protection*, 130, 238–249.

Wang, S., Gao, M., Li, Z., She, Z., Wu, J., Zheng, D., . . . Wang, X. (2016). Performance evaluation, microbial enzymatic activity and microbial community of a sequencing batch reactor under long-term exposure to cerium dioxide nanoparticles. *Bioresource Technology*, 220, 262–270.

Wang, S., Li, Z., Gao, M., She, Z., Guo, L., Zheng, D., . . . Wang, X. (2017). Long-term effects of nickel oxide nanoparticles on performance, microbial enzymatic activity, and microbial community of a sequencing batch reactor. *Chemosphere*, 169, 387–395.

Wang, S., Li, Z., Gao, M., She, Z., Guo, L., . . . Gao, F. (2017). Long term effects of cupric oxide nanoparticles (CuO NPs) on the performance, microbial community and enzymatic activity of activated sludge in a sequencing batch reactor. *Journal of Environmental Management*, 187, 330–339.

Wu, J., Chang, Y., Gao, H., Liang, G., Yu, R., & Ding, Z. (2018). Responses and recovery assessment of continuously cultured *Nitrosomonas europaea* under chronic ZnO nanoparticle stress: Effects of dissolved oxygen. *Chemosphere*, 195, 693–701.

Wu, Q., Huang, K., Sun, H., Ren, H., Zhang, X. X., & Ye, L. (2018). Comparison of the impacts of zinc ions and zinc nanoparticles on nitrifying microbial community. *Journal of Hazardous Materials*, 343, 166–175.

Yang, X., Chen, Y., Guo, F., Liu, X., Su, X., & He, Q. (2019). Metagenomic analysis of the biotoxicity of titanium dioxide nanoparticles to microbial nitrogen transformation in constructed wetlands. *Journal of Hazardous Materials*, 3, 121376.

Yang, Y., Zhu, H. G., Colvin, V. L., & Alvarez, P. J. (2011). Cellular and transcriptional response of *Pseudomonas stutzeri* to quantum dots under aerobic and denitrifying conditions. *Environmental Science & Technology*, 45(11), 4988–4994.

Yazdanbakhsh, A. R., Rafiee, M., Daraei, H., & Amoozegar, M. A. (2019). Responses of flocculated activated sludge to bimetallic AgFe nanoparticles toxicity: Performance, activity enzymatic, and bacterial community shift. *Journal of Hazardous Materials*, 366, 114–123.

Yu, R., Wu, J., Liu, M., Chen, L., Zhu, G., & Lu, H. (2016). Physiological and transcriptional responses of *Nitrosomonas europaea* to $TiO_2$ and ZnO nanoparticles and their mixtures. *Environmental Science and Pollution Research International*, 23(13), 13023–13034.

Yu, R., Wu, J., Liu, M., Zhu, G., Chen, L., Chang, Y., & Lu, H. (2016). A toxicity of binary mixtures of metal oxide nanoparticles to *Nitrosomonas europaea*. *Chemosphere*, 153, 187–197.

Zeng, G., Zhang, L., Dong, H., Chen, Y., Zhang, J., Zhu, Y., . . . Fang, W. (2018). Pathway and mechanism of nitrogen transformation during composting: Functional enzymes and genes under different concentrations of PVP-AgNPs. *Bioresource Technology*, 253, 112–120.

Zhang, C., Liang, Z., & Hu, Z. (2014). Bacterial response to a continuous long-term exposure of silver nanoparticles at sub-ppm silver concentrations in a membrane bioreactor activated sludge system. *Water Research*, 50, 350–358.

Zhang, X., Chen, Z., Ma, Y., Zhang, N., Wei, D., Zhang, H., & Zhang, H. (2019). Response of partial nitrification sludge to the single and combined stress of CuO nanoparticles and sulfamethoxazole antibiotic on microbial activity, community and resistance genes. *Science of the Total Environment*, 712, 135759. doi:https://doi.org/10.1016/j.scitotenv.2019.135759

Zhang, X., Chen, Z., Zhang, N., Ma, Y., Song, Y., Li, Y., & Zhang, H. (2020). Resistance to copper oxide nanoparticle and oxytetracycline of partial nitrification sludge. *Chemical Engineering Journal*, 381, 122661.

Zhang, Z., Gao, P., Li, M., Cheng, J., Liu, W., & Feng, Y. (2016). Influence of silver nanoparticles on nutrient removal and microbial communities in SBR process after long-term exposure. *Science of the Total Environment*, 569-570, 234–243.

Zhao, J., Wu, J., Li, X., Wang, S., Hu, B., & Ding, X. (2017). The denitrification characteristics and microbial community in the cathode of an MFC with aerobic denitrification at high temperatures. *Frontiers in Microbiology*, 8, 9.

Zhao, S., Su, X., Wang, Y., Yang, X., Bi, M., He, Q., & Chen, Y. (2019). Copper oxide nanoparticles inhibited denitrifying enzymes and electron transport system activities denitrification and $N_2O$ emission to influence soil. *Chemosphere*, 245, 125394. doi:https://doi.org/10.1016/j.chemosphere.2019.125394

Zheng, X., Chen, Y., & Wu, R. (2011a). Long term effects of titanium dioxide nanoparticles on nitrogen and phosphorus removal from wastewater and bacterial community shift in activated sludge. *Environmental Science & Technology*, 45(17), 7284–7290.

Zheng, X., Su, Y., Chen, Y., Wan, R., Liu, K., Li, M., & Yin, D. (2014). Zinc oxide nanoparticles cause inhibition of microbial denitrification by affecting transcriptional regulation and enzyme activity. *Environmental Science & Technology*, 48, 13800–13807.

Zheng, X., Wu, R., & Chen, Y. (2011b). Effects of ZnO nanoparticles on wastewater biological nitrogen and phosphorus removal. *Environmental Science & Technology*, 45(7), 2826–2832.

Zheng, Y. L., Hou, L. J., Liu, M., Newell, S. E., Yin, G. Y., Yu, C. D., . . . Liu, C. (2017). Effects of silver nanoparticles on nitrification and associated nitrous oxide production in aquatic environments. *Science Advances*, 3, e1603229.

## Glossary

**Activated sludge:** A mass of aerated precipitated sewage added to untreated sewage to bring about purification by hastening decomposition by microorganisms.

**Anammox microorganism:** It is an abbreviation for anaerobic ammonium oxidation microorganism, which converts ammonium into dinitrogen gas anaerobically.

**Anatase:** Metastable mineral form of titanium dioxide ($TiO_2$).

**Antagonistic response:** Opposite response produced from contrasting action of two or more chemical groups.

**Bimetallic nanoparticle:** Nanoparticle comprised of two different metals.
**Bioavailability:** Portion of any toxicants from its bulk concentration that enters the living cell.
**Cell viability:** Refers to number of healthy living cells based on a total cell sample.
**Denitrifying microorganism:** Denitrifying microorganisms reduce nitrate or nitrite into dinitrogen gas or nitrogen oxide.
**Diauxic:** Two different responses by same microorganism in sequential manner upon exposure to two or more environmental stimuli.
**Engineered nanoparticle:** Particles with at least one dimension between 1 and 100 nm in size. These are artificially synthesized from bulk material to offer unique characteristics at cellular, atomic, and molecular levels and, also, artificially designed to impart new functional behavior.
**Enzyme:** Proteinaceous substance that is able to catalyses a particular biochemical reaction.
**Gene cluster:** Group of two or more genes found that encode for similar polypeptides, or proteins, which collectively share a generalized function.
**Gene expression:** Process by which genetic information in DNA is used in the synthesis of functional gene product.
**Lipid peroxidation:** Oxidative degradation of lipids leading to cell damage.
**Membrane bioreactor:** It is a combination of membrane process like microfiltration and ultrafiltration integrated to biological wastewater treatment process.
**Membrane fouling:** Membrane fouling is the accumulation of substances on the membrane surface and/or within the membrane pores, which results in deterioration of membrane performance.
**Membrane integrity:** Continuous plasma membrane protecting the cell from any damage.
**Nanofertilizer:** A nanofertilizer is any product that is made with nanoparticles or uses nanotechnology to improve nutrient efficiency for plants and it is categorized into three types – nanoscale fertilizer, nanoscale coating, and nanoscale additive.
**Nanoparticle aggregation:** Formation of clusters of nanoparticles that affects the sorption, retention, and speciation of metals differently based on the mechanism of aggregation (drying, freezing, pH, ionic strength).
**Nanopesticide:** These are pesticide formulations in the nanometer size range and claimed to have novel properties associated with the small size.
**Nitrifying microorganism:** These microorganisms oxidize ammonia into nitrite.
**Operon:** It is a regulatory unit of DNA that consists of group of genes together with promoter sequence and operator sequence that control transcription of a gene.

**Reactive oxygen species:** These are metabolic by-products of aerobic respiration, for which excessive accumulation can lead to oxidative stress in cells, for example – superoxide and hydroxyl radicle.

**Sequential batch reactor:** A sequencing batch reactor is fill and draw type activated sludge-type wastewater treatment system that can carry out various treatment operations in one tank.

**Synergistic response:** Similar response arising from two or more chemical substances that produces an effect greater than the sum of their individual effects.

# 8

# Silver-Based Nanoparticles for Antibacterial Activity

## Recent Development and Mechanistic Approaches

*Arpita Roy*[1,2], *Papia Basuthakur*[1,2], *Shagufta Haque*[1,2], *and Chitta Ranjan Patra*[1,2]

[1] *Department of Applied Biology, CSIR-Indian Institute of Chemical Technology, Hyderabad, Telangana State, India*
[2] *Academy of Scientific and Innovative Research (AcSIR), Ghaziabad, Uttar Pradesh, India*

## Objectives

- Explore the background of silver nanoparticles in biomedical applications.
- Understand basic methodologies for the synthesis of silver nanoparticles.
- Investigation of various mechanisms of action of silver nanoparticles in antibacterial therapy.
- Discuss the recent applications of silver nanoparticles in commercial products.
- Future perspective and challenges of silver nanoparticles before use in clinics.

## 8.1 Introduction

Recently, numerous inorganic metal nanoparticles have gained major attention in the field of nanomedicine. For example, metal oxide/hydroxide nanoparticles (CONP: cerium oxide nanoparticles, ZONP: zinc oxide nanoparticles, EHN: europium hydroxide nanorods, THN: terbium hydroxide nanorods, zinc-doped titanium oxide nanoparticles, etc.) (S. Das et al., 2012; Barui et al., 2012; C. R. Patra et al., 2008; Nethi, Barui, Bollu, Rao, & Patra, 2017a; Nethi, Rico-Oller, Rodríguez-Diéguez, Gómez-Ruiz, & Patra, 2017b) and carbon-based nanoparticles (like fullerene, graphene oxide nanoparticles, etc.) have been used for angiogenesis study and for their applications in biological research for the treatment of different diseases like cancer, vascular, ischemic, antibacterial, etc. (Chaudhuri,

*Microbial Interactions at Nanobiotechnology Interfaces: Molecular Mechanisms and Applications*, First Edition. Edited by R. Navanietha Krishnaraj and Rajesh K. Sani.

Harfouche, Soni, Hentschel, & Sengupta, 2009; Mukherjee et al., 2015; Chakraborty, Ponrasu, Chandel, Dixit, & Muthuvijayan, 2018). Noble metal nanoparticles like silver, gold, and platinum nanoparticles have been used as antimicrobial agents, drug delivery vehicles, cancer therapeutics, and diagnostics (Kotcherlakota et al., 2019; Zhou, Gao, Liu, & Chen, 2015; B. R. Rao et al., 2018; Rolim et al., 2019; Shoshan, Vonderach, Hattendorf, & Wennemers, 2019). Among all noble metal nanoparticles, silver nanoparticles have drawn attention in various fields of biological application due to their simple synthesis procedures, ease of characterization, capacity to attach various targeting moieties on their surface, and therapeutic efficacy especially in the treatment of microbial infections (Ivanova et al., 2018, X.-F. Zhang, Liu, Shen, & Gurunathan, 2016; J.-S. Lee, Lytton-Jean, Hurst, & Mirkin, 2007). Again, silver nanoparticle is recognized as highly toxic toward microorganisms due to the release of silver ions (Siddiqi, Husen, & Rao, 2018). Furthermore, these nanoparticles possess greater bioavailability, biocompatibility, and fast clearance from the system of host (Ivanova et al., 2018; Behra et al., 2013). Since there have been reports of several biological applications of silver nanoparticles, (Ishwarya et al., 2017; S. W. Kim et al., 2012; S. S. D. Kumar, Rajendran, Houreld, & Abrahamse, 2018; T. Yang et al., 2016), it is essential to compile all the underlying mechanisms of the action of silver nanoparticles as antibacterial agent in a comprehensive book chapter. Therefore, within the scope of this book chapter, we focus on silver-based nanoparticles, their synthesis, and antibacterial activities. Additionally, the recent advances of mechanistic approaches and commercial applications of silver-based nanomaterial are also discussed along with their future perspective and challenges.

## 8.2 Historical Background of Silver

Silver is known since time immemorial to have excellent medicinal properties. Since 3000 BC, silver has been used for jewelry, cutlery, weaponries, and various medical conditions (Barillo & Marx, 2014; Raub, 1989; Toynbee & Painter, 1986; Tozzi, Al-Darweesh, Vogt, & Stumpe, 2001). Mostly, silver was known to treat ailments much prior to the discovery of microbial infections. It has been reported that primordial Romans, Phoenicians, Greeks, and Egyptians utilized silver in water and food preservation (Wanhill, Steijaart, Leenheer, & Koens, 1998). Hippocrates and Macedonians used silver in the treatment of wound healing for the first time. Silver was used in the form of silver nitrate by Gabor in 702–705. Later, Avicenna in 980 AD, Paracelsus in 1520, and Angelo Sala in 1614 utilized silver for wound healing, blood purification, and treat various infections. In 1800, silver vessels were used to store water, wine, vinegar, etc. for prolonged usage (Davies & Etris, 1997; Levey, 1967). Doctor Carl in the 1880s established the use of silver nitrate in eye drops for the treatment of ophthalmia neonatorium in newborns. It was reported in the year 1889 about the citrate stabilized colloidal silver by M. C. Lea (1889) and the size formed in the nanoscale range with average diameter of 7–9 nm (Frens & Overbeek, 1969). Since 1897, nanosilver has been manufactured under the name "Collargol" and marketed for various medical purposes (Boese, 1921). Starting from then, colloidal nanosilver was first used as antiseptic for wounds by Dr. Crede in 1891 (Alexander, 2009). In 1915, Roe A.L. reported treating infected corneal ulcers, interstitial keratitis, blepharitis, and dacryocystitis using colloidal silver (Roe, 1915). Various investigations in the 1900s revealed the effective treatment strategies of colloidal silver against tonsillitis, sepsis, epididymitis, and other infectious diseases (Duhamel, 1912), (Sanderson-Wells, 1918; G. V. A. Brown, 1916). Another method was introduced to synthesize gelatin-stabilized nanosilver by Moudry, and it was patented in the year 1953 (Vaclav, 1960). The nanosilver bearing the medicinal properties were commercially sold under the brand names Collargol, Protargol, and Argyrol. The widespread marketing of these products were conducted over 50 years period and was recommended to treat different diseases related to bacterial infections (Fung & Bowen, 1996).

## 8.3 Synthesis Procedures of Silver Nanoparticles

Silver nanoparticles exhibit unique physical and functional properties, and hence find applications for use in day-to-day life. Hence it is very important to synthesize silver nanoparticles with different shapes and sizes for various applications. Since this book chapter mainly focuses on the antibacterial activity of silver nanoparticles, we briefly mention the synthesis and characterization of this nanoparticle. The synthesis techniques have mainly been divided into three foremost categories:

- Chemical synthesis (J. Chen, Wang, Zhang, & Jin, 2008; W. Zhang, Qiao, & Chen, 2007)
- Physical synthesis (Jung, Cheol Oh, Soo Noh, Ji, & Soo Kim, 2006; D. K. Lee & Kang, 2004)
- Biological synthesis (Korbekandi, Ashari, Iravani, & Abbasi, 2013; Saifuddin, Wong, & Yasumira, 2009)

### 8.3.1 Chemical Synthesis

Chemical synthesis involves the synthesis of silver nanoparticles with the help of several chemical agents using reduction processes: microwave-assisted synthesis, thermal reduction, polyol process, electrochemical synthetic method, microemulsion, photochemical reduction methods, and photoreduction method (Ma et al., 2004;

M. Chen et al., 2007; Yin et al., 2002; Soto et al., 2011; Abid, Wark, Brevet, & Girault, 2002; Shchukin, Radtchenko, & Sukhorukov, 2003; Krutyakov, Olenin, Kudrinskii, Dzhurik, & Lisichkin, 2008; W. Zhang et al., 2007; Huang & Yang, 2008; Seku, Gangapuram, Pejjai, Kadimpati, & Golla, 2018; Shchukin et al., 2003; Socol et al., 2002; J. Chen et al., 2008). The advantages of using this type of synthesis are that it requires very less time for reaction and high amount of yield can be obtained with the help of very less energy consumption. For example, our group synthesized silver nitroprusside (AgNNPs) nanocomposites using silver nitrate and sodium nitroprusside where the nitroprusside present provides increased stability to the formed nanocomplex. Figure 8.1 depicts the synthesis of AgNNPs by stirring method and its characterization through different techniques like XRD, TEM, FESEM, and DLS (B. R. Rao et al., 2018). The results together demonstrate the crystalline nature of monodispersed nanomaterials.

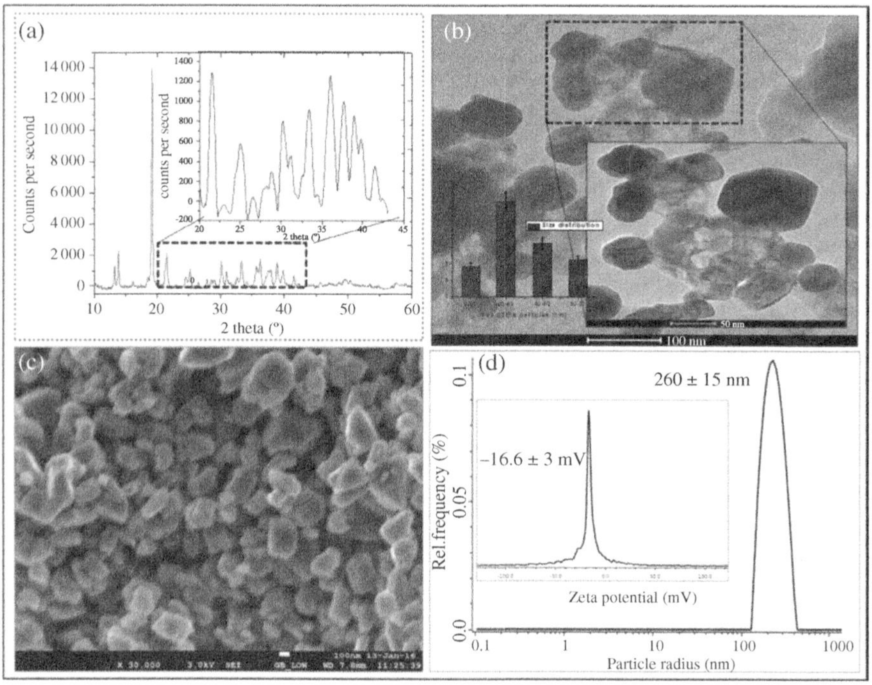


**Figure 8.1** Physicochemical characterizations of AgNNPs: (a) XRD analysis depicting the crystalline nature of AgNNPs, (b) TEM images of AgNNPs demonstrate monodispersed nature of nanoparticles (inset: size distribution), (c) FESEM image of AgNNPs indicating their morphology of nanocrystals, (d) DLS analysis of AgNNPs shows the hydrodynamic radius and negative zeta potential (inset) of nanoparticles. *Source:* Reproduced with permission from B. R. Rao et al. (2018) Copyright © (2018) American Chemical Society.

### 8.3.2 Physical Methods

Physical methods involve the synthesis of silver nanoparticles without the assistance of any chemical or biological agents. The different physical methods involve laser ablation, evaporation and condensation, thermal decomposition procedures, arc discharge, etc. (D. K. Lee & Kang, 2004; S. Kim et al., 2005; K. Kim & Kim, 2019; Alaqad & Saleh, 2016). Physical methods are usually employed because they are devoid of any contamination occurring from the solvent and the nanoparticles produced have particle size distribution in a narrow range (Jung et al., 2006; Kruis, Fissan, & Rellinghaus, 2000).

### 8.3.3 Biological Methods

One of the most economically cheap and eco-friendly methods for the synthesis of silver nanoparticles is through biological approach. This method is attributed to the use of natural products and organisms for the synthesis of silver nanoparticles such as plants, bacteria, fungi, etc. There have been various drawbacks regarding the chemical synthesis procedures of silver nanoparticles owing to the amount of contamination, not being environment friendly and toxic. Hence scientists are searching for eco-friendly ways for the formation of silver nanoparticles using several microorganisms including bacteria, fungi, and also many plants extracts (A. Roy, Bulut, Some, Mandal, & Yilmaz, 2019; Gahlawat & Choudhury, 2019; Kaur & Kumar, 2019). Silver nanoparticles can be synthesized using either intracellular or extracellular processes (V. L. Das et al., 2014). In extracellular process, the bacterial or fungal extracellular enzymes, proteins or any secretions are used for the synthesis of the silver nanoparticles. In this process, only the microbial supernatants are used excluding the microbial biomass. There are no living microorganisms involved to confer resistance. However, in the intracellular synthesis process, the metal salts enter inside the bacterial or fungal cells where several enzymes or reductases like nitrate reductase reduce silver ions ($Ag^{+}$) to metallic silver nanoparticles ($Ag^{0}$) (Kalimuthu, Suresh Babu, Venkataraman, Bilal, & Gurunathan, 2008). The synthesized silver nanoparticles were released by processing the microbial cells carrying them. For the intracellular synthesis, resistant microbes are chosen where the resistant mechanism varies with different species (Ali, Ali, Wang, Waseem, & Pan, 2019). Use of bacteria for the synthesis of silver nanoparticles has been reported by many scientists (Kalimuthu et al., 2008; Saifuddin et al., 2009). The mechanism behind the use of bacteria as a precursor to synthesize nanoparticles involves that bacteria have the ability for the reduction of metallic ions into nanoparticles. Some of the bacteria use NADPH reductase enzyme for synthesis of nanoparticles (Srivastava & Constanti, 2012). They are easy to handle and have good growth rate. Researchers have utilized algae in silver nanoparticle formation. For example, marine algae such as *Chlorella salina* and many more have been used for the synthesis activity (Govindaraju, Basha, Kumar, & Singaravelu, 2008; Shahverdi, Minaeian, Shahverdi, Jamalifar, & Nohi, 2007). Generally, silver-resistant bacteria that have intrinsic and evolutionary resistance mechanism to survive in the silver environment are isolated from the silver mines for the synthesis of silver nanoparticles (Rajora, Kaushik, Jyoti, & Kothari, 2016). Also, bacteria that grow in harsh conditions like heavy metal-contaminated area or sewage wastes (V. L. Das et al., 2014), (Kalimuthu et al., 2008) are suitable for the synthesis of silver nanoparticles. It is reported that *Bacillus licheniformis* are resistant to silver at 1 mM concentration to synthesize silver nanoparticles intracellularly. However, at higher concentration (5–10 mM), the bacterial cell undergoes apoptosis resulting in death of the bacteria. (Pandian, Deepak, Kalishwaralal, Viswanathan, & Gurunathan, 2010). The exact resistance mechanism against silver is not well-explored.

In this context, other than these, fungi are also among the organisms used for the synthesis of silver nanoparticles (Anil Kumar et al., 2007; Korbekandi et al., 2013). Fungi have increased bioaccumulation capability and exhibit increased metal tolerance (Kitching, Ramani, & Marsili, 2015). The fungal cell wall can bind high amounts of metal (Singh, Kim, Zhang, & Yang, 2016b; Khandel & Shahi, 2018). Other than the use of microorganisms, there have been various reports on the use of plants for the synthesis of silver nanoparticles. Since this is a very economical approach, it can be used to produce silver nanoparticles in large scale. Several plants such as *Camellia sinensis* commonly known as green tea, *Pinus densiflora,* and *Magnolia kobus* are used for the formation of silver nanoparticles (Kesharwani, Yoon, Hwang, & Rai, 2009; Song & Kim, 2009; Vilchis-Nestor et al., 2008). The different compounds present in plants namely flavones, terpenoids, and polysaccharides are responsible for the reduction of the metal ions as well as for acting as capping agents for the nanoparticles (Moodley, Krishna, Pillay, & Govender, 2018).

The microorganisms used for the synthesis for silver nanoparticles even act as stabilizing and capping agents. The stabilizing or capping agents help the formed nanoparticles in several ways such as by reducing toxicity, preventing the nanoparticles from being agglomerated, and by increasing their antimicrobial action. Even the stabilizing or capping agents attached to the nanoparticle contribute synergistically to its antimicrobial activity (Khanra, Roy, & Bhattacharyya, 2013) (N. Roy, Gaur, Jain, Bhattacharya, & Rani, 2013). To this context, Kiran et al., synthesized silver nanoparticles with the help of biosurfactants formed from *Brevibacterium casei* MSA19, which has been derived from marine sponge. The formed nanoparticles exhibited smaller size and were stabilized for nearly two months (Kiran, Sabu, & Selvin, 2010). In another work, Hamouda et al. biosynthesized silver nanoparticles by using aqueous extract of cyanobacteria *Oscillatoria limnetica*. The cyanobacteria served the dual purpose of acting as a reducing as well as stabilizing agent. The characterization study demonstrated that the presence of the free amino groups along with the sulfur groups contributed to the stability of the nanoparticles. Even the formed nanoparticles exhibited excellent antibacterial activity against multidrug-resistant bacteria (Hamouda, Hussein, Abo-elmagd, & Bawazir, 2019).

## 8.4 Biological Application of Silver Nanoparticles

Silver nanoparticles have been widely studied for various biomedical applications. Numerous research groups have reported versatile applications of AgNPs in combating different diseases such as bacterial, fungal, and viral infections; cancer; and diabetes. Few of the studies and the mechanism of action of AgNPs are summarized in Table 8.1.

Elaborate discussions of the above applications of AgNPs are beyond the scope of this book chapter as this chapter focuses only on the antibacterial activity of AgNPs and their mechanism of actions.

## 8.5 Bacterial Infection and Antibiotic Resistance

Humans and animals not only harbor different bacterial species within the body but are also exposed to various types of bacteria in the outside environment. Diseases occur when bacteria damage and infect the cells of our body. Some bacteria can colonize and multiply on the surface of the host cells such as mucosal surface of the urogenital tract, respiratory tract, etc. Others simply invade the host cells, hijack the normal cellular machinery, and multiply. Again, some bacteria secrete toxins that paralyze and destroy cell's metabolic functions or evoke heavy immune reactions that themselves are toxic (Arciola, Campoccia, & Montanaro, 2018; Cohen et al., 2016; Didelot, Walker, Peto, Crook, & Wilson, 2016; Poole, Day, von Itzstein, Paton, & Jennings, 2018).

Normally, the immune system provides the first line of defense against the infecting bacterial pathogens. Bacteria also possess multiple mechanisms to evade body's immune response. Standard curative options include treatment with antibacterial drugs called antibiotics. These antibiotics are common agents (molecules or drugs) that are used to inhibit bacterial infections, which act through perturbation of cellular processes (protein synthesis, DNA replication, impairment in cell membrane function, etc.) in bacterial cells, thereby resulting in inhibition of cell growth and cell division (Epand, Walker, Epand, & Magarvey, 2016; Polikanov, Aleksashin, Beckert, & Wilson, 2018; van Eijk, Wittekoek, Kuijper, & Smits, 2017).

**Table 8.1** Various biological application of silver nanoparticles.

| Applications | Silver nanoparticles | Mechanism(s) | References |
|---|---|---|---|
| Antifungal | Silver colloidal solution | Membrane damage, hyphae damage | S. W. Kim et al. (2012) |
| | Silver nanoparticles and silver ions | Inhibit colony formation | Jo, Kim, and Jung (2009) |
| Antiviral | Biosynthesized silver nanoparticles | Larvicidal effect on Zika virus mosquito larvae | Ishwarya et al. (2017) |
| | Electrochemical synthesized silver nanoparticles | Anti-poliovirus activity | Huy et al. (2017) |
| Anticancer | Silver nanoparticles in specific polysaccharide | Increase in ROS and induce apoptosis in cancer cells | Buttacavoli et al. (2018) |
| | Biosynthesized silver nanoparticles | DNA damage and ROS generation | Mukherjee, Chowdhury, Kotcherlakota, and Patra (2014) |
| Anti-inflammatory | Liquid ibuprofen-silver nanoparticles | Increased local anesthetic effect | Jiang, Yu, Li, Ma, and Li (2018) |
| | Chitosan silver nanoparticles | Anti-inflammatory | Karthik et al. (2018) |
| Anti-angiogenic | Silver nanoparticles | Disrupt HIF signaling, inhibit angiogenesis | T. Yang, Li, et al. (2016b), |
| | Biosynthesized silver nanoparticles | Inhibit angiogenesis and DNA diffusion | Hullikere, Joshi, Vijay, Ananda, and Nivya (2015) |
| Anti-diabetic | Herbal mediated silver nanoparticles | Inhibition of protein tyrosine phosphatase 1B | Shanker, Mohan, Hussain, Jayarambabu, and Pravallika (2017) |
| | Biosynthesized silver nanoparticles | Inhibition of $\alpha$-amylase | Prabhu, Vinodhini, Elanchezhiyan, and Rajeswari (2018) |
| Antioxidant | Biosynthesized silver nanoparticles | Superoxide, radical scavenging | Cyril, George, Joseph, Raghavamenon, and Sylas (2019) |
| | Biosynthesized silver nanoparticles | Increased scavenging activity | Ahn, Jin, and Park (2019) |
| Wound healing | Biopolymer-based silver nanoparticles | Inhibiting infection | S. S. D. Kumar et al. (2018) |
| | Silver nitroprusside nanoparticles | Alteration of M1 macrophage to M2 | B. R. Rao et al. (2018) |

Side by side the extensive use of antibiotics, the microbes have also evolved skills to counter the dire effects of antibiotics (Blair, Webber, Baylay, Ogbolu, & Piddock, 2015). Pathogens try to evade the effects of antibiotics by four main mechanisms: (i) mutations in genes that affect the drug activity; (ii) horizontal transfer of resistance-determining genes; (iii) modification of internal molecular mechanisms for resistance, like altering the targets for antibiotics; (iv) enzyme-catalyzed modification of antibiotics (E. D. Brown & Wright, 2016; Blair et al., 2015).

Within the bacterial population, that is susceptible to antibiotics, some pathogens may undergo mutations in endogenous genes that may alter the drug uptake, efflux mechanisms of bacterial cells, modulation of regulatory networks, etc. Modification in all these mechanisms may favor the survival of bacteria, even under the influence of antibiotics (Munita & Arias, 2016). Moreover, horizontal gene transfer also contributes to the development of antibiotics resistance in bacteria. Bacteria acquire exogenous genetic material by transformation (uptake of naked DNA from environment), transduction (phage-mediated transfer of DNA), or conjugation (transfer of DNA by direct contact with another bacterium) (Huddleston, 2014). On the other hand, more or less all pathogens harbor some innate mechanisms for confronting detrimental effects of antibiotics. To begin with, most of the antibiotics enter the bacterial cells by crossing the cell membrane. Overexposure of drugs can induce mutations in cell membrane channel proteins, thus barring the drugs from entering the cells. Overexpression of genes encoding bacterial efflux pumps can export the antibiotics that have already entered the cells (Blair et al., 2015). Some endogenous enzymes of bacteria induce modifications like acetylation, phosphorylation, adenylation, glycosylation, hydroxylation, etc. on the drugs, often rendering them ineffective. Some enzymes are even capable of destroying the antimicrobial drug molecules by hydrolysis of ester or amide bonds (Kong, Schneper, & Mathee, 2010). In addition to that, bacteria also hinder the interaction of antibiotics to their targets by modifying different molecules inside the cells (Billal, Feng, Leprohon, Légaré, & Ouellette, 2011; Holmes et al., 2016).

Apart from the abovementioned mechanisms for antibiotic resistance, bacteria also manifest biofilm formation in order to circumvent antibiotics attack. Biofilms consist of interdependent consortium of microbial cells adhering to both biotic or abiotic surfaces and being sheathed in a self-produced extracellular matrix. This extracellular matrix is antibiotic recalcitrant. Hence, the biofilms impart increased tolerance to antibiotics and also help evade the host immune defenses. Taken together, simple antibiotic treatment options do not imply convincing reprisal to bacterial infections. Hence, stringent approaches are necessary for impeding the spread of bacterial infections. With the advancement of nanotechnology, different types of nanoparticles or combinations of nanoparticles are being explored for employment as antibiotics. In the subsequent sections, we describe the synthesis and applications of different forms of silver nanoparticles in combating bacterial infections (Hall & Mah, 2017; S. Singh, Singh, Chowdhury, & Singh, 2017; R. Roy, Tiwari, Donelli, & Tiwari, 2018).

## 8.6 Nanosilver for Antibacterial Therapy

Silver nanoparticles are studied in different forms (metallic, biosynthesized, nanocomposite, nanoscaffolds, etc.) for their antibacterial activity, which are described in the following sections.

### 8.6.1 Metallic Silver Nanoparticles

Among the metallic nanoparticles that have wide spectrum of application in the field of biomedical science, silver nanoparticle is the most imperative (Abbas, Naeem, Iftikhar, & Latif, 2018). The noble metal silver is considered to be useful in various prospect of mankind since ancient times. Its use in coin, cutlery, and weapons, in medicines (as antibacterial agent) are well known. The release of silver ions from the metallic silver contributes to their antimicrobial efficacy (Chernousova & Epple, 2013). Silver is used in various forms such as metals, salts, or nanoparticles. In recent times, the use of silver in the form of nanoparticle has been receiving greater attention due to its size, shape, better biodistribution, and bioavailability (Acharya et al., 2018; Hong, Wen, Xiong, & Hu, 2016; Mata, Nakkala, Chandra, Raja, & Sadras, 2018). Typically, chemical synthesis of silver nanoparticles involves the reduction of silver salts into silver nanoparticles using reducing agents like sodium borohydride, citrate, ethylene glycol, etc. in the presence of water or organic solvents. Further stabilization may be done to avoid aggregation and growth by following different functionalization methods (Chernousova & Epple, 2013). The synthesis of the silver nanoparticles could be navigated using different techniques to obtain variable sizes and morphology of the nanoparticle like spheres, rods, tubes, wires, and fibers (C. Rao & Cheetham, 2001).

### 8.6.2 Biosynthesized Silver Nanoparticles

Biosynthesis of nanoparticles can be defined as the process of synthesis of nanoparticles using biological entities using plant extracts (like alkaloids, anthraquinones, polyphenols, terpenoids, proteins, enzymes reductases, etc.) or microorganisms (bacteria, fungi, algae, etc.). This is an emerging branch, in the field of nanotechnology and can be called "bio-nanotechnology." This approach has gained much attention in the last two decades because biosynthesized nanoproducts are nontoxic, eco-friendly, cost-effective, and easy to synthesize (Keat, Aziz, Eid, & Elmarzugi, 2015; Sheoran & Kaur, 2018; Lateef et al., 2016; Lakshmanan, Sathiyaseelan, Kalaichelvan, & Murugesan, 2018; Singh, Kim, & Yang, 2016a; X. Zhang, Yan, Tyagi, & Surampalli, 2011; Amooaghaie, Saeri, & Azizi, 2015; Velmurugan et al., 2015; Mukherjee et al., 2014; Castro-Longoria, Vilchis-Nestor, & Avalos-Borja, 2011; S. A. Kumar et al., 2007; Singh et al., 2016b). This approach of using green technology in the synthesis of nanoparticles has revolutionized the field of nanotechnology.

Applying such green chemistry approaches, Behravan et al. synthesized silver nanoparticles using *Berberis vulgaris* leaf and root extract for their antibacterial activity. The authors have shown that the as-synthesized silver nanoparticles exhibited better antibacterial activity compared to plant extracts against both Gram-positive and Gram-negative bacteria (Behravan et al., 2019). Lateef et al. utilized cocoa pod husk extract to synthesize silver nanoparticles (4–32 nm in size). The antibacterial potential of the silver nanoparticles was found to be effective at judicious concentration against multidrug-resistant isolates of *Klebsiella pneumonia* and *Escherichia coli*. Also, these nanoparticles showed synergistic antibacterial activity in combination with conventional drugs cefuroxime and ampicillin, respectively. Along with that, the nanoparticle is reported to possess antioxidant activity, anti-larvicidal activity (against *Anopheles* mosquitoes), and antibacterial properties in paint additives (Lateef et al., 2016). Likewise, our group already demonstrated the green chemistry approach for the synthesis of colloidal silver nanoparticles by reducing silver nitrate using leaf extract of *Olax scandens*. The nanoparticles exhibited antibacterial activity against *E. coli* by damaging the bacterial cell membrane. Along with that the nanoparticle also has anticancer and fluorescence properties with biocompatible nature. Figure 8.2 depicts the growth inhibition kinetics and colony formation of *E. coli* due to biosynthesized silver nanoparticles using different concentrations. Also the figure shows the SEM images (Figure 8.2f–h) of the silver nanoparticles-treated *E. coli* cells (Mukherjee et al., 2014). Another group synthesized silver nanoparticles using the stem bark extracts of *Diospyros montana*. The particles successfully constrained the growth of both Gram-positive (*Bacillus subtilis* and *Staphylococcus aureus*) and Gram-negative (*E. coli* and *Klebsiella aerogenes*) bacteria. The author found that the particles also exhibited significant free radical scavenging activity by increasing the percentage of DPPH and $H_2O_2$ in these cells (Bharathi, Josebin, Vasantharaj, & Bhuvaneshwari, 2018). Likewise, Masum and group green synthesized nanosilver using fruit extract of *Phyllanthus emblica* plant. The synthesized particles effectively inhibited the booming of rice bacterial brown stripe pathogen *Acidovorax oryzae*. The authors also demonstrated that the silver nanoparticles could inhibit the biofilm formation as well as swarming of the pathogens. The molecular mechanism behind the inhibitory property of the nanoparticles was found to increase

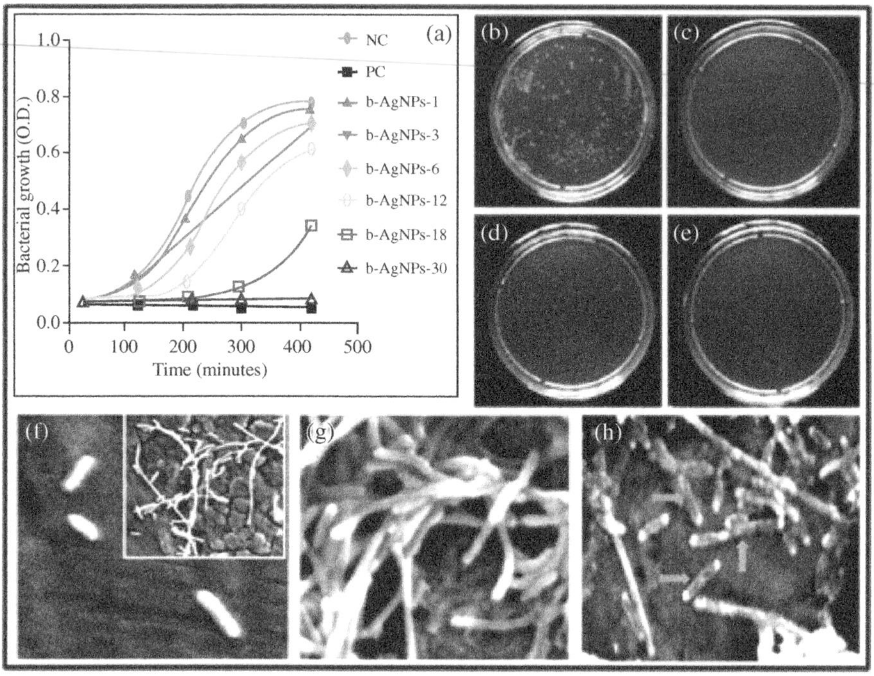


**Figure 8.2** Study of antibacterial activities: (a) liquid growth inhibition kinetics of *E. coli* using different concentrations of b-AgNPs, b-AgNP-30 (at 30 μM) shows almost 100% growth inhibition. Ampicillin has been used as a positive control (PC) and NC: negative control or untreated *E. coli*. The numerical number indicates the concentration of b-AgNPs in μM, (b–e) optical images of bacterial colonies formed by *E. coli* cells, i.e. colony counting assay (after 24 hours); (b) control; (c) ampicillin (100 μg/ml); (d) b-AgNPs (18 μM); (e) b-AgNPs (30 μM); and (f–h) SEM images of *E. coli* cells (f) without being treated (control), (g) treated with *Olax* for one hour, (h) treated with b-AgNPs (30 μM) for one hour. The SEM images show that the silver nanoparticles damage the bacterial cell membrane (marked in **black** arrow), whereas the bacterial membranes of untreated and treated *E. coli* with *Olax* are intact. *Source:* Reproduced with permission from Mukherjee et al. (2014) Copyright © (2019) Ivyspring International Publisher.

the Hcp protein secretion in the strain RS-2 resulting in cell membrane damage as confirmed by TEM images and live/dead cell staining assay (Masum et al., 2019). Recently, Pirtarighat et al. described another method to synthesize silver nanoparticle using the plant extract *Salvia spinosa*. In this reaction, the carboxylic acid functional group present in the carnosic acid and flavonoids of plant extract reduced the silver salt and stabilized the nanoparticles' formation. The group showed that the antibacterial activity of the nanosilver against both Gram-positive and Gram-negative bacteria can be attributed to alterations in cell membrane permeability leading to ATP leakage and free radical generation, resulting in cell death

(Pirtarighat, Ghannadnia, & Baghshahi, 2019). Rautela et al. used seed extract of *Tectona grandis* for synthesis of silver nanoparticles. These nanoparticles showed antibacterial activity against *E. coli, B. cereus,* and *S. aureus* strains, as determined by well-diffusion, broth dilution assays. The mode of action was found to be disruption of cell membrane, leading to leakage of reducing sugars and proteins, as detected by DNS and Bradford's method (Rautela, Rani, & Das, 2019). In another approach, Shaik et al. produced silver nanoparticles by reducing silver with the help of aqueous solution of Saudi *Origanum vulgare* L. extract. The as-synthesized nanoparticles were tested for antimicrobial activity against bacterial as well as fungal species namely *Shigella sonnei, Micrococcus luteus, E. coli, Aspergillus flavus, Alternaria alternata, Paecilomyces variotii,* and *Phialophora alba,* respectively employing well diffusion method (Shaik et al., 2018). Recently, Anwar et al. reported green synthesis of silver nanoparticle by glycosidic polymers of plant gum and further functionalized with citrus fruit flavonoids hesperidin (HDN) and naringin (NRG). These nanoparticles were studied for their antimicrobial properties against brain-eating amoebae (*Acanthamoeba castellanii* and *Naegleria fowleri* ) as well as multidrug-resistant and neuropathogenic bacteria (MRSA and *E. coli* K1 respectively). The nano-formulation successfully inhibited the encystation and excystation of the amoebae *A. castellanii* and reduced the host cell cytotoxicity initiated by the pathogens (Anwar et al., 2019).

In a different approach, Divya et al. synthesized silver nanoparticles using coral associated bacteria. From diverse coral isolates, MGL-D10 strain of bacteria was used for the biosynthesis of nanosilver. This biogenic nanosilver was checked for their antimicrobial action against urinary tract infection causing pathogens like *Pseudomonas aeruginosa, E. coli,* and *S. aureus,* as well as *Candida albicans.* Additionally, the group studied its antibiofilm activity using the catheter-mediated biofilm formed by *S. aureus* evidenced by confocal microscopy. The MGL-D10 reduced silver nanoparticles successfully inhibited the nosocomial infection causing pathogens and the biofilm formation in catheters and medical devices (Divya, Kiran, Hassan, & Selvin, 2019). Hu et al. employed mycelial extract of endophytic fungus *Talaromyces purpureogenus* to synthesize silver nanoparticles and termed as Tp-AgNPs. The biosynthesized silver nanoparticles showed antibacterial activity against both Gram-positive (*S. aureus, B. cereus*) and Gram-negative (*Salmonella enterica, P. aeruginosa*, and *E. coli*) bacteria as well as wound-healing capability and anticancer activity (Hu, Saravanakumar, Jin, & Wang, 2019).

### 8.6.3 Silver Nanocomposites

Nanocomposites are multipotent materials consisting of different phase-separated materials where at least one phase is in the nanoscale range (Müller et al., 2017). Nanocomposites have contributed immensely in the field of biomedicine and material science industry. In this section of the book chapter, we put light on the nanocomposites containing silver associated with functional materials and their application as antimicrobial agents. Generally, within the nanocomposites, silver ions remain assimilated with the polymers or matrices. The manifestation of the effects of these nanocomposites depends upon the release of silver ions in liquid suspension. These released silver ions cause death of bacterial cells by various mechanisms like increase in cell membrane permeability, disruption of essential cellular metabolic processes, etc. (Egger, Lehmann, Height, Loessner, & Schuppler, 2009).

Fu et al. reported on the efficient antibacterial activity of a newly designed silver nanocomposite containing cellulose. The nanocomposite was prepared by eco-friendly hydrothermal method where hemicellulose was the reductant for cellulose (natural polymer). AgNPs were incorporated into the cellulose substrates during the synthesis. The as-synthesized nanocomposites were evaluated for their antibacterial activities against two common pathogens: *S. aureus* and *E. coli* using some genuine methods like agar disc diffusion assays, **m**inimum **i**nhibitory **c**oncentration (MIC) assay. The results of these assays showed considerable zone of inhibition against both the species of bacteria. The author also revealed that the antibacterial activity of the nanocomposites was due to the releasing behavior of silver in the nanocomposite (Fu, Gao, Qi, Ma, & Li, 2018).

Our group also developed a nanocomposite, silver nitroprusside (AgNNPs), using sodium nitroprusside and silver nitrate. The as-synthesized biocompatible (observed by various in vitro and in vivo assays) silver nanocomposite possessed antibacterial properties against both Gram-positive (*B. subtilis, Staphylococcus epidermidis*) and Gram-negative *(E. coli, K. pneumonia* and *P. aeruginosa*) bacteria. Bacterial growth kinetics study revealed that the nanocomposite exhibited inhibitory effect at the log phase stage of the growth curve. While investigating the molecular mechanisms of antibacterial activity, it was found that the nanocomposite exerted antibacterial activity by membrane disruption and oxidative stress-induced DNA damage. Additionally, the nanocomposite was found to exhibit wound-healing property by accelerating re-epithelialization and endothelial cell migration. The nanocomposite enhanced the polarization of M1–M2 macrophages as observed by increased expression of Arginase-1, an M2 macrophage marker in skin tissues. Figure 8.3 reflects the immunohistochemical analysis of skin tissues of mice demonstrating the M1 macrophage polarization to M2 macrophages as induced by AgNNP treatment upon skin tissues using Arg-1 and NOS markers (B. R. Rao et al., 2018). Mei et al. synthesized such nanocomposites using PDMAEMA [polymerized monomer 2-(dimethylamino) ethyl methacrylate] stabilized with AgNPs (AgNPs@PDMAEMA-C4), which have strong antimicrobial activity. The combination of these polymers and the antimicrobial silver nanoparticles effectively induce antimicrobial activity against two bacterial strains,

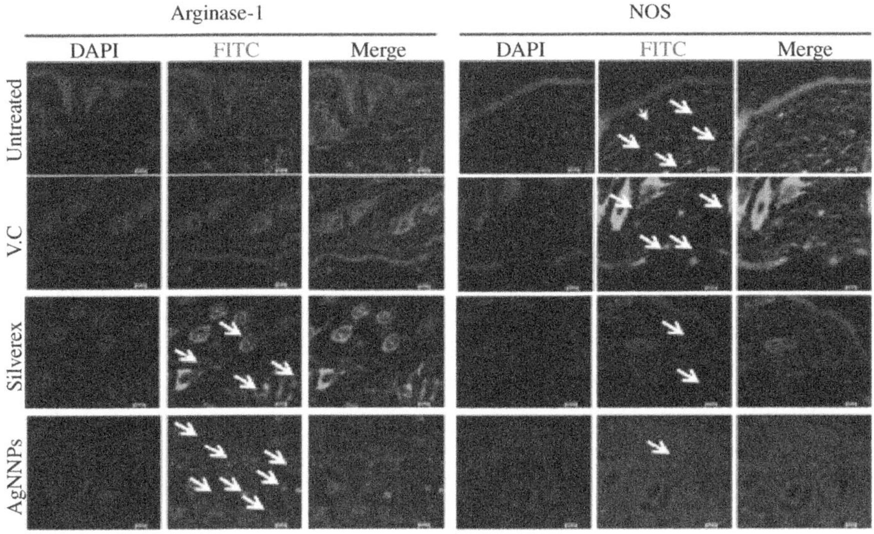


**Figure 8.3** Immunohistochemistry study of skin tissues of C57BL6 female mice sacrificed on day 9 post-treatment with AgNNPs. The presence of macrophage polarized forms such as M2 and M1 are identified using antibodies (1 : 1000). Nuclear staining was performed using DAPI. Arg-1 positive cells depicted as arrows in M2 macrophages present in skin tissues. Similar kind of observation for M1 macrophages polarization is shown in arrows in the stained skin tissues. First row: untreated skin, second row: vehicle control (Vaseline), third row: Silverex 0.2% (w/W) treated, fourth row: AgNNPs-0.1% (w/w) treated skin tissues, respectively. Cells were observed under 60× objective. Scale bar: 5 μm. *Source:* Reproduced with permission from B. R. Rao et al. (2018) Copyright © (2018) American Chemical Society.

*S. aureus* and *P. aeruginosa* without eliciting bacterial resistance. Molecular mechanisms have revealed that the PDMAEMA-C4 nanoparticles interfered with the cell membrane fluidity and hence killed the cells by causing membrane disruption resulting in cell lysis. In addition to membrane fluidity, hindrance of activity of intracellular substance can also disrupt the normal physiology of the cell. Hence, inhibitory effect of AgNPs@PDMAEMA-C4 on intracellular enzyme, $\beta$-galactosidase, nanocomposite, and silver nanoparticle separately was also measured. The results indicated that the nanoparticle was responsible for killing bacteria by interfering with cellular functions (Mei, Lu, Zhang, Li, & Jia, 2014).

Nowadays, researchers are aiming at developing antimicrobial agents with materials that are not harmful to living beings. In this context, development of polymer-based metal nanocomposites has gained a lot of attention. This is because polymers are biocompatible, abundantly available in nature, possesses biological activities like antimicrobial activity, and enhance antimicrobial effect by several folds, when linked with antimicrobial metal nanoparticles. Futrya et al.

synthesized a nanocomposite using cationic polymers, chitosan and AgNP, derived from ascorbic acid (Regiel-Futyra et al., 2017). The group investigated the antimicrobial activity against some common pathogens like *S. aureus*, *P. aeruginosa*, and *E. coli*. The nanocomposites were found to be toxic against these strains during both stationary and growth phase of the pathogens. The promising result of this study highlights the potential for application in biomedical field.

### 8.6.4 Silver Nanoscaffolds

Nanoscaffolds are multipotent materials where at least one phase is polymeric and another phase is in the nanoscale range (Bhattacharyya, 2019). Such scaffolds are being utilized for manufacturing various medical devices like medical implants and consumer goods (Doolette et al., 2013; Zheng, Song, Wang, & Wang, 2014; Z. Li, Zhang, Cheng, & Yang, 2016).

In this regard, Lu et al. designed a titanium/silver containing nanoscaffold material, nano-hydroxyapatite/polyamide-66 composite (TA-nHP66), that exhibits antimicrobial activity (Lu et al., 2016). The formed nanoscaffold successfully inhibited the bone infection caused by *S. aureus* and *E. coli* during osteomyelitis. TA-nHP66 scaffolds show significantly larger and prolonged inhibitory effect on the two aforementioned bacterial strains than only silver and titanium in disc diffusion assay. This proves that Ag and Ti nanoparticles complement each other to exhibit synergistic antibacterial activity. Additionally, in vivo tests proved that the as-synthesized scaffolds did not evoke inflammation in osteomyelitis-induced rabbit models.

In another study, a research group from Southwest Jiaotong University, China reported on the development of medical implants formed by electrochemical deposition of Ag and electrostatic immobilization of BMP-2 on HA-coated Ti surfaces. Standard in vitro and in vivo assays revealed that these BMP/CS/Ag/HA-coated Ti bars not only execute osteoconductivity by promoting differentiation of bone marrow stromal cells (BMSCs), but also show prolonged bactericidal effect against both *S. epidermidis* and *E. coli*. Overall, the authors showed that the implant has the potential for enhancing bone growth in case of bone disorders, without the concern for bacterial infection (Xie et al., 2014).

## 8.7 Influence of Size and Shape of Silver Nanoparticles as Antibacterial Agents

The antibacterial activity of the silver nanoparticles is highly dependent upon the shape and size of the nanoparticles. Many studies have reported that silver nanoparticles with smaller sizes were more cytotoxic than the larger ones

(X. Yang et al., 2012). The clarification to this point is attributed to the fact that small size of silver nanoparticles helps in easy uptake, prompt release of silver ions, increased surface area, and ready dissipation (Powers, Badireddy, Ryde, Seidler, & Slotkin, 2011). In this regard, Morones et al. exhibited the bactericidal property of the silver nanoparticles. They showed that the silver nanoparticles within the size range of 1–10 nm were most effective as they were able to attach to the bacteria directly (Morones et al., 2005). The abovementioned findings are similar to the results obtained by Choi et al. In this study, the authors incubated nitrifying bacteria with silver nanoparticles of different sizes (9–21 nm). The results revealed that nanosilver measuring less than 5 nm in size were most effective in inhibiting these nitrifying bacteria. Direct attachment to the bacterial cell and subsequent damage to cell membrane permeability, cytosolic redox cycle, and dissipation of proton motive force were the possible mechanisms of action for cell death. In this regard, silver nanoparticles were more compelling than silver ions themselves (Choi & Hu, 2008). Another group of scientists from IIT, Bombay, India, demonstrated that the activity of silver nanoparticles against bacterial strains *E. coli* (MTCC 443, MTCC 739), *S. aureus* (NCIM 5021), and *B. subtilis* (MTCC 441) increased with decrease in size of nanosilver (Agnihotri, Mukherji, & Mukherji, 2014).

In addition to size, efficiency of silver nanoparticles also depended on the shape of nanosilver. Acharya et al. reported that spherically shaped silver nanoparticle was more lethal than rod-shaped nanosilver against Gram-negative bacteria *K. pneumonia*. The results were confirmed by various in vitro assays. Figure 8.4 shows the FE-SEM images of the effect of the sphere- and rod-shaped silver nanoparticles on *K. pneumonia* bacteria and how the nanoparticles ruptures its cell membrane (Acharya et al., 2018). In another study, Kim et al. reported that plate-shaped and small spherical silver nanoparticles exhibited enhanced killing of Gram-negative bacteria *E. coli* than cube-shaped silver nanoparticles (D. H. Kim, Park, Jeon, Kim, & Seo, 2017).

Changes in the surface modification of the silver nanoparticles also play an important role in the antibacterial activity of the nanoparticle since they change the shape and size of the formed nanoparticle (Montes-Burgos et al., 2009).

## 8.8 Nanosilver and Its Mechanism of Action for Antibacterial Therapy

Silver nanoparticles (Ag-NPs) have been used as antibacterial, antifungal, antiviral, anti-inflammatory, and anti-angiogenic agents due to their unique physical, chemical, and biological properties (You et al., 2012). The most commonly

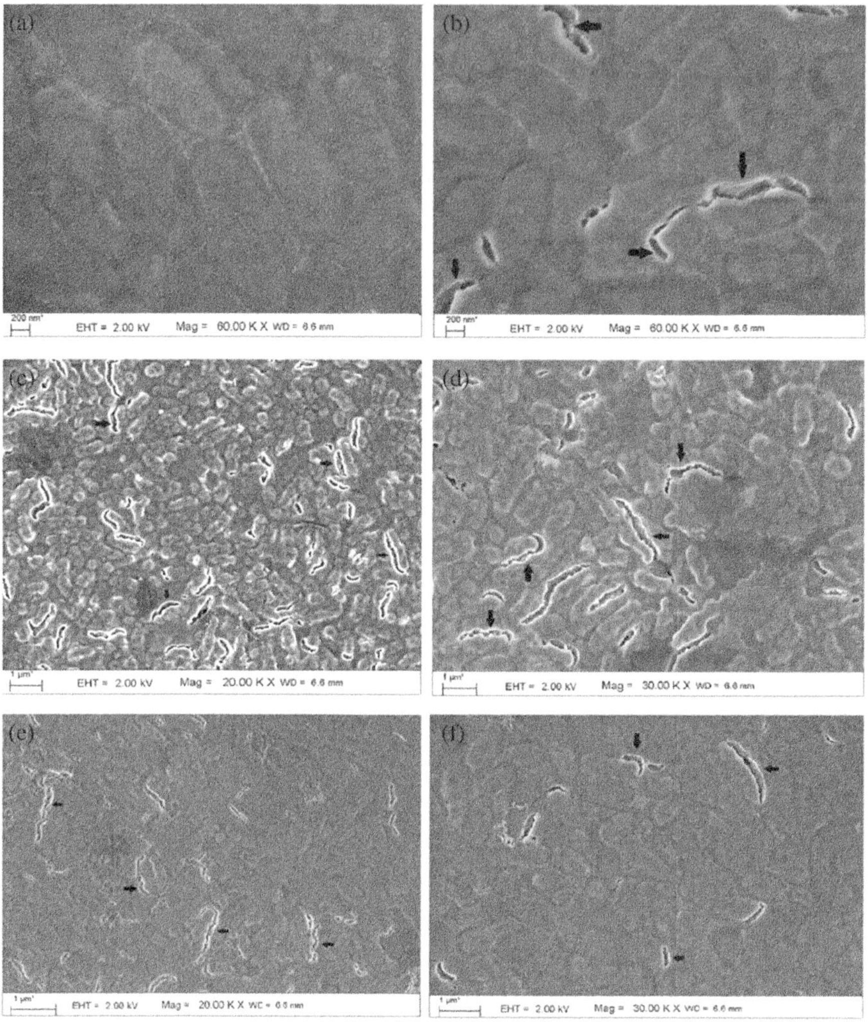


**Figure 8.4** FE-SEM images of nanoparticle-untreated *K. pneumonia* AWD5 (a) with intact cell surfaces, in contrast to ANP-sp treated *K. pneumonia* AWD5 (b). The cells of *K. pneumonia* AWD5 were disrupted due to interaction with AgNP-sp (c, d) and AgNR (e, f). Arrow indicates disrupted regions on cell surface. *Source:* Reproduced with permission from Acharya et al. (2018) Copyright © (2019) Springer Nature Limited.

reported mechanism of the silver nanoparticles is membrane damage leading to the leakage of cellular contents and consequent cell lysis (Seong & Lee, 2017), (Khalandi et al., 2017). The affectivity of the silver nanoparticle in bacterial cell membrane damage also depends on the bacterial cell wall composition, i.e. layer of the peptidoglycan, based on which bacteria are classified as Gram-positive and Gram-negative bacteria (Chatterjee, Chatterjee, Majumdar, & Chakrabarti, 2015). Figure 8.5 shows the comparative study of the silver nanoparticles and the silver ions on Gram-positive and Gram-negative bacteria. It shows the complete pathway of leakage of metabolites, inactivation of enzymes, and interaction with different organelles leading to the ultimate death of the bacteria (Kędziora et al., 2018). The silver nanoparticles with positive charge can interact well with the negatively charged bacterial membrane (Abbaszadegan et al., 2015; Mandal et al., 2016). Additionally, the modulation of the size and morphology of the silver nanoparticles facilitates their enhanced cellular internalization and subsequent interaction with the intracellular biomolecules like nucleic acids, lipid, proteins, etc. leading to their denaturation or inactivation, which ultimately lead to bacterial cell death (You et al., 2012). Further, the underlying mechanism of bactericidal action of silver nanoparticles could be the generation of excessive reactive oxygen species (ROS) and release of free radicals like hydrogen peroxide, hydroxyl radical, superoxide anion, singlet oxygen, etc. (Gomaa, 2017; Siritongsuk et al., 2016; Zhao et al., 2017). The increased ROS reduces the production of antioxidant enzymes such as glutathione (GSH), catalase, superoxide dismutase as well as induce DNA damage and lipid peroxidation. All these activities trigger death of the bacterial cell (W. Lee, Kim, & Lee, 2014; Yun'an Qing et al., 2018; Yuan, Peng, & Gurunathan, 2017; Korshed, Li, Liu, & Wang, 2016). Another important aspect of the bacterial toxicity of the silver nanoparticles is the release of silver ions. According to Stabryla et al., only silver ions have decreased antibacterial activity as compared to the engineered silver nanoparticles. This indicates that the silver nanoparticles have different silver ion release and binding pathways in the environment external and internal to the bacterial cells as shown in Figure 8.6 (Stabryla et al., 2018). In this context, a proteomics approach by Yan et al. revealed that one of the bactericidal effects of silver nanoparticles against *P. aeruginosa* was due to upregulation of 27 and downregulation of 32 silver controlled proteins. This study also uncovered five silver binding proteins within the cells. The study also uncovered that, along with the abovementioned proteins, interference with cell membrane and intracellular ROS generation were also crucial pathways, in this bactericidal activity (Yan et al., 2018). Due to the increasing antibiotic resistance, Surwade et al. suggested the combination therapy of silver nanoparticles with the conventional antibiotic ampicillin against methicillin-resistant *S. aureus* (MRSA). The study showed synergistic effect of the nanoparticle and the antibiotic proving its effectiveness at a lower concentration (Surwade et al., 2019).

Another major hurdle in the way of effective antibacterial therapy is the biofilm formation. Biofilms are the extracellular matrix formed by the bacteria mostly on solid surfaces like tissues, medical devices, catheters, and pacemakers to protect the bacterial community from any antibiotics (Neethirajan, Clond, & Vogt, 2014;

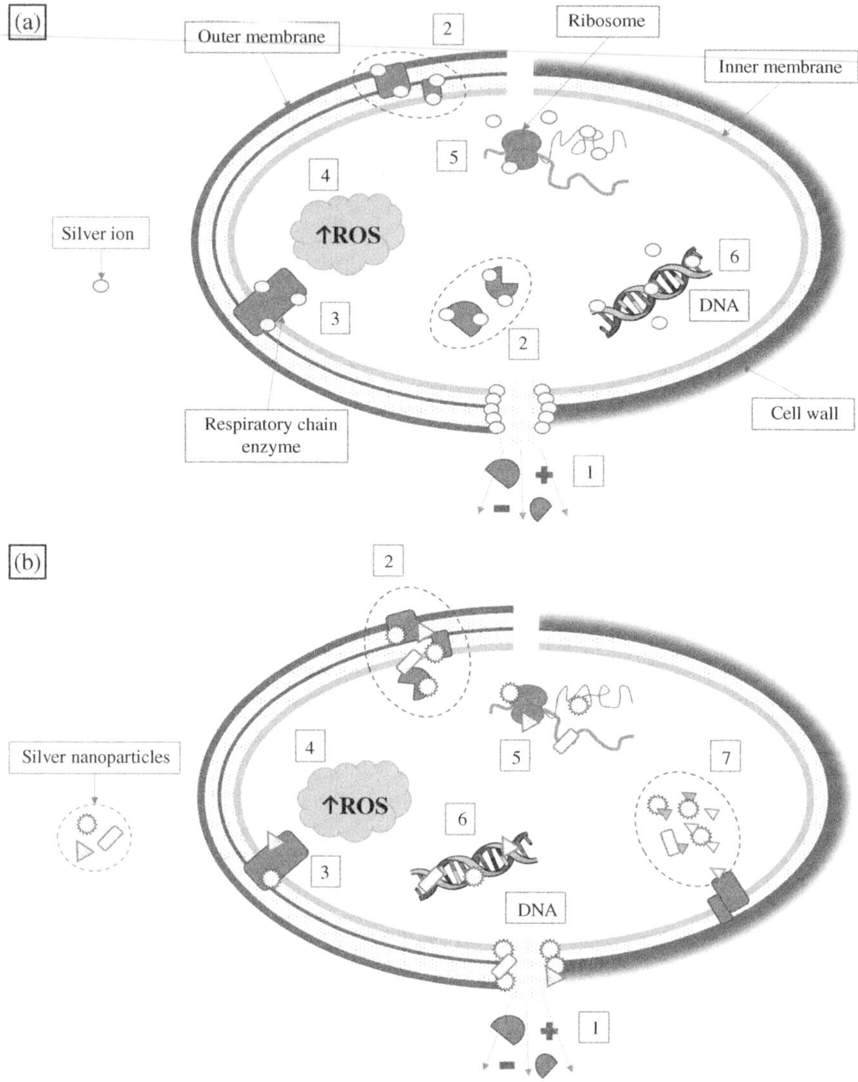


**Figure 8.5** A comparison of the silver ions (a) and silver nanoparticles (b) mode of action to Gram-negative (left) and Gram-positive (right) bacteria. (1) Pore formation: metabolites and ions leakage (shown as plus and minus in the figure above). (2) Denaturation of structural and cytoplasmic proteins; enzyme inactivation. (3) Inactivation of respiratory chain enzymes. (4) Increase of intracellular reactive oxygen species (ROS) concentration. (5) Interaction with ribosome. (6) Interaction with nucleic acids. (7) Inhibition of signal transduction. *Source:* Reproduced with permission Kędziora et al. (2018) Copyright © (2018) by the authors. Licensee MDPI, Basel, Switzerland.

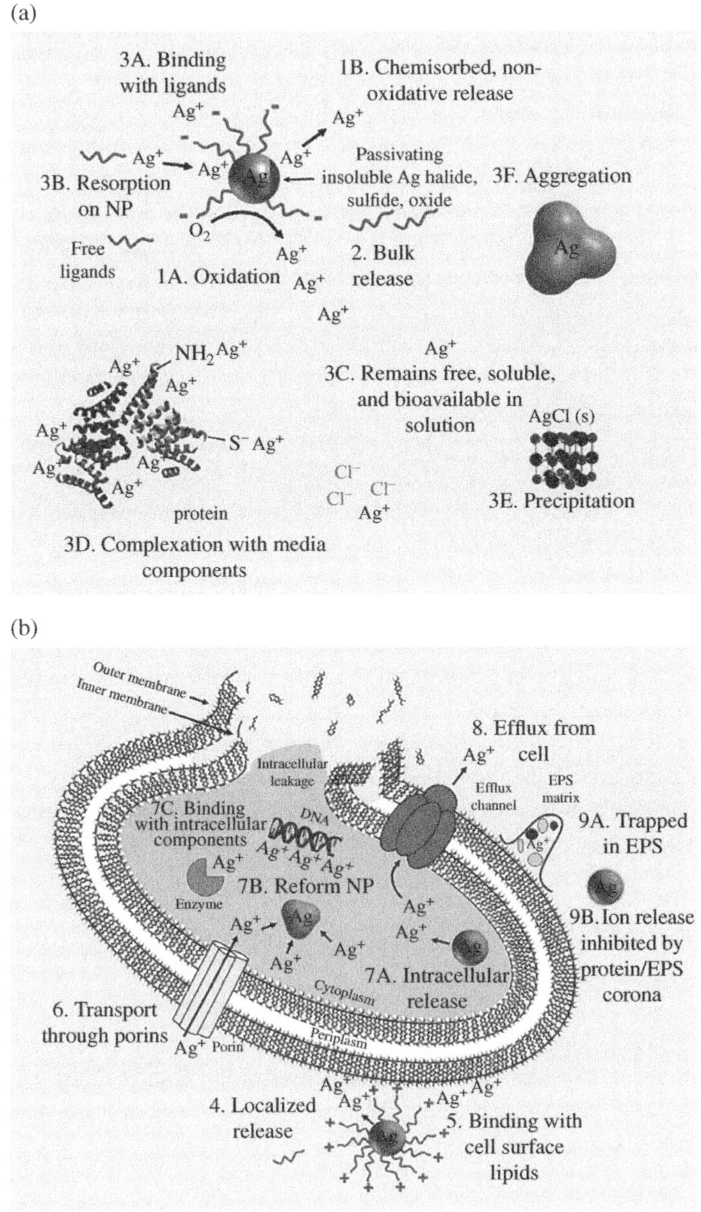


**Figure 8.6** Potential Ag(I) ion release pathways, binding, and partitioning events within the AgNP exposure condition. (a) State of Ag in the media, external to the bacterial cell. (b) Interaction of Ag with the cell and state of Ag inside the cell. Note: Figure is not to scale as the particle is enlarged to demonstrate effect. *Source:* Reproduced with permission from Stabryla, Johnston, Millstone, and Gilbertson (2018) Copyright © (2018) The Royal Society of Chemistry.

Stoodley, Sauer, Davies, & Costerton, 2002). Recently, Hashimoto et al. demonstrated the size-dependent effect of silver nanoparticles against biofilm-forming *Streptococcus mutants*. Silver nanoparticles were synthesized by coating with polyvinylpyrrolidone of different size range (74, 33.9, and 57.0 nm). The TEM analysis revealed that the nanoparticles bearing least size were most effective in interacting with bacteria and inhibiting the biofilm formation (Hashimoto & Honda, 2019). Similarly, silver nanoparticles could successfully diminish the biofilm formation in the marine environment where the marine biofilm was formed on glass and polydimthylsiloxane (PDMS). The effect of the silver nanoparticle on biofilm formation was evaluated using MiSeq sequencing technology (J.-L. Yang et al., 2016). Another group reported that quercetin-decorated silver nanoparticles showed elevated antibacterial as well as antibiofilm activities against *E. coli* (strain ECDCM1) isolated from cows with mastitis. Such a report was confirmed by the results obtained in antibacterial assays, SEM image. Further, deeper molecular study using real-time PCR analysis revealed that nanosilver reduced transcription of biofilm forming genes (Yu et al., 2018). Very recently, Singh et al. demonstrated a genome-wide transcriptome analysis of the *P. aeruginosa* biofilm via RNA-Seq using silver nanoparticles. Along with RNA-Seq evaluation, qRT-PCR, SEM, and confocal microscopy revealed the modification of the cellular morphology with substantial inhibition of the biofilm formation. The group highlighted the fact that although silver nanoparticles are effective in the reduction of biofilms, thus minimizing the predicament of antibiotic resistance. However, overuse of these nanoparticles in the products of daily use may adversely affect the ecosystem (P. Singh, Paknikar, & Rajwade, 2019). Figure 8.7 illustrates the overall mechanistic pathways for antibacterial activity of the silver nanoparticles, which includes membrane damage, destabilization of the proteins and organelles along with interaction with DNA (Dakal et al., 2016).

## 8.9 Application of Silver Nanoparticle in Commercial Products

Silver nanoparticles have gained commercial acceptance due to wide range of biological and non-biological activities. Silver nanoparticles are employed to prepare large number products of commercial importance and for our daily use. In the following sections, we discuss some of the aspects.

### 8.9.1 Silver Nanoparticles in Wound Dressing Materials and Devices

Wounded areas on the skin are prone to microbial infections. These microbial infections impede the wound-healing process by colonizing and thriving on the wounded regions. Hence, competent treatment strategies are required to deal

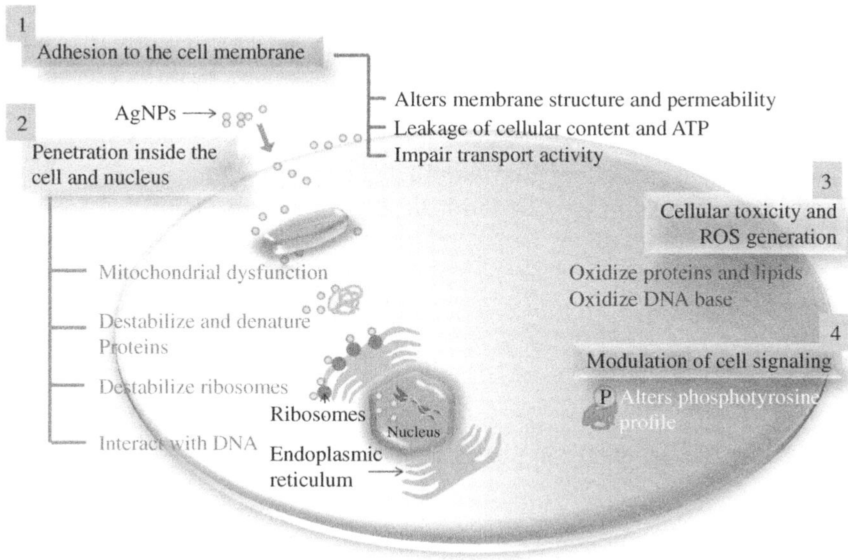


**Figure 8.7** The four most prominent routes of antimicrobial action of AgNPs: 1, AgNPs adhere to microbial cell surface and result in membrane damage and altered transport activity; 2, AgNPs penetrate inside the microbial cells and interact with cellular organelles and biomolecules, and thereby, affect respective cellular machinery; 3, AgNPs cause increase in ROS inside the microbial cells leading to cell damage; and 4, AgNPs modulate cellular signal system ultimately causing cell death. *Source:* Reproduced with permission from Dakal, Kumar, Majumdar, and Yadav (2016) Copyright © (2016) Dakal, Kumar, Majumdar and Yadav.

with such pathological conditions. One viable approach to reduce wound infections is the use of wound dressings loaded with antimicrobial agents. Use of AgNPs in wound dressings can help eradicate the problem related to wound infections. A number of researchers have developed various types of such wound dressing materials (Augustine et al., 2018; GhavamiNejad et al., 2015; Thomas, Soumya, Mathew, & Radhakrishnan, 2015). In a report published by Chen et al., the author and his team synthesized a novel Konjac glucomannan/silver nanoparticle (KGM/AgNP) composite for use as wound dressing device. The composite sponge was synthesized by simple method and was shown to execute cytocompatibility on normal mammalian cell line, L929. Besides that, the composite material exhibited excellent antibacterial activity owing to the presence of silver nanoparticles. By further screening of the product for potential applications, it was found by histological analysis of wound tissues that the novel composite sponge also promoted fibroblast growth and accelerated epithelialization. Taken together, the as-synthesized wound dressing material showed excellent potential for inducing faster healing of wounds (H. Chen et al., 2018). In another study, a research group reported the use of sericin- and chitosan-capped silver nanoparticle (S/C-SNP)-loaded wound dressing hydrogel. In this composite, silver and chitosan exhibited antibacterial activity against *E. coli.* (Gram-negative bacteria) and *S. aureus* (Gram-positive bacteria). On the other hand, sericin was useful in promoting wound healing. In addition, histopathological analysis revealed that, this wound dressing device augmented keratin and blood vessel formation, leading to rapid wound closure. Finally, the hydrogel showed no significant sign of dermal irritation. Overall, the nanocomposite demonstrated significant potential to be used as a wound dressing material (Verma, Kanoujia, Parashar, Tripathi, & Saraf, 2017). Thus in various studies, the importance of use of silver as an antibacterial agent in the fabrication of new wound dressing materials has been approved.

### 8.9.2 Silver Nanoparticles in Soaps and Detergents

Nanotechnology-based companies have focused on the role of nanoparticles in topical therapy for the management of dermatological conditions. Currently, various types of nanoparticle-based soaps, detergents, lotions, shampoo, ointment, cream, gel products, etc. are available in the market. The role of the nanoparticles here is to remove the pathogens and allergens on the skin surface without disrupting the skin epidermal lipid barrier. In this context, nanoparticles like carbon, silver (S. Tang & Zheng, 2018; C. Tang, Hu, Cao, Yan, & Xing, 2017; J. K. Patra & Baek, 2017), titanium (Alavi & Karimi, 2018; Subhapriya & Gomathipriya, 2018) zinc, (Raja et al., 2018; Kadiyala, Turali-Emre, Bahng, Kotov, & VanEpps, 2018), silica (Gounani, Asadollahi, Meyer, & Arpanaei, 2018; Park et al., 2016) copper, (Khatami, Heli, Jahani, Azizi, & Nobre, 2017), and graphene (Yousefi et al., 2017), which have robust antimicrobial activities, are often used singly or in combination with other polymers or nanoparticles (Alavi & Karimi, 2018; Slavin, Asnis, Häfeli, & Bach, 2017). To this end, scientists have carried out research works for developing nano-functionalized antimicrobial products. In one study, Bansod et al prepared AgNP functionalized oil for the development of herbal soaps, shampoos, and ointments for topical application. The group synthesized AgNPs from *Azadirachta indica* and combined them with herbal oils extracted from medicinal plants including *Cymbopogon citratus, Cymbopogon martini, Eucalyptus globules, A. indica,* and *Ocimum sanctum.* The formulation successfully exhibited antimicrobial activity against animal skin pathogens, *S. aureus* and *E. coli* bacteria. With regard to formulations for soaps and shampoos, the optimum range of functional conditions like pH balance, temperature, etc. must be compatible with the surface of animal skins. In this formulation, the pH of the soap and shampoo were found to be 6.5 and 6.7

respectively, which is considered to be a safe range. The antimicrobial property of the ointment also helped in preventing infections from occurring in wounds, thus accelerating healing of wounds (Bansod, Bawaskar, Gade, & Rai, 2015). The antimicrobial products impede the survivability of the pathogens by general mechanisms like alteration in cell membrane permeability, disruption of cell membrane potential. As a result, there was a release of membrane lipopolysaccharides, generation of free radicals, disintegration of membrane potential by dissipating the proton motive force, etc. (Bansod et al., 2015).

### 8.9.3 Silver Nanoparticles in Fabrics

There are a variety of uses of silver nanoparticles in the field of fabrics (Ballottin et al., 2017). To this context, Su et al., synthesized silver nanoparticles with the help of sodium alginate (SA@AgNPs) as capping agent and incorporated into cotton fabrics that exhibit antibacterial activity against nosocomial pathogens (*S. aureus, P. aeruginosa, E. coli*). Figure 8.8 shows the SEM images (Figure 8.8a) of the SA@AgNPs-incorporated cotton fabrics and antibacterial activity of the nanoparticles by agar diffusion (Figure 8.8b) (Su et al., 2017). Often, garments are made by integrating silver nanoparticles with cotton fabrics to impart the antibacterial properties (Ibrahim & Hassan, 2016; Marákova et al., 2017; Wu, Ma, Pan, Chen, & Sun, 2016). Medical practitioners like doctors, surgeons, etc. are always exposed to infections. Hence, wearing surgical aprons with antibacterial properties can help prevent the spread of infections from patients to doctors and surgeons. There have been reports of development of surgical aprons impregnated with silver nanoparticles to provide protection from infection (Sadanand, Tian, Rajulu, & Satyanarayana, 2017; Walser, Demou, Lang, & Hellweg, 2011). In this context, Freeman et al. used silver nanoparticles and functionalized them in fabrics to prevent bacterial contamination (Freeman, Halladay, & Cripps, 2012; Q. Li et al., 2008).

There are many ways to fabricate silver nanoparticles. Some include blending the silver nanoparticles with the clothing materials and others include immobilization of the material with the silver nanoparticles. Meng et al. fabricated silver nanoparticles with silk fibers, which were multilayered coated film. The silk fibers exhibited antibacterial property. The silver ions in the fabric were found to be released in a sustainable manner, thus prolonging antimicrobial activity (Meng et al., 2016). In another study, Zulfiqar Ali Raza et al. designed cotton fabric with silver nanoparticles incorporated into it with the help of starch. The starch acted as reducing agent and also helped to stabilize the silver nanoparticle-incorporated cotton fabrics. The material exhibited good antimicrobial activity against the *E. coli* and *S. aureus* strains (Raza et al., 2015).

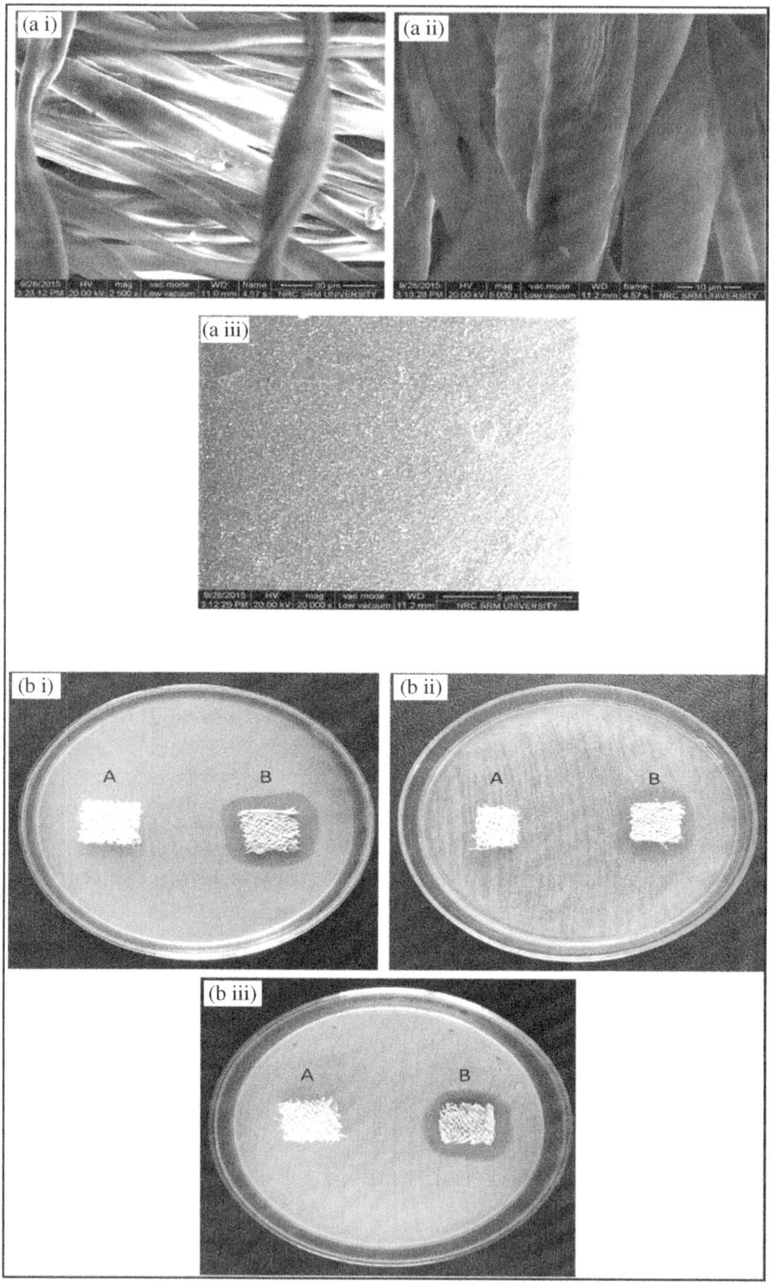


**Figure 8.8** (a) SEM images of cotton fabrics with SA@AgNPs at different magnifications: (i) 2500×, (ii) 5000×, and (iii) 20 000×. (b) Antibacterial activity of untreated and SA@AgNP-treated cotton fabrics against (i) *E. coli,* (ii) *P. aeruginosa,* and (iii) *S. aureus* measured by the agar diffusion method. A, Untreated cotton fabrics; B, cotton fabrics treated with SA@NPs. *Source:* Reproduced with permission from Su et al. (2017) Copyright © (2016) Elsevier B.V.

### 8.9.4 Silver Nanoparticles in Cosmetics

Nanotechnology is one of the key technologies that is now being greatly explored in the field of cosmetics (Kaul, Gulati, Verma, Mukherjee, & Nagaich, 2018; Kraeling et al., 2018). Silver nanoparticles owing to their antibacterial properties are now being widely used in cosmetics. Several toxicity studies have been made on animals with the silver nanoparticles to establish them as cosmetic agents and their use for topical applications (Sweet & Singleton, 2011). In this regard, Jain et al. through their studies confirmed that silver nanoparticles had no toxic effect on rats. This proved that silver nanoparticles could be safely used for topical applications (Jain et al., 2009).

Nowadays, lots of products of daily use like toothpastes, creams, baby products, soaps, etc. are formulated with silver nanoparticles because of their excellent antimicrobial properties. Ointments used for the treatment of skin diseases like acne, folliculitis, cellulitis, etc. are often formulated with nanosilver for enhanced activity (Samberg, Oldenburg, & Monteiro-Riviere, 2009; Z. Zhang, Tsai, Ramezanli, & Michniak-Kohn, 2013; Lohani, Verma, Joshi, Yadav, & Karki, 2014). Hand disinfectants are prepared using silver nanoparticles that are effective in preventing disease transmission in a limited period of time and minimizing the effect of infectious diseases (Agnihotri et al., 2014; Wei et al., 2010). Other utilities of silver nanoparticles include their use in nail paints. This includes mixing of silver with other metal and additional chemicals and compounds (Domeradzka-Gajda et al., 2017).

### 8.9.5 Silver Nanoparticles in Food Packaging

Food borne diseases are formidable menace in public health worldwide. Most of the foodborne diseases arise from contamination of food, which can occur at any stage, from farm to table. Most common contributing factors are related to unhygienic growth condition, food handling, preparation methods, and food delivery (Angelo, Nisler, Hall, Brown, & Gould, 2017). One approach to reduce microbial contamination during food delivery is the use of antimicrobial food containers or food packages. Hence, manufacturers of consumable products are trying relentlessly to meet the increasing demand for food delivery in an attractive and hygienic way. Consequently, the food packaging industries are working on innovative packaging strategies that will help reduce the risk of food spoilage by pathogens and at the same time extend the shelf life of food (Carbone, Donia, Sabbatella, & Antiochia, 2016). With tremendous advancement in nanotechnology, food packaging industries have recognized the benefits of incorporating nanomaterials in food packaging materials. In this regard, nanoparticles with antibacterial, antimicrobial, water-resistant, and other protecting properties have gained attention. Among the different nanomaterials, possessing these properties, silver has found the maximum usage in the manufacture of food containers or packages. Different polymers such as nontoxic, biodegradable, and abundantly available natural polymers (starch, cellulose, agar, chitosan, gluten, gelatin) and highly durable, flexible, and thermostable synthetic polymers (like PLA, PGA, PVA, and PBS) are being exploited. Amalgamation of antibacterial silver with biopolymers is an area of emerging interest (Rezić, Haramina, & Rezić, 2017). Several research groups around the globe are working on developing new types of food packing materials or coatings that are biodegradable and at the same time possess antimicrobial activity. In this regard, Sarwar et al. synthesized a nanocomposite film using polyvinyl alcohol (PVA), nanocellulose, starch-capped AgNPs, and glycerin by solution casting method. In this film, PVA was complexed with nanocellulose, which formed a percolating network. The AgNPs were used as antimicrobial fillers. The AgNPs show excellent antibacterial activity against Gram-negative *E. coli* and Gram-positive methicillin-resistant *S. aureus* (MRSA) strains. In Gram-negative bacteria, the positively charged AgNPs interacted with the negative cell membrane and penetrated through the thin peptidoglycan layer, thus disrupting the cell membrane leading to cell death. In the case of Gram-positive bacteria, the positively charged AgNPs accumulated at the cell membrane but could not penetrate the thick peptidoglycan layer and hence showed slightly fluctuating cell damaging activity with varying concentration of silver. Structural analysis showed that the nanocomposite possessed efficient moisture retention capability and thermal stability. Based on the mechanical and antibacterial properties, the author envisaged that the PVA/NC/Ag nanocomposite films can be used for food packaging (Sarwar, Niazi, Jahan, Ahmad, & Hussain, 2018). Scientists are putting a lot of effort to develop biodegradable films or coatings that are biocompatible, nontoxic, and cost-effective. For this purpose, different types of bionanocomposite packaging materials are being synthesized, and these bionanocomposite packaging materials are fast replacing the conventional plastic-based packaging materials. These are mostly formed by amalgamation of nanoparticles and biopolymers. On the other hand, the nanoparticles impart additional strength, and other suitable properties like oxygen scavenging capability, antimicrobial activity, temperature maintenance ability, etc. (Youssef & El-Sayed, 2018). In a similar way, scientists are focusing on the blending, layering, and filling natural polymer derivatives with different biopolymers or synthetic materials to impart enhanced mechanical and physical strength to ultimately develop best films to meet the ASTM standards (Malhotra, Keshwani, & Kharkwal, 2015). One naturally occurring protein, zein, found in the endosperm of corn is often considered as a waste protein due to its poor water solubility and inconsistent amino acid profile. However, the hydrophobicity of zein makes it less water permeable and, hence, has been regarded as important property for making films for use as food packaging materials. Zein films are being extensively used for making antimicrobial food packaging materials (Kashiri et al., 2017). Mei et al., for the first time, produced silver nanoclusters embedded zein films. This novel coating material exhibits potent antimicrobial activity and low cytotoxicity and hence forms proficient food packaging material. Moreover, it was observed that silver nanoclusters show higher antimicrobial efficacy than silver nanoparticles. This was vindicated by systemic evaluation of antibacterial activity and cytotoxicity of AgNC, AgNP, and $AgNO_3$. Standard antibacterial activity assays revealed that AgNCs are far more effective at lower doses than AgNPs, thus allowing for a reduction in administration doses for antimicrobial applications. Corroborative results were also obtained from bacterial growth kinetics assay, proving that AgNCs and $AgNO_3$ had antimicrobial efficiency in a dose-dependent manner, while AgNPs had no antimicrobial efficiency at even high doses. The nanoclusters have high surface-to-volume ratio, which allows for their proficient release from composites and also grants better contact with bacteria (Mei et al., 2017).

### 8.9.6 Silver Nanoparticles in Paints

Microbial growth on painted walls contributes to weathering and destruction of the paints (Wangler et al., 2012). The microorganisms destroy the surface of the paint by forming a mesh with production of spores and dirt retention. As a result of that, the paint gets discolored. In order to counter all these shortcomings, paint industries have focused on the production of paints with new formulations having antimicrobial properties (Schoknecht, Gruycheva, Mathies, Bergmann, & Burkhardt, 2009). Thus, the paint industry is now focusing on antimicrobial silver nanoparticles to improve the antimicrobial activity of the paints (Koponen, Jensen, & Schneider, 2011).

The antimicrobial activity of the silver nanoparticles has been reported in several articles. When the silver nanoparticles are being incorporated into paints, they act as a reservoir of silver ions. Depending upon the size and structure of the nanoparticles, silver ions are released from the paints. The silver nanoparticles containing paints can have antimicrobial effect that induces ROS production and generates stress leading to cell death (Matsumura, Yoshikata, Kunisaki, & Tsuchido, 2003). The silver nanoparticles release silver ions, which can interfere with the antioxidant property of the cell by interacting with sulfur and phosphor molecules (Pappa et al., 2007). The silver nanoparticles are considered to be least toxic in comparison to other biocidal products used in paints (Kaiser, Diener, & Wick, 2013).

## 8.10 Toxicity of Silver Nanoparticles

Owing to various biomedical applications, silver nanoparticles or silver-based nanoparticles have gained commercial acceptance. Consequently, assessment of their health risks toward their nontarget species also becomes imperative.

However, not much information is available regarding their toxicity mechanisms toward nontarget organisms.

Recently, Habas et al. investigated the mechanism of cytotoxic and genotoxic effects of AgNP nanoparticles on human skin keratinocytes (HaCaT). Results from their studies indicated that AgNPs caused severe oxidative damage, accompanied by increased production of superoxide anion levels as well as significant decrease in endogenous antioxidant enzyme of SOD, CAT, GPX expression involved in HaCaT cells in vitro (Habas & Shang, 2019). Figure 8.9 represents the overall toxicity of silver nanoparticles, which leads to cytotoxicity and induces genotoxic as well as inflammatory responses in the mammalian and human cells. These consequences lead to fibrosis and apoptotic cell death (Dakal et al., 2016).

Cho et al. demonstrated that small-sized nanoparticles were found to be more effective than larger ones. The group found that following intraperitoneal administration of AgNPs, reduced activity, piloerection, and lowered body temperature were observed. Further histopathological analysis revealed increased apoptosis in

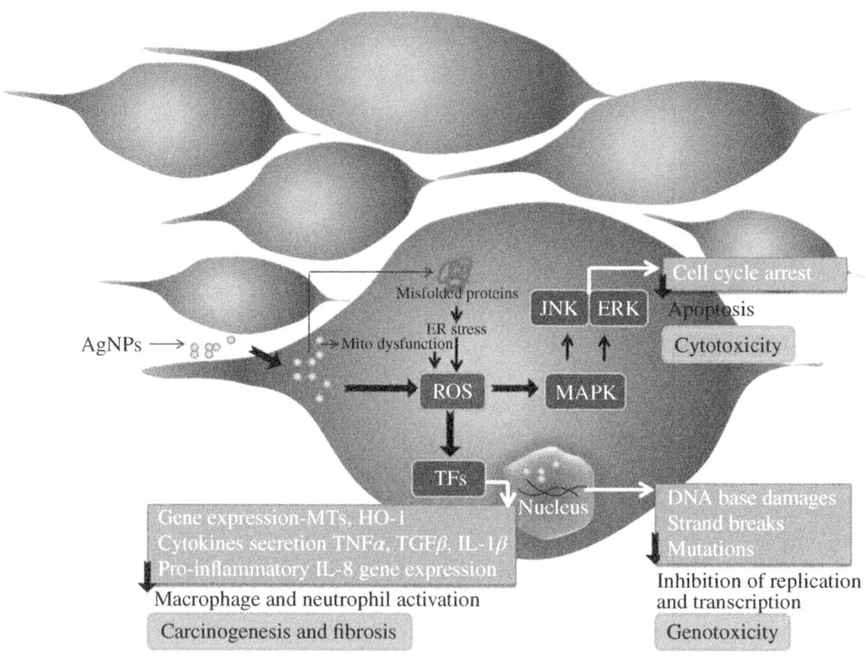


**Figure 8.9** AgNPs exposure to human or mammalian cells. AgNPs induce cytotoxic, genotoxic, and inflammatory response in human and mammalian cells and consequently trigger apoptotic cell death, carcinogenesis, and fibrosis. *Source:* Reproduced with permission from Dakal et al. (2016) Copyright © (2016) Dakal, Kumar, Majumdar and Yadav.

thymus cortex, spleen congestion, and necrosis in liver tissues in mice (Cho et al., 2018). Another research group from Tehran University of Medical Sciences reported that the nanoparticle interfered with blood–brain barrier permeability, and sperm quality and quantity (Moradi-Sardareh et al., 2018). Ema et al. investigated the reproductive and developmental toxicity of AgNPs using both male and female rats. In male rats, AgNPs were found to be deposited in testes. On the other hand, in female rats, exposure of the nanoparticles during early gestation leads to their deposition in visceral yolk sac, while late-gestation exposure to AgNPs led to their accumulation in breast milk, placenta, and pre- and postnatal offspring. AgNPs also effectively impaired physical development and cognitive behavior in offsprings born to AgNP-exposed mothers (Ema, Okuda, Gamo, & Honda, 2017). Release of silver nanoparticles into the environment can cause disruption of ecosystem (Tlili, Jabiol, Behra, Gil-Allué, & Gessner, 2017).

## 8.11 Future Prospective and Challenges

The concept of nanotechnology is reforming in day-to-day life with the research advances and improvements in the field of nanotechnology. Many companies and organizations are taking the help of nanoparticles for making new formulations. Silver nanoparticles in this prospect have been one of the most promising candidates in the field of antibacterial resistance due to their unique characterization and properties. But there remain many future options that need to be addressed. One of the foremost options is more accurate toxicity measurements, accumulation of the nanoparticles, and the amount of dosage given. In response to addressing these problems, more in vitro and in vivo data analyses should be performed. Also there should be proper procedures for synthesis that can increase the yield of the silver nanoparticles and increase their stability (Calderon-Jimenez et al., 2017). The silver nanoparticles have varying sizes and shapes, surface modifications; hence it is very necessary to understand the actual underlying mechanisms in order to enhance the antibacterial properties of silver. Further increase in antibacterial activity can be done by enhancing the dispersibility of the silver nanoparticles by changing the surface modifications. Decreased chemical stability is also required in order to allow the release of silver ions slowly and render long-term effect; finally, the nanoparticles must possess positive charge for a proper connection with the cell membranes. Also the nanoparticles should be made with proper shape and size so that they resemble platelets and interact better with bacteria (S. Tang & Zheng, 2018).

Silver nanoparticles have a wide range of application in the field of antibacterial activity. But there are many drawbacks regarding the usage of silver nanoparticles since there have been several bacterial strains that are resistant to silver nanoparticles (Holland et al., 2011). There have been various mechanisms reported for the cause of bacterial resistance to silver nanoparticles (Parikh et al., 2008). The other problem associated with the silver nanoparticles is that the regular ingestion can cause accumulation in the tissues of the skin and this condition can lead to argyrosis in reaction to sunlight. There have been several reports of silver nanoparticles being toxic to bone marrow cells, hepatocytes, osteoblasts, and fibroblasts (Gaiser et al., 2013). Hence if these challenges could be addressed by improving the synthesis procedures and dosing mechanisms, it would be helpful in the production of a new technology for antibacterial treatments.

## 8.12 Conclusion

This chapter has illustrated the remarkable antibacterial activity of diverse forms of silver nanoparticles and described their widespread applications in the field of biomedicine. Silver nanoparticles can be synthesized through an array of methods. Moreover, metallic silver nanoparticles have been assimilated with different biological or synthetic polymers to develop silver nanocompounds. These silver nanocompounds showed enhanced antimicrobial activity and are economically viable. These nanoparticles impart bactericidal effect via various mechanisms like prompting cell membrane leakage by interacting with membrane proteins; induction of DNA damage; disruption of normal cellular machinery by interfering with DNA synthesis, protein synthesis, or regular cellular metabolism; and generation of oxidative stress via ROS production. As a result of these antibacterial properties, silver, in the form of nanoparticle has gained enormous credibility as therapeutic. More so, this antibacterial potential has been used for the formulation of various commercially important products like wound dressing materials, cosmetics, washing detergents, fabrics, paints, etc. Nevertheless, there have been reports of potential toxic effects of silver nanoparticles in both human and animal hosts. But, the exact mechanism of toxicity is still not clearly known. Hence, it is essential to evaluate the toxicity matrix of silver nanoparticles and establish a safety profile of silver or silver nanoparticle-based agents to translate their therapeutic potential from bench to bedside.

## Take Home Message

- Due to the misuse and overuse of antibiotics, bacteria frequently adopt resistance mechanism by manipulating the mode of action of the antibiotics or by discarding the antibiotics through the efflux pump.
- Therefore, nanotechnology plays a pivotal role in dealing with these kinds of crisis where bacteria get lesser chance to develop resistance. Silver

nanoparticles are one such material to have intrinsic antibacterial activity with least chances to developing resistance.

- Silver nanoparticles with their size in the nanoscale range can easily disrupt or permeabilize through the bacterial cell membrane to inhibit various cellular mechanisms.
- With the evolution of silver nanoparticles as an effective alternative antibacterial therapy in biomedical field, they have been used in many commercial products for daily consumer's application.
- However, the toxicity issues related to silver nanoparticles are a major concern. In-depth molecular mechanisms are needed to be explored comprehensively in order for the products to qualify the clinical trials and reach the market.

## Questions and Answers

**1** How do silver nanoparticles alter cell membrane permeability?
Silver nanoparticles attach to the bacterial cell surfaces and interact with the cell membrane proteins like lipopolysaccharides to disrupt the integrity of the membrane, thus leaking the cellular contents and resulting in death of the bacteria.

**2** Which size range of silver nanoparticles exhibits highest cytotoxicity?
Silver nanoparticles less than 20 nm in diameter exhibit the most efficient cytotoxicity. These nanoparticles can easily attach to the bacterial cell membrane surfaces and alter the cell membrane permeability. Further, nanosilver of this size can penetrate inside the bacterial cells and induce various types of damage, resulting in cell death.

**3** What is the role of silver nanoparticles in wound dressing materials?
Wounds are the target sites for the growth of bacteria as well as other microbes. These microbes augment infection at the wound site and thus delay healing. Under these circumstances, antimicrobial nanoparticles like nanosilver prevent bacterial growth and help accelerate wound healing.

**4** How do silver nanoparticles influence spoilage-free food delivery?
During transfer of food in food packages, the food is at high risk of getting contaminated by various microbes. The presence of silver nanoparticles in food packaging materials can prevent the growth of microbes including bacteria and thus help in spoilage-free food delivery.

**5** How do bacteria overcome antibiotics attack?
Bacteria have evolved various mechanisms to evade the detrimental effects of antibiotics. Some of the mechanisms include the following: (i) alterations of cell membrane channel proteins to prevent the entry of antibiotics, (ii) enhanced expression of efflux pumps to increase the export of antibiotics that have already entered the cells, (iii) chemical modifications like acetylation, phosphorylation, adenylation, glycosylation, hydroxylation on the drugs to render them ineffective, (iv) modification of the intracellular targets of antibiotics etc.

## Abbreviation

AIDS: Acquired immune deficiency syndrome
ALP: Alkaline phosphatase
ALT: Alanine transaminase
Ag-NP: Silver nanoparticle
$AgNO_3$: Silver nitrate
AST: Aspartate aminotransferase
ASTM: American Society for Testing and Materials
ATP: Adenosine triphosphate
BMP: Bone morphogenic proteins
BMSC: Bone marrow stromal cells
BUN: Blood urea nitrogen
CEA: Chick embryo angiogenesis
CONP: Cerium oxide nanoparticles
CREA: Creatinine
CS: Chitosan
DMF: *N,N*-Dimethylformamide
DNA: Deoxyribonucleic acid
DNS: Dinitrosalicylic acid
DPH: Diphenyl hexatriene
DPPH: 2,2-Diphenyl-1-picrylhydrazyl
EHN: Europium hydroxide nanorods
FE-SEM: Field emission scanning electron microscope
GFP: Green fluorescent protein
GSH: Glutathione
HA: Hydroxyapatite
HCP: Host cell protein
HDN: Hesperidin

| | |
|---|---|
| HIV: | Human immune deficiency virus |
| IL-6,8: | Interleukin-6,8 |
| KGM: | Konjac glucomannan |
| MBC: | Minimum bactericidal concentration |
| MIC: | Minimum inhibitory concentration |
| MRSA: | Methicillin-resistance *Staphylococcus aureus* |
| MTCC: | Microbial Type Culture Collection |
| MTT: | 3-(4,5-Dimethylthiazol-2-yl)-2,5-Diphenyltetrazolium Bromide |
| NC: | Nanocellulose |
| NRG: | Naringin |
| PBS: | Polybutylene succinate |
| PGA: | Polyglycolic acid |
| PLA: | Polylactic acid |
| PLS: | Polybutylene succinate |
| PVA: | Polyvinyl alcohol |
| PCL: | Polycaprolactone |
| PDMAEMA: | Polymerized monomer 2-(dimethylamino)ethyl methacrylate |
| qRT-PCR: | Qualitative reverse transcriptase polymerase chain reaction |
| PDMS: | Polydimthylsiloxane |
| RNA: | Ribonucleic acid |
| ROS: | Reactive oxygen species |
| SEM: | Scanning electron microscope |
| S/C-SN: | Sericin- and chitosan-capped silver nanoparticle |
| THN: | Terbium hydroxide nanorods |
| TEM: | Transmission electron microscope |
| XRD: | X-ray diffraction |
| ZONP: | Zinc oxide nanoparticles |

## Acknowledgments

Financial support from Nano Mission-DST (SR/NM/NS-1252/2013; GAP 570), New Delhi, to C.R.P. is duly acknowledged. A.R. is thankful to ICMR, New Delhi, for granting Senior Research Fellowship. P.B.T. and S.H are thankful to CSIR, New Delhi for awarding Senior Research Fellowship. The authors are thankful to the Director, CSIR-IICT for his support, encouragement, and keen interest in this work. IICT Manuscript No. IICT/Pubs./2019/326 dated 6 September 2019 for this book chapter is duly acknowledged.

## References

Abbas, M., Naeem, N., Iftikhar, H., & Latif, U. (2018). Synthesis, characterization and antimicrobial properties of silver nanocomposites. *Silver Nanoparticles: Fabrication, Characterization and Applications*, 71. doi:10.5772/intechopen.74623

Abbaszadegan, A., Ghahramani, Y., Gholami, A., Hemmateenejad, B., Dorostkar, S., Nabavizadeh, M., & Sharghi, H. (2015). The effect of charge at the surface of silver nanoparticles on antimicrobial activity against gram-positive and gram-negative bacteria: A preliminary study. *Journal of Nanomaterials*, 16(1), 53. doi:10.1155/2015/720654

Abid, J.-P., Wark, A., Brevet, P.-F., & Girault, H. (2002). Preparation of silver nanoparticles in solution from a silver salt by laser irradiation. *Chemical Communications*, 7, 792–793. doi:10.1039/B200272H

Acharya, D., Singha, K. M., Pandey, P., Mohanta, B., Rajkumari, J., & Singha, L. P. (2018). Shape dependent physical mutilation and lethal effects of silver nanoparticles on bacteria. *Scientific Reports*, 8(1), 201–201. doi:10.1038/s41598-017-18590-6

Agnihotri, S., Mukherji, S., & Mukherji, S. (2014). Size-controlled silver nanoparticles synthesized over the range 5–100 nm using the same protocol and their antibacterial efficacy. *RSC Advances*, 4(8), 3974–3983. doi:10.1039/C3RA44507K

Ahn, E.-Y., Jin, H., & Park, Y. (2019). Assessing the antioxidant, cytotoxic, apoptotic and wound healing properties of silver nanoparticles green-synthesized by plant extracts. *Materials Science and Engineering.C*, 101, 204–216. doi:10.1016/j.msec.2019.03.095

Alaqad, K., & Saleh, T. A. (2016). Gold and silver nanoparticles: Synthesis methods, characterization routes and applications towards drugs. *Journal of Environmental & Analytical Toxicology*, 6(4), 525–2161. doi:10.4172/2161-0525.1000384

Alavi, M., & Karimi, N. (2018). Characterization, antibacterial, total antioxidant, scavenging, reducing power and ion chelating activities of green synthesized silver, copper and titanium dioxide nanoparticles using *Artemisia haussknechtii* leaf extract. *Artificial Cells, Nanomedicine, and Biotechnology*, 46(8), 2066–2081. doi:10.1080/21691401.2017.1408121

Alexander, J. W. (2009). History of the medical use of silver. *Surgical Infections*, 10(3), 289–292. doi:10.1089/sur.2008.9941

Ali, J., Ali, N., Wang, L., Waseem, H., & Pan, G. (2019). Revisiting the mechanistic pathways for bacterial mediated synthesis of noble metal nanoparticles. *Journal of Microbiological Methods*, 159, 18–25. doi:10.1016/j.mimet.2019.02.010

Amooaghaie, R., Saeri, M. R., & Azizi, M. (2015). Synthesis, characterization and biocompatibility of silver nanoparticles synthesized from *Nigella sativa* leaf extract in comparison with chemical silver nanoparticles. *Ecotoxicology and Environmental Safety*, 120, 400–408. doi:10.1016/j.ecoenv.2015.06.025

Angelo, K., Nisler, A., Hall, A., Brown, L., & Gould, L. (2017). Epidemiology of restaurant-associated foodborne disease outbreaks, United States, 1998–2013. *Epidemiology and Infection*, 145(3), 523–534. doi:10.1017/S0950268816002314

Anil Kumar, S., Abyaneh, M. K., Gosavi, S. W., Kulkarni, S. K., Pasricha, R., Ahmad, A., & Khan, M. I. (2007). Nitrate reductase-mediated synthesis of silver nanoparticles from $AgNO_3$. *Biotechnology Letters*, 29(3), 439–445. doi:10.1007/s10529-006-9256-7

Anwar, A., Masri, A., Rao, K., Rajendran, K., Khan, N. A., Shah, M. R., & Siddiqui, R. (2019). Antimicrobial activities of green synthesized gums-stabilized nanoparticles loaded with flavonoids. *Scientific Reports*, 9(1), 3122. doi:10.1038/s41598-019-39528-0

Arciola, C. R., Campoccia, D., & Montanaro, L. (2018). Implant infections: Adhesion, biofilm formation and immune evasion. *Nature Reviews Microbiology*, 16(7), 397. doi:10.1038/s41579-018-0019-y

Augustine, R., Hasan, A., Nath, V. Y., Thomas, J., Augustine, A., Kalarikkal, N., . . . Thomas, S. (2018). Electrospun polyvinyl alcohol membranes incorporated with green synthesized silver nanoparticles for wound dressing applications. *Journal of Materials Science: Materials in Medicine*, 29(11), 163. doi:10.1007/s10856-018-6169-7

Ballottin, D., Fulaz, S., Cabrini, F., Tsukamoto, J., Duran, N., Alves, O. L., & Tasic, L. (2017). Antimicrobial textiles: Biogenic silver nanoparticles against *Candida* and *Xanthomonas*. *Materials Science & Engineering, C: Materials for Biological Applications*, 75, 582–589. doi:10.1016/j.msec.2017.02.110

Bansod, S. D., Bawaskar, M. S., Gade, A. K., & Rai, M. K. (2015). Development of shampoo, soap and ointment formulated by green synthesised silver nanoparticles functionalised with antimicrobial plants oils in veterinary dermatology: Treatment and prevention strategies. *IET Nanobiotechnology*, 9(4), 165–171. doi:10.1049/iet-nbt.2014.0042

Barillo, D. J., & Marx, D. E. (2014). Silver in medicine: A brief history bc 335 to present. *Burns*, 40, S3–S8. doi:10.1016/j.burns.2014.09.009

Barui, A. K., Veeriah, V., Mukherjee, S., Manna, J., Patel, A. K., Patra, S., . . . Chatterjee, S. (2012). Zinc oxide nanoflowers make new blood vessels. *Nanoscale*, 4(24), 7861–7869. doi:10.1039/C2NR32369A

Behra, R., Sigg, L., Clift, M. J., Herzog, F., Minghetti, M., Johnston, B., . . . Rothen-Rutishauser, B. (2013). Bioavailability of silver nanoparticles and ions: From a chemical and biochemical perspective. *Journal of Royal Society Interface*, 10(87), 20130396. doi:10.1098/rsif.2013.0396

Behravan, M., Panahi, A. H., Naghizadeh, A., Ziaee, M., Mahdavi, R., & Mirzapour, A. (2019). Facile green synthesis of silver nanoparticles using *Berberis vulgaris* leaf and root aqueous extract and its antibacterial activity. *International Journal of Biological Macromolecules*, 124, 148–154. doi:10.1016/j.ijbiomac.2018.11.101

Bharathi, D., Josebin, M. D., Vasantharaj, S., & Bhuvaneshwari, V. (2018). Biosynthesis of silver nanoparticles using stem bark extracts of *Diospyros montana* and their antioxidant and antibacterial activities. *Journal of Nanostructure Chemistry*, 8(1), 83–92. doi:10.1007/s40097-018-0256-7

Bhattacharyya, T. (2019). A polarization based study of gold nanoparticles entrapped in single-wall carbon nanotube doped nanoscaffold. *Colloids and Surfaces A: Physicochemical and Engineering Aspects*, 570, 444–448. doi:10.1016/j.colsurfa.2019.03.043

Billal, D. S., Feng, J., Leprohon, P., Légaré, D., & Ouellette, M. (2011). Whole genome analysis of linezolid resistance in *Streptococcus pneumoniae* reveals resistance and compensatory mutations. *BMC Genomics*, 12(1), 512. doi:10.1186/1471-2164-12-512

Blair, J. M., Webber, M. A., Baylay, A. J., Ogbolu, D. O., & Piddock, L. J. (2015). Molecular mechanisms of antibiotic resistance. *Nature Reviews Microbiology*, 13(1), 42. doi:10.1038/nrmicro3380

Boese, K. (1921). Über Collargol, seine Anwendung und seine Erfolge in der Chirurgie und Gynäkologie. *Deutsche Zeitschrift f. Chirurgie*, 163, 62–84. https://doi.org/10.1007/BF02801881

Brown, E. D., & Wright, G. D. (2016). Antibacterial drug discovery in the resistance era. *Nature*, 529(7586), 336. doi:10.1038/nature17042

Brown, G. V. A. (1916). Colloidal silver in sepsis. *Transactions of the American Association of Obstetricians and Gynecologists for the Year*, 28, 262.

Buttacavoli, M., Albanese, N. N., Di Cara, G., Alduina, R., Faleri, C., Gallo, M., . . . Baldi, F. (2018). Anticancer activity of biogenerated silver nanoparticles: an integrated proteomic investigation. *Oncotarget*, 9(11), 9685. doi:10.18632/oncotarget.23859

Calderon-Jimenez, B., Johnson, M. E., Montoro Bustos, A. R., Murphy, K. E., Winchester, M. R., & Vega Baudrit, J. R. (2017). Silver nanoparticles: Technological advances, societal impacts, and metrological challenges. *Frontiers in Chemistry*, 5, 6. doi:10.3389/fchem.2017.00006

Carbone, M., Donia, D. T., Sabbatella, G., & Antiochia, R. (2016). Silver nanoparticles in polymeric matrices for fresh food packaging. *Journal of King Saud University – Science*, 28(4), 273–279. doi:10.1016/j.jksus.2016.05.004

Castro-Longoria, E., Vilchis-Nestor, A. R., & Avalos-Borja, M. (2011). Biosynthesis of silver, gold and bimetallic nanoparticles using the filamentous fungus *Neurospora crassa*. *Colloids and Surfaces B: Biointerfaces*, 83(1), 42–48. doi:10.1016/j.colsurfb.2010.10.035

Chakraborty, S., Ponrasu, T., Chandel, S., Dixit, M., & Muthuvijayan, V. (2018). Reduced graphene oxide-loaded nanocomposite scaffolds for enhancing angiogenesis in tissue engineering applications. *Royal Society Open Science*, 5(5), 172017. doi:10.1098/rsos.172017

Chatterjee, T., Chatterjee, B. K., Majumdar, D., & Chakrabarti, P. (2015). Antibacterial effect of silver nanoparticles and the modeling of bacterial growth kinetics using a modified Gompertz model. *Biochimica et Biophysica Acta-General Subjects*, 1850(2), 299–306. doi:10.1016/j.bbagen.2014.10.022

Chaudhuri, P., Harfouche, R., Soni, S., Hentschel, D. M., & Sengupta, S. (2009). Shape effect of carbon nanovectors on angiogenesis. *ACS Nano*, 4(1), 574–582. doi:10.1021/nn901465h

Chen, H., Lan, G., Ran, L., Xiao, Y., Yu, K., Lu, B., . . . Lu, F. (2018). A novel wound dressing based on a Konjac glucomannan/silver nanoparticle composite sponge effectively kills bacteria and accelerates wound healing. *Carbohydrate Polymers*, 183, 70–80. doi:10.1016/j.carbpol.2017.11.029

Chen, J., Wang, J., Zhang, X., & Jin, Y. (2008). Microwave-assisted green synthesis of silver nanoparticles by carboxymethyl cellulose sodium and silver nitrate. *Materials Chemistry and Physics*, 108(2), 421–424. doi:10.1016/j.matchemphys.2007.10.019

Chen, M., Feng, Y.-G., Wang, X., Li, T.-C., Zhang, J.-Y., & Qian, D.-J. (2007). Silver nanoparticles capped by oleylamine: Formation, growth, and self-organization. *Langmuir*, 23(10), 5296–5304. doi:10.1021/la700553d

Chernousova, S., & Epple, M. (2013). Silver as antibacterial agent: Ion, nanoparticle, and metal. *Angewandte Chemie, International Edition*, 52(6), 1636–1653. doi:10.1002/anie.201205923

Cho, Y.-M., Mizuta, Y., Akagi, J.-I., Toyoda, T., Sone, M., & Ogawa, K. (2018). Size-dependent acute toxicity of silver nanoparticles in mice. *Journal of Toxicologic Pathology*, 31(1), 73–80. doi:10.1293/tox.2017-0043

Choi, O., & Hu, Z. (2008). Size dependent and reactive oxygen species related nanosilver toxicity to nitrifying bacteria. *Environmental Science & Technology*, 42(12), 4583–4588. doi:10.1021/es703238h

Cohen, T. S., Hilliard, J. J., Jones-Nelson, O., Keller, A. E., O'Day, T., Tkaczyk, C., . . . Wang, Q. (2016). *Staphylococcus aureus* α toxin potentiates opportunistic bacterial lung infections. *Science Translational Medicine*, 8(329), 329ra331–329ra331. doi:10.1126/scitranslmed.aad9922

Cyril, N., George, J. B., Joseph, L., Raghavamenon, A., & Sylas, V. (2019). Assessment of antioxidant, antibacterial and anti-proliferative (lung cancer cell line A549) activities of green synthesized silver nanoparticles from Derris trifoliata. *Toxicology Research*, 8(2), 297–308. doi:10.1039/c8tx00323

Dakal, T. C., Kumar, A., Majumdar, R. S., & Yadav, V. (2016). Mechanistic basis of antimicrobial actions of silver nanoparticles. *Frontiers in Microbiology*, 7, 1831. doi:10.3389/fmicb.2016.01831

Das, S., Singh, S., Dowding, J. M., Oommen, S., Kumar, A., Sayle, T. X., . . . Sayle, D. C. (2012). The induction of angiogenesis by cerium oxide nanoparticles through the modulation of oxygen in intracellular environments. *Biomaterials*, 33(31), 7746–7755. doi:10.1016/j.biomaterials.2012.07.019

Das, V. L., Thomas, R., Varghese, R. T., Soniya, E., Mathew, J., & Radhakrishnan, E. (2014). Extracellular synthesis of silver nanoparticles by the *Bacillus* strain CS 11 isolated from industrialized area. *3 Biotech*, 4(2), 121–126. doi:10.1007/s13205-013-0130-8

Davies, R. L., & Etris, S. F. (1997). The development and functions of silver in water purification and disease control. *Catalysis Today*, 36(1), 107–114. doi:10.1016/S0920-5861(96)00203-9

Didelot, X., Walker, A. S., Peto, T. E., Crook, D. W., & Wilson, D. J. (2016). Within-host evolution of bacterial pathogens. *Nature Reviews Microbiology*, 14(3), 150. doi:10.1038/nrmicro.2015.13

Divya, M., Kiran, G. S., Hassan, S., & Selvin, J. (2019). Biogenic synthesis and effect of silver nanoparticles (AgNPs) to combat catheter-related urinary tract infections. *Biocatalysis and Agricultural Biotechnology*, 18, 101037. doi:10.1016/j.bcab.2019.101037

Domeradzka-Gajda, K., Nocuń, M., Roszak, J., Janasik, B., Quarles, C. D., Jr., Wąsowicz, W., . . . Ranoszek-Soliwoda, K. (2017). A study on the in vitro percutaneous absorption of silver nanoparticles in combination with aluminum chloride, methyl paraben or *di-n*-butyl phthalate. *Toxicology Letters*, 272, 38–48. doi:10.1016/j.toxlet.2017.03.006

Doolette, C. L., McLaughlin, M. J., Kirby, J. K., Batstone, D. J., Harris, H. H., Ge, H., & Cornelis, G. (2013). Transformation of PVP coated silver nanoparticles in a simulated wastewater treatment process and the effect on microbial communities. *Chemistry Central Journal*, 7(1), 46. doi:10.1186/1752-153X-7-46

Duhamel, B. (1912). Electric metallic colloids and their therapeutical applications. *The Lancet*, 179(4611), 89–90. doi:10.1016/S0140-6736(01)67045-4

Egger, S., Lehmann, R. P., Height, M. J., Loessner, M. J., & Schuppler, M. (2009). Antimicrobial properties of a novel silver-silica nanocomposite material. *Applied and Environmental Microbiology*, 75(9), 2973–2976. doi:10.1128/AEM.01658-08

Ema, M., Okuda, H., Gamo, M., & Honda, K. (2017). A review of reproductive and developmental toxicity of silver nanoparticles in laboratory animals. *Reproductive Toxicology*, 67, 149–164. doi:10.1016/j.reprotox.2017.01.005

Epand, R. M., Walker, C., Epand, R. F., & Magarvey, N. A. (2016). Molecular mechanisms of membrane targeting antibiotics. *Biochimica et Biophysica Acta-Biomembranes*, 1858(5), 980–987. doi:10.1016/j.bbamem.2015.10.018

Freeman, A. I., Halladay, L. J., & Cripps, P. (2012). The effect of silver impregnation of surgical scrub suits on surface bacterial contamination. *Veterinary Journal*, 192(3), 489–493. doi:10.1016/j.tvjl.2011.06.039

Frens, G., & Overbeek, J. T. G. (1969). Carey Lea's colloidal silver. *Colloid Polymer Sciences*, 233(1–2), 922–929. doi:10.1007/BF01508016

Fu, L.-H., Gao, Q.-L., Qi, C., Ma, M.-G., & Li, J.-F. (2018). Microwave-hydrothermal rapid synthesis of cellulose/Ag nanocomposites and their antibacterial activity. *Nanomaterials*, 8(12), 978. doi:10.3390/nano8120978

Fung, M. C., & Bowen, D. L. (1996). Silver products for medical indications: Risk-benefit assessment. *Journal of Toxicology – Clinical Toxicology*, 34(1), 119–126. doi:10.3109/15563659609020246

Gahlawat, G., & Choudhury, A. R. (2019). A review on the biosynthesis of metal and metal salt nanoparticles by microbes. *RSC Advances*, 9(23), 12944–12967. doi:10.1039/C8RA10483B

Gaiser, B. K., Hirn, S., Kermanizadeh, A., Kanase, N., Fytianos, K., Wenk, A., . . . Stone, V. (2013). Effects of silver nanoparticles on the liver and hepatocytes in vitro. *Toxicological Sciences*, 131(2), 537–547. doi:10.1093/toxsci/kfs306

GhavamiNejad, A., Rajan Unnithan, A., Ramachandra Kurup Sasikala, A., Samarikhalaj, M., Thomas, R. G., Jeong, Y. Y., . . . Hee Park, C. (2015). Mussel-inspired electrospun nanofibers functionalized with size-controlled silver nanoparticles for wound dressing application. *ACS Applied Materials & Interfaces*, 7(22), 12176–12183. doi:10.1021/acsami.5b02542

Gomaa, E. Z. (2017). Silver nanoparticles as an antimicrobial agent: A case study on *Staphylococcus aureus* and *Escherichia coli* as models for Gram-positive and Gram-negative bacteria. *The Journal of General and Applied Microbiology*, 63(1), 36–43. doi:10.2323/jgam.2016.07.004

Gounani, Z., Asadollahi, M. A., Meyer, R. L., & Arpanaei, A. (2018). Loading of polymyxin B onto anionic mesoporous silica nanoparticles retains antibacterial activity and enhances biocompatibility. *International Journal of Pharmaceutics*, 537(1-2), 148–161. doi:10.1016/j.ijpharm.2017.12.039

Govindaraju, K., Basha, S. K., Kumar, V. G., & Singaravelu, G. (2008). Silver, gold and bimetallic nanoparticles production using single-cell protein (*Spirulina platensis*) Geitler. *Journal of Materials Science*, 43(15), 5115–5122. doi:10.1007/s10853-008-2745-4

Habas, K., & Shang, L. (2019). Silver nanoparticle-mediated cellular responses in human keratinocyte cell line HaCaT in vitro. *Nanoscale Reports*, 2(2), 1–9. doi:10.26524/nr1921

Hall, C. W., & Mah, T.-F. (2017). Molecular mechanisms of biofilm-based antibiotic resistance and tolerance in pathogenic bacteria. *FEMS Microbiology Review*, 41(3), 276–301. doi:10.1093/femsre/fux010

Hamouda, R. A., Hussein, M. H., Abo-elmagd, R. A., & Bawazir, S. S. (2019). Synthesis and biological characterization of silver nanoparticles derived from the cyanobacterium *Oscillatoria limnetica*. *Scientific Reports*, 9(1), 1–17. doi:10.1038/s41598-019-49444-y

Hashimoto, M., & Honda, Y. (2019). Effect of silver nanoparticle size on biofilm formation. *Journal of Biomaterials and Tissue Engineering*, 9(4), 450–455. doi:10.1166/jbt.2019.2002

Holland, S. L., Dyer, P. S., Bond, C. J., James, S. A., Roberts, I. N., & Avery, S. V. (2011). *Candida argentea* sp. nov., a copper and silver resistant yeast species. *Fungal Biology*, 115(9), 909–918. doi:10.1016/j.funbio.2011.07.004

Holmes, A. H., Moore, L. S., Sundsfjord, A., Steinbakk, M., Regmi, S., Karkey, A., . . . Piddock, L. J. (2016). Understanding the mechanisms and drivers of antimicrobial resistance. *The Lancet*, 387(10014), 176–187. doi:10.1016/S0140-6736(15)00473-0

Hong, X., Wen, J., Xiong, X., & Hu, Y. (2016). Shape effect on the antibacterial activity of silver nanoparticles synthesized via a microwave-assisted method. *Environemental Science and Pollution Research*, 23(5), 4489–4497. doi:10.1007/s11356-015-5668-z

Hu, X., Saravanakumar, K., Jin, T., & Wang, M.-H. (2019). Mycosynthesis, characterization, anticancer and antibacterial activity of silver nanoparticles from endophytic fungus *Talaromyces purpureogenus*. *International Journal of Nanomedicine*, 14, 3427. doi:10.2147/IJN.S200817

Huang, H., & Yang, Y. (2008). Preparation of silver nanoparticles in inorganic clay suspensions. *Composites Science and Technology*, 68(14), 2948–2953. doi:10.1016/j.compscitech.2007.10.003

Huddleston, J. R. (2014). Horizontal gene transfer in the human gastrointestinal tract: Potential spread of antibiotic resistance genes. *Infection and Drug Resistance*, 7, 167. doi:10.2147/IDR.S48820

Hullikere, M. M., Joshi, C. G., Vijay, R., Ananda, D., & Nivya, M. (2015). Antiangiogenic, cytotoxic and antimicrobial activity of plant mediated silver nano particle from Tragia involucrate. *Research Journal of Nanoscience and Nanotechnology*, 5, 16–26. doi:10.3923/rjnn.2015.16.26

Huy, T. Q., Thanh, N. T. H., Thuy, N. T., Van Chung, P., Hung, P. N., Le, A.-T., & Hanh, N. T. H. (2017). Cytotoxicity and antiviral activity of electrochemical–synthesized silver nanoparticles against poliovirus. *Journal of Virological Methods*, 241, 52–57. doi:10.1016/j.jviromet.2016.12.015

Ibrahim, H. M., & Hassan, M. S. (2016). Characterization and antimicrobial properties of cotton fabric loaded with green synthesized silver nanoparticles. *Carbohydrate Polymers*, 151, 841–850. doi:10.1016/j.carbpol.2016.05.041

Ishwarya, R., Vaseeharan, B., Anuradha, R., Rekha, R., Govindarajan, M., Alharbi, N. S., . . . Benelli, G. (2017). Eco-friendly fabrication of Ag nanostructures using the seed extract of *Pedalium murex*, an ancient Indian medicinal plant: Histopathological effects on the Zika virus vector *Aedes aegypti* and inhibition of biofilm-forming pathogenic bacteria. *Journal of Photochemistry and Photobiology B*, 174, 133–143. doi:10.1016/j.jphotobiol.2017.07.026

Ivanova, N., Gugleva, V., Dobreva, M., Stefanov, S., Pehlivanov, I., & Andonova, V. (2018). Silver nanoparticles as multy-functional drug delivery systems. *Nanomedicines: IntechOpen*. doi:10.5772/intechopen.80238

Jain, J., Arora, S., Rajwade, J. M., Omray, P., Khandelwal, S., & Paknikar, K. M. (2009). Silver nanoparticles in therapeutics: Development of an antimicrobial gel formulation for topical use. *Molecular Pharmaceutics*, 6(5), 1388–1401. doi:10.1021/mp900056g

Jiang, Q., Yu, S., Li, X., Ma, C., & Li, A. (2018). Evaluation of local anesthetic effects of lidocaine-ibuprofen ionic liquid stabilized silver nanoparticles in male swiss mice. *Journal of Photochemistry and Photobiology B*, 178, 367–370. doi:10.1016/j.jphotobiol.2017.11.028

Jo, Y.-K., Kim, B. H., & Jung, G. (2009). Antifungal activity of silver ions and nanoparticles on phytopathogenic fungi. *Plant Disease*, 93(10), 1037–1043. doi:10.1094/PDIS-93-10-1037

Jung, J. H., Cheol Oh, H., Soo Noh, H., Ji, J. H., & Soo Kim, S. (2006). Metal nanoparticle generation using a small ceramic heater with a local heating area. *Journal of Aerosol Science*, 37(12), 1662–1670. doi:https://doi.org/10.1016/j.jaerosci.2006.09.002

Kadiyala, U., Turali-Emre, E. S., Bahng, J. H., Kotov, N. A., & VanEpps, J. S. (2018). Unexpected insights into antibacterial activity of zinc oxide nanoparticles against methicillin resistant Staphylococcus aureus (MRSA). *Nanoscale*, 10(10), 4927–4939. doi:10.1039/c7nr08499d

Kaiser, J. P., Diener, L., & Wick, P. (2013). Nanoparticles in paints: A new strategy to protect façades and surfaces? *Journal of Physics Conference Series*, 429, 012036. doi:10.1088/1742-6596/429/1/012036

Kalimuthu, K., Suresh Babu, R., Venkataraman, D., Bilal, M., & Gurunathan, S. (2008). Biosynthesis of silver nanocrystals by *Bacillus licheniformis*. *Colloids Surface B Biointerfaces*, 65(1), 150–153. doi:10.1016/j.colsurfb.2008.02.018

Karthik, C., Manukumar, H., Ananda, A., Nagashree, S., Rakesh, K., Mallesha, L., . . . Krishnamurthy, N. (2018). Synthesis of novel benzodioxane midst piperazine moiety decorated chitosan silver nanoparticle against biohazard pathogens and as potential anti-inflammatory candidate: A molecular docking studies. *International Journal of Biological Macromolecules*, 108, 489–502. doi:10.1016/j.ijbiomac.2017.12.045

Kashiri, M., Cerisuelo, J. P., Domínguez, I., López-Carballo, G., Muriel-Gallet, V., Gavara, R., & Hernández-Muñoz, P. (2017). Zein films and coatings as carriers and release systems of *Zataria multiflora* Boiss. essential oil for antimicrobial food packaging. *Food Hydrocolloids*, 70, 260–268. doi:10.1016/j.foodhyd.2017.02.021

Kaul, S., Gulati, N., Verma, D., Mukherjee, S., & Nagaich, U. (2018). Role of nanotechnology in cosmeceuticals: A review of recent advances. *Journal of Pharmaceutics*, 2018. doi:10.1155/2018/3420204

Kaur, A., & Kumar, R. (2019). Enhanced bactericidal efficacy of polymer stabilized silver nanoparticles in conjugation with different classes of antibiotics. *RSC Advances*, 9(2), 1095–1105. doi:10.1039/C8RA07980C

Keat, C. L., Aziz, A., Eid, A. M., & Elmarzugi, N. A. (2015). Biosynthesis of nanoparticles and silver nanoparticles. *Bioresources and Bioprocessing*, 2(1), 47. doi:10.1186/s40643-015-0076-2

Kędziora, A., Speruda, M., Krzyżewska, E., Rybka, J., Łukowiak, A., & Bugla-Płoskońska, G. (2018). Similarities and differences between silver ions and silver in nanoforms as antibacterial agents. *International Journal of Molecular Sciences*, 19(2), 444. doi:10.3390/ijms19020444

Kesharwani, J., Yoon, K.-Y., Hwang, J., & Rai, M. (2009). Phytofabrication of silver nanoparticles by leaf extract of *Datura metel*: Hypothetical mechanism involved in synthesis. *Journal of Bionanoscience*, 3, 39–44. doi:10.1166/jbns.2009.1008

Khalandi, B., Asadi, N., Milani, M., Davaran, S., Abadi, A. J. N., Abasi, E., & Akbarzadeh, A. (2017). A review on potential role of silver nanoparticles and possible mechanisms of their actions on bacteria. *Drug Research*, 11(02), 70–76. doi:10.1055/s-0042-113383

Khandel, P., & Shahi, S. K. (2018). Mycogenic nanoparticles and their bio-prospective applications: current status and future challenges. *Journal of Nanostructure Chemistry*, 8(4), 369–391. doi:10.1007/s40097-018-0285-2

Khanra, K., Roy, A., & Bhattacharyya, N. (2013). Evaluation of antibacterial activity and cytotoxicity of green synthesized silver nanoparticles using *Hemidesmus indicus* R. *The Beilstein Journal of Nanoscience and Nanotechnology*, 1(1), 1–6.

Khatami, M., Heli, H., Jahani, P. M., Azizi, H., & Nobre, M. A. L. (2017). Copper/copper oxide nanoparticles synthesis using *Stachys lavandulifolia* and its antibacterial activity. *IET Nanobiotechnology*, 11(6), 709–713. doi:10.1049/iet-nbt.2016.0189

Kim, D. H., Park, J. C., Jeon, G. E., Kim, C. S., & Seo, J. H. (2017). Effect of the size and shape of silver nanoparticles on bacterial growth and metabolism by monitoring optical density and fluorescence intensity. *Biotechnology and Bioprocess Engineering*, 22(2), 210–217. doi:10.1007/s12257-016-0641-3

Kim, K., & Kim, T. (2019). Nanofabrication by thermal plasma jets: From nanoparticles to low-dimensional nanomaterials. *Journal of Applied Physics*, 125(7), 070901. doi:10.1063/1.5060977

Kim, S., Yoo, B. K., Chun, K., Kang, W., Choo, J., Gong, M.-S., & Joo, S.-W. (2005). Catalytic effect of laser ablated Ni nanoparticles in the oxidative addition reaction for a coupling reagent of benzylchloride and bromoacetonitrile. *Journal of Molecular Catalysis A: Chemical*, 226(2), 231–234. doi:10.1016/j.molcata.2004.10.038

Kim, S. W., Jung, J. H., Lamsal, K., Kim, Y. S., Min, J. S., & Lee, Y. S. (2012). Antifungal effects of silver nanoparticles (AgNPs) against various plant pathogenic fungi. *Mycobiology*, 40(1), 53–58. doi:10.5941/MYCO.2012.40.1.05

Kiran, G. S., Sabu, A., & Selvin, J. (2010). Synthesis of silver nanoparticles by glycolipid biosurfactant produced from marine *Brevibacterium casei* MSA19. *Journal of Biotechnology*, 148(4), 221–225. doi:10.1016/j.jbiotec.2010.06.012

Kitching, M., Ramani, M., & Marsili, E. (2015). Fungal biosynthesis of gold nanoparticles: Mechanism and scale up. *Microbial Biotechnology*, 8(6), 904–917. doi:10.1111/1751-7915.12151

Kong, K. F., Schneper, L., & Mathee, K. (2010). Beta-lactam antibiotics: From antibiosis to resistance and bacteriology. *APMIS*, 118(1), 1–36. doi:10.1111/j.1600-0463.2009.02563.x

Koponen, I. K., Jensen, K. A., & Schneider, T. (2011). Comparison of dust released from sanding conventional and nanoparticle-doped wall and wood coatings.

*Journal of Exposure Science & Environmental Epidemiology*, 21(4), 408–418. doi:10.1038/jes.2010.32

Korbekandi, H., Ashari, Z., Iravani, S., & Abbasi, S. (2013). Optimization of biological synthesis of silver nanoparticles using *Fusarium oxysporum*. *Iranian Journal of Pharmaceutical Research*, 12(3), 289–298. PMID:24250635

Korshed, P., Li, L., Liu, Z., & Wang, T. (2016). The molecular mechanisms of the antibacterial effect of picosecond laser generated silver nanoparticles and their toxicity to human cells. *Plos One*, 11(8), e0160078. doi:10.1371/journal.pone.0160078

Kotcherlakota, R., Vydiam, K., Jeyalakshmi Srinivasan, D., Mukherjee, S., Roy, A., Kuncha, M., . . . Patra, C. R. (2019). Restoration of p53 function in ovarian cancer mediated by gold nanoparticles based EGFR targeted gene delivery system. *ACS Biomaterials Science & Engineering*, 7, 3631–3644. doi:10.1021/acsbiomaterials.9b00006

Kraeling, M. E., Topping, V. D., Keltner, Z. M., Belgrave, K. R., Bailey, K. D., Gao, X., & Yourick, J. J. (2018). in vitro percutaneous penetration of silver nanoparticles in pig and human skin. *Regulatory Toxicology and Pharmacology*, 95, 314–322. doi:10.1016/j.yrtph.2018.04.006

Kruis, F. E., Fissan, H., & Rellinghaus, B. (2000). Sintering and evaporation characteristics of gas-phase synthesis of size-selected PbS nanoparticles. *Materials Science and Engineering*, 69–70, 329–334. doi:10.1016/S0921-5107(99)00298-6

Krutyakov, Y. A., Olenin, A. Y., Kudrinskii, A., Dzhurik, P., & Lisichkin, G. (2008). Aggregative stability and polydispersity of silver nanoparticles prepared using two-phase aqueous organic systems. *Nanotechnologies in Russia*, 3(5-6), 303–310. doi:10.1134/S1995078008050054

Kumar, S. A., Abyaneh, M. K., Gosavi, S., Kulkarni, S. K., Pasricha, R., Ahmad, A., & Khan, M. (2007). Nitrate reductase-mediated synthesis of silver nanoparticles from $AgNO_3$. *Biotechnology Letters*, 29(3), 439–445. doi:10.1007/s10529-006-9256-7

Kumar, S. S. D., Rajendran, N. K., Houreld, N. N., & Abrahamse, H. (2018). Recent advances on silver nanoparticle and biopolymer-based biomaterials for wound healing applications. *International Journal of Biological Macromolecules*, 115, 165–175. doi:10.1016/j.ijbiomac.2018.04.003

Lakshmanan, G., Sathiyaseelan, A., Kalaichelvan, P., & Murugesan, K. (2018). Plant-mediated synthesis of silver nanoparticles using fruit extract of *Cleome viscosa* L.: Assessment of their antibacterial and anticancer activity. *Karbala International Journal of Modern Science*, 4(1), 61–68. doi:10.1016/j.kijoms.2017.10.007

Lateef, A., Azeez, M. A., Asafa, T. B., Yekeen, T. A., Akinboro, A., Oladipo, I. C., . . . Beukes, L. S. (2016). Cocoa pod husk extract-mediated biosynthesis of silver nanoparticles: Its antimicrobial, antioxidant and larvicidal activities. *Journal of Nanostructure Chemistry*, 6(2), 159–169. doi:10.1007/s40097-016-0191-4

Lea, M. C. (1889). ART. L. – on allotropic forms of silver. *American Journal of Science (1880–1910)*, 37(222), 476. doi:10.2475/ajs.s3-38.223.47

Lee, D. K., & Kang, Y. S. (2004). Synthesis of silver nanocrystallites by a new thermal decomposition method and their characterization. *ETRI Journal*, 26(3), 252–256. doi:10.4218/etrij.04.0103.0061

Lee, J.-S., Lytton-Jean, A. K., Hurst, S. J., & Mirkin, C. A. (2007). Silver nanoparticle – oligonucleotide conjugates based on DNA with triple cyclic disulfide moieties. *Nano Letters*, 7(7), 2112–2115. doi:10.1021/nl071108g

Lee, W., Kim, K.-J., & Lee, D. G. (2014). A novel mechanism for the antibacterial effect of silver nanoparticles on *Escherichia coli*. *BioMetals*, 27(6), 1191–1201. doi:10.1007/s10534-014-9782-z

Levey, M. (1967). Medieval arabic minting of gold and silver coins. *Chymia*, 12, 3–14.

Li, Q., Mahendra, S., Lyon, D. Y., Brunet, L., Liga, M. V., Li, D., & Alvarez, P. J. J. (2008). Antimicrobial nanomaterials for water disinfection and microbial control: Potential applications and implications. *Water Research*, 42(18), 4591–4602. doi:10.1016/j.watres.2008.08.015

Li, Z., Zhang, M., Cheng, D., & Yang, R. (2016). Preparation of silver nano-particles immobilized onto chitin nano-crystals and their application to cellulose paper for imparting antimicrobial activity. *Carbohydrate Polymers*, 151, 834–840. doi:10.1016/j.carbpol.2016.06.012

Lohani, A., Verma, A., Joshi, H., Yadav, N., & Karki, N. (2014). Nanotechnology-based cosmeceuticals. *ISRN Dermatology*, 2014, 14. doi:10.1155/2014/843687

Lu, M., Liao, J., Dong, J., Wu, J., Qiu, H., Zhou, X., . . . Quan, Z. (2016). An effective treatment of experimental osteomyelitis using the antimicrobial titanium/silver-containing nHP66 (nano-hydroxyapatite/polyamide-66) nanoscaffold biomaterials. *Scientific Reports*, 6, 39174. doi:10.1038/srep39174

Ma, H., Yin, B., Wang, S., Jiao, Y., Pan, W., Huang, S., . . . Meng, F. (2004). Synthesis of silver and gold nanoparticles by a novel electrochemical method. *ChemPhysChem*, 5(1), 68–75. doi:10.1002/cphc.200300900

Malhotra, B., Keshwani, A., & Kharkwal, H. (2015). Natural polymer based cling films for food packaging. *International Journal of Pharmacy and Pharmaceutical Sciences*, 7(4), 10–18.

Mandal, D., Dash, S. K., Das, B., Chattopadhyay, S., Ghosh, T., Das, D., & Roy, S. (2016). Bio-fabricated silver nanoparticles preferentially targets Gram positive depending on cell surface charge. *Biomedicine & Pharmacotherapy*, 83, 548–558. doi:10.1016/j.biopha.2016.07.011

Maráková, N., Humpolíček, P., Kašpárková, V., Capáková, Z., Martinková, L., Bober, P., . . . Stejskal, J. (2017). Antimicrobial activity and cytotoxicity of cotton fabric coated with conducting polymers, polyaniline or polypyrrole, and with deposited silver nanoparticles. *Applied Surface Science*, 396, 169–176. doi:10.1016/j.apsusc.2016.11.024

Masum, M. M. I., Siddiqa, M. M., Ali, K. A., Zhang, Y., Abdallah, Y., Ibrahim, E., . . . Li, B. (2019). Biogenic synthesis of silver nanoparticles using *Phyllanthus emblica* fruit extract and its inhibitory action against the pathogen *Acidovorax oryzae* strain RS-2 of rice bacterial brown stripe. *Frontiers in Microbiology*, 10. doi:10.3389/fmicb.2019.00820

Mata, R., Nakkala, J. R., Chandra, V. K., Raja, K., & Sadras, S. R. (2018). in vivo bio-distribution, clearance and toxicity assessment of biogenic silver and gold nanoparticles synthesized from *Abutilon indicum* in Wistar rats. *Journal of Trace Elements in Medicine and Biology*, 48, 157–165. doi:10.1016/j.jtemb.2018.03.015

Matsumura, Y., Yoshikata, K., Kunisaki, S.-I., & Tsuchido, T. (2003). Mode of bactericidal action of silver zeolite and its comparison with that of silver nitrate. *Applied and Environmental Microbiology*, 69(7), 4278–4281. doi:10.1128/aem.69.7.4278-4281.2003

Mei, L., Lu, Z., Zhang, X., Li, C., & Jia, Y. (2014). Polymer–Ag nanocomposites with enhanced antimicrobial activity against bacterial infection. *ACS Applied Materials & Interfaces*, 6(18), 15813–15821. doi:10.1021/am502886m

Mei, L., Teng, Z., Zhu, G., Liu, Y., Zhang, F., Zhang, J., . . . Chen, X. (2017). Silver nanocluster-embedded zein films as antimicrobial coating materials for food packaging. *ACS Applied Materials & Interfaces*, 9(40), 35297–35304. doi:10.1021/acsami.7b08152

Meng, M., He, H., Xiao, J., Zhao, P., Xie, J., & Lu, Z. (2016). Controllable in situ synthesis of silver nanoparticles on multilayered film-coated silk fibers for antibacterial application. *Journal of Colloid and Interface Science*, 461, 369–375. doi:10.1016/j.jcis.2015.09.038

Montes-Burgos, I., Walczyk, D., Hole, P., Smith, J., Lynch, I., & Dawson, K. (2009). Characterisation of nanoparticle size and state prior to nanotoxicological studies. *Journal of Nanoparticle Research*, 12, 47–53. doi:10.1007/s11051-009-9774-z

Moodley, J. S., Krishna, S. B. N., Pillay, K., & Govender, P. (2018). Green synthesis of silver nanoparticles from *Moringa oleifera* leaf extracts and its antimicrobial potential. *Advances in Natural Sciences: Nanoscience*, 9(1), 015011. doi:10.1088/2043-6254/aaabb2

Moradi-Sardareh, H., Basir, H. R. G., Hassan, Z. M., Davoudi, M., Amidi, F., & Paknejad, M. (2018). Toxicity of silver nanoparticles on different tissues of Balb/C mice. *Life Sciences*, 211, 81–90. doi:10.1016/j.lfs.2018.09.001

Morones, J. R., Elechiguerra, J. L., Camacho, A., Holt, K., Kouri, J. B., Ramirez, J. T., & Yacaman, M. J. (2005). The bactericidal effect of silver nanoparticles. *Nanotechnology*, 16(10), 2346–2353. doi:10.1088/0957-4484/16/10/059

Mukherjee, S., Chowdhury, D., Kotcherlakota, R., & Patra, S. (2014). Potential theranostics application of bio-synthesized silver nanoparticles (4-in-1 system). *Theranostics*, 4(3), 316. doi:10.7150/thno.7819

Mukherjee, S., Sriram, P., Barui, A. K., Nethi, S. K., Veeriah, V., Chatterjee, S., . . . Patra, C. R. (2015). Graphene oxides show angiogenic properties. *Advanced Healthcare Materials*, 4(11), 1722–1732. doi:10.1002/adhm.201500155

Müller, K., Bugnicourt, E., Latorre, M., Jorda, M., Echegoyen Sanz, Y., Lagaron, J., . . . Bölz, U. (2017). Review on the processing and properties of polymer nanocomposites and nanocoatings and their applications in the packaging, automotive and solar energy fields. *Nanomaterials*, 7(4), 74. doi:10.3390/nano7040074

Munita, J. M., & Arias, C. A. (2016). Mechanisms of antibiotic resistance. *Microbiology Spectrum*, 4(2). doi:10.1128/microbiolspec

Neethirajan, S., Clond, M. A., & Vogt, A. (2014). Medical biofilms—Nanotechnology approaches. *Journal of Biomedical Nanotechnology*, 10(10), 2806–2827. doi:10.1166/jbn.2014.1892

Nethi, S. K., Barui, A. K., Bollu, V. S., Rao, B. R., & Patra, C. R. (2017a). Pro-angiogenic properties of terbium hydroxide nanorods: Molecular mechanisms and therapeutic applications in wound healing. *ACS Biomaterials Science & Engineering*, 3(12), 3635–3645. doi:10.1021/acsbiomaterials.7b00457

Nethi, S. K., Rico-Oller, B., Rodríguez-Diéguez, A., Gómez-Ruiz, S., & Patra, C. R. (2017b). Design, synthesis and characterization of doped-titanium oxide nanomaterials with environmental and angiogenic applications. *Science of the Total Environment*, 599, 1263–1274. doi:10.1016/j.scitotenv.2017.05.005

Pandian, S. R. K., Deepak, V., Kalishwaralal, K., Viswanathan, P., & Gurunathan, S. (2010). Mechanism of bactericidal activity of silver nitrate—A concentration dependent bi-functional molecule. *Brazilian Journal of Microbiology*, 41(3), 805–809. doi:10.1590/S1517-83822010000300033. PMCID:PMC3768641

Pappa, A., Franco, R., Schoneveld, O., Galanis, A., Sandaltzopoulos, R., & Panayiotidis, M. I. (2007). Sulfur-containing compounds in protecting against oxidant-mediated lung diseases. *Current Medicinal Chemistry*, 14(24), 2590–2596. doi:10.2174/092986707782023262

Parikh, R. Y., Singh, S., Prasad, B. L., Patole, M. S., Sastry, M., & Shouche, Y. S. (2008). Extracellular synthesis of crystalline silver nanoparticles and molecular evidence of silver resistance from *Morganella* sp.: Towards understanding biochemical synthesis mechanism. *ChemBioChem*, 9(9), 1415–1422. doi:10.1002/cbic.200700592

Park, S.-H., Ko, Y.-S., Park, S.-J., Lee, J. S., Cho, J., Baek, K.-Y., . . . Lee, J.-H. (2016). Immobilization of silver nanoparticle-decorated silica particles on polyamide thin film composite membranes for antibacterial properties. *Journal of Membrane Science*, 499, 80–91. doi:10.1016/j.memsci.2015.09.060

Patra, C. R., Bhattacharya, R., Patra, S., Vlahakis, N. E., Gabashvili, A., Koltypin, Y., . . . Mukhopadhyay, D. (2008). Pro-angiogenic properties of europium (III) hydroxide nanorods. *Advanced Materials*, 20(4), 753–756. doi:10.1002/adma.200701611

Patra, J. K., & Baek, K.-H. (2017). Antibacterial activity and synergistic antibacterial potential of biosynthesized silver nanoparticles against foodborne pathogenic bacteria along with its anticandidal and antioxidant effects. *Frontiers in Microbiology*, 8, 167. doi:10.3389/fmicb.2017.00167

Pirtarighat, S., Ghannadnia, M., & Baghshahi, S. (2019). Green synthesis of silver nanoparticles using the plant extract of *Salvia spinosa* grown in vitro and their antibacterial activity assessment. *Journal of Nanostructure Chemisty*, 9(1), 1–9. doi:10.1007/s40097-018-0291-4

Polikanov, Y. S., Aleksashin, N. A., Beckert, B., & Wilson, D. N. (2018). The mechanisms of action of ribosome-targeting peptide antibiotics. *Frontiers in Molecular Biosciences*, 5, 48. doi:10.3389/fmolb.2018.00048

Poole, J., Day, C. J., von Itzstein, M., Paton, J. C., & Jennings, M. P. (2018). Glycointeractions in bacterial pathogenesis. *Nature Reviews Microbiology*, 16(7), 440. doi:10.1038/s41579-018-0007-2

Powers, C. M., Badireddy, A. R., Ryde, I. T., Seidler, F. J., & Slotkin, T. A. (2011). Silver nanoparticles compromise neurodevelopment in PC12 cells: Critical contributions of silver ion, particle size, coating, and composition. *Environmental Health Perspectives*, 119(1), 37–44. doi:10.1289/ehp.1002337

Prabhu, S., Vinodhini, S., Elanchezhiyan, C., & Rajeswari, D. (2018). Evaluation of antidiabetic activity of biologically synthesized silver nanoparticles using Pouteria sapota in streptozotocin-induced diabetic rats. *Journal of Diabetes*, 10(1), 28–42. doi:10.1111/1753-0407.12554

Raja, A., Ashokkumar, S., Marthandam, R. P., Jayachandiran, J., Khatiwada, C. P., Kaviyarasu, K., . . . Swaminathan, M. (2018). Eco-friendly preparation of zinc oxide nanoparticles using *Tabernaemontana divaricata* and its photocatalytic and antimicrobial activity. *Journal of Photochemistry and Photobiology B: Biology*, 181, 53–58.

Rajora, N., Kaushik, S., Jyoti, A., & Kothari, S. L. (2016). Rapid synthesis of silver nanoparticles by *Pseudomonas stutzeri* isolated from textile soil under optimised conditions and evaluation of their antimicrobial and cytotoxicity properties. *IET Nanobiotechnology*, 10(6), 367–373. doi:10.1049/iet-nbt.2015.0107

Rao, B. R., Kotcherlakota, R., Nethi, S. K., Puvvada, N., Sreedhar, B., Chaudhuri, A., & Patra, C. R. (2018). $Ag_2$ [Fe (CN) 5NO] nanoparticles exhibit antibacterial activity and wound healing properties. *ACS Biomaterials Science & Engineering*, 4(9), 3434–3449. doi:10.1021/acsbiomaterials.8b00759

Rao, C., & Cheetham, A. (2001). Science and technology of nanomaterials: Current status and future prospects. *Journal of Materials Chemistry*, 11(12), 2887–2894. doi:10.1039/B105058N

Raub, C. (1989). *Use of silver in jewelry*. Paper presented at the The Santa Fe Symposium on Jewelry Manufacturing Technology 1989. May 20. Tennessee.

Rautela, A., Rani, J., & Das, M. D. (2019). Green synthesis of silver nanoparticles from Tectona grandis seeds extract: Characterization and mechanism of antimicrobial action on different microorganisms. *Journal of Analytical Science and Technology*, 10(1), 5. doi:10.1186/s40543-018-0163-z

Raza, Z. A., Rehman, A., Mohsin, M., Bajwa, S. Z., Anwar, F., Naeem, A., & Ahmad, N. (2015). Development of antibacterial cellulosic fabric via clean impregnation of silver nanoparticles. *Journal of Cleaner Production*, 101, 377–386. doi:10.1016/j.jclepro.2015.03.091

Regiel-Futyra, A., Kus-Liśkiewicz, M., Sebastian, V., Irusta, S., Arruebo, M., Kyzioł, A., & Stochel, G. (2017). Development of noncytotoxic silver–chitosan nanocomposites for efficient control of biofilm forming microbes. *RSC Advances*, 7(83), 52398–52413. doi:10.1039/C7RA08359A

Rezić, I., Haramina, T., & Rezić, T. (2017). Metal nanoparticles and carbon nanotubes—Perfect antimicrobial nano-fillers in polymer-based food packaging materials. In *Food packaging* (pp. 497–532). Elsevier. doi:10.1016/B978-0-12-804302-8.00015-7

Roe, A. L. (1915). Collosol argentum and its ophthalmic uses. *British Medical Journal*, 1(2820), 104. doi:10.1136/bmj.1.2820.104

Rolim, W. R., Pieretti, J. C., Renó, D. L., Lima, B. A., Nascimento, M. N. H., Ambrosio, F. N., . . . Seabra, A. B. (2019). Antimicrobial activity and cytotoxicity to tumor cells of nitric oxide donor and silver nanoparticles containing PVA/PEG films for topical applications. *ACS Applied Materials & Interfaces*, 11(6), 6589–6604. doi:10.1021/acsami.8b19021

Roy, A., Bulut, O., Some, S., Mandal, A. K., & Yilmaz, M. D. (2019). Green synthesis of silver nanoparticles: Biomolecule-nanoparticle organizations targeting antimicrobial activity. *RSC Advances*, 9(5), 2673–2702. doi:10.1039/C8RA08982E

Roy, N., Gaur, A., Jain, A., Bhattacharya, S., & Rani, V. (2013). Green synthesis of silver nanoparticles: An approach to overcome toxicity. *Environmental Toxicology and Pharmacology*, 36(3), 807–812. doi:10.1016/j.etap.2013.07.005

Roy, R., Tiwari, M., Donelli, G., & Tiwari, V. (2018). Strategies for combating bacterial biofilms: A focus on anti-biofilm agents and their mechanisms of action. *Virulence*, 9(1), 522–554. doi:10.1080/21505594.2017.1313372

Sadanand, V., Tian, H., Rajulu, A. V., & Satyanarayana, B. (2017). Antibacterial cotton fabric with in situ generated silver nanoparticles by one-step hydrothermal method. *International Journal of Polymer Analysis and Characterization*, 22(3), 275–279. doi:10.1080/1023666X.2017.1287828

Saifuddin, N., Wong, C. W., & Yasumira, A. A. N. (2009). Rapid biosynthesis of silver nanoparticles using culture supernatant of bacteria with microwave irradiation. *E-Journal of Chemistry*, 6(1), 61–70. doi:10.1155/2009/734264

Samberg, M. E., Oldenburg, S. J., & Monteiro-Riviere, N. A. (2009). Evaluation of silver nanoparticle toxicity in skin in vivo and keratinocytes in vitro. *Environmental Health Perspectives*, 118(3), 407–413.

Sanderson-Wells, T. (1918). A case of puerperal septicæmia successfully treated with intravenous injections of collosol argentum. *The Lancet*, 191(4929), 258–259. doi:10.1016/S0140-6736(01)23965-8

Sarwar, M. S., Niazi, M. B. K., Jahan, Z., Ahmad, T., & Hussain, A. (2018). Preparation and characterization of PVA/nanocellulose/Ag nanocomposite films for antimicrobial food packaging. *Carbohydrate Polymers*, 184, 453–464. doi:10.1016/j.carbpol.2017.12.068

Schoknecht, U., Gruycheva, J., Mathies, H., Bergmann, H., & Burkhardt, M. (2009). Leaching of biocides used in facade coatings under laboratory test conditions. *Environmental Science & Technology*, 43(24), 9321–9328. doi:10.1021/es9019832

Seku, K., Gangapuram, B. R., Pejjai, B., Kadimpati, K. K., & Golla, N. (2018). Microwave-assisted synthesis of silver nanoparticles and their application in catalytic, antibacterial and antioxidant activities. *Journal of Nanostructure Chemistry*, 8(2), 179–188. doi:10.1007/s40097-018-0264-7

Seong, M., & Lee, D. G. (2017). Silver nanoparticles against *Salmonella enterica* serotype typhimurium: role of inner membrane dysfunction. *Current Microbiology*, 74(6), 661–670. doi:10.1007/s00284-017-1235-9

Shahverdi, A. R., Minaeian, S., Shahverdi, H. R., Jamalifar, H., & Nohi, A.-A. (2007). Rapid synthesis of silver nanoparticles using culture supernatants of Enterobacteria: A novel biological approach. *Process Biochemistry*, 42(5), 919–923. doi:10.1016/j.procbio.2007.02.005

Shaik, M., Khan, M., Kuniyil, M., Al-Warthan, A., Alkhathlan, H., Siddiqui, M., . . . Khan, M. (2018). Plant-extract-assisted green synthesis of silver nanoparticles using *Origanum vulgare* L. extract and their microbicidal activities. *Sustainability*, 10(4), 913. doi:10.3390/su10040913

Shanker, K., Mohan, G. K., Hussain, M. A., Jayarambabu, N., & Pravallika, P. L. (2017). Green biosynthesis, characterization, in vitro antidiabetic activity, and investigational acute toxicity studies of some herbal-mediated silver nanoparticles on animal models. *Pharmacognosy Magazine*, 13(49), 188.

Shchukin, D. G., Radtchenko, I. L., & Sukhorukov, G. B. (2003). Photoinduced reduction of silver inside microscale polyelectrolyte capsules. *ChemPhysChem*, 4(10), 1101–1103. doi:10.1002/cphc.200300740

Sheoran, N., & Kaur, P. (2018). Biosynthesis of nanoparticles using eco-friendly factories and their role in plant pathogenicity: A review. *Biotechnology Research & Innovation*, 24, 473–484. doi:10.1016/j.biori.2018.09.003

Shoshan, M. S., Vonderach, T., Hattendorf, B., & Wennemers, H. (2019). Peptide-coated platinum nanoparticles with selective toxicity against liver cancer cells. *Angewandte Chemie International Edition*, 58(15), 4901–4905. doi:10.1002/anie.201813149

Siddiqi, K. S., Husen, A., & Rao, R. A. (2018). A review on biosynthesis of silver nanoparticles and their biocidal properties. *Journal of Nanobiotechnolgy*, 16(1), 14. doi:10.1186/s12951-018-0334-5

Singh, N., Paknikar, K. M., & Rajwade, J. (2019). RNA-sequencing reveals a multitude of effects of silver nanoparticles on *Pseudomonas aeruginosa* biofilms. *Environmental Science Nano*, 6(6), 1812–1828. doi:10.1039/C8EN01286E

Singh, P., Kim, Y. J., & Yang, D. C. (2016a). A strategic approach for rapid synthesis of gold and silver nanoparticles by *Panax ginseng* leaves. *Artificial Cells, Nanomedicine, and Biotechnology*, 44(8), 1949–1957. doi:10.3109/21691401.2015.1115410

Singh, P., Kim, Y.-J., Zhang, D., & Yang, D.-C. (2016b). Biological synthesis of nanoparticles from plants and microorganisms. *Trends in Biotechnology*, 34(7), 588–599. doi:10.1016/j.tibtech.2016.02.006

Singh, S., Singh, S. K., Chowdhury, I., & Singh, R. (2017). Understanding the mechanism of bacterial biofilms resistance to antimicrobial agents. *The Open Microbiology Journal*, 11, 53. doi:10.2174/1874285801711010053

Siritongsuk, P., Hongsing, N., Thammawithan, S., Daduang, S., Klaynongsruang, S., Tuanyok, A., & Patramanon, R. (2016). Two-phase bactericidal mechanism of silver nanoparticles against *Burkholderia pseudomallei*. *Plos One*, 11(12), e0168098. doi:10.1371/journal.pone.0168098

Slavin, Y. N., Asnis, J., Häfeli, U. O., & Bach, H. (2017). Metal nanoparticles: Understanding the mechanisms behind antibacterial activity. *Journal of Nanobiotechnology*, 15(1), 65. doi:10.1186/s12951-017-0308-z

Socol, Y., Abramson, O., Gedanken, A., Meshorer, Y., Berenstein, L., & Zaban, A. (2002). Suspensive electrode formation in pulsed sonoelectrochemical synthesis of silver nanoparticles. *Langmuir*, 18(12), 4736–4740. doi:10.1021/la015689f

Song, J. Y., & Kim, B. S. (2009). Rapid biological synthesis of silver nanoparticles using plant leaf extracts. *Bioprocess and Biosystems Engineering*, 32(1), 79–84. doi:10.1007/s00449-008-0224-6

Soto, G., Tiznado, H., Contreras, O., Pérez-Tijerina, E., Cruz-Reyes, J., Del Valle, M., & Portillo, A. (2011). Preparation of a $Ag/SiO_2$ nanocomposite using a fluidized bed microwave plasma reactor, and its hydrodesulphurization and Escherichia coli bactericidal activities. *Powder Technology*, 213(1–3), 55–62. doi:10.1016/j.powtec.2011.07.005

Srivastava, S. K., & Constanti, M. (2012). Room temperature biogenic synthesis of multiple nanoparticles (Ag, Pd, Fe, Rh, Ni, Ru, Pt, Co, and Li) by Pseudomonas aeruginosa SM1. *Journal of Nanoparticle Research*, 14(4), 831. doi:101007/s11051-012-0831-7

Stabryla, L. M., Johnston, K. A., Millstone, J. E., & Gilbertson, L. M. (2018). Emerging investigator series: It's not all about the ion: Support for particle-specific contributions to silver nanoparticle antimicrobial activity. *Environmental Science Nano*, 5(9), 2047–2068. doi:10.1039/C8EN00429C

Stoodley, P., Sauer, K., Davies, D. G., & Costerton, J. W. (2002). Biofilms as complex differentiated communities. *Annual Review of Microbiology*, 56(1), 187–209. doi:10.1146/annurev.micro.56.012302.160705

Su, C.-H., Kumar, G. V., Adhikary, S., Velusamy, P., Pandian, K., & Anbu, P. (2017). Preparation of cotton fabric using sodium alginate-coated nanoparticles to protect against nosocomial pathogens. *Biochemical Engineering Journal*, 117, 28–35. doi:10.1016/j.bej.2016.10.020

Subhapriya, S., & Gomathipriya, P. (2018). Green synthesis of titanium dioxide ($TiO_2$) nanoparticles by *Trigonella foenum*-graecum extract and its antimicrobial properties. *Microbial Pathogenesis*, 116, 215–220. doi:10.1016/j.micpath.2018.01.027

Surwade, P., Ghildyal, C., Weikel, C., Luxton, T., Peloquin, D., Fan, X., & Shah, V. (2019). Augmented antibacterial activity of ampicillin with silver nanoparticles against methicillin-resistant *Staphylococcus aureus* (MRSA). *The Journal of Antibiotics*, 72(1), 50. doi:10.1038/s41429-018-0111-6

Sweet, M., & Singleton, I. (2011). Silver nanoparticles: A microbial perspective. *Advances in Applied Microbiology*, 77, 115–133). Wiley Online Library. doi:10.1016/B978-0-12-387044-5.00005-4

Tang, C., Hu, D., Cao, Q., Yan, W., & Xing, B. (2017). Silver nanoparticles-loaded activated carbon fibers using chitosan as binding agent: Preparation, mechanism, and their antibacterial activity. *Applied Surface Science*, 394, 457–465. doi:10.1016/j.apsusc.2016.10.095

Tang, S., & Zheng, J. (2018). Antibacterial activity of silver nanoparticles: Structural effects. *Advanced Healthcare Materials*, 7(13), 1701503. doi:10.1002/adhm.201701503

Thomas, R., Soumya, K., Mathew, J., & Radhakrishnan, E. (2015). Electrospun polycaprolactone membrane incorporated with biosynthesized silver nanoparticles as effective wound dressing material. *Applied Biochemistry and Biotechnology*, 176(8), 2213–2224. doi:10.1007/s12010-015-1709-9

Tlili, A., Jabiol, J. R. m., Behra, R., Gil-Allué, C., & Gessner, M. O. (2017). Chronic exposure effects of silver nanoparticles on stream microbial decomposer communities and ecosystem functions. *Environmental Science & Technology*, 51(4), 2447–2455. doi:10.1021/acs.est.6b05508

Toynbee, J. M. C., & Painter, K. S. (1986). II. Silver picture plates of late antiquity: ad 300 to 700. *Archaeologia*, 108, 15–65. doi:10.1017/S002613409001170X

Tozzi, P., Al-Darweesh, A., Vogt, P., & Stumpe, F. (2001). Silver-coated prosthetic heart valve: A double-bladed weapon. *European Journal of Cardio-Thoracic Surgery*, 19(5), 729–731. doi:10.1016/s1010-7940(01)00675-3

Vaclav, M. Z. (1960). *Process of producing oligodynamic metal biocides*. Google Patents.

van Eijk, E., Wittekoek, B., Kuijper, E. J., & Smits, W. K. (2017). DNA replication proteins as potential targets for antimicrobials in drug-resistant bacterial pathogens. *The Journal of Antimicrobial Chemotherapy*, 72(5), 1275–1284. doi:10.1093/jac/dkw548

Velmurugan, P., Park, J.-H., Lee, S.-M., Jang, J.-S., Lee, K.-J., Han, S.-S., . . . Oh, B.-T. (2015). Synthesis and characterization of nanosilver with antibacterial properties using *Pinus densiflora* young cone extract. *Journal of Photochemistry and Photobiology B*, 147, 63–68. doi:10.1016/j.jphotobiol.2015.03.008

Verma, J., Kanoujia, J., Parashar, P., Tripathi, C. B., & Saraf, S. A. (2017). Wound healing applications of sericin/chitosan-capped silver nanoparticles incorporated hydrogel. *Drug Delivery and Translational Research*, 7(1), 77–88. doi:10.1007/s13346-016-0322-y

Vilchis-Nestor, A. R., Sánchez-Mendieta, V., Camacho-López, M. A., Gómez-Espinosa, R. M., Camacho-López, M. A., & Arenas-Alatorre, J. A. (2008). Solventless synthesis and optical properties of Au and Ag nanoparticles using *Camellia sinensis* extract. *Materials Letters*, 62(17), 3103–3105. doi:10.1016/j.matlet.2008.01.138

Walser, T., Demou, E., Lang, D. J., & Hellweg, S. (2011). Prospective environmental life cycle assessment of nanosilver T-shirts. *Environmental Science & Technology*, 45(10), 4570–4578. doi:10.1021/es2001248

Wangler, T. P., Zuleeg, S., Vonbank, R., Bester, K., Boller, M., Carmeliet, J., & Burkhardt, M. (2012). Laboratory scale studies of biocide leaching from façade coatings. *Building and Environment*, 54, 168–173. doi:10.1016/j.buildenv.2012.02.021

Wanhill, R., Steijaart, J., Leenheer, R., & Koens, J. (1998). Damage assessment and preservation of an Egyptian silver vase (300–200 bc). *Archaeometry*, 40(1), 123–137. doi:10.1111/j.1475-4754.1998.tb00828.x

Wei, K., Wu, Y., Xi, Y., Luo, Y., Shu, L., J., & Zeng, X., F. (2010). Research on disinfection efficacy of the compound antimicrobial emulsion containing silver nanoparticles [J]. *Chinese Journal of Disinfection*, 5.

Wu, M., Ma, B., Pan, T., Chen, S., & Sun, J. (2016). Silver-nanoparticle-colored cotton fabrics with tunable colors and durable antibacterial and self-healing superhydrophobic properties. *Advanced Functional Materials*, 26(4), 569–576. doi:10.1002/adfm.201504197

Xie, C.-M., Lu, X., Wang, K.-F., Meng, F.-Z., Jiang, O., Zhang, H.-P., . . . Fang, L.-M. (2014). Silver nanoparticles and growth factors incorporated hydroxyapatite coatings on metallic implant surfaces for enhancement of osteoinductivity and antibacterial properties. *ACS Applied Materials & Interfaces*, 6(11), 8580–8589. doi:10.1021/am501428e

Yan, X., He, B., Liu, L., Qu, G., Shi, J., Hu, L., & Jiang, G. (2018). Antibacterial mechanism of silver nanoparticles in *Pseudomonas aeruginosa*: Proteomics approach. *Metallomics*, 10(4), 557–564. doi:10.1039/c7mt00328e

Yang, J.-L., Li, Y.-F., Liang, X., Guo, X.-P., Ding, D.-W., Zhang, D., . . . Dobretsov, S. (2016). Silver nanoparticles impact biofilm communities and mussel settlement. *Scientific Reports*, 6, 37406. doi:10.1038/srep37406

Yang, T., Yao, Q., Cao, F., Liu, Q., Liu, B., & Wang, X.-H. (2016). Silver nanoparticles inhibit the function of hypoxia-inducible factor-1 and target genes: Insight into the cytotoxicity and antiangiogenesis. *International Journal of Nanomedicine*, 11, 6679. doi:10.2147/IJN.S109695

Yang, X., Gondikas, A. P., Marinakos, S. M., Auffan, M., Liu, J., Hsu-Kim, H., & Meyer, J. N. (2012). Mechanism of silver nanoparticle toxicity is dependent on dissolved silver and surface coating in *Caenorhabditis elegans*. *Environmental Science & Technology*, 46(2), 1119–1127. doi:10.1021/es202417t

Yin, Y., Li, Z.-Y., Zhong, Z., Gates, B., Xia, Y., & Venkateswaran, S. (2002). Synthesis and characterization of stable aqueous dispersions of silver nanoparticles through the Tollens process. *Journal of Materials Chemistry*, 12(3), 522–527. doi:10.1039/B107469E

You, C., Han, C., Wang, X., Zheng, Y., Li, Q., Hu, X., & Sun, H. (2012). The progress of silver nanoparticles in the antibacterial mechanism, clinical application and cytotoxicity. *Molecular Biology Reports*, 39(9), 9193–9201. doi:10.1007/s11033-012-1792-8

Yousefi, M., Dadashpour, M., Hejazi, M., Hasanzadeh, M., Behnam, B., de la Guardia, M., . . . Mokhtarzadeh, A. (2017). Anti-bacterial activity of graphene oxide as a new weapon nanomaterial to combat multidrug-resistance bacteria. *Materials Science and Engineering: C*, 74, 568–581. doi:10.1016/j.msec.2016.12.125

Youssef, A. M., & El-Sayed, S. M. (2018). Bionanocomposites materials for food packaging applications: Concepts and future outlook. *Carbohydrate Polymers*, 193, 19–27. doi:10.1016/j.carbpol.2018.03.088

Yu, L., Shang, F., Chen, X., Ni, J., Yu, L., Zhang, M., . . . Xue, T. (2018). The anti-biofilm effect of silver-nanoparticle-decorated quercetin nanoparticles on a multi-drug resistant *Escherichia coli* strain isolated from a dairy cow with mastitis. *PeerJ*, 6, e5711. doi:10.7717/peerj.5711

Yuan, Y.-G., Peng, Q.-L., & Gurunathan, S. (2017). Effects of silver nanoparticles on multiple drug-resistant strains of *Staphylococcus aureus* and *Pseudomonas aeruginosa* from mastitis-infected goats: An alternative approach for antimicrobial therapy. *International Journal of Molecular Sciences*, 18(3), 569. doi:10.3390/ijms18030569

Yun'an Qing, L. C., Li, R., Liu, G., Zhang, Y., Tang, X., Wang, J., . . . Qin, Y. (2018). Potential antibacterial mechanism of silver nanoparticles and the optimization of orthopedic implants by advanced modification technologies. *International Journal of Nanomedicine*, 13, 3311. doi:10.2147/IJN.S165125

Zhang, W., Qiao, X., & Chen, J. (2007). Synthesis of nanosilver colloidal particles in water/oil microemulsion. *Colloids and Surfaces A: Physicochemical and Engineering Aspects*, 299(1), 22–28. doi:10.1016/j.colsurfa.2006.11.012

Zhang, X., Yan, S., Tyagi, R., & Surampalli, R. (2011). Synthesis of nanoparticles by microorganisms and their application in enhancing microbiological reaction rates. *Chemosphere*, 82(4), 489–494. doi:10.1016/j.chemosphere.2010.10.023

Zhang, X.-F., Liu, Z.-G., Shen, W., & Gurunathan, S. (2016). Silver nanoparticles: Synthesis, characterization, properties, applications, and therapeutic approaches. *International Journal of Molecular Sciences*, 17(9), 1534. doi:10.3390/ijms17091534

Zhang, Z., Tsai, P. C., Ramezanli, T., & Michniak-Kohn, B. B. (2013). Polymeric nanoparticles-based topical delivery systems for the treatment of dermatological diseases. *Wiley Interdisciplinary Reviews: Nanomedicine and Nanobiotechnology*, 5(3), 205–218. doi:10.1002/wnan.1211

Zhao, R., Lv, M., Li, Y., Sun, M., Kong, W., Wang, L., . . . Qiu, S. (2017). Stable nanocomposite based on PEGylated and silver nanoparticles loaded graphene oxide for long-term antibacterial activity. *ACS Applied Materials & Interfaces*, 9(18), 15328–15341. doi:10.1021/acsami.7b03987

Zheng, J., Song, F., Wang, X.-L., & Wang, Y.-Z. (2014). In-situ synthesis, characterization and antimicrobial activity of viscose fiber loaded with silver nanoparticles. *Cellulose*, 21(4), 3097–3105. doi:10.1007/s10570-014-0324-1

Zhou, W., Gao, X., Liu, D., & Chen, X. (2015). Gold nanoparticles for in vitro diagnostics. *Chemical Reviews*, 115(19), 10575–10636. doi:10.1021/acs.chemrev.5b00100

## Glossary

**Abiotic:** Nonliving part of an ecosystem.

**Argyrosis:** A condition caused as a result of excessive exposure to silver or silver dust-based chemical compounds.

**Allergens:** Any substance (antigen), most often eaten or inhaled, that is recognized by the immune system and causes an allergic reaction.

**Angiogenesis:** The formation of new blood vessels from the pre-existing ones.

**Antibiotics:** The agents that inhibit the growth of bacteria.

**Antidiabetic:** Agents associated with the decrease in the level of glucose in the patient's body.

**Antiviral:** Section of drugs used for the treatment of viral infections.

**Bacterial efflux pumps:** The mechanism by which bacteria remove unwanted substance from the cell.

**Bioavailability:** The part of the drug or any treatment that enters the body circulation and remains unchanged.

**Biotic:** Living components of an ecosystem.

**Biocompatible:** The quality of being nontoxic to cells.

**Biofilm:** A collection of microorganisms consisting of extracellular matrix, proteins, DNA, and lipids. Here, the microorganisms grow and communicate with each other.

**Bio-nanotechnology:** The convergence of biology and nanotechnology.

**Cellulose:** An organic compound composed of carbon, hydrogen, and oxygen atoms that helps in maintaining the structural integrity of plant cell walls.

**Cell lysis:** The process by which cell membranes are broken down such that the cellular integrity is compromised and the inner contents of the cell come out

**Cell membrane fluidity:** The extent of viscosity of the lipid bilayer of the cell membrane.

**Curative:** Ability to cure diseases.

**Cytotoxicity:** The potential of being toxic to cells.

**Dispersibility:** Uniform distribution of a solute in a solution.

**Drug-resistant bacteria:** Bacteria that have developed mechanisms to circumvent toxic effect of antibacterial drugs.

**Enzymes:** Protein molecules that take part in certain reactions and influence the rate of the chemical reaction.

**Epithelialization:** The process by which epithelial tissue expands and replaces a denuded surface.

**Extracellular matrix:** A noncellular component present in all tissues and containing collagen, enzyme, glycoproteins, etc. that offers structural and biochemical support to surrounding cells.

**Flavonoids:** Plant secondary metabolites that play a role in defense system.

**Fluorescence properties:** The physical property of an object to absorb light of one wavelength and re-emit into another wavelength.

**Gestation:** The period during which the fetus develops inside the uterus of the biological mother.

**Immune system:** The defense system of a living organism that comprises certain cells that protect the host from infectious agents.

**Intraperitoneal:** Administered through the peritoneum.

**Log phase of growth:** The stage of life cycle of cells when the cells grow by doubling themselves rapidly.

**Mutation:** Alterations in nucleic acids of the genome of an organism.

**Nanocomposite:** A solid material composed of different materials where one of the materials has less than 100 nm diameter.

**Nosocomial:** An infection or a disease that is acquired as a result of exposure to infection or toxin from certain locations like hospital.

**Pathogens:** Infectious agents (mostly microorganisms) that cause diseases to multicellular organisms like plants and animals.

**Peptidoglycan layer:** A polymer consisting of sugars and amino acids that forms a mesh-like layer outside the plasma membrane of most bacteria, forming the cell wall.

**Stationary phase of growth:** The stage of the life cycle of cells when they stop growing but remain metabolically active.

**Synergism:** Cumulative effect of more than one agent.

**Toxins:** Small molecules or proteins that can cause diseases upon coming in contact with body tissues.

**Transcriptome:** The complete set of all transcripts, i.e. RNA molecules present in one cell or a population of cells at a specific developmental stage or physiological condition.

**Wound:** An injury that occurs when a skin is cut, torn, burnt, or punctured.

# 9

# Microbial Gold Nanoparticles and Their Biomedical Applications

*Dindyal Mandal, Rohit Kumar Singh, Uday Suryakant Maharana, Bijayananda Panigrahi, and Sourav Mishra*

*School of Biotechnology, KIIT University, Bhubaneswar, Odisha, India*

## Objectives

- Understand the importance of green synthesis of gold nanoparticles.
- Get a general idea about the application of microbes in gold nanoparticle synthesis.
- Understand the difference between extracellular and intracellular synthesis of gold nanoparticles.
- Understand the mechanism of microbial gold nanoparticles synthesis.
- Know how to apply gold nanoparticles in the field of sensors, antimicrobial and anticancer research

## 9.1 Introduction

Recently, various metal nanoparticles such as titanium, aluminum, iron, copper, silver, and gold have drawn enormous attention in the field of materials science owing to their special physicochemical properties (Giljohann et al., 2010; Haber & Buhro, 1998; Jana, Gearheart, & Murphy, 2001; Kuemmeth, Bolotin, Shi, & Ralph, 2008; Lee, Choi, Lee, Stott, & Kim, 2008; Ling, Lee, & Hyeon, 2015; Narayanan & Sakthivel, 2011; Peng, Zhao, Dai, Shi, & Hirao, 2005; Sun & Xia, 2002). Among them, gold nanoparticles (Au NPs) are well-recognized in materials science as well as biological research specially as catalysts, sensors, bio-labeling agents, drug delivery vehicles, antibacterial agents, and tissue engineering materials due to their distinctive characteristics (Gao, Wei, Yang, Chen, & Yang, 2015; Howes, Chandrawati, & Stevens, 2014; Shi, Votruba, Farokhzad, & Langer, 2010; Wang & Wang, 2013).

*Microbial Interactions at Nanobiotechnology Interfaces: Molecular Mechanisms and Applications*, First Edition. Edited by R. Navanietha Krishnaraj and Rajesh K. Sani.

From ancient times, gold nanoparticles (Au NPs)/compounds have been playing a major role in the history of medicine for the treatment of joints, bones, facial paralysis, and various inflammatory skin disorders (Naraginti et al., 2016; Sanvicens & Marco, 2008; Shedbalkar, Singh, Wadhwani, Gaidhani, & Chopade, 2014; Yun & Kwok, 2017). The key factor contributing toward the popularity of Au NPs in biomedical and clinical applications is their stability against oxidation (Gupta, Naregalkar, Vaidya, & Gupta, 2007).

Traditional chemical synthesis has many disadvantages, which include the toxicity and hazardous nature of the reagents used in the synthesis (Ahmed, Ahmad, Swami, & Ikram, 2016). Furthermore, toxic chemicals adsorbed on the surface of the nanoparticles restrict their use in biomedical research. Therefore, the development of biocompatible, environment-friendly, and nontoxic methods is required for the synthesis and stabilization of nanoparticles. Thus, biological synthesis or green synthesis of nanoparticles came into the picture. In green synthesis, plant extract, different biomolecules such as polyphenols, carbohydrates, and microbes including algae, fungi, viruses, and bacteria are used as a reducing agent (Dauthal & Mukhopadhyay, 2016; Iravani, 2011; Mandal, Bolander, Mukhopadhyay, Sarkar, & Mukherjee, 2006; Shah, Fawcett, Sharma, Tripathy, & Poinern, 2015; Singh, Mishra, et al., 2018; Singh, Panigrahi, et al., 2018). A major advantage of microbes is that they contain a significant number of enzymes that degrade the toxic chemicals during the synthesis of nanoparticles as a defense mechanism. Green synthesis is now also a more demanding approach for bio-production of nanoparticles at large scale.

Herein, the objective of the chapter is to extensively explore the current research scenario on the microbial gold nanoparticles and its potential biomedical applications. This chapter focuses on microbial synthesis of gold nanoparticles, mechanism of synthesis of Au NPs, and potential biomedical applications of Au NPs.

## 9.2 Microbial Gold Nanoparticles Synthesis

Microbial gold nanoparticles synthesis is broadly divided into two categories: intracellular and extracellular modes. In the case of intracellular synthesis, nanoparticles are formed inside the cellular matrix (Samanta, Singh, & Adholeya, 2017), whereas the reduction of chloroaurate ions occurs by the proteins/enzymes present on the cell wall leads to extracellular synthesis of gold nanoparticles as described in Figure 9.1 (Sengani, Grumezescu, & Rajeswari, 2017).

Various methods of nanoparticles synthesis are summarized in Figure 9.2. In the area of biological synthesis, plants generally form nanoparticles in the solution; in contrast, microorganisms such as bacteria, algae, and fungi are reported to generate gold nanoparticles (Au NPs) intracellularly or extracellularly with well-defined phase and morphology. However, the downstream processing of intracellular Au

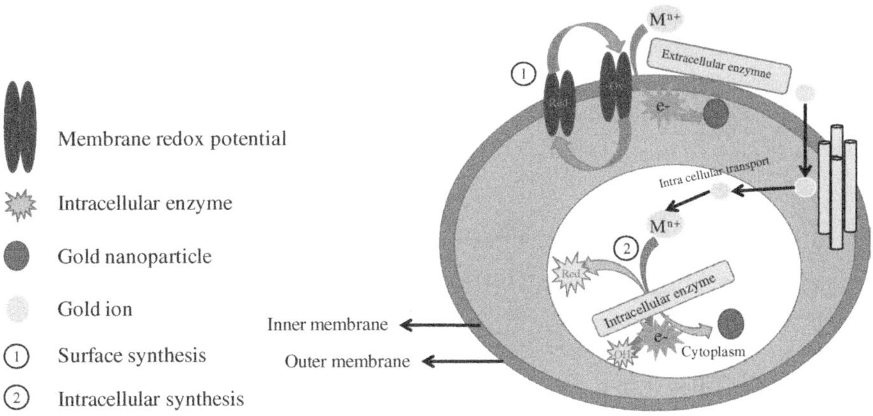


**Figure 9.1** Schematic diagram for the mechanism of extracellular and intracellular synthesis of Au NPs. *Source*: Reprinted with permission from Shah et al. (2015) Copyright © 2017 Elsevier.

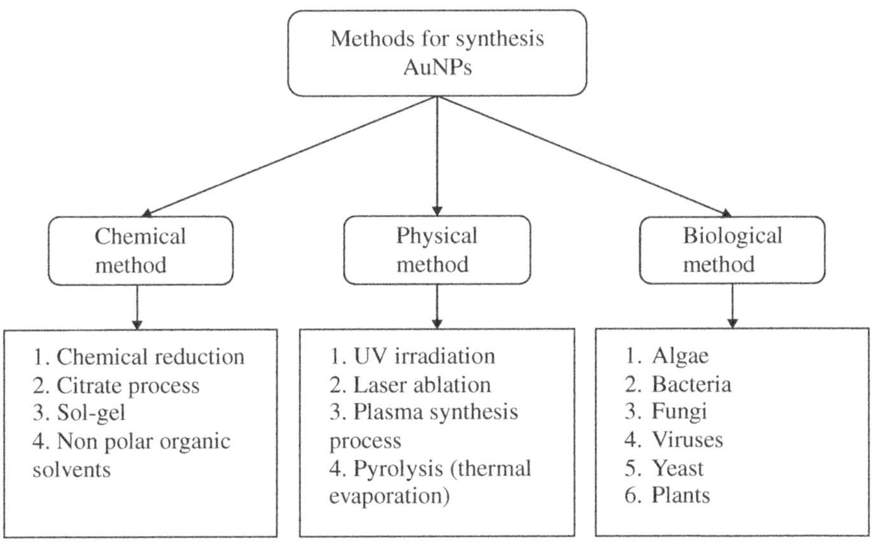


**Figure 9.2** Schematic diagram of various methods of nanoparticles synthesis.

NPs is very hard, which restricts their use in the real applications. Intracellular synthesis requires detergents and ultrasonication step to recover the nanoparticles, which can be omitted in the extracellular synthesis of Au NPs (Castro, Blazquez, Gonzalez, & Ballester, 2014). Thus, microbes with extracellular constituents have been explored mostly for screening the biosynthesis of Au NPs. As it is already known that microbes cause detoxification, reduction of lethal salts to nanoparticles occurs by reductase enzymes present within the microbes (Salunke, Sawant, Lee, & Kim, 2016). Therefore, the microbes have been used as alternative manufacturing tool for the synthesis of Au NPs, which has created a new and promising area of research in the materials synthesis.

### 9.2.1 Bacteria-Mediated Gold Nanoparticles

Among the microorganisms, prokaryotic bacteria have drawn tremendous attention in the field of gold nanoparticles synthesis, because they have the ability to acclimatize to intense circumstances and the bacterial subculture is cheap and fast growing. Moreover, the optimum parameters such as aeration, cultivation time interval, and temperature can be simply managed. The mechanism involved in the biosynthesis of Au NPs is that the cell wall of bacteria can be easily manipulated on the basis of their culture conditions, thus crucial metals mostly penetrate from the cell wall to the cytoplasm and reverse through the wall of the network-like structure of the extracellular membrane (Gericke & Pinches, 2006a). The cell wall contains peptidoglycan, which consists of several numbers of anionic sites, which further help in the interlinkage between metal and functional groups to deposit metal (French, Puddephatt, Habash, & Glasauer, 2013).

Primarily, the synthesis of gold nanoparticles is visualized by the color change, which works on the principle of surface plasmon resonance (SPR). SPR is the combined excitation of the conduction electrons in metal with the wavelength of irradiated light. Depending upon the nanoparticle size, color of the solution is changed. For instance, in the case of gold nanoparticles the color change occurs from ruby red to purple indicates the increase in the size of nanoparticles. The color variation is attributed to the localized surface plasmon resonance (LSPR), which they exhibit and the corresponding wavelength lies in the visible region of the electromagnetic spectrum where absorption occurs in a particular wavelength and the complementary color is emitted. The absorbance of this color change is measured by UV–visible spectroscopy. For example, bacterial solution is milky white and the solution of gold chloride is pale yellow. However, when bacterial solution is incubated with gold chloride, it turns red, indicating the formation of gold nanoparticles.

Early studies depict that *Bacillus subtilis* 168 is capable of reducing $Au^{3+}$ ions to generate octahedral gold nanoparticles of nanoscale dimensions (5–25 nm) within bacterial cells when gold chloride is incubated with bacterial cells under ambient temperature and pressure conditions (Beveridge & Murray, 1980; Southam & Beveridge, 1994). It is presumed that organic phosphate compounds play a role in the development of octahedral Au NPs, probably by the formation of bacteria–Au complex formation. It is found that Fe(III)-reducing bacteria *Shewanella algae* is capable of reducing gold ions in the anaerobic condition. The bacteria can reduce gold ions in the presence of hydrogen gas, which results in the formation of nano-sized Au NPs (Konishi, 2004).

Another bacterial strain *Pseudomonas stutzeri* AG259 was first reported for the extracellular synthesis of gold nanoparticles (Khan & Pethe, 2016), where the size range of biosynthesis of gold nanoparticles was reported from 50 to 60 nm and absorbance was found to be at ca. 530 nm, known as the surface plasmon resonance (SPR) of gold nanoparticles. Apart from whole cell biomass, bacterial extract was also used for the nanoparticles synthesis. For example, bacterial cell extract of *Rhodopseudomonas capsulate* was used for the extracellular synthesis of gold nanoparticles at room temperature (He et al., 2007).

*Lactobacillus* strain was reported for the synthesis of Au NPs with well-defined morphology (Nair & Pradeep, 2002). Husseiny et al. studied the synthesis of extracellular Au NPs using the sediment of *Pseudomonas aeruginosa*. In this work, they have demonstrated that cell filtrate can control the dispersity and shape of the nanoparticles (Husseiny, El-Aziz, Badr, & Mahmoud, 2007). This work was helpful to establish the molecular and chemical mechanism of nanoparticles synthesis. It has been reported that in alkaline condition *Thermomonospora* sp. produces gold nanoparticles within size ranges of 8–10 nm (Ahmad et al., 2003a). Further, Ahmad, Senapati, Khan, Kumar, and Sastry (2003b) have validated the efficiency of *Rhodococcus* sp. toward the production of gold nanoparticles. Lengke, Fleet, and Southam (2006) found the formation of cube-shaped Au NPs using filamentous cyanobacterium *Plectonema boryanum* UTEX 485 upon incubation with different precursors of gold ions such as $HAuCl_4$ and aqueous $Au(S_2O_3)_2$. In order to achieve extracellular gold nanoparticles, Thirumurugan et al. have used the supernatant of *B. subtilis 168*, where bioreduction of $Au^{3+}$ to $Au^0$ occurred through the modification of carboxyl group (Thirumurugan, Ramachandran, Tomy, Jiflin, & Rajagomathi, 2012)

In order to understand the mechanism of synthesis, nanoparticles' surface should be well-characterized. Tanzil et al. (2016) established from FTIR analysis that external proteins, phospholipids, and lipopolysaccharides are responsible for the synthesis of gold nanoparticles when they use different bacterial strains. Morphology of gold nanoparticles was found to be tuned with the help of various components of cell surface of *Escherichia coli* (Du, Jiang, Liu, & Wang, 2007). Later, Li et al. (2016a) explored the extreme bacterium *Deinococcus radiodurans*, which is resistant to oxidants and stresses toward the synthesis of Au NPs. They have also shown from FTIR analysis that the presence of protein molecules on the surface of Au NPs is responsible for the synthesis.

In order to eliminate the necessity of aqueous extract of microbial strain toward nanoparticles synthesis, nontraditional experimental methods should be explored. To address this issue, dried bacterial strain *Bacillus megaterium* D01 has been used by Li et al. for the biosynthesis of spherically shaped Au NPs. They further have shown that the shape, size, and monodispersity of NPs can be adjusted by dodecanethiol (Wen et al., 2009).

Sometimes it is not important to use bacterial biomass for their synthesis. In order to simplify the downstream processing, Nadaf and Kanase (2016) reported the synthesis of Au NPs by cell-free extract of *Bacillus marisflavi* YCIS MN 5 strain. The synthesized nanoparticles were found to be spherical, crystalline, and of size ~14 nm. The generated Au NPs exhibit efficiency for catalytic biodegradation of Congo red and methylene blue. Another species *Bacillus licheniformis* has been investigated by Sriram, Kalishwaralal, and Gurunathan (2012) toward the intracellular synthesis of Au NPs. Li et al. (2016b) demonstrated the extracellular Au NPs synthesis using *Bacillus niabensis 45* bacterial strain and they speculated that the cyclic peptides present in the bacteria are responsible for the NPs synthesis.

In similar studies, other bacterial strains such as *Klebsiella pneumonia* (Malarkodi et al., 2013) and *Brevibacterium casei* have been used for the biosynthesis of Au NPs (Kalishwaralal et al., 2010). Rajasree and Suman (2012) reported the extracellular biosynthesis of Au NPs using *Pseudomonas fluorescens* and observed the size of nanoparticle in the range of 20–80 nm. In order to study the effect of concentration of gold ions, Wadhwani et al. studied the effect of bacterium cells (*Acinetobacter* sp.) with different concentrations of the gold ion. The Au NPs were observed to be spherical, monodispersed, and of size 19 nm. The FTIR spectra at 3221–3196 $cm^{-1}$ revealed the existence and role of amino acids (Wadhwani et al., 2016). Correa-Llanten et al. studied bacterium *Geobacillus* sp. for the synthesis of Au NPs, in which color of the bacterial biomass changes from colorless to purple upon the exposure of bacterial cells to $Au^{3+}$ ions, indicating the accumulation of intracellular gold nanoparticles. Synthesized Au NPs were found to be of quasihexagonal shape, and FTIR revealed the presence of protein-like molecules (NADH) on the surface of the nanoparticles (Correa-Llanten, Muñoz-Ibacache, Castro, Munoz, & Blamey, 2013).

De Corte et al. explored *Shewanella oneidensis* for the synthesis of Au NPs. Using biosorption and slow reduction process, the reaction of sodium acetate with $HAuCl_4$ initially leads to the formation of gold ions. In the next step, these gold ions are adsorbed on the external membrane of the cell and, finally, Au NPs are generated. Further, this process has been used in industrial waste for recovery of gold ions (De Corte et al., 2011). An overview of bacterial synthesis of Au NPs is given in Table 9.1.

### 9.2.2 Algae-Mediated Gold Nanoparticles

Recently, researchers have explored marine organisms for the synthesis of Au NPs. Because marine algae contain variety of biomolecules like pigments, peptides, and polysaccharides (Rajesh Kumar et al., 2013). Therefore, adverse marine groups like Euglenoids, Chlorophyceae, Rhodophyceae, Cyanophyceae, Diatoms, and Phaeophyceae have been evolved for the synthesis of gold nanoparticles (Table 9.2).

**Table 9.1** Synthesis of gold nanoparticles by bacteria.

| Bacterial strain | Size (nm) | Shape | References |
|---|---|---|---|
| *Pseudomonas stutzeri* | 50–60 | Spherical | Khan and Pethe (2016) |
| *Rhodopseudomonas capsulata* | 10–12 | Spherical | He et al. (2007) |
| *Lactobacillus* sp. | 100–300 | Hexagonal/triangular/other | Nair and Pradeep (2002) |
| *Pseudomonas aeruginosa* | 15–30 | — | Husseiny et al. (2007) |
| *Rhodococcus* sp. | 5–15 | Spherical | Ahmad, Senapati, Khan, Kumar, and Sastry (2003) |
| *Plectonema boryanum* | 20–25 | Cubic | Lengke et al. (2006) |
| *Bacillus subtilis 168* | 5–25 | — | Thirumurugan et al. (2012) |
| *Escherichia coli DH5α* | $25 \pm 8$ | Spherical | Du et al. (2007) |
| *Deinococcus radiodurans* | 43.75 | Spherical/triangular/irregular | Li et al. (2016a) |
| *Bacillus marsiflavi* | ~14 | Spherical | Nadaf and Kanase (2016) |
| *Bacillus licheniformis* | 38 | Spherical | Sriram et al. (2012) |
| *Bacillus niabensis 45* | 10–20 | Spherical | Li et al. (2016a) |
| *Klebsiella pneumoniae* | 35–65 | Spherical | Malarkodi et al. (2013) |
| *Brevibacterium casei* | 10–50 | Spherical | Kalishwaralal et al. (2010) |
| *Pseudomonas fluorescens* | 20–80 | Spherical | Rajasree and Suman (2012) |
| *Acinetobacter* sp. *SW30* | 19 | Spherical | Wadhwani et al. (2016) |
| *Geobacillus* sp. | 5–20 | Hexagonal | Correa-Llanten et al. (2013) |
| *Shewanella oneidensis* | $12 \pm 5$ | Spherical | De Corte et al. (2011) |

**Table 9.2** Synthesis of gold nanoparticles by algae.

| Algae species | Size (nm) | Shape | References |
|---|---|---|---|
| *Turbinaria conoides* | 6–10 | Spherical | Rajesh Kumar et al. (2013) |
| *Fucus vesiculosus* | — | Spherical | Mata et al. (2009) |
| *Padina gymnospora* | 53–67 | Spherical | M. Singh, Kalaivani, Manikandan, Sangeetha, and Kumaraguru (2013) |
| *Tetraselmis kochinensis* | 5–35 | Spherical | Senapati, Syed, Moeez, Kumar, and Ahmad (2012) |
| *Sargassum muticum* | 27–35 | Spherical | Namvar et al. (2015) |
| *Chlorella vulgaris* | 2–10 | Spherical | Annamalai and Nallamuthu (2015) |
| *Sargassum wightii* | | Spherical | Singaravelu, Arockiamary, Kumar, and Govindaraju (2007) |
| *Ecklonia cava* | 20–50 | Spherical | Venkatesan et al. (2014) |
| *Stoechospermum marginatum* | 18.7–93.7 | Spherical | Rajathi, Parthiban, Kumar, and Anantharaman (2012) |
| *Calothrix algae* | 30–120 | Spherical | B. Kumar, Smita, Sánchez, Guerra, and Cumbal (2016) |

Interestingly, significant studies have been reported by Mata et al. using algae *Fucus vesiculosus* for the synthesis of Au NPs. Herein, synthesis of Au NPs has been performed based upon the redox potential and pH of the solution (Mata et al., 2009). From the mechanistic insight they have shown that hydroxyl groups present in the algal polysaccharides might be responsible for the reduction of gold ions. Various strains such as brown marine macroalga (M. Singh et al., 2013), *Tetraselmis kochinensis* (Senapati et al., 2012), *Sargassum muticum* (Namvar et al., 2015), *Chlorella vulgaris* (Annamalai & Nallamuthu, 2015) *Sargassum wightii Greville* (Singaravelu et al., 2007) have been investigated with regard to Au NPs synthesis. In most of the studies, it is concluded that the enzyme present in cell membrane is responsible for the reduction of gold ions and formation of Au NPs.

Later, algal extract has been used for the synthesis of gold nanoparticles using *Galaxaura elongate* at room temperature. Mechanistic analysis from FTIR indicates that nanoparticles are conjugated with algal compounds mainly via glutamic acid and hexadecanoic acid, which act as both reducing and stabilizing agents. The generated Au NPs showed better antibacterial activity against *P. aeruginosa, K. pneumoniae, Staphylococcus aureus,* and *E. coli,* respectively (Abdel-Raouf, Al-Enazi, & Ibraheem, 2017; Feng et al., 2000). Depending upon the protecting agents and nucleation process, sometimes nanoparticles develop anisotropic shapes. Recently, Venkatesan et al. (2014) reported that marine brown alga *Ecklonia cava* can produce cubical gold nanoparticles. Generated anisotropic gold nanoparticles exhibit interesting biomedical activity. In another mechanistic analysis, Rajathi et al. (2012) demonstrated that presence of hydroxyl groups in the form of diterpenoids inside algae caused the reduction of gold ions. In a different approach, B. Kumar et al., (2016) demonstrated the effect of ultrasound methods for the synthesis of Au NPs using *Calothrix algae.*

### 9.2.3 Fungi-Mediated Gold Nanoparticles

Fungi appear to be more promising for large-scale generation of nanoparticles owing to their simpler growing method in the laboratory and industrial scale (Sastry, Ahmad, Khan, & Kumar, 2003; Senapati, Ahmad, Khan, Sastry, & Kumar, 2005; Table 9.3). In general, fungi tolerate higher metal concentrations than bacteria and secrete abundant extracellular redox proteins to reduce soluble metal ions to their insoluble form and eventually to nanocrystals. Fungi harbor untapped biological diversity and may provide novel metal reductases for metal detoxification and bioreduction. A thorough understanding of the biosynthetic mechanism of Au NPs in fungi is needed to reduce the time of biosynthesis and to scale up the Au NP production process (Kitching, Ramani, & Marsili, 2015).

**Table 9.3** Synthesis of gold nanoparticles by fungi.

| Fungi species | Size (nm) | Shape | References |
|---|---|---|---|
| *Verticillium* sp. | 20 ± 8 | Spherical, Quasihexagonal | Mukherjee et al. (2001) |
| *Yarrowia lipolytica NCIM 3589* | 15 | Hexagonal, triangular | Agnihotri, Joshi, Kumar, Zinjarde, and Kulkarni (2009) |
| *Rhizopus oryzae* | 10 | — | Das, Das, and Guha (2009) |
| *Rhizopus stolonifer* | 1–5 | — | Binupriya, Sathishkumar, and Yun (2010) |
| *Aspergillus niger* | 10–20 | Polydispersed | Xie, Lee, Wang, and Ting (2007) |
| *Aspergillus niger* | 12.79 ± 5.61 | — | Bhambure, Bule, Shaligram, Kamat, and Singhal (2009) |
| *Fusarium oxysporum* | 20–40 | Spherical, triangular | Mukherjee et al. (2002) |
| *Colletotrichum* sp. | 20–40 | Decahedral and icosahedral | Shankar, Ahmad, Pasricha, and Sastry (2003) |
| *Fusarium semitectum* | 25 | — | Sawle et al. (2008) |
| *Neurospora crassa* | 32 | Spherical | Castro-Longoria, Vilchis-Nestor, and Avalos-Borja (2011) |
| *Helminthosporum solani* | 2–70 | Polydispersed | S. A. Kumar, Peter, and Nadeau (2008) |
| *Penicillium brevicompactum* | 10~50 | Spherical | Mishra et al. (2011) |
| *Trichoderma koningii* | 10–14 | Spheres | Maliszewska, Aniszkiewicz, and Sadowski (2009) |
| *Verticillium luteoalbum* | <10 | Spherical spheres and rods | Gericke and Pinches (2006b) |
| *Phanerochaete chrysosporium* | 10–100 | Spherical | Sanghi, Verma, and Puri (2011) |
| Extremophilic yeasts (isolated from acid mine drainage) | 30–100 | — | Arumugam and Berchmans (2011) |

Moreover, fungi yield a large number of proteins that are further helpful in synthesis of Au NPs. Fungi-mediated synthesis helps in production of nanoparticles having good monodispersity along with well-defined dimensions (Das & Marsili, 2010). Mukherjee et al. (2001), (2002) are pioneers in exploring the fungi as nano-factories, where they first used *Fusarium oxysporum* and *Verticillium* sp. for extracellular and intracellular synthesis of nanoparticles, respectively. This group reported the first extracellular synthesis of gold nanoparticles using a eukaryotic system such as fungi, which is much smaller and also monodisperse than the nanoparticles that are synthesized intracellularly with bacteria. Further, they have shown the intracellular synthesis of Au NPs using *Verticillium* sp. and it was presumed that bioreduction of $AuCl_4^-$ metal ions occurred by enzymes that are present on the cell walls of microorganisms. These types of gold nanoparticles can be utilized for catalysis purposes and can be used in synthesis of coatings for electronic applications as precursors (Alghuthaymi, Almoammar, Rai, Said-Galiev, & Abd-Elsalam, 2015; G. P. Kumar et al., 2007; Mukherjee et al., 2001).

Different shapes such as gold nanoplates have been reported by Shankar et al. (2004) using fungal extracts. Yeast like *Yarrowia lipolytica* and *Pichia jadinii* showed really good potential for the synthesis of Au NPs (Agnihotri et al., 2009; Gericke & Pinches, 2006b). Yeast strains are more advantageous than bacterial strains due to production of NPs in large amounts as yeasts are easier to maintain in laboratory environments, producing enzymes in large quantities and growth occurs rapidly engaging simple nutrients.

Based on the functional groups present on the surface of the nanoparticles, nanoparticles exhibit affinity toward some specific biomolecules. For example, when gold ions are exposed to *Rhizopus oryzae* a nanogold-bioconjugate (NGBC) is formed. The new conjugate exhibits excellent adsorption ability to various dissimilar organophosphorus pesticides. The treatment of NGBC upon microorganisms leads to the damage of the cell membrane, which further causes cell death. NGBC helps in removing pesticides and *E. coli* from polluted water (Das et al., 2009). Binupriya et al. explored a green synthesis process for the crystallization of soluble ionic silver and gold. Inactive cell filtration (ICF) of the filamentous fungus *Rhizopus stolonifera* confirmed as biomaterials having good potential to bioreduce the ionic silver and gold particles to nanosized crystalline particles at room temperature. Au NPs synthesized via ICF produced very small-sized particles, less than 5 nm (Binupriya et al., 2010).

Xie et al. (2007) reported the generation of gold nanocrystals consisting of various morphologies (nanowalls, spiral nanoplates, regular nanoplates, lamellar nanoagglomerates, spherical, and spherical nanoagglomerates) upon the variation of cellular contents and spent medium. The gold nanoparticles generated by *Phanerochaete chrysosporium* reveal plate-like structure formation with the basal plane (Sanghi et al., 2011). The experimental report suggested that the proteins present on the fungal cell walls and in fungal extract were primary biomolecules responsible for synthesis of gold nanocrystals. Furthermore, they have also reported the synthesis of gold nanoplates in presence of spent medium with help of an enzymatic reaction. The anisotropic structure formation was found to be kinetically controlled. Additionally, the shapes and sizes were dependent on environmental factors like pH and temperature of the solution. Another fungal strain *Aspergillus niger* produces stable extracellular gold nanoparticles while exposing the gold ions to the cell filtrate (Bhambure et al., 2009).

The mechanism of fungi-mediated gold nanoparticles synthesis is not clear yet. However, according to literature reports the biotransformation mechanisms of gold ions into gold nanoparticles by fungi can be described by two reduction routes: (a) binding of Au(III) on the cell wall through electrostatic interaction followed by reduction to gold nanoparticles and (b) transportation of Au(III) into the cytoplasm and reduction to form Au nanoparticles. For example, negatively charged gold chloride binds to the positively charged mycelia of *R. oryzae* through electrostatic interaction with phosphoproteins and then reduce to Au(I) species as a result of the high redox potential of Au(III) (Das, Dickinson, Lafir, Brougham, & Marsili, 2012).

However, Zhang et al. investigated the mechanism using proteomic analysis. It was observed that *Aureobasidium. pullulans, Fusarium* sp., and *F. oxysporum* produced intracellular gold nanoparticles. They speculated that probably intracellular reducing sugar and proteins of *A. pullulans, Fusarium* sp., and *F. oxysporum* were the active biomolecules involved in the biosynthesis of gold nanoparticles (Mukherjee et al., 2002).They identified the proteins from the extract of *Fusarium* sp. and molecular weights of three fungal proteins were found to be around100, 25, and 19 kDa, respectively, as identified by using the capillary LC/MS method. Identified proteins were found to be plasma membrane 3-glucan binding protein, ATPase, and glyceraldehyde-3-phosphate dehydrogenase, respectively (Zhang, He, Wang, & Yang, 2011).

Another class of fungi, i.e. endophytic fungi, known to infect living plant tissues is also capable of reducing the gold ions to produce Au NPs. When *Colletotrichum* was exposed to gold ions, extracellular rapid synthesis of Au NPs was found with variable sizes of nanoparticles. A plausible mechanism of synthesis was found to be the enzymes/polypeptides present in the fungal strain (Shankar et al., 2003).

Apart from gold nanoparticles, gold–silver alloy (Au–Ag) is also another important material in nanotechnology. Limited reports are available for the synthesis of Au–Ag alloy using fungal strain as reducing agents. Sawle et al. reported the formation of spherically shaped Au–Ag alloy and Au NPs using the aqueous extract of fungus *Fusarium semitectum* in Sawle et al. (2008). Another strain *Neurospora crassa* is also known to generate Au–Ag alloy (Castro-Longoria et al., 2011).

Kumar et al. have reported that plant pathogenic fungus *Helminthosporum solani* when incubated with chloroaurate ions can produce gold nanoparticles. The nanoparticles size was found to be in the range from 2 to 70 nm. It consists of mostly polydispersed nanoparticles; however, a good population is found to be equally sized triangles, rods, pyramids, stars, and pentagons shape (S. A. Kumar et al., 2008). Mishra et al. reported a synthesis of gold nanoparticles using an industrial strain *Penicillium brevicompactum*. In order to compare the production rate of Au NPs, they investigated both supernatant broth (SB) and live-cell filtrate (LCF) and the former method was observed to be relatively fast (Mishra et al., 2011). Similar methodology was adopted later using *Trichoderma koningii*. Herein, the cell-free extract was exploited for the nanoparticles synthesis, where gold nanospheres were obtained. Further, it was shown that size-dependent Au NPs could be separated by sucrose-based density gradient method (Gericke & Pinches, 2006b; Maliszewska et al., 2009).

In a mechanistic study, Sanghi et al. found that the different fungal proteins were responsible for the extracellular and intracellular synthesis of Au NPs, respectively, when *Phanerochaete chrysosporium* was incorporated with gold ions. It was noticed that the laccase enzyme was responsible for the extracellular synthesis of nanoparticles, whereas the ligninase enzyme was responsible for the intracellular synthesis of nanoparticles on the mycelium surface (Sanghi et al., 2011). Further detailed study revealed that the metal ions $AuCl^{-4}$ were trapped on the surface of fungal biomass by electrostatic interaction with their charged amino acid residues present in fungal enzymes in the cell wall of the fungal mycelia (Khandel & Shahi, 2018).

### 9.2.4 Yeast-Mediated Gold Nanoparticles

The presence of characteristic proteases or reductases leads to the reduction of gold salt into nanoparticles. In one study, P.S. Pimprikar et al. reported a nice arrangement of nanoplates and nanoparticles synthesized by the tropical marine yeast isolate *Y. lipolytica* NCIM 3589, when incubated with different concentrations of $HAuCl_4$. They established that depending upon the cell numbers and gold concentrations, shape of Au NPs could be tuned (Pimprikar, Joshi, Kumar, Zinjarde, & Kulkarni, 2009). Sometimes, external stabilizing agents can be used in the biological synthesis of nanoparticles. Au NPs can be synthesized using yeast species *Hansenula anomala* in the presence of amine-terminated polyamidopamine dendrimer, which is extracted from the same yeast. This method established that some characteristics like size, aggregation, and monodispersity of metal nanoparticles could be controlled by PAMAM dendrimers of different generations, especially in the case of gold nanoparticles (Arumugam & Berchmans, 2011). Extremophilic yeasts are also reported to reduce metal ions such as gold and silver toward the formation of nanoparticles (Mourato, Gadanho, Lino, & Tenreiro, 2011).

### 9.2.5 Mechanism Involved in Microbial Nanoparticles Synthesis

The specific mechanism of nanoparticle synthesis varies from one microbial strain to another strain. Nevertheless, the synthesis pattern follows a general principle, which involves the entrapment of metal ions into the microbial strain or on the surface of the microbial strain in the presence of an enzyme and followed by the reduction of metal ions to their corresponding NPs. Most probably, electrostatic attraction is responsible for the entrapment of metal ions mainly due to the negative charge of carboxylate group of enzymes present on the surface of the microbes. Then, metal ions are reduced by the enzymes to form nuclei of NPs synthesis.

In literature microbial gold nanoparticles synthesis and the mechanism of their synthesis by various bioreducing agents such as enzymes, peptides, proteins, electron shuttle quinones, and exopolysaccharides have been reported. The production of Au NPs essentially requires an enzyme (NADH-dependent reductase) and cell surface proteins, which help in transfer of electron from NADH, which further results in the bioreduction of gold ions into gold nanoparticles. During nanoparticles synthesis, electrons can be moved by means of low-molecular weight redox mediators like ubiquinol, NADH, or oxygen/superoxide or by direct association between c-type cytochromes, redox proteins, and the metal ion moving electrons across cell membranes to the surface (Gahlawat & Choudhury, 2019). The different cofactors like NADH or NADH-dependent enzymes are also released extracellularly. In the reduction process, the initial step is the transfer of an electron from NADH through an electron carrier (NADH-dependent reductase), while in next step electrons are transferred to gold ions ($Au^{3+}$) and finally they are reduced to elemental gold ($Au^0$) or Au NPs (Ali, Ali, Wang, Waseem, & Pan, 2019). Some authors have suggested the involvement of NADPH-dependent reductase enzyme in the reduction of $Au^{3+}$ to $Au^0$ and stabilization via capping molecules (Das & Marsili, 2010).

Choudhury et al. built a technique for the synthesis of gold nanoparticles utilizing pullulan exopolysaccharide (EPS) as reducing agent. The FTIR investigation clearly demonstrated that the essential structure of pullulan remained unaffected during synthesis of gold nanoparticles by pullulan polymer, and a-1,4 and a-1,6 linkages of polymer were also intact even after the reaction. It was estimated that the biomineralization of gold ions was because of the oxidation of side chain aliphatic alcoholic groups of pullulan molecule. In this way, it might be presumed that the free $CH_2OH$ groups of pure pullulan molecules were oxidized to $COO^-$ carboxyl group, while simultaneously there is reduction of $Au^{3+}$ to $Au^0$ resulting in the formation of gold nanoparticles (Choudhury, Malhotra, Bhattacharjee, & Prasad, 2014).

Also, the electrostatic interactions between metal cationic ions and anionic groups, such as carboxylic and phosphoric functional groups of EPS, have been mentioned to be a favorable position for the synthesis of metal nanoparticles. Among these functional groups, hydroxyl, carboxyl, phosphoric, hemiacetals, and amino end groups have been proposed for the reduction of metal ions from the precursor salts to obtain the respective nanoparticles (Emam & Ahmed, 2016). Additionally, hydroxyl groups have been known to have strong binding affinity with metal ions (Pandey, Goswami, & Nanda, 2012). Actually, during the reduction process involving hydroxyl, alcoholic, and aldehyde groups, oxidation of functional groups take place to their corresponding oxidized state, which is an important factor during synthesis of metal nanoparticles (Mata et al., 2009).

## 9.3 Applications of Microbial Gold Nanoparticles

### 9.3.1 Biosensing

Microbes-mediated gold nanoparticles are considered to be the best candidates for the application of biosensors due to their unique abilities of electron transferring and adsorptive capacity. One of the microbial Au NPs shows sensing activity against toxin A (TOA) of *Clostridium difficile.* This mechanism involves the self-assembly of thiolated single-stranded DNA with Nafion–thionine–GNPS-modified electrode via Au–thiol interaction, which leads to hybridization to form an aptamer–DNA duplex. The principle of the reported electrochemical biosensor for TOA is based upon AP probe and an enzymatic amplification by simply immobilizing thiolated CP on Nafion–thionine–GNPS-modified SPE with an Au–thiol interaction. The HRP-labeled AP (aptamer probe) was modified upon electrode surface by hybridizing to CP to produce an aptamer–DNA duplex (Figure 9.3a). In the absence of TOA, HRP fixed to the electrode helps in catalysis of the oxidation of thionine by hydrogen peroxide. Figure 9.3c shows the oxidation of HRP by electrochemical method when target TOA is added. The binding between TOA and aptamer leads to the formation of aptamer–TOA complex, replacing the aptamer–DNA complex leading to the detachment of the HRP-labeled AP from the electrode (Figure 9.3b). This leads to the reduction of the catalytic activity of HRP along with response of current. As the concentrations of thionine and hydrogen peroxide were constant, the decrease in the response of current was totally dependent on the concentration of TOA. These outcomes were used to design the biosensor for the detection of Toxin A, which exhibits good selectivity, stability, reproducibility, and accuracy. The limit of detection was found to be 1 nM for TOA (Luo, Liu, Xia, Xu, & Xie, 2014).

Biosynthesized Au NPs can be used as sensor for the detection of organophosphate pesticides like dichlorvos, dimethoate, and phorate. For example, Au NPs are derived from cell-free supernatant of marine *Staphylococcus lentus* exhibit this type of activity. Among the pesticides only dichlorvos shows the accumulation of the gold nanoparticles, which leads to the change in color from pink to blue with

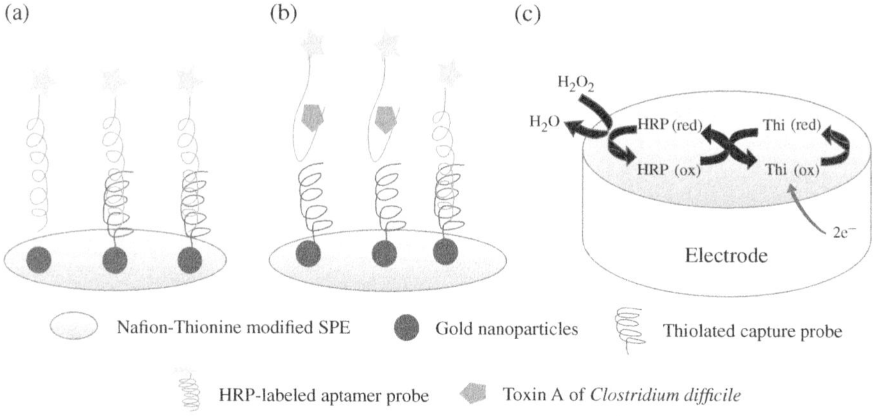


**Figure 9.3** Schematic representation of the aptasensor for TOA. (a) HRP-labeled aptamer probe hybridizes with the complementary part of the capture probe, (b) HRP-labeled aptamer probe recognizes toxin A, and (c) three enzymatic mechanisms based on HRP reactivity for $H_2O_2$. *Source*: Reprinted with permission from Malarkodi et al. (2013) Copyright © 2014 Elsevier.

a redshift in absorption spectra (Figure 9.4a). However, A control reaction mixture and lower concentration values of the pesticides appear pink without any color change and display a peak at 540 nm corresponding to SPR band of Au NPs. On the other hand, beyond a concentration of 2.5 μg of dichlorvos (5.0, 7.5, and 10.0), there is a change in color with gradual shift in absorption spectra, suggesting affinity of Au NPs toward pesticides. TEM results also confirm the difference in pre- and post-interaction of Au NPs with pesticide leading to the formation of agglomeration (Figure 9.4b, c). The results indicate that the synthesized Au NPs can be used for the colorimetric detection of dichlorvos based on color change and SPR band shifting of Au NPs (Apte et al., 2016).

### 9.3.2 Antibacterial Activity of Au NPs

There are many effective chemicals-based antimicrobial agents reported so far for pharmaceutical and biomedical applications. However, the major hurdle with the usage of chemical antimicrobial materials is multidrug resistance (MDR). To address this problem, there is a need to develop an alternative way to formulate new antimicrobial drugs. For decades, it has been observed that Au-based compounds have been showing antimicrobial activity for their growth-inhibitory ability against microorganisms. The cell wall of Gram-positive bacteria is composed of a thick peptidoglycan (PGN) layer, which consists of linear polysaccharide chains cross-linked by short peptides, thus forming more rigid structure, which

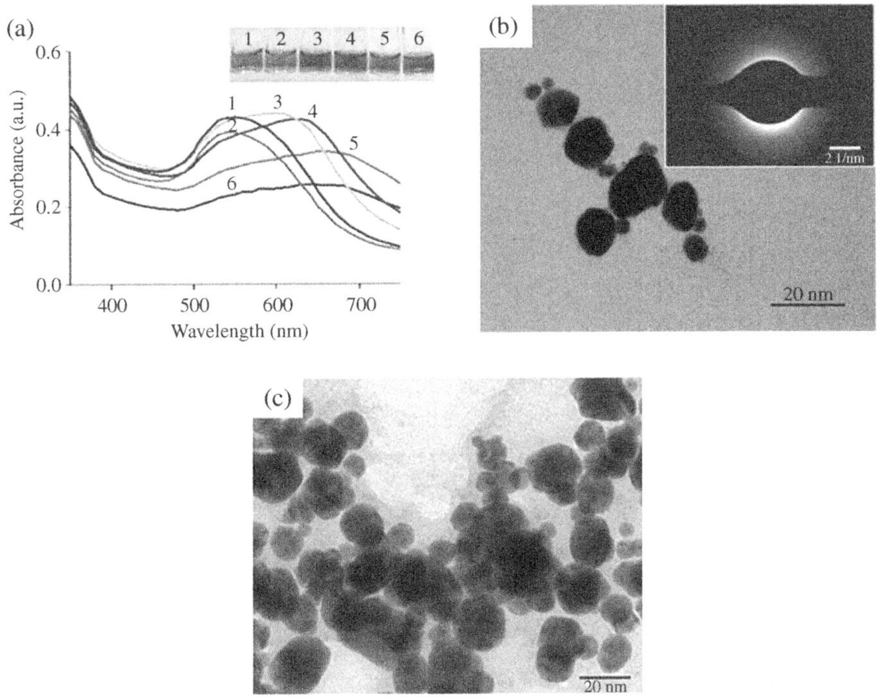


**Figure 9.4** (a) UV–vis spectra of Au NPs with different concentrations of dichlorvos for five minutes (lines and tubes 1–6: control, 2.5, 5.0, 10.0, 15.0, and 20 mg/l, respectively. (b) TEM image of Au NPs. (c) TEM image of AuNPs after interaction with dichlorvos (scale bar = 20 nm and 2.00 l/nm in figure and inset, respectively). *Source*: Reprinted with permission from Apte, Chaudhari, Vaidya, Kumar, and Zinjarde (2016) Copyright © 2016 Elsevier.

may prevent the action of Au NPs by blocking the process of gold ion infiltration into the cell and resides the gold ion in the cell wall area. Depending on the shape, size, and surface properties nanoparticles interact with the PGN layer of cell wall and the plasma membrane. They also bind to the cytoplasmic DNA of bacteria and prevent DNA replication and protein synthesis. Gram-negative bacteria, on the other hand, contain a thin PGN layer on the cell wall. The antimicrobial activity by nanoparticles on Gram-negative bacteria might be related to the pits formation on the cell wall of the bacteria, causing cell death. Gold nanoparticles derived from *P. fluorescens* 417 inhibiting *Coffea arabica* exhibit the bactericidal activity against *Pseudomonas aeruginosa* followed by *B. subtilis, S. aureus, E. coli,* and *K. pneumoniae* (Baker & Satish, 2015). In order to evaluate the antimicrobial effect, the zone of inhibition is measured. The results indicate higher zone of inhibition in the case of *P. aeruginosa* (MTCC 7903). Similar observations were recorded in broth dilution assay, where a decrease in OD (measured at 600 nm) was observed with an increase in concentration of gold nanoparticles. Bactericidal activity also was investigated by CFU plate method where gradual decrease in the colonies was found as the gold nanoparticles' concentration was increased. The result exhibits the highest activity against clinically important Gram-negative *P. aeruginosa*. In another study, the antimicrobial activity of Au NPs synthesized by *Pseudomonas veronii* AS 41G was evaluated via broth dilution assay against the methicillin-resistant bacterial strain *S. aureus* (MTCC 7443) and *E. coli* (MTCC 7443). In order to understand the mode of antimicrobial action of the nanoparticles, DNA was extracted from *S. aureus*. Then extracted DNA was treated with biosynthesized Au NPs and incubated at 37 °C. Strand breaking activity was noticed when observed in agarose gel in the case of nanoparticles treated DNA, in contrast, an intact band was clearly visible in the case of control (Baker & Satish, 2015). Gold nanoparticles synthesized by *Trichoderma viride* and *Hypocrea lixii* also show antibacterial activity against human and plant pathogens such as *Shigella sonnei*, *E. coli*, and *Pseudomonas syringae*, respectively (Mishra et al., 2014). Gold nanoparticles derived from *Deinococcus radiodurans* exhibit antibacterial activity against *S. aureus* and *E. coli*, and substantial antibacterial activity is observed with destruction of the cytoplasmic membrane. Inhibition in *E. coli* is found to be more significant than *S. aureus* (Li et al., 2016a).

Later, Jena et al. also showed the antimicrobial activity of Au NPs derived from *B. megaterium*. FTIR study indicates that a flavin-containing reductase domain of cytochrome P450 of *B. megaterium* is involved in the biosynthesis of nanoparticles. The extracellular formation of gold nanoparticles was confirmed by the use of bacterial cell-free extract. Further, the antibacterial properties were investigated by well diffusion process against *P. aeruginosa*, *S. aureus*, *Staphylococcus flexneri*, and *Micrococcus luteus*, respectively, with distinct zone of inhibition. Minimum inhibitory concentration (MIC) of Au NPs was determined by the broth dilution method against above strains and values were determined to be 15.10, 14.93, 15.33, and 16.75 μg/ml, respectively. The data indicate that these biosynthesized gold nanoparticles can be used against various multidrug-resistant bacteria (Jena, Das, Bosu, Suar, & Mandal, 2015).

Although the exact mechanism of antibacterial activity of Au NPs is still unknown, some of the researchers have proposed the possible mechanism. Cui et al. (2012) have proposed that the bactericidal activity in presence of Au NPs takes place because of inhibition of ATPase activities due to rupture of potential membrane gradient. Further, this leads to the decrease in the level of ATP and stops the binding between tRNA and subunit of ribosome. Taken together, it can be concluded that Au NPs can be used as antimicrobial agent as they do not possess risk to human health (Cui et al., 2012).

### 9.3.3 Anticancer Activity of Microbial Gold Nanoparticles

Au NPs are considered as novel and potential nanoparticles that can be explored in the field of biomedical applications such as anticancer agents. In order to explore the capability of Au NPs as anticancer agents, several research groups have been involved in the evaluation of cytotoxicity of the nanoparticles (Connor, Mwamuka, Gole, Murphy, & Wyatt, 2005; Patra, Banerjee, Chaudhuri, Lahiri, & Dasgupta, 2007). According to Pan et al. smaller size of gold nanoparticles induces more cytotoxicity in HeLa cells than the bigger size (Pan et al., 2007). Kumar et al. explored the cytotoxicity of microbial Au NPs against HEK293 cells and performed dose-dependent study. The study indicates that the nanoparticles do not exhibit any significant cytotoxicity against the tested cells. In contrast, Au NPs conjugated with doxorubicin [Dox] showed approximately 100-fold greater concentration than free Dox, which could be attributed to the enhanced solubility of Au NPs in presence of Dox (S. A. Kumar et al., 2008). In another study, Mishra et al. analyzed the cytotoxic effects of Au NPs obtained from the supernatant broth of *P. brevicompactum* KCCM 60390, against mouse mayo blast cancer $C_2C_{12}$ cells at different time intervals (24, 48, and 72 hours). Their study reveals that cytotoxicity of these nanoparticles is time-dependent and concentration-dependent against the cancer cells (Mishra et al., 2011).

The surface properties of Au NPs are also important for the induction of cytotoxicity in the cancer cells. Maliszewska et al. synthesized Au NPs by using cell-free extract of *T. koningii* and separated nanoparticles of smaller size (Au NPs-30) by sucrose gradient. Further, they compared the cytotoxicity of different concentrations of Au NPs, Au NPs-30 (contains 30% sucrose fraction), and CF (cell-free filtrate) against human colon cancer cell line LoVo and multidrug-resistant subline LoVo/DX. The cytotoxicity studies were done by colorimetric assay using SRB (sulphorhodamine B is a bright pink aminoxanthine dye) method. The assay showed that Au NPs were able to suppress proliferation of Lovo/DX cells more effectively than LoVo; this is attributed to the endocytosis caused by the proteins present in the cell filtrate (Maliszewska, 2013).

Kalpana et al. described in vitro; cytotoxicity of gold nanoparticles synthesized by using the cultural filtrate of *K. pneumoniae* against 3T3L1, H9c2, and HepG2 cell lines. Instead of cytotoxicity these synthesized Au NPs are found to be non-toxic and biocompatible. Instead of blocking cell proliferation they all promote cell proliferation of above cell lines at high concentration of 1000 μg/ml gold nanoparticles. Further, the cell proliferation was found to be dose-dependent.

The nontoxic nature of gold nanoparticles could be due to the large size and capping molecules, which help the cells to adhere to nanoparticles, and, thus, further lead to enhancement of cell viability instead of toxicity (Kalpana, Srikanth, Pichiah, Cha, & Lee, 2014).

Venkatesan et al. successfully used marine brown alga *E. cava* for the synthesis of Au NPs and investigated the cytotoxicity effect against HaCaT cells (human keratinocyte cell) line. However, Au NPs did not show any toxicity to HaCaT cell lines, suggesting the biocompatibility of generated Au NPs. Thus, they can be used safely in biomedical applications (Venkatesan et al., 2014). Singh et al. reported poly shaped gold nanoparticles synthesized using aqueous macroalgae extract *Padina gymnospora*. The cytotoxicity of the different concentrations of nanoparticles was examined by MTT assay, which showed $IC_{50}$ value 51.9 and 76.40 nM for Hep G2 and A549, respectively, which was further confirmed by LDH leakage. The LDH assay indicates that biomolecules capped Au NPs offers an outstanding platform for attachment of biological drugs for controlled and sustained release of drugs and can be used in in vivo; application (M. Singh et al., 2015). Rajesh Kumar explored the synthesis of Au NPs by using eco-friendly marine bacteria *Enterococcus* sp. Further, they evaluated in vitro; cytotoxicity activity against Hep-G2 and A549 cell lines at different concentrations of Au NPs such as 1, 10, 25, 50, and 100 μg, respectively. The results show significant cytotoxicity against the cancer cells. The anticancer activity of Au NPs probably originated from the interaction of gold atoms with the functional groups of intracellular proteins or the nitrogen bases and phosphate groups present in DNA, which is further confirmed by the suppression of abnormally expressed signaling proteins, such as Akt and Ras and cytokine (Rajesh Kumar, 2016).

The mechanism of anticancer activity of Au NPs is not clear. It is reported that there is increased oxidative stress, which further leads to mitochondrial damage and increased amount of ROS (reactive oxygen species) can cause DNA damage and apoptosis. However, further research work is required to understand the in-depth mechanism of anticancer activities of Au NPs on cancer cells (Pan et al., 2009).

## 9.4 Conclusion

The present chapter summarizes the biosynthesis of gold nanoparticles by microbes (bacteria, algae, fungi, and yeast) thought to be safe, clean, nonhazardous, and eco-friendly. Thus, it can be considered as alternative to various chemical methods of synthesis of gold nanoparticles. These microbes perform metabolic activities that secrete enzymes and proteins and they can be used as reducing agents and stabilizing agents for gold ions to their subsequent gold colloids of different sizes, shapes, and compositions. Thus, microbes mediated gold nanoparticles with ongoing efforts in improving particle synthesis efficiency and exploring their biomedical applications, biosensor, an antibacterial and catalytic activity which further implemented in the health care industry in recent years. These are the few future aspects for the generation of nanoparticles using green synthesis and yet many more are yet to be explored by the researchers.

## Acknowledgements

The authors gratefully acknowledge the financial support from Department of Atomic Energy (grant no. 37(2)/14/22/2015/BRNS), Department of Biotechnology (grant no. BT/PR10019/NNT/28/708/ 2013), and Department of Science & Technology (grant no. SR/NM/NS-1075/2013), Govt of India to DM. RKS thanks the Council of Scientific and Industrial Research (CSIR), Govt. of India, for providing Senior Research Fellowship (SRF) award (09/1035(0018)/2019-EMR-1).

## Take Home Message

- Gold nanoparticles are known as nontoxic substance and that is why they can be used in biomedical research.
- Because of their unique optical properties such as absorption and scattering properties, gold nanoparticles are used in many applications.
- Green synthesis such as microbial synthesis of gold nanoparticles is important for the applications of nanoparticles in biomedical research because traditional synthesis methods will produce chemically stabilized nanoparticles, whereas biosynthesis will lead to biomolecule-capped nanoparticles.
- Extracellular biosynthesis of nanoparticles always has advantage over intracellular synthesis because of easy downstream processing.
- In general, gold nanoparticles do not show any antimicrobial activity. However, depending on the surface properties, they may show some activity against some pathogenic strains.

## Questions and Answers

1 Why is microbial nanoparticles synthesis called green synthesis?
Ans. Traditional chemical synthesis has lot of disadvantages, which include the toxicity and hazardous nature of the reagents used in the synthesis. Furthermore, toxic chemicals adsorbed on the nanoparticles' surface restrict their use in the biomedical research. Green synthesis is considered as a cost-effective, safe, environmentally friendly sustainable method by which harsh conditions and/or toxic chemicals may be avoided. By using

microbes in nanoparticle synthesis, involvement of any toxicity or hazardous reagents can be avoided.

2 Why are gold nanoparticles red in color, whereas the metallic gold is yellow?
The characteristic optical property shown by gold nanoparticles is due to the oscillations of the conduction band of electrons present on the surface of nanoparticles. It happens based on the principle of surface plasmon resonance (SPR).

3 Why fungi produce large number of nanoparticles compared to bacteria?
Fungi secrete large number of enzymes that are responsible for the nanoparticle synthesis.

4 How can gold nanoparticles used as sensors?
The optical property of gold nanoparticles can be used as basis of sensor. When they interact with any external agent, there is a color change of the nanoparticles' solution. In this way, colorimetric sensor can be developed using gold nanoparticles.

5 What is surface plasmon resonance (SPR)?
SPR is the combined excitation of the conduction electrons in metal with the wavelength of irradiated light.

6 How can gold nanoparticles be characterized?
The formation of gold nanoparticles can be characterized mainly by UV–visible spectroscopy, transmission electron microscopy (TEM), and scanning electron microscopy (SEM).

7 Among intracellular synthesis and extracellular synthesis, which method is advantageous and why?
Extracellular synthesis of nanoparticles always has advantage because, here, nanoparticles can be separated easily from the biomass by filtration.

8 If absorbance of gold nanoparticles solution is shifted to higher wavelength with the addition of any external agent in the same solvent, what does it indicate?
If SPR band of gold nanoparticles solution is shifted to higher wavelength, this indicates that there is a possibility of change in size or shape of nanoparticles.

## References

Abdel-Raouf, N., Al-Enazi, N. M., & Ibraheem, I. B. (2017). Green biosynthesis of gold nanoparticles using *Galaxauraelongata* and characterization of their antibacterial activity. *Arabian Journal of Chemistry*, 10, S3029–S3039. doi:10.1016/j.arabjc.2013.11.044

Agnihotri, M., Joshi, S., Kumar, A. R., Zinjarde, S., & Kulkarni, S. (2009). Biosynthesis of gold nanoparticles by the tropical marine yeast *Yarrowialipolytica* NCIM 3589. *Materials Letters*, 63(15), 1231–1234. doi:10.1016/j.matlet.2009.02.042

Ahmad, A., Senapati, S., Khan, M. I., Kumar, R., Ramani, R., Srinivas, V., & Sastry, M. (2003a). Intracellular synthesis of gold nanoparticles by a novel alkalotolerant actinomycete, *Rhodococcus species. Nanotechnology*, 14(7), 824. doi:10.1088/0957-4484/14/7/323

Ahmad, A., Senapati, S., Khan, M. I., Kumar, R., & Sastry, M. (2003b). Extracellular biosynthesis of monodisperse gold nanoparticles by a novel extremophilic actinomycete, *Thermomonospora* sp. *Langmuir*, 19(8), 3550–3553. doi:10.1021/la026772l

Ahmed, S., Ahmad, M., Swami, B. L., & Ikram, S. (2016). A review on plants extract mediated synthesis of silver nanoparticles for antimicrobial applications: A green expertise. *Journal of Advanced Research*, 7(1), 17–28. doi:10.1016/j.jare.2015.02.007

Alghuthaymi, M. A., Almoammar, H., Rai, M., Said-Galiev, E., & Abd-Elsalam, K. A. (2015). Myco nanoparticles: Synthesis and their role in phytopathogens management. *Biotechnology & Biotechnological Equipment*, 29(2), 221–236. doi:10.1080/13102818.2015.1008194

Ali, J., Ali, N., Wang, L., Waseem, H., & Pan, G. (2019). Revisiting the mechanistic pathways for bacterial mediated synthesis of noble metal nanoparticles. *Journal of Microbiological Methods*, 159, 18–25. doi:10.1016/j.mimet.2019.02.010

Annamalai, J., & Nallamuthu, T. (2015). Characterization of biosynthesized gold nanoparticles from aqueous extract of *Chlorella vulgaris* and their anti-pathogenic properties. *Applied Nanoscience*, 5(5), 603–607. doi:10.1007/s13204-014-0353-y

Apte, M., Chaudhari, P., Vaidya, A., Kumar, A. R., & Zinjarde, S. (2016). Application of nanoparticles derived from marine *Staphylococcus lentus* in sensing dichlorvos and mercury ions. *Colloids and Surfaces A: Physicochemical and Engineering Aspects*, 501, 1–8. doi:10.1016/j.colsurfa.2016.04.055

Arumugam, P., & Berchmans, S. (2011). Synthesis of gold nanoparticles: An ecofriendly approach using *Hansenulaanomala. ACS Applied Materials & Interfaces*, 3(5), 1418–1425. doi:10.1021/am200443j

Baker, S., & Satish, S. (2015). Biosynthesis of gold nanoparticles by *Pseudomonas veronii* AS41G inhabiting *Annona squamosa* L. *Spectrochimica Acta Part A: Molecular and Biomolecular Spectroscopy*, 150, 691–695. doi:10.1016/j.saa.2015.05.080

Baker, S., Volova, T., Prudnikova, S. V., Satish, S., & Prasad, N. (2017). Nano agro particles emerging trends and future prospect in modern agriculture system. *Environmental Toxicology and Pharmacology*, 53, 10–17. doi:10.1016/j.etap.2017.04.012

Beveridge, T. J., & Murray, R. G. (1980). Sites of metal deposition in the cell wall of *Bacillus subtilis*. *Journal of Bacteriology*, 141(2), 876–887. PMID: 6767692

Bhambure, R., Bule, M., Shaligram, N., Kamat, M., & Singhal, R. (2009). Extracellular biosynthesis of gold nanoparticles using *Aspergillus niger* – its characterization and stability. *Chemical Engineering & Technology: Industrial Chemistry–Plant Equipment–Process Engineering–Biotechnology*, 32(7), 1036–1041. doi:10.1002/ceat.200800647

Binupriya, A. R., Sathishkumar, M., & Yun, S. I. (2010). Biocrystallization of silver and gold ions by inactive cell filtrate of *Rhizopus stolonifer*. *Colloids and Surfaces B: Biointerfaces*, 79(2), 531–534. doi:10.1016/j.colsurfb.2010.05.021

Castro, L., Blazquez, M. L., Gonzalez, F. G., & Ballester, A. (2014). Mechanism and applications of metal nanoparticles prepared by bio-mediated process. *Reviews in Advanced Sciences and Engineering*, 3(3), 199–216. doi:10.1166/rase.2014.1064

Castro-Longoria, E., Vilchis-Nestor, A. R., & Avalos-Borja, M. (2011). Biosynthesis of silver, gold and bimetallic nanoparticles using the filamentous fungus *Neurospora crassa*. *Colloids and Surfaces B: Biointerfaces*, 83(1), 42–48. doi:10.1016/j.colsurfb.2010.10.035

Choudhury, A. R., Malhotra, A., Bhattacharjee, P., & Prasad, G. S. (2014). Facile and rapid thermo-regulated biomineralization of gold by pullulan and study of its thermodynamic parameters. *Carbohydrate Polymers*, 106, 154–159. doi:10.1016/j.carbpol.2014.01.072

Connor, E. E., Mwamuka, J., Gole, A., Murphy, C. J., & Wyatt, M. D. (2005). Gold nanoparticles are taken up by human cells but do not cause acute cytotoxicity. *Small*, 1(3), 325–327. doi:10.1002/smll.200400093

Correa-Llanten, D. N., Muñoz-Ibacache, S. A., Castro, M. E., Munoz, P. A., & Blamey, J. M. (2013). Gold nanoparticles synthesized by *Geobacillus* sp. strain ID17 a thermophilic bacterium isolated from Deception Island, Antarctica. *Microbial Cell Factories*, 12(1), 75. doi:10.1186/1475-2859-12-75

Cui, Y., Zhao, Y., Tian, Y., Zhang, W., Lü, X., & Jiang, X. (2012). The molecular mechanism of action of bactericidal gold nanoparticles on *Escherichia coli*. *Biomaterials*, 33(7), 2327–2333. doi:10.1016/j.biomaterials.2011.11.057

Das, S. K., Das, A. R., & Guha, A. K. (2009). Gold nanoparticles: Microbial synthesis and application in water hygiene management. *Langmuir*, 25(14), 8192–8199. doi:10.1021/la900585p

Das, S. K., Dickinson, C., Lafir, F., Brougham, D. F., & Marsili, E. (2012). Synthesis, characterization and catalytic activity of gold nanoparticles biosynthesized with *Rhizopus oryzae* protein extract. *Green Chemistry*, 14(5), 1322–1334. doi:10.1039/C2GC16676C

Das, S. K., & Marsili, E. (2010). A green chemical approach for the synthesis of gold nanoparticles: Characterization and mechanistic aspect. *Reviews in Environmental Science and Bio/Technology*, 9(3), 199–204. doi:10.1007/s11157-010-9188-5

Dauthal, P., & Mukhopadhyay, M. (2016). Noble metal nanoparticles: Plant-mediated synthesis, mechanistic aspects of synthesis, and applications. *Industrial & Engineering Chemistry Research*, 55(36), 9557–9577. doi:10.1021/acs.iecr.6b00861

De Corte, S., Hennebel, T., Verschuere, S., Cuvelier, C., Verstraete, W., & Boon, N. (2011). Gold nanoparticle formation using *Shewanella oneidensis*: A fast biosorption and slow reduction process. *Journal of Chemical Technology & Biotechnology*, 86(4), 547–553. doi:10.1002/jctb.2549

Du, L., Jiang, H., Liu, X., & Wang, E. (2007). Biosynthesis of gold nanoparticles assisted by *Escherichia coli DH5α* and its application on direct electrochemistry of hemoglobin. *Electrochemistry Communications*, 9(5), 1165–1170. doi:10.1016/j.elecom.2007.01.007

Emam, H. E., & Ahmed, H. B. (2016). Polysaccharides templates for assembly of nanosilver. *Carbohydrate Polymers*, 135, 300–307. doi:10.1016/j.carbpol.2015.08.095

Feng, Q. L., Wu, J., Chen, G. Q., Cui, F. Z., Kim, T. N., & Kim, J. O. (2000). A mechanistic study of the antibacterial effect of silver ions on *Escherichia coli* and *Staphylococcus aureus*. *Journal of Biomedical Materials Research*, 52(4), 662–668. doi:10.1002/1097-4636(20001215)52:4%3C662::AID-JBM10%3E3.0.CO;23

French, S., Puddephatt, D., Habash, M., & Glasauer, S. (2013). The dynamic nature of bacterial surfaces: Implications for metal–membrane interaction. *Critical Reviews in Microbiology*, 39(2), 196–217. doi:10.3109/1040841X.2012.702098

Gahlawat, G., & Choudhury, A. R. (2019). A review on the biosynthesis of metal and metal salt nanoparticles by microbes. *RSC Advances*, 9(23), 12944–12967. doi:10.1039/C8RA10483B

Gao, C., Wei, D., Yang, H., Chen, T., & Yang, L. (2015). Nanotechnology for treating osteoporotic vertebral fractures. *International Journal of Nanomedicine*, 10, 5139. doi:10.2147/IJN.S85037

Gericke, M., & Pinches, A. (2006a). Biological synthesis of metal nanoparticles. *Hydrometallurgy*, 83(1–4), 132–140. doi:10.1016/j.hydromet.2006.03.019

Gericke, M., & Pinches, A. (2006b). Microbial production of gold nanoparticles. *Gold Bulletin*, 39(1), 22–28. doi:10.1007/BF03215529

Giljohann, D. A., Seferos, D. S., Daniel, W. L., Massich, M. D., Patel, P. C., & Mirkin, C. A. (2010). Gold nanoparticles for biology and medicine. *Angewandte Chemie International Edition*, 49(19), 3280–3294. doi:10.1002/anie.200904359

Gupta, A. K., Naregalkar, R. R., Vaidya, V. D., & Gupta, M. (2007). Recent advances on surface engineering of magnetic iron oxide nanoparticles and their biomedical applications. 2 (1) 23–39. doi: 10.2217/17435889.2.1.23

Haber, J. A., & Buhro, W. E. (1998). Kinetic instability of nanocrystalline aluminum prepared by chemical synthesis; facile room-temperature grain growth. *Journal of the American Chemical Society*, 120(42), 10847–10855. doi:10.1021/ja981972y

He, S., Guo, Z., Zhang, Y., Zhang, S., Wang, J., & Gu, N. (2007). Biosynthesis of gold nanoparticles using the bacteria *Rhodopseudomonascapsulata*. *Materials Letters*, 61(18), 3984–3987. doi:10.1016/j.matlet.2007.01.018

Howes, P. D., Chandrawati, R., & Stevens, M. M. (2014). Colloidal nanoparticles as advanced biological sensors. *Science*, 346(6205), 1247390. doi:10.1126/science.1247390

Husseiny, M. I., El-Aziz, M. A., Badr, Y., & Mahmoud, M. A. (2007). Biosynthesis of gold nanoparticles using *Pseudomonas aeruginosa*. *Spectrochimica Acta Part A: Molecular and Biomolecular Spectroscopy*, 67(3-4), 1003–1006. doi:10.1016/j.saa.2006.09.028

Iravani, S. (2011). Green synthesis of metal nanoparticles using plants. *Green Chemistry*, 13(10), 2638–2650. doi:10.1039/C1GC15386B

Jana, N. R., Gearheart, L., & Murphy, C. J. (2001). Wet chemical synthesis of high aspect ratio cylindrical gold nanorods. *The Journal of Physical Chemistry B*, 105(19), 4065–4067. doi:10.1021/jp0107964

Jena, S., Das, B., Bosu, R., Suar, M., & Mandal, D. (2015). Bacteria generated antibacterial gold nanoparticles and potential mechanistic insight. *Journal of Cluster Science*, 26(5), 1707–1721. doi:10.1007/s10876-015-0869-7

Kalishwaralal, K., Deepak, V., Pandian, S. R. K., Kottaisamy, M., BarathManiKanth, S., Kartikeyan, B., & Gurunathan, S. (2010). Biosynthesis of silver and gold nanoparticles using *Brevibacterium casei*. *Colloids and Surfaces B: Biointerfaces*, 77(2), 257–262. doi:10.1016/j.colsurfb.2010.02.007

Kalpana, D., Srikanth, K., Pichiah, P. T., Cha, Y. S., & Lee, Y. S. (2014). Synthesis, characterization and in vitro; cytotoxicity of gold nanoparticles using cultural filtrate of low shear modeled microgravity and normal gravity cultured *K. pneumoniae*. *Macromolecular Research*, 22(5), 487–493. doi:10.1007/s13233-014-2072-3

Khan, W., & Pethe, A. S. (2016). Extracellular synthesis of gold nano particles from *Pseudomonas stutzeri* (MTCC 8362) and its antimicrobial activity. *International Journal of Innovative Research in Science, Engineering and Technology*, 5(9), 16482–16489. doi:10.15680/IJIRSET.2016.0509134

Khandel, P., & Shahi, S. K. (2018). Mycogenic nanoparticles and their bio-prospective applications: Current status and future challenges. *Journal of Nanostructure in Chemistry*, 8(4), 369–391. doi:10.1007/s40097-018-0285-2

Kitching, M., Ramani, M., & Marsili, E. (2015). Fungal biosynthesis of gold nanoparticles: Mechanism and scale up. *Microbial Biotechnology*, 8(6), 904–917. doi:10.1111/1751-7915.12151

Konishi, Y. (2004). Microbial preparation of gold nanoparticles by anaerobic bacterium. *Transactions of the Materials Research Society of Japan*, 29, 2341–2343. NII Article ID (NAID)10017630539

Kuemmeth, F., Bolotin, K. I., Shi, S. F., & Ralph, D. C. (2008). Measurement of discrete energy-level spectra in individual chemically synthesized gold nanoparticles. *Nano Letters*, 8(12), 4506–4512. doi:10.1021/nl802473n

Kumar, B., Smita, K., Sánchez, E., Guerra, S., & Cumbal, L. (2016). Ecofriendly ultrasound-assisted rapid synthesis of gold nanoparticles using *Calothrix algae*. *Advances in Natural Sciences: Nanoscience and Nanotechnology*, 7(2), 025013. doi:10.1088/2043-6262/7/2/025013

Kumar, G. P., Shruthi, S., Vibha, B., Reddy, B. A., Kundu, T. K., & Narayana, C. (2007). Hot spots in Ag core–Au shell nanoparticles potent for surface-enhanced Raman scattering studies of biomolecules. *The Journal of Physical Chemistry C*, 111(11), 4388–4392. doi:10.1021/jp068253n

Kumar, S. A., Peter, Y. A., & Nadeau, J. L. (2008). Facile biosynthesis, separation and conjugation of gold nanoparticles to doxorubicin. *Nanotechnology*, 19(49), 495101. doi:10.1088/0957-4484/19/49/495101

Lee, Y., Choi, J. R., Lee, K. J., Stott, N. E., & Kim, D. (2008). Large-scale synthesis of copper nanoparticles by chemically controlled reduction for applications of inkjet-printed electronics. *Nanotechnology*, 19(41), 415604. doi:10.1088/0957-4484/19/41/415604

Lengke, M. F., Fleet, M. E., & Southam, G. (2006). Morphology of gold nanoparticles synthesized by filamentous cyanobacteria from gold (I)–thiosulfate and gold (III)–chloride complexes. *Langmuir*, 22(6), 2780–2787. doi:10.1021/la052652c

Li, J., Li, Q., Ma, X., Tian, B., Li, T., Yu, J., & Hua, Y. (2016a). Biosynthesis of gold nanoparticles by the extreme bacterium *Deinococcusradiodurans* and an evaluation of their antibacterial properties. *International Journal of Nanomedicine*, 11, 5931. doi:10.2147/IJN.S119618

Li, Y., Li, Y., Li, Q., Fan, X., Gao, J., & Luo, Y. (2016b). Rapid biosynthesis of gold nanoparticles by the extracellular secretion of *Bacillus niabensis* 45: Characterization and anti biofilm activity. *Journal of Chemistry*, 2016. doi:10.1155/2016/2781347

Ling, D., Lee, N., & Hyeon, T. (2015). Chemical synthesis and assembly of uniformly sized iron oxide nanoparticles for medical applications. *Accounts of Chemical Research*, 48(5), 1276–1285. doi:10.1021/acs.accounts.5b00038

Luo, P., Liu, Y., Xia, Y., Xu, H., & Xie, G. (2014). Aptamer biosensor for sensitive detection of toxin A of *Clostridium difficile* using gold nanoparticles synthesized by *Bacillus stearothermophilus*. *Biosensors and Bioelectronics*, 54, 217–221. doi:10.1016/j.bios.2013.11.013

Malarkodi, C., Rajesh Kumar, S., Vanaja, M., Paulkumar, K., Gnanajobitha, G., & Annadurai, G. (2013). Eco-friendly synthesis and characterization of gold nanoparticles using *Klebsiella pneumoniae*. *Journal of Nanostructure in Chemistry*, 3(1), 30. doi:10.1186/2193-8865-3-30

Maliszewska, I. (2013). Microbial mediated synthesis of gold nanoparticles: Preparation, characterization and cytotoxicity studies. *Digest Journal of Nanomaterials & Biostructures (DJNB)*, 8(3), 1123–1131.

Maliszewska, I., Aniszkiewicz, L., & Sadowski, Z. (2009). Biological synthesis of gold nanostructures using the extract of *Trichoderma koningii*. *Acta Physica Polonica, A.*, 116, S163–S165. doi:10.12693/APhysPolA.116.S-163

Mandal, D., Bolander, M. E., Mukhopadhyay, D., Sarkar, G., & Mukherjee, P. (2006). The use of microorganisms for the formation of metal nanoparticles and their application. *Applied Microbiology and Biotechnology*, 69(5), 485–492. doi:10.1007/s00253-005-0179-3

Mata, Y. N., Torres, E., Blazquez, M. L., Ballester, A., González, F. M. J. A., & Munoz, J. A. (2009). Gold (III) biosorption and bioreduction with the brown alga *Fucusvesiculosus*. *Journal of Hazardous Materials*, 166(2–3), 612–618. doi:10.1016/j.jhazmat.2008.11.064

Mishra, A., Kumari, M., Pandey, S., Chaudhry, V., Gupta, K. C., & Nautiyal, C. S. (2014). Biocatalytic and antimicrobial activities of gold nanoparticles synthesized by *Trichoderma* sp. *Bioresource Technology*, 166, 235–242. doi:10.1016/j.biortech.2014.04.085

Mishra, A., Tripathy, S. K., Wahab, R., Jeong, S. H., Hwang, I., Yang, Y. B., . . . Yun, S. I. (2011). Microbial synthesis of gold nanoparticles using the fungus *Penicillium brevicompactum* and their cytotoxic effects against mouse mayo blast cancer C 2 C 12 cells. *Applied Microbiology and Biotechnology*, 92(3), 617–630. doi:10.1007/s00253-011-3556-0

Mourato, A., Gadanho, M., Lino, A. R., & Tenreiro, R. (2011). Biosynthesis of crystalline silver and gold nanoparticles by extremophilic yeasts. *Bioinorganic Chemistry and Applications*, 2011, 1–8. doi:10.1155/2011/546074

Mukherjee, P., Ahmad, A., Mandal, D., Senapati, S., Sainkar, S. R., Khan, M. I., . . . Sastry, M. (2001). Bioreduction of $AuCl_4^-$ ions by the fungus *Verticillium* sp. and surface trapping of the gold nanoparticles formed. *Angewandte Chemie International Edition*, 40(19), 3585–3588. doi:10.1002/1521-3773(20011001)40:19

Mukherjee, P., Senapati, S., Mandal, D., Ahmad, A., Khan, M. I., Kumar, R., & Sastry, M. (2002). Extracellular synthesis of gold nanoparticles by the fungus *Fusarium oxysporum*. *ChemBioChem*, 3(5), 461463. doi:10.1002/14397633(20020503)3:5%3C461::AID-CBIC461%3E3.0.CO;2-X

Nadaf, N. Y., & Kanase, S. S. (2016). Biosynthesis of gold nanoparticles by *Bacillus marisflavi* and its potential in catalytic dye degradation. *Arabian Journal of Chemistry.*, 12(8), 4806–4814. doi:10.1016/j.arabjc.2016.09.020

Nair, B., & Pradeep, T. (2002). Coalescence of nanoclusters and formation of submicron crystallites assisted by *Lactobacillus strains*. *Crystal Growth & Design*, 2(4), 293–298. doi. doi:10.1021/cg0255164

Namvar, F., Azizi, S., Ahmad, M. B., Shameli, K., Mohamad, R., Mahdavi, M., & Tahir, P. M. (2015). Green synthesis and characterization of gold nanoparticles using the marine macroalgae *Sargassum muticum*. *Research on Chemical Intermediates*, 41(8), 5723–5730. doi:10.1007/s11164-014-1696-4

Naraginti, S., Kumari, P. L., Das, R. K., Sivakumar, A., Patil, S. H., & Andhalkar, V. V. (2016). Amelioration of excision wounds by topical application of green synthesized, formulated silver and gold nanoparticles in albino Wistar rats. *Materials Science and Engineering: C*, 62, 293–300. doi:10.1016/j.msec.2016.01.069

Narayanan, K. B., & Sakthivel, N. (2011). Synthesis and characterization of nano-gold composite using *Cylindrocladium floridanum* and its heterogeneous catalysis in the degradation of 4-nitrophenol. *Journal of Hazardous Materials*, 189(1–2), 519–525. doi:10.1016/j.jhazmat.2011.02.069

Pan, Y., Leifert, A., Ruau, D., Neuss, S., Bornemann, J., Schmid, G., . . . Jahnen-Dechent, W. (2009). Gold nanoparticles of diameter 1.4 nm trigger necrosis by oxidative stress and mitochondrial damage. *Small*, 5(18), 2067–2076. doi:10.1002/smll.200900466

Pan, Y., Neuss, S., Leifert, A., Fischler, M., Wen, F., Simon, U., . . . Jahnen-Dechent, W. (2007). Size-dependent cytotoxicity of gold nanoparticles. *Small*, 3(11), 1941–1949. doi:10.1002/smll.200700378

Pandey, S., Goswami, G. K., & Nanda, K. K. (2012). Green synthesis of biopolymer–silver nanoparticle nanocomposite: An optical sensor for ammonia detection. *International Journal of Biological Macromolecules*, 51(4), 583–589. doi:10.1016/j.ijbiomac.2012.06.033

Patra, H. K., Banerjee, S., Chaudhuri, U., Lahiri, P., & Dasgupta, A. K. (2007). Cell selective response to gold nanoparticles. *Nanomedicine: Nanotechnology, Biology and Medicine*, 3(2), 111–119. doi:10.1016/j.nano.2007.03.005

Peng, T., Zhao, D., Dai, K., Shi, W., & Hirao, K. (2005). Synthesis of titanium dioxide nanoparticles with mesoporous anatase wall and high photocatalytic activity. *The Journal of Physical Chemistry B*, 109(11), 4947–4952. doi:10.1021/jp044771r

Pimprikar, P. S., Joshi, S. S., Kumar, A. R., Zinjarde, S. S., & Kulkarni, S. K. (2009). Influence of biomass and gold salt concentration on nanoparticle synthesis by the tropical marine yeast *Yarrowia lipolytica* NCIM 3589. *Colloids and Surfaces B: Biointerfaces*, 74(1), 309–316. doi:10.1016/j.colsurfb.2009.07.040

Rajasree, S. R., & Suman, T. Y. (2012). Extracellular biosynthesis of gold nanoparticles using a gram negative bacterium *Pseudomonas fluorescens*. *Asian Pacific Journal of Tropical Disease*, 2, S796–S799. doi:10.1016/S2222-1808(12)60267-9

Rajathi, F. A. A., Parthiban, C., Kumar, V. G., & Anantharaman, P. (2012). Biosynthesis of antibacterial gold nanoparticles using brown alga, *Stoechospermum marginatum* (kützing). *Spectrochimica Acta Part A: Molecular and Biomolecular Spectroscopy*, 99, 166–173. doi:10.1016/j.saa.2012.08.081

Rajesh Kumar, S. (2016). Anticancer activity of eco-friendly gold nanoparticles against lung and liver cancer cells. *Journal of Genetic Engineering and Biotechnology*, 14(1), 195–202. doi:10.1016/j.jgeb.2016.05.007

Rajesh Kumar, S., Malarkodi, C., Gnanajobitha, G., Paulkumar, K., Vanaja, M., Kannan, C., & Annadurai, G. (2013). Seaweed-mediated synthesis of gold nanoparticles using *Turbinaria conoides* and its characterization. *Journal of Nanostructure in Chemistry*, 3(1), 44. doi:10.1186/2193-8865-3-44

Salunke, B. K., Sawant, S. S., Lee, S. I., & Kim, B. S. (2016). Microorganisms as efficient biosystem for the synthesis of metal nanoparticles: Current scenario and future possibilities. *World Journal of Microbiology and Biotechnology*, 32(5), 88. doi:10.1007/s11274-016-2044-1

Samanta, S., Singh, B. R., & Adholeya, A. (2017). Intracellular synthesis of gold nanoparticles using an ectomycorrhizal strain *EM-1083* of *Laccaria fraterna* and its nanoanti-quorum sensing potential against *Pseudomonas aeruginosa*. *Indian Journal of Microbiology*, 57(4), 448–460. doi:10.1007/s12088-017-0662-4

Sanghi, R., Verma, P., & Puri, S. (2011). Enzymatic formation of gold nanoparticles using *Phanerochaete chrysosporium*. *Advances in Chemical Engineering and Science*, 1(03), 154. doi:10.4236/aces.2011.13023

Sanvicens, N., & Marco, M. P. (2008). Multifunctional nanoparticles–properties and prospects for their use in human medicine. *Trends in Biotechnology*, 26(8), 425–433. doi:10.1016/j.tibtech.2008.04.005

Sastry, M., Ahmad, A., Khan, M. I., & Kumar, R. (2003). Biosynthesis of metal nanoparticles using fungi and actinomycete. *Current Science*, 85(2), 162–170. http://repository.ias.ac.in/47133/1/92_pub.pdf

Sawle, B. D., Salimath, B., Deshpande, R., Bedre, M. D., Prabhakar, B. K., & Venkataraman, A. (2008). Biosynthesis and stabilization of Au and Au–Ag alloy nanoparticles by fungus, *Fusarium semitectum*. *Science and Technology of Advanced Materials*, 9(3), 035012. doi:10.1088/1468-6996/9/3/035012

Senapati, S., Ahmad, A., Khan, M. I., Sastry, M., & Kumar, R. (2005). Extracellular biosynthesis of bimetallic Au–Ag alloy nanoparticles. *Small*, 1(5), 517–520. doi:10.1002/smll.200400053

Senapati, S., Syed, A., Moeez, S., Kumar, A., & Ahmad, A. (2012). Intracellular synthesis of gold nanoparticles using alga *Tetraselmis kochinensis*. *Materials Letters*, 79, 116–118. doi:10.1016/j.matlet.2012.04.009

Sengani, M., Grumezescu, A. M., & Rajeswari, V. D. (2017). Recent trends and methodologies in gold nanoparticle synthesis—A prospective review on drug delivery aspect. *OpenNano*, 2, 37–46. doi:10.1016/j.onano.2017.07.001

Shah, M., Fawcett, D., Sharma, S., Tripathy, S., & Poinern, G. (2015). Green synthesis of metallic nanoparticles via biological entities. *Materials*, 8(11), 7278–7308. doi:10.3390/ma8115377

Shankar, S. S., Ahmad, A., Pasricha, R., & Sastry, M. (2003). Bioreduction of chloroaurate ions by geranium leaves and its endophytic fungus yields gold nanoparticles of different shapes. *Journal of Materials Chemistry*, 13(7), 1822–1826. doi:10.1039/B303808B

Shankar, S. S., Rai, A., Ankamwar, B., Singh, A., Ahmad, A., & Sastry, M. (2004). Biological synthesis of triangular gold nanoprisms. *Nature Materials*, 3(7), 482. doi:10.1038/nmat1152

Shedbalkar, U., Singh, R., Wadhwani, S., Gaidhani, S., & Chopade, B. A. (2014). Microbial synthesis of gold nanoparticles: Current status and future prospects. *Advances in Colloid and Interface Science*, 209, 40–48. doi:10.1016/j.cis.2013.12.011

Shi, J., Votruba, A. R., Farokhzad, O. C., & Langer, R. (2010). Nanotechnology in drug delivery and tissue engineering: From discovery to applications. *Nano Letters*, 10(9), 3223–3230. doi:10.1021/nl102184c

Singaravelu, G., Arockiamary, J. S., Kumar, V. G., & Govindaraju, K. (2007). A novel extracellular synthesis of monodisperse gold nanoparticles using marine alga, *Sargassum wightii* Greville. *Colloids and Surfaces B: Biointerfaces*, 57(1), 97–101. doi:10.1016/j.colsurfb.2007.01.010

Singh, M., Kalaivani, R., Manikandan, S., Sangeetha, N., & Kumaraguru, A. K. (2013). Facile green synthesis of variable metallic gold nanoparticle using *Padina gymnospora*, a brown marine macroalga. *Applied Nanoscience*, 3(2), 145–151. doi:10.1007/s13204-012-0115-7

Singh, M., Saurav, K., Majouga, A., Kumari, M., Kumar, M., Manikandan, S., & Kumaraguru, A. K. (2015). The cytotoxicity and cellular stress by temperature-fabricated polyshaped gold nanoparticles using marine macroalgae, *Padina gymnospora*. *Biotechnology and Applied Biochemistry*, 62(3), 424–432. doi:10.1002/bab.1271

Singh, R. K., Mishra, S., Jena, S., Panigrahi, B., Das, B., Jayabalan, R., & Mandal, D. (2018). Rapid colorimetric sensing of gadolinium by EGCG-derived AgNPs: The development of a nanohybrid bioimaging probe. *Chemical Communications*, 54(32), 3981–3984. doi:10.1039/C8CC01777H

Singh, R. K., Panigrahi, B., Mishra, S., Das, B., Jayabalan, R., Parhi, P. K., & Mandal, D. (2018). pH triggered green synthesized silver nanoparticles toward selective colorimetric detection of kanamycin and hazardous sulfide ions. *Journal of Molecular Liquids*, 269, 269–277. doi:10.1016/j.molliq.2018.08.056

Southam, G., & Beveridge, T. J. (1994). The in vitro; formation of placer gold by bacteria. *Geochimica et Cosmochimica Acta*, 58(20), 4527–4530. doi:10.1016/0016-7037(94)90355-7

Sriram, M. I., Kalishwaralal, K., & Gurunathan, S. (2012). Biosynthesis of silver and gold nanoparticles using *Bacillus licheniformis*. In *Nanoparticles in Biology and Medicine* (pp. 33–43). Totowa, NJ: Humana Press. doi:10.1007/978-1-61779-953-2_3

Sun, Y., & Xia, Y. (2002). Shape-controlled synthesis of gold and silver nanoparticles. *Science*, 298(5601), 2176–2179. doi:10.1126/science.1077229

Tanzil, A. H., Sultana, S. T., Saunders, S. R., Dohnalkova, A. C., Shi, L., Davenport, E., . . . Beyenal, H. (2016). Production of gold nanoparticles by electrode-respiring *Geobacter sulfurreducens* biofilms. *Enzyme and Microbial Technology*, 95, 69–75. doi:10.1016/j.enzmictec.2016.07.012

Thirumurugan, A., Ramachandran, S., Tomy, N. A., Jiflin, G. J., & Rajagomathi, G. (2012). Biological synthesis of gold nanoparticles by *Bacillus subtilis* and evaluation of increased antimicrobial activity against clinical isolates. *Korean Journal of Chemical Engineering*, 29(12), 1761–1765. doi:10.1007/s11814-012-0055-7

Venkatesan, J., Manivasagan, P., Kim, S. K., Kirthi, A. V., Marimuthu, S., & Rahuman, A. A. (2014). Marine algae-mediated synthesis of gold nanoparticles using a novel *Ecklonia cava*. *Bioprocess and Biosystems Engineering*, 37(8), 1591–1597. doi:10.1007/s00449-014-1131-7

Wadhwani, S. A., Shedbalkar, U. U., Singh, R., Vashisth, P., Pruthi, V., & Chopade, B. A. (2016). Kinetics of synthesis of gold nanoparticles by *Acinetobacter* sp. *SW30* isolated from environment. *Indian Journal of Microbiology*, 56(4), 439–444. doi:10.1007/s12088-016-0598-0

Wang, E. C., & Wang, A. Z. (2013). Nanoparticles and their applications in cell and molecular biology. *Integrative Biology*, 6(1), 9–26. doi:10.1002/anie.200904359

Wen, L., Lin, Z., Gu, P., Zhou, J., Yao, B., Chen, G., & Fu, J. (2009). Extracellular biosynthesis of monodispersed gold nanoparticles by a SAM capping route. *Journal of Nanoparticle Research*, 11(2), 279–288. doi:10.1007/s11051-008-9378-z

Xie, J., Lee, J. Y., Wang, D. I., & Ting, Y. P. (2007). High-yield synthesis of complex gold nanostructures in a fungal system. *The Journal of Physical Chemistry C*, 111(45), 16858–16865. doi:10.1021/jp0752668

Yun, S. H., & Kwok, S. J. (2017). Light in diagnosis, therapy and surgery. *Nature Biomedical Engineering*, 1(1), 0008. doi:10.1038/s41551-016-0008

Zhang, X., He, X., Wang, K., & Yang, X. (2011). Different active biomolecules involved in biosynthesis of gold nanoparticles by three fungus species. *Journal of Biomedical Nanotechnology*, 7(2), 245–254.

# Glossary

**Intracellular synthesis:** This is the mode of nanoparticle synthesis where particles are formed inside the cells. Here, microbes become cultured depending on the nanoparticles.

**Extracellular synthesis:** Here, nanoparticles are formed outside the cellular matrix. By filtration nanoparticles can be separated from the reaction mixture.

**Sol–gel method:** The method is used for preparing specially metal oxide glasses and ceramics through hydrolysis of a chemical precursor that passes through solution state first, then a gel state before being dehydrated to a glass or ceramic.

**Citrate–gel process:** Citrate process is almost comparable to sol–gel method, small molecules can be converted to solid materials. In this process, metal salts or alkoxides are incorporated into a citric acid solution.

**Laser ablation synthesis:** It is a commonly used method to obtain colloidal solution of nanoparticles. In this method, nanoparticles are formed during the condensation of a plasma plume generated by laser ablation of a bulk metal plate, which is inserted in a liquid solution.

**Plasma synthesis:** Here, synthesis occurs through evaporation of the bulk metal precursor followed by the condensation.

**Pyrolysis:** This is a chemical decomposition method of organic materials at higher temperatures in absence of oxygen. This process takes place under high temperature above 430 °C and under pressure.

**Electrochemical synthesis:** This method uses electrical energy to drive chemical change; it controls reagent activity by varying electrode potential.

**Catalyst:** A substance that increases the rate of a chemical reaction without itself undergoing any permanent chemical change.

**Sensor:** A device that detects or measures a physical property and records, indicates, or otherwise responds to it.

**Antibacterial activity:** Defined as a collective term for all active principles (agents) that inhibit the growth of bacteria, prevent the formation of microbial colonies, and may destroy microorganisms.

**Ultrasonication:** The act of applying frequencies (>20 kHz) to agitate particles in a sample, for various purposes such as the extraction of multiple compounds from plants, microalgae, and seaweeds.

**Detoxification:** The physiological or medicinal removal of toxic substances from a living organism, including the human body, which is mainly carried out by the liver.

**Surface plasmon resonance (SPR):** It is a physical process that can occur when plane-polarized light hits a thin metal film under total internal reflection conditions.

**Electromagnetic spectrum:** The range of wavelengths or frequencies over which electromagnetic radiation extends.

**Biosorption:** It is a physiochemical process that occurs naturally in certain biomass that allows it to passively concentrate and bind contaminants onto its cellular structure.

**Redox potential:** The measure of the tendency of a chemical species to acquire electrons from or lose electrons to an electrode and thereby be reduced or oxidized, respectively.

**Nucleation:** This is the first step of any thermodynamic phase or new structure that is developed through self-assembly or self-organization.

**Anisotropy:** The property of a substance because of which it shows various physical properties such as absorbance, refractive index, conductivity, tensile strength, etc. along different molecular axes. It is usually observed in crystals.

**Reductase:** An enzyme that promotes the chemical reduction of a specified substance.

**Endophyte:** It is a endosymbiont, often a bacterium or fungus, that lives within a plant for at least part of its life cycle without causing apparent disease.

**Mycelium:** The vegetative part of a fungus, consisting of a network of fine white filaments (hyphae).

**Protease:** An enzyme that breaks down proteins and peptides.

**Reactive oxygen species (ROS):** One type of unstable molecule that contains oxygen and that easily reacts with other molecules in a cell. In the cells it causes damage to DNA, RNA, and proteins, and may cause cell death.

**Peptidoglycan:** A substance forming the cell walls of many bacteria, consisting of glycosaminoglycan chains interlinked with short peptides.

# 10

# Nano-Bio Interactions and Their Practical Implications in Agriculture

*Achintya N. Bezbaruah*[1] *and Ann-Marie Fortuna*[2]

[1]*Department of Civil and Environmental Engineering, North Dakota State University, Fargo, ND, USA*
[2]*United States Department of Agriculture, Grazingland Research Laboratory, Agricultural Research Service, El Reno, OK, USA*

## 10.1 Introduction

Engineered nanomaterials (ENMs, size ≤100 nm at least in one dimension) are increasingly being studied for use in agriculture as fertilizers and pesticides. The most commonly studied nanomaterials for agricultural applications are iron, iron oxides, zinc oxide, silver, gold, copper oxide, titanium dioxide, silica, aluminum oxide, carbon nanotubes, and graphene (Sabir, Arshad, & Chaudhari, 2014). This chapter is devoted to the discussion of some of the most relevant findings in the area of nanomaterial applications in agriculture and how the different components of a soil system interact with the ENMs. A holistic overview is presented here without specific nanomaterials being extensively covered. The potential areas for further research have been identified in this chapter.

*Importance of plant–soil–microbial interactions:* The microbial community serves as a critical driver of nutrient cycling within the soil, plant, water, and atmospheric continuum of agricultural production systems. Microorganisms also play a critical role in the formation and function of soil organic matter (SOM), the organic component of soil that consists of plant material, soil organisms, and actively decomposing as well as stable organic matter (humus) (Weil & Brady, 2017). Soil typically has a large and diverse population of microorganisms that include bacteria, archaea, fungi, algae, and protozoa. The nutrient supplying capacity of soil is continuously managed and augmented by microorganisms that contribute to plant nutrition by way of fixation, solubilization, and/or transformation in the soil matrix. Plant growth-promoting soil

*Microbial Interactions at Nanobiotechnology Interfaces: Molecular Mechanisms and Applications*, First Edition. Edited by R. Navanietha Krishnaraj and Rajesh K. Sani.

microorganisms enhance the bioavailability of macronutrients (measured in g/kg) like nitrogen (N), phosphorus (P), and potassium (K), secondary nutrients (g/kg–mg/kg) like calcium (Ca), magnesium (Mg), sulfur (S), and micronutrients (mg/kg) like zinc (Zn), iron (Fe), manganese (Mn), boron (B), chlorine (Cl), copper (Cu), molybdenum (Mo), cobalt (Co), vanadium (V), sodium (Na), and silicon (Si) to crops. Microorganisms also influence structural nutrients, which include carbon (C) via mineralization and carbon fixation, hydrogen (H) by changing the pH, and oxygen (O) by energy transformations that result in changes in redox potentials. Microbes also contribute to the creation and maintenance of soil structure fostering root growth and crop production. Through an effective maintenance of the soil biogeochemical cycles (Figure 10.1) and suppression of ill-effects of pathogens and natural and anthropogenic toxins, the microbes literally act as the managers of soil health (Fierer, 2017; Fukami, Cerezini, & Hungria, 2018; Lindstroem & Mousavi, 2019; Puri, Padda, & Chanway, 2020; Sammauria, Kumawat, Kumawat, Singh, & Jatwa, 2020; Tien, Gaskins, & Hubbell, 1979). Sustaining healthy and diverse populations of soil microorganisms is vital to meeting the nutrient demands of our agricultural crops. A consortium of microorganisms can help in maintaining soil's usable capacity for future generations (Sammauria et al., 2020).

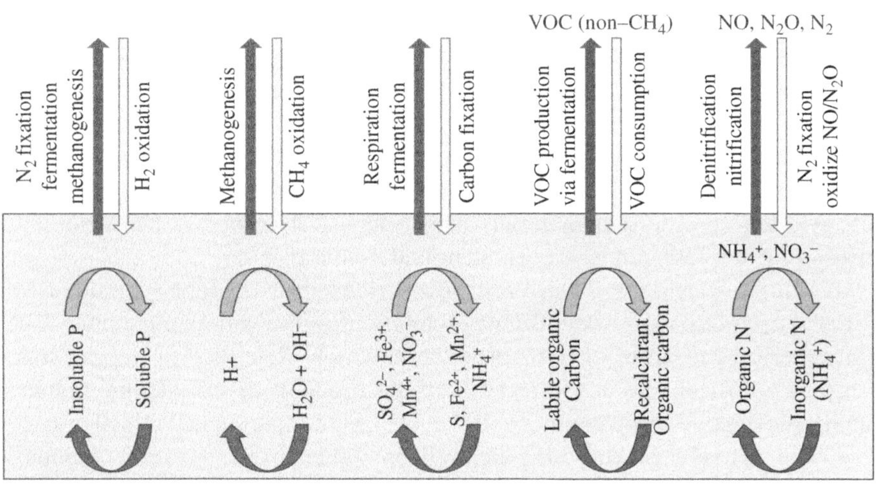


**Figure 10.1** Major soil microorganisms modulate biogeochemical processes. The key microbial processes that occur within soil regulate soil acidity, and N, P, and the availability of other nutrients. The microorganisms also consume soil organic carbon. Note: The volatile organic compounds (VOCs) include acetone, methanol, formaldehyde, isoprene, and other low-molecular weight organic compounds. *Source*: Figure modified after Fierer, (2017).

*Soil and microbial health:* Soil functions include sustaining biological diversity, activity, and productivity, regulating water and solute flow, filtering, buffering, degrading organic and inorganic materials, storing and cycling nutrients and carbon, and providing physical stability and support (Gupta, Neate, & Leonard, 1997). Increased and unscientific applications of agrochemicals have resulted in inhibition of microbial growth as well as reductions in diversity leading to the deterioration of soil health (Sammauria et al., 2020), the outcome of which leads to reduced agricultural production as well as the potential to pollute water bodies and contribute to genetic mutations of ubiquitous soil microbial populations and/or their gradual reduction due to this and other deleterious conditions. These changes in soil health will make our agri-production system susceptible and such instability may consequently lead to serious adverse effects on the global economy and public health.

### 10.1.1 Agriculturally Beneficial Soil Microorganisms

Plant growth-promoting microorganisms (PGPMs) or agriculturally beneficial microorganisms (ABMs) are important to our agri-production system (Achari & Kowshik, 2018). Different populations of PGPMs fix inorganic forms of N and C from the atmosphere, and convert these inorganic forms to organic forms that are bioavailable to themselves and a range of other organisms. These include symbiotic N-fixing rhizobium and free-living rhizobium some of which fix both N and C. Other PGPMs solubilize macro- and micronutrients increasing nutrient bioavailability, improving food quality via biofortification, and producing phyto-stimulants that modulate plant hormone levels (Asghari, Khademian, & Sedaghati, 2020; Fukami et al., 2018; Herrera, Novotna, Ortiz, Soto, & Arriagada, 2020; Mehmood et al., 2020; Tien et al., 1979). These microbes also play a role in biological control of soilborne plant pathogens by producing antibiotics and enzymes that lyse pathogen cell walls. Also, microbe-produced siderophores reduce the bioavailability of micronutrients to pathogens by chelation of iron (Fe) and making the nutrient available to the PGPM and crop plants. These and similar activities improve plant health (Azabou et al., 2020; Moya et al., 2020; Zhang & Zhuang, 2020). PGPMs are known to compete with plant pathogens, thus, reducing the potential for infections. PGPMs stimulate induced systemic resistance (ISR) in plants through the release of compounds such as ethylene that prepare plants to respond more quickly to plant pathogen infections (Glick, 2012).

Additionally, PGPMs moderate the effects of environmental stresses that include temperature, drought, flooding, and presence of toxic compounds such as metals and organic contaminants (Glick, 2012). Several mechanisms enable PGPMs to reduce growth inhibition in plants resulting from the abovementioned stresses. Examples include the production of the enzyme 1-aminocyclopropane-1-carboxylate (ACC) deaminase that reduces plant growth inhibition resulting from plant stress-induced production of ethylene, growth hormones such as auxins like indole-3-acetic acid (IAA), cytokinins that promote cell division in plants, antioxidants that degrade reactive oxygen species (ROS), and osmoprotectors such as trehalose, a nonreducing disaccharide and exopolysaccharides both of which decrease plant damage from draught (i.e. extreme osmotic stress) (Mishra, Singh, & Arora, 2017).

## 10.2 Engineered Nanomaterials and Agriculture

### 10.2.1 Pathways for ENM to Soil

Engineered nanomaterials (ENMs) enter soils through applications of nano-based fertilizers, insecticides/pesticides, and soil amendments. The primary uncontrolled (unmanaged) pathways for ENMs' entry to soil are via the application of biosolids (50% land applied but 1% is applied as a fertilizer to cropland (USEPA (United States Environmental Protection Agency), 2020)) from municipal wastewater treatment plants, use of partially treated or untreated municipal wastewater in agricultural fields, and waste streams from manufacturing units where nanomaterials are used (Fadeel et al., 2018; Fijalkowski, Rorat, Grobelak, & Kacprzak, 2017; Mortimer, Li, Wang, & Holden, 2020; Sinha, Das, Bezbaruah, & Fortuna, 2018).

### 10.2.2 Fate of ENMs in Soil

The fate of both inorganic and organic ENMs is dependent upon the interaction of properties specific to an ENM and that of a given soil type. There have been efforts to represent the various processes involved in soil–ENM interactions that determine the behaviors of the ENMs in soil matrices (Figure 10.2). The properties that determine the interactions and fate of ENMs in soil are soil mineralogy, organic matter quantity and quality (degree of humification), structure (aggregate stability), pH, salinity, and oxidation state (Cornelis, Hund-Rinke, Kuhlbusch, Van den Brink, & Nickel, 2014). The types of soil minerals present in a soil, their shrink/swell potential and charge are all dependent on the type of parent material of the soil (i.e. minerals from which the soil was formed) and the degree to which the soil has weathered (Weil & Brady, 2017). Soil colloids contain silicate clays (structured aluminosilicate minerals), humic substances, and aluminum (Al) and iron (Fe) oxides, hydroxides, oxyhydroxides, sesquioxides and amorphous gels (Weil & Brady, 2017).

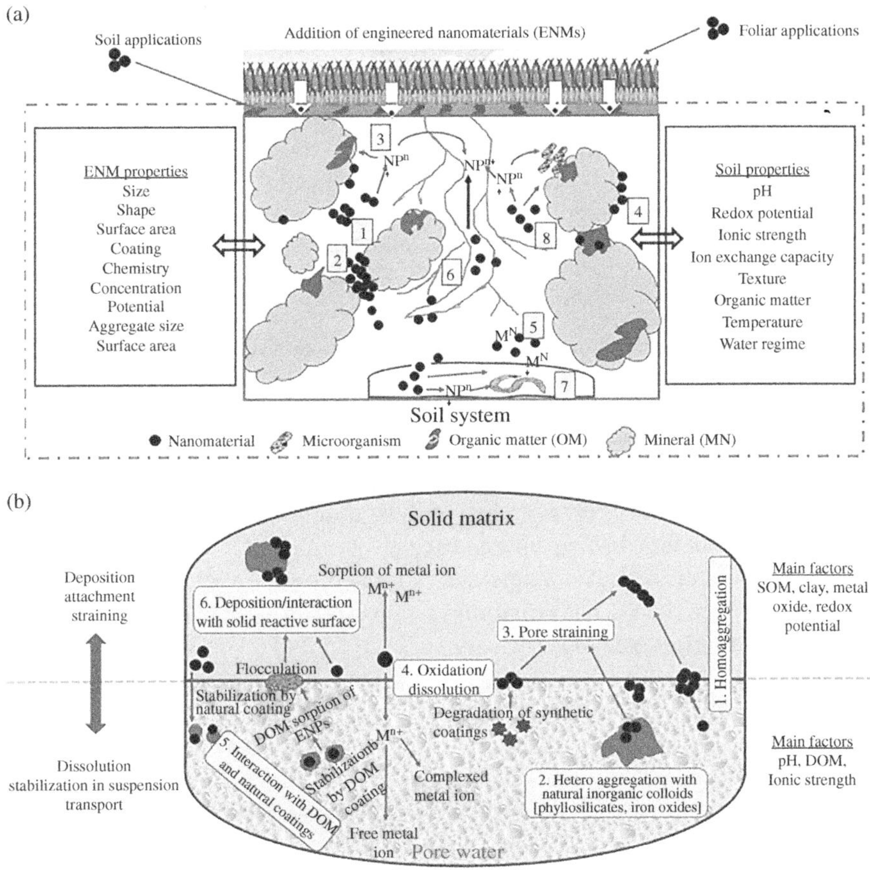


**Figure 10.2** Representations of possible processes and factors that determine fate and transport of ENMs in soil matrix. (a) A soil system with plants present. Reading keys: 1. Agglomeration; 2. straining; 3. dissolution and release of free ionic metals from metal-based nanomaterials ($NP^{n+}$); 4. sorption onto soil particles; 5. electrostatic stabilization of nanomaterials with major cation ($M^{N+}$) present in solid and aqueous phases; 6. uptake and translocation by plant roots; 7. uptake by invertebrates likes earthworms and nematodes; and 8. uptake by soil microorganisms. Figure modified after de Santiago-Martin et al. (2016). (b) Soil matrix and pore water subsystems. *Source*: Figure modified after Rodrigues et al. (2016).

Agricultural systems typically require aerobic conditions. Optimum oxygen ($O_2$) concentrations for aerobic processes range from the atmospheric level (21%) to ~5%. Soil has 10- to100-fold more $CO_2$ than the atmosphere. Soil $O_2$ is a measure of soil redox potential (i.e. oxidation-reduction potential or ORP). Increased $O_2$ concentrations result in a loss of electrons, which is equal to positive millivolts (i.e. oxidation). Loss of $O_2$ results in a gain in electrons, which is equal to negative millivolts (i.e. reduction). Soil redox conditions range from highly oxidized (+900 mV) to highly reduced (−300 mV). Nitrate ($NO_3^-$) is reduced to $N_2O$ and $N_2$ gases by denitrifiers at 250 mV. Manganese ($Mn^{4+}$) is reduced to $Mn^{2+}$ by manganese-reducing bacteria at 200 mV. Iron ($Fe^{3+}$) is reduced to $Fe^{2+}$ by iron-reducing bacteria at 100 mV. Sulfate ($SO_4^{2-}$) is reduced to $S^{2-}$ by sulfate-reducing bacteria at −150 mV. Carbon dioxide ($CO_2$) is reduced to methane ($CH_4$) by methanogens at −400 mV. This range of redox conditions determines the transformations, forms, and bioavailability of non-nano and nano forms of N, Mn, Fe, S, and C (Figure 10.1, also see Fierer (2017)).

Soil's cation exchange capacity (CEC) is a result of the mineral structure of clays, sheets of crystalline silica and alumina. One tetrahedral silica sheet and one octahedral alumina sheet is a 1:1 clay. One octahedral sheet sandwiched between two tetrahedral sheets is a 2:1 clay. Clay minerals (<2 μm) provide the majority of the cation (positively charged ion) exchange sites and a small amount of pH-dependent charge comes from amorphous Fe and Al oxide coatings (Weil & Brady, 2017). Mineral-associated anion (negatively charged ion) exchange capacity (AEC) results in negative charge and is located on the broken edges of clay alumina and silica sheets. The remaining cation exchange capacity (CEC) and the majority of the anion exchange capacity are supplied by soil organic matter (Weil & Brady, 2017). A soil's CEC and AEC determine the amount of exchangeable cations (+) and anions (−) in the soil and, thus, determine the different forms of naturally occurring ions and ENMs that can be held in the soil.

Formation of soil colloids and aggregates occur when environmental variables interact with soil water, plant roots, soil microorganisms, inorganic binding agents such as hydroxides and polyvalent cations and organic substances of plant and microbial origins. The interactions of ENMs with soil constituents determine the ENMs' bioavailability and/or toxicity in part. Inorganic binding agents contribute to the aggregation of primary particles (sand, silt, and clay) into colloids (particles < 0.001 mm), microaggregates (05–0.21 mm) and macroaggregate (size classes of >2.00, 0.50–2.00, and 0.21–0.50 mm), all of which can have variable shape and charge. Due to their small size, colloids are most likely to interact with and result in the transport of ENMs. Aluminosilicates, iron oxides, and calcium carbonates are important for the production of stable microaggregates that are created through largely chemical processes. Aluminum (Al) and other metallic ions, such as iron (Fe), magnesium (Mn), and calcium (Ca), are adsorbed on to the surface of amorphous silicates, creating a stable silicate coating. This increases the surface reactivity of layered silicate clays. Highly weathered silicate clays and modified clays form the most stable aggregates, especially in the presence of organic matter. In addition, Al has the capacity to bind humus into Al–humus complexes, resulting in strong stabilization of organic matter. Additional primary binding agents in soils are iron oxides that bind electrostatically with positively charged oxides and negatively charged clay minerals. Oxides of any kind coat mineral surfaces forming bridges between primary and secondary particles. Oxides on mineral surfaces bind and adsorb organic materials, contributing to the formation of organo–mineral complexes. The formation of clay–polyvalent cation–organic matter complexes is critical for the stabilization of microaggregates. Polyvalent cations such as $Ca^{2+}$ reduce repulsive forces between negatively charged clay particles, facilitate flocculation, and strengthen the aggregates. Stable Ca-organic matter in aggregates can represent >20% of the soil organic carbon content. Application of calcium carbonate alters the soil pH that contributes to macroaggregation, stimulates microbial activity, and reacts with compounds that contain N-linked oligosaccharides and iron, facilitating their binding to soil particles (Clough & Skjemstad, 2000). These processes may play major roles when ENMs are introduced to the soil system.

Organic binding agents that aid in the formation of micro- and macroaggregates include polysaccharides, proteins, lipids, and humic materials that bind clays and other primary particles into aggregates. These organic substances originate from microbial cells and plant roots or they are the by-products of organic matter decomposition. Binding occurs through van der Waals and weak hydrogen bonds between charged groups and the siloxane oxygen on clay surfaces. Adsorption of these binding agents onto clay surfaces is affected by the chemical composition and structural complexity of the macromolecule, the surface charge of the clay mineral, pH (charge varies with pH), and the presence of polyvalent cations. Stable macroaggregates often contain fungal hyphae and fine plant roots, illustrating the importance of these materials for aggregation (Clough & Skjemstad, 2000).

### 10.2.3 Chemical Interactions of ENM in Soil

ENMs enter the soil via the applications of nano-based fertilizers, insecticides/pesticides, or soil amendments (Figure 10.2). Upon entering the soil, nanoparticles can heteroaggregate (aggregate with dissimilar colloids), react with soil constituents, and are most often adsorbed to soil colloids or dissolved organic matter (DOM) (Cornelis et al., 2014). Homoaggregation (ENMs aggregating with each other) can also be expected depending on the pH and ionic strength in the soil matrix. Soil colloids, ENMs, and soil colloids with adsorbed ENMs are transported in soil pore water when ion concentrations in pore solution drop and/or a hydrodynamic event (i.e. rainfall, irrigation) occurs. Adsorption of phosphate, amphiphilic proteins (attracted to both water and fat or lipid molecules), and humic acids aid in dispersing ENMs in soil pores (Cornelis et al., 2014). High dispersion leads to added mobility of the nanomaterials. Lower redox conditions (limited $O_2$) will result in a reduction in iron oxides that bind colloid particles together and that will lead to increased mobility of ENMs, colloids, or colloid-adsorbed ENMs (Cornelis et al., 2014).

### 10.2.4 Mechanisms Controlling Heteroaggregation

Aggregation of ENMs with soil colloids occurs when the charges on each particle are opposite in nature. Typically, ENMs are negatively charged and soil colloids are positively charged (Cornelis et al., 2014). Similar to iron oxides, ENMs and natural colloids contain pH-dependent variable charges. Such particles have what is known as a point-of-zero-charge (PZC), which is the pH at which the absorbent's surface has no charge. The aggregation forces are particularly favorable for the ENMs and colloids to interact when the pH is between the ZPC and the point at which the two sets of particles have the greatest positive and negatively charged surfaces (Cornelis et al., 2014). The mobility of any molecule including a heteroaggregate is dependent upon its molecular size and charge (a pH-dependent parameter) as well as the viscosity and electrical field in the fluid through which the molecule is moving. Therefore, as the size of a soil colloids increases, the potential for heteroaggregation with an ENM decreases (Cornelis et al., 2014). Smaller particles in the range of <1 μm (0.001 mm) have high diffusion coefficients and are highly mobile, and their mobility can be predicted by Brownian diffusion that is controlled by temperature and particle size (Cornelis et al., 2014). Deposition of colloids and ENMs is controlled by Brownian diffusion, direct interception, and gravitational sedimentation. Deposition reduces the mobility of ENMs and often prevents heteroaggregation. The larger the particle the more likely it is to be strained by size exclusion while some smaller particles may be trapped in discontinuous pores in soil aggregates (Cornelis et al., 2014). These mechanisms are critical in determining the mobility and fate of ENMs in terrestrial soil and agricultural soils in particular.

### 10.2.5 Mobility of Colloids and ENMs in Soil

ENMs adsorb (or coat themselves) with dissolved organic matter (DOM) such as humic and fulvic acids. This occurs via three mechanisms, namely ligand exchange with surface hydroxyl groups, hydrophilic interactions (mixes with water), and hydrophobic interactions (repels water) with aromatic groups (Cornelis et al., 2014). Humic acids and $Ca^{2+}$ stabilize ENM heteroaggregation and may lead to reduced ENM mobility in soil. Carbon-based ENMs act similar to soil organic matter (SOM) adsorbing onto soil minerals. Higher pH creates greater negative charge on the DOM favoring adsorption by positively charged ENMs. In contrast, lower pH favors positive charges repelling negatively charged

ENMs. As DOM concentration increases, adsorption of DOM by ENMs also increases. At this point, the adsorbed DOM dominates the surface properties of the ENMs (even if they are surface coated) and creates negatively charged surfaces that reduce heteroaggregation and deposition. Such conditions lead to higher mobility of ENMs in soil (Cornelis et al., 2014).

### 10.2.6 Nanoagriculture

Nanotechnology is a promising tool for enhanced and sustainable agricultural production. Some ENMs have shown unique physiochemical properties like aiding plant growth-promoting microorganisms (PGPMs) for plant growth and increased stress tolerance (Adisa et al., 2019; He & Deng, 2019; Kah, Tufenkji, & White, 2019; Rai et al., 2018; Rastogi et al., 2019; Rodriguez-Gonzalez, Terashima, & Fujishima, 2019; Shang et al., 2019; Zhao et al., 2020). ENMs have been reported to positively impact seed germination (Almeelbi & Bezbaruah, 2014; Kim, Kim, & Chung, 2020; Zhu et al., 2019), root, shoot, and biomass growth (Acharya, Jayaprakasha, Crosby, Jifon, & Patil, 2019; Almeelbi & Bezbaruah, 2014; Jacob, Borchardt, Navaratnam, Otte, & Bezbaruah, 2013; Kim et al., 2020; Zhu et al., 2019), and crop yield (Kumari et al., 2019; Zahedi, Karimi, & da Silva, 2020). Almeelbi and Bezbaruah (2014) have demonstrated that iron nanomaterials can not only be used as a phosphate delivery vehicle for *Spinacia oleracea* (spinach) growth but also to fortify the edible part of the fresh crop with iron (Fe). The use of nanotechnologies for control of plant pests and diseases is a relatively recent but promising area that continues to evolve with numerous potential field applicable technologies (Antonoglou et al., 2018; Kah et al., 2019; Kah, Beulke, Tiede, & Hofmann, 2013; Kah & Hofmann, 2014; Kumar et al., 2019; 2020; Rai et al., 2018; Shang et al., 2019). Foliar and soil applications of nanomaterials have been tried to augment photosynthesis in plants in an effort to improve crop production (Alidoust & Isoda, 2013; Cao et al., 2017; Faizan, Faraz, Yusuf, Khan, & Hayat, 2018; Liu, Yue, Wang, & Xing, 2019; Miranda-Villagomez et al., 2019; Swift, Oliver, Galan, & Whitney, 2019). Researchers and industry experts are strategizing as to how to design specific nanomaterials for agricultural needs while ensuring the sustainability of agriculture practices (Zhao et al., 2020).

Fortuna (2019) and (Sinha et al., 2018) determined that low concentrations (1–5 mg/l) of nanoscale zero-valent iron (NZVI) particles were made bioavailable to a PGPM (*Pseudomonas fluorescens*) in nutrient broth solution (Figure 10.3). It is known that some microorganisms including *P. fluorescens* produce siderophores to chelate insoluble forms of Fe that are then rendered bioavailable to the microorganisms as well as plants in some instances. Siderophore production is an energy-intensive process and its production is an indication that concentrations of bioavailable iron are insufficient in the growth medium. Iron is typically limited

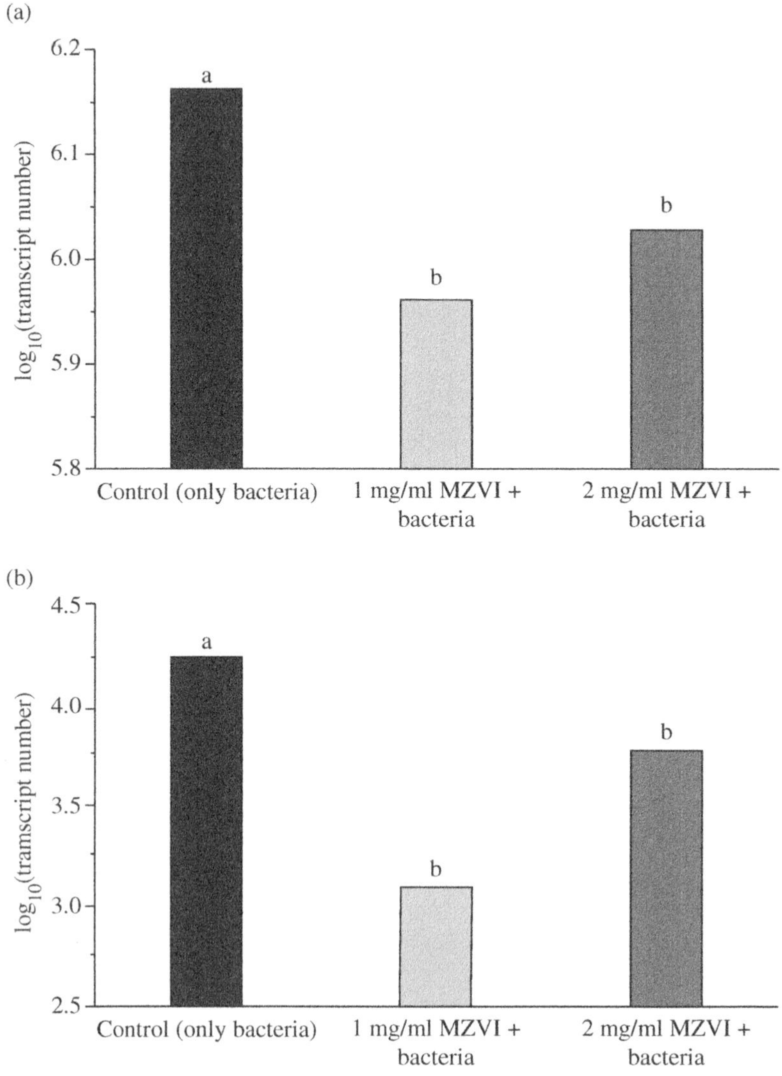


**Figure 10.3** Bioavailability of iron (Fe) from NZVI and MZVI. mRNA expression levels determined by qRT-PCR that represent (a) the bacterioferritin-associated ferredoxin gene (PFL 4858), and (b) *pvdS* gene (PFL 4190). PFL 4858 is responsible for mobilization and storage of Fe from bacterioferritin B, and PFL 4190 is a cluster of genes that regulate siderophore production. The control is the *P. fluorescens* culture with no MZVI addition. The figures do not show any NZVI treatment as there was no measurable gene expression when 1 g/l NZVI was used even through there was healthy microbial growth and higher concentrations of NZVI were toxic to *P. fluorescens* and resulted in no or insignificant bacteria growth. The genes were not expressed in the *P. fluorescens* with 1 g/l NZVI treatment because iron was easily bioavailable from the ENMs used in this treatment. *Source*: Figure modified after Sinha et al. (2018).

and siderophore production rarely ceases. The above team equated siderophore production by the bacterium with lowered bioavailability of Fe. Their research was one of the few experiments to compare a nano-sized micronutrient (NZVI) with that of a microscale nutrient (microscale zero-valent iron or MZVI). MZVI application does not produce reactive oxygen species (ROS) while NZVI applications do. Therefore, additions of both NZVI and MZVI were monitored to determine their effects on *P. fluorescens* viability, growth, and production of siderophores. The team reported that Fe species derived from low concentrations (1 g/l) of NZVI applications were bioavailable and could be utilized by *P. fluorescens* without the production of siderophores (Figure 10.3). However, higher concentrations of NZVI were toxic to the bacterium. MZVI particles were not toxic to the bacterium but the Fe from the microsized particles was not readily bioavailable requiring siderophore production by *P. fluorescens* for the acquisition of the Fe. The team hypothesized that NZVI application as a fertilizer may lead to better utilization of the iron by plants.

*Nanofertilizers:* One of the most promising applications of green nanotechnologies is the adoption of nano-based macronutrient (N, P, K, Ca, Mg, S) and some micronutrient (Fe, Mn, Zn, Cu) fertilizers to replace or supplement conventional fertilizer applications to crops. The concept behind the technology is that nutrient use efficiency (NUE) will be improved in part due to the greater surface area of nano-sized nutrients and in some instances, the potential of nano-sized materials to penetrate roots may lead to lower fertilizer demand (Adisa et al., 2019). The effect of root penetration by nanoparticles requires further investigation to verify that penetration does not compromise root function (Chichiricco & Poma, 2015). In addition, nanocoatings on N fertilizers such as "nano-enabled urea-coated zeolite chips and urea-modified hydroxyapatite" containing both N and P may prove to be more efficient than current slow-release N fertilizers such as polymer-coated urea, methylene urea, sulfur-coated urea, and polymer-coated sulfur-coated urea (Adisa et al., 2019; Liu & Lal, 2014) These technologies are designed to reduce N runoff/leaching and denitrification, and most importantly, to time N availability close to crop uptake. Combining nano-chitosan in a composite containing NPK fertilizer has been shown to have a similar effect as the above mentioned coatings leading to greater NUE and has been shown to shorten crop life cycle as well as significantly increase yield (Abdel-Aziz, Hasaneen, & Omer, 2016; Adisa et al., 2019). Chitosan is a naturally occurring sugar contained in the exoskeletons of insects and shellfish. Transport of nutrients such as Ca in the nano form has been shown to change the flow of the nutrient in some plants allowing Ca to be transported through the phloem of groundnuts and enhanced root development (Adisa et al., 2019; Deepa et al., 2015).

Research and development of micronutrient nanofertilizers have advanced rapidly and have potential to positively impact nutrient acquisition of metals by plants. Particularly, nutrients such as Fe, Zn, Mn, and Cu that are often not bioavailable to crop plants. Iron deficiencies are prevalent particularly in calcareous soils with high pH. Nano-humic Fe fertilizers have been shown to release bioavailable Fe continuously, increasing biomass and Fe content in soybeans (*Glycine max*) grown in pots containing soil (Cieschi et al., 2019). Almeelbi and Bezbaruah (2014) have demonstrated for the first time that nanoscale zero-valent iron (NZVI) can be used as a phosphate delivery vehicle for *Spinacia oleracea* (spinach) and *Selenastrum capricornutum* (a microalga) and fortify the edible part of the fresh crop with Fe. They reported up to three- to fourfold increase in biomass growth (in spinach and the alga) and up to 11-fold increase in Fe content in the spinach leaves due the application of NZVI as the sole source of Fe. Similar to nano-Fe fertilizers, nano-Mn fertilizers have been shown to have higher reactivity and provide greater continuous availability of soluble Mn relative to micro-Mn fertilizers in a greenhouse experiment in which wheat (*Triticum aestivum* L.) was grown in soil (Dimkpa et al., 2018). Less desirable effects of nano-Mn included higher Mn concentrations in shoot and grain, decreased plant available soil N, and higher concentrations of soil and shoot P. Zinc nanofertilizer significantly enhanced pearl millet (*Pennisetum americanum* L.) yield, shoot and root length, chlorophyll content, total soluble leaf protein, plant biomass, and enzyme activities of phosphatase, phytase, and dehydrogenase (Tarafdar, Raliya, Mahawar, & Rathore, 2014). Zinc and phosphorus typically have an antagonistic effect on each other reducing uptake of each other and uptake of other nutrients (Weil & Brady, 2017). In contrast, nano-Zn increases utilization of N, B, Fe, and Mg (Weil & Brady, 2017). Other researchers have reported similar positive effects of nutrient utilization when a number of crop plants were fertilized with nano-Zn (Adisa et al., 2019). Nano-copper (Cu) has also been shown to have positive effects on plant growth and enzyme activities, and both positive and negative effects on nutrient utilization in crops (Adisa et al., 2019; Weil & Brady, 2017).

### 10.2.7 Nanopesticides

Some plant micronutrients such as Cu and Zn have been used as biocides along with sulfides, silver (Ag), and heavy metals/metalloids such as mercury (Hg), cadmium (Cd), arsenic (As), and lead (Pb) as far back as ancient Sumeria, Egypt, China, Greece, and Rome (Jentsch, 2020). Bordeaux mix (a mixer of copper sulfate [$CuSO_4$] and slaked lime ($Ca[OH]_2$) were used as a fungicide and Paris Green (copper acetoarsenite, $C_4H_6As_6Cu_4O_{16}$) as an insecticide and fungicide (also as rodenticide) in the 1860s. While arsenates, lead, and mercury-based pesticides were largely phased out by the end of the twentieth century, copper and zinc are still widely used in many countries. Nanopesticide applications are expected, like their nanofertilizer counterparts, to be as effective or more effective than their non-nano counterparts (Adisa et al., 2019). Currently, several different types of nano-Zn and nano-Cu have been found to be highly effective in eliminating both fungal and bacterial pathogens, as has nano-chitosan alone or as a composite with nano-Cu or nano-Ag (Adisa et al., 2019). Silver (Ag) in nano and non-nano forms is a known biocide that binds to proteins on bacterial plasma membranes. Much of the published literature equates nanopesticides' effectiveness "with improved solubility, increased adhesion and absorption to plant leaves, controlled and responsive release of active ingredients that can be engineered for targeted delivery, enhanced bioavailability and biodegradability, and improved stability of active ingredients in the environment" (Adisa et al., 2019).

### 10.2.8 ENMs and Agriculturally Beneficial Microorganisms

While there are numerous studies that explore the effects of ENMs on plant growth and microorganisms in general (Mortimer et al., 2020), there are limited reports addressing the interactive effects of PGPM/ABM and ENMs. Fortuna et al., (2019) (Sinha et al., 2018) used nanoscale zero-valent iron (NZVI) particles to study their impact on the growth of a PGPM (*P. fluorescens*) and the bacterium's ease of Fe uptake (Figure 10.3). In a recent publication, Mortimer et al. (2020) highlighted the need for additional research to determine the future ramifications of ENMs on global food security.

Mortimer et al. (2020) reported the use of nanomaterials to assess the role of the plant itself (Soybean *(Glycine Max)*) in regulating the ENM–bacteria interactions. They used a group of emerging ENMs, which included nanoceria (i.e. $CeO_2$ nanoparticles), multi-walled carbon nanotubes (MWCNT), graphene nanoplatelets (GNP or graphene sheets), and carbon black (CB). These ENMs are of relevance because of their potential direct use in agricultural soils. Further, the carbon nanomaterials are used as matrices for the development of nanohybrids that are designed for delivery of plant macronutrients, micronutrients, and growth stimulants. *Bradyrhizobium diazoefficiens*, a symbiotic N-fixing bacteria in soybeans, was used to evaluate the role of soybean root exudates on ENM–bacteria interactions, specifically the effects ENMs on *B. diazoefficiens* growth and gene expression. Higher concentrations (50 mg/l) of all ENMs with the exception of graphene sheets inhibited bacterial growth. This was likely due to the interference of plant-bacterial signaling leading to downregulation of genes related to transport of RNA that facilitates nodulation signaling, and that resulted in reduced nodulation competitiveness *of B. diazoefficiens*. However, lower concentrations (0.1–10 mg/l) of ENMs and root exudates added as separate treatments stimulated global transcriptomic responses in the bacteria. Thus, lower doses of ENMs mimicked the effects of root exudates. However, the presence of low concentrations of

ENMs inhibited root exudate gene regulation expression resulting in reduced plant signaling, reduced infection of roots by rhizobium, and lower root nodule counts. This research indicated that the physical properties of ENMs, not chemical, affected the bacteria as evidenced by the number of differently regulated genes (see data related to MWCNT and $CeO_2$ ENMs in Figure 10.4. Also see Figure 10.5).

The results reported by Mortimer et al. (2020) are promising in a sense because the detrimental effects of ENMs appear to go down with dose with plant exudates present in the growth media. Typically, plant exudates are present in soil playing a major role in overall health and growth of the plant. However, more research is needed to determine if ENMs in soil interferes with plant growth particularly if ENMs persist in soil. Xenobiotic compounds often mimic biotic compounds that are used in metabolic pathways without providing microorganisms the benefit of obtaining energy. ENMs may induce a similar effect in plants by mimicking signal molecules in individual microorganisms that regulate a specific gene function such as motility or signaling in a biofilm (a colony of microorganisms that grow together and are surrounded by the extra extracellular polymeric substances they produce). Mortimer et al. (2020) verified that the low presence of ENMs (0.1–10 mg/l) increased the activity of flagellar genes that enhance the motility of the rhizobium. These genes aid a bacterium in its free-living stage when it is in search of a symbiotic host plant. Root exudates that are released from plant roots signal the presence of a host. Thus, the ENMs' mimicking host signal molecules that are not adjacent

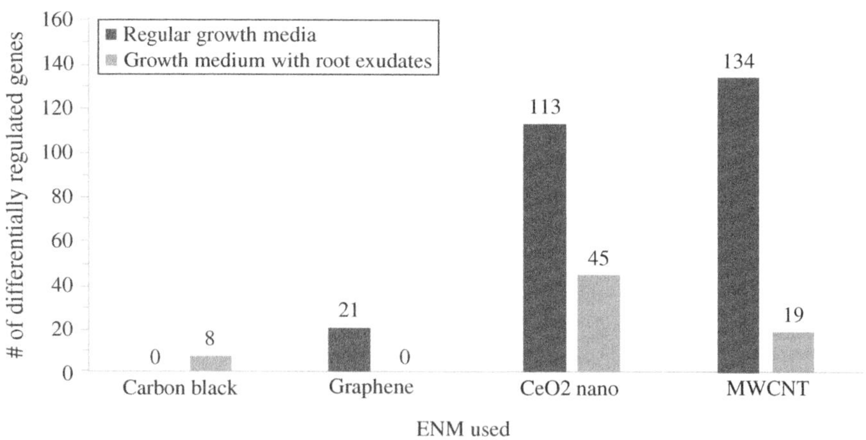


**Figure 10.4** Exposure of *B. diazoefficiens* to ENMs resulted in differentially regulated genes in the bacteria. The presence of root exudates in the growth medium reduced total numbers of genes that were differentially regulated. *Source*: Figure modified after Mortimer et al. (2020).

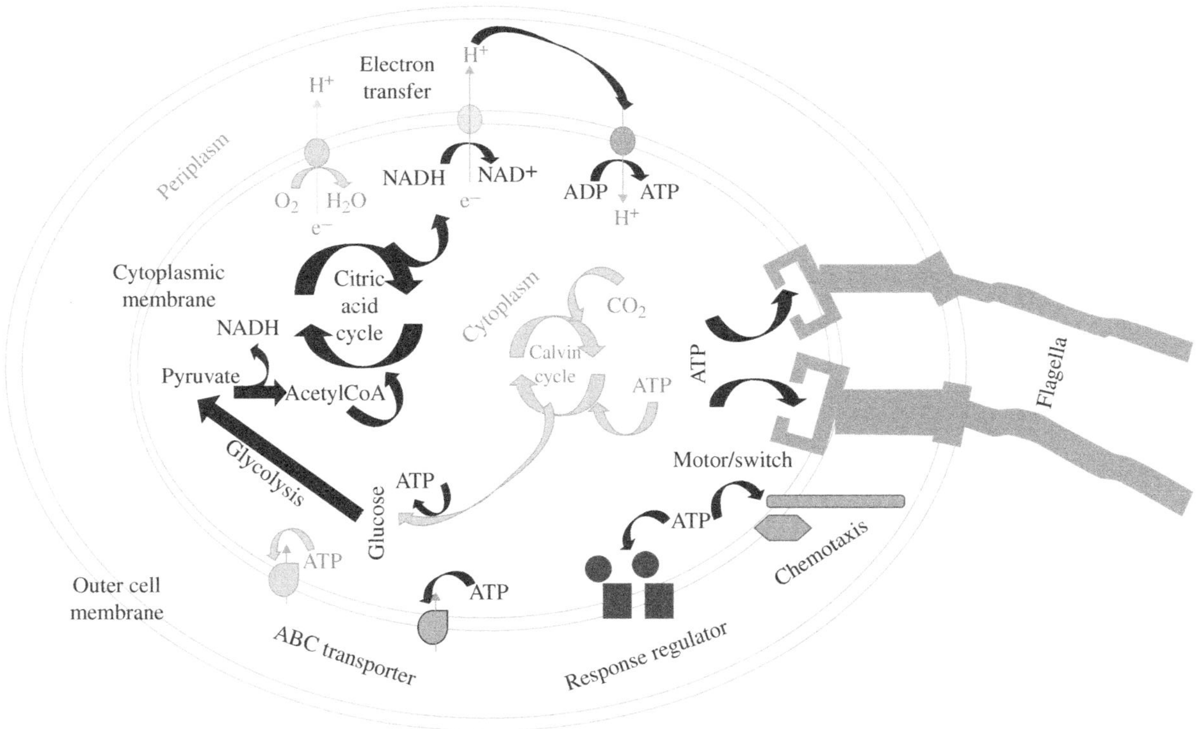


**Figure 10.5** Upregulation (dark-colored symbols) and downregulated (gray symbols) genetic pathways in *B. diazoefficiens* when exposed to low concentration of ENMs (0.1–10 mg/l of MWCNT and $CeO_2$ nanoparticles). Both MWCNT and $CeO_2$ nanoparticles upregulated flagellar assembly, motor- and chemotaxis-associated genes, and several ABC transporters among others. The Calvin cycle, several ABC transporters, and electron transport chains (containing cytochrome c family proteins as electron transporters) were downregulated. *Source*: Figure modified after Mortimer et al. (2020).

to a rhizobium's symbiotic host could interfere with the potential of rhizobia to infect the roots of a host plant leading to a potential reduction in N fixation.

Biofilms play a critical role in the rhizosphere or plant root interface. Although biofilms associated with plant roots are typically nonpathogenic plant growth-promoting rhizobacteria (PGPRs) that enhance acquisition of water and nutrients and phytohormone production, biofilms can contain potential plant pathogens (Rudrappa, Biedrzycki, & Bais, 2008). A biofilm protects microorganisms against desiccation and other adverse environmental factors and allows for the transport of nutrients and signal molecules between members of the colony. Ouyang et al. (2020) reported on the effects of ZnO nanoparticles on the biofilm formation mechanism of *Pseudomonas putida* KT2440. The *P. putida* used in this study is a plant-beneficial bacterium (ABM). Low concentrations (0.5–30 mg/l) of the ENMs significantly enhanced the bacterial colonization and biofilm formation process but the process was inhibited at ENM concentration above 30 mg/l. The protein and sugar contents in the biofilm increased at low concentrations of ZnO ENMs. The changes in expression of the genes associated with oxidative stress possibly led to enhanced biofilm development. Upregulation of gene expressions associated with quorum sensing (releasing, sensing, and responding to signal molecules); lipopolysaccharide biosynthesis; and antibiotic resistance was observed at low ENM concentration (0.5 mg/l). However, at high ENM concentration (250 mg/l), the biofilm formation-related genes were downregulated while anti-oxidative stress-related genes were upregulated. It has been argued that excess biofilm formation due to the presence of ENMs may change the established ecological balances in soil systems. So, long-term release of low concentrations of ENMs to the environment may not be of benefit to ABMs. It should, however, be noted that the long-term effects of ENMs on biofilm development have not been addressed in major research efforts so far. We feel that the biofilms may be a means of trapping ENMs and, thus, reducing their toxicity to microorganisms and plant roots.

## 10.3 Summary

This chapter is an effort to discuss some important aspects of nonmaterial applications in agriculture. While there is only limited understanding at this time, the elucidation of interactions of the ENM–plant–soil–microbiome will contribute to addressing our existing knowledge gaps with respect to the mechanisms of interactions and potential ENM dose or application rates for a given soil, crop, and climate. Such information will provide a foundation for the future design and development of smart and safe nanomaterials (Ouyang et al., 2020). The long-term application and retention of ENMs in soil necessitate the development of toxicology studies. Further research is also needed to aid agronomists in the development of guidelines and regulations for field applications of nano-based fertilizers, pesticides, and growth stimulants based on the properties and fate of nanoscale materials.

## References


Abdel-Aziz, H. M. M., Hasaneen, M. N. A., & Omer, A. M. (2016). Nano chitosan-NPK fertilizer enhances the growth and productivity of wheat plants grown in sandy soil. *Spanish Journal of Agricultural Research*, 14(1), 2171–9292.

Achari, G. A., & Kowshik, M. (2018). Recent developments on nanotechnology in agriculture: Plant mineral nutrition, health, and interactions with soil microflora. *Journal of Agricultural and Food Chemistry*, 66(33), 8647–8661.

Acharya, P., Jayaprakasha, G. K., Crosby, K. M., Jifon, J. L., & Patil, B. S. (2019). Green-synthesized nanoparticles enhanced seedling growth, yield, and quality of onion (*Allium cepa* L.). *Acs Sustainable Chemistry & Engineering.*, 7(17), 14580–14590.

Adisa, I. O., Pullagurala, V. L. R., Peralta-Videa, J. R., Dimkpa, C. O., Elmer, W. H., Gardea-Torresdey, J. L., et al. (2019). Recent advances in nano-enabled fertilizers and pesticides: A critical review of mechanisms of action. *Environmental Science-Nano*, 6(7), 2002–2030.

Alidoust, D., & Isoda, A. (2013). Effect of gamma Fe2O3 nanoparticles on photosynthetic characteristic of soybean (*Glycine max* (L.) Merr.): Foliar spray versus soil amendment. *Acta Physiologiae Plantarum*, 35(12), 3365–3375.

Almeelbi, T., & Bezbaruah, A. (2014). Nanoparticle-sorbed phosphate: Iron and phosphate bioavailability studies with *Spinacia oleracea* and *Selenastrum capricornutum*. *ACS Sustainable Chemistry & Engineering.*, 2(7), 1625–1632.

Antonoglou, O., Moustaka, J., Adamakis, I.-D. S., Sperdouli, I., Pantazaki, A. A., Moustakas, M., et al. (2018). Nanobrass CuZn nanoparticles as foliar spray nonphytotoxic fungicides. *Acs Applied Materials & Interfaces*, 10(5), 4450–4461.

Asghari, B., Khademian, R., & Sedaghati, B. (2020). Plant growth promoting rhizobacteria (PGPR) confer drought resistance and stimulate biosynthesis of secondary metabolites in pennyroyal (*Mentha pulegium* L.) under water shortage condition. *Scientia Horticulturae*, 15, 263.

Azabou, M. C., Gharbi, Y., Medhioub, I., Ennouri, K., Barham, H., Tounsi, S., et al. (2020). The endophytic strain *Bacillus velezensis* OEE1: An efficient biocontrol agent against *Verticillium wilt* of olive and a potential plant growth promoting bacteria. *Biological Control*, 142, 104168.

Cao, Z., Stowers, C., Rossi, L., Zhang, W., Lombardini, L., & Ma, X. (2017). Physiological effects of cerium oxide nanoparticles on the photosynthesis and water use efficiency of soybean (*Glycine max* (L.) Merr.). *Environmental Science-Nano*, 4(5), 1086–1094.

Chichiricco, G., & Poma, A. (2015). Penetration and toxicity of nanomaterials in higher plants. *Nanomaterials*, 5(2), 851–873.

Cieschi, M. T., Polyakov, A. Y., Lebedev, V. A., Volkov, D. S., Pankratov, D. A., Veligzhanin, A. A., et al. (2019 Apr). Eco-friendly iron-humic nanofertilizers synthesis for the prevention of iron chlorosis in soybean (*Glycine max*) grown in calcareous soil. *Frontiers in Plant Science*, 10.

Clough, A., & Skjemstad, J. O. (2000). Physical and chemical protection of soil organic carbon in three agricultural soils with different contents of calcium carbonate. *Australian Journal of Soil Research*, 38(5), 1005–1016.

Cornelis, G., Hund-Rinke, K., Kuhlbusch, T., Van den Brink, N., & Nickel, C. (2014). Fate and bioavailability of engineered nanoparticles in soils: A review. *Critical Reviews in Environmental Science and Technology*, 44(24), 2720–2764.

de Santiago-Martin, A., Constantin, B., Guesdon, G., Kagambega, N., Raymond, S., & Cloutier, R. G. (2016). Bioavailability of engineered nanoparticles in soil systems. *Journal of Hazardous Toxic and Radioactive Waste*, 20(1), B4015001-1-B4015001-15.

Deepa, M., Sudhakar, P., Nagamadhuri, K. V., Reddy, K. B., Krishna, T. G., & Prasad, T. (2015). First evidence on phloem transport of nanoscale calcium oxide in groundnut using solution culture technique. *Applied Nanoscience*, 5(5), 545–551.

Dimkpa, C. O., Singh, U., Adisa, I. O., Bindraban, P. S., Elmer, W. H., Gardea-Torresdey, J. L., et al. (2018). Effects of manganese nanoparticle exposure on nutrient acquisition in wheat (*Triticum aestivum* L.). *Agronomy-Basel*, 8(9), 158.

Fadeel, B., Farcal, L., Hardy, B., Vazquez-Campos, S., Hristozov, D., Marcomini, A., et al. (2018). Advanced tools for the safety assessment of nanomaterials. *Nature Nanotechnology*, 13(7), 537–543.

Faizan, M., Faraz, A., Yusuf, M., Khan, S. T., & Hayat, S. (2018). Zinc oxide nanoparticle-mediated changes in photosynthetic efficiency and antioxidant system of tomato plants. *Photosynthetica*, 56(2), 678–686.

Fierer, N. (2017). Embracing the unknown: Disentangling the complexities of the soil microbiome. *Nature Reviews Microbiology*, 15(10), 579–590.

Fijalkowski, K., Rorat, A., Grobelak, A., & Kacprzak, M. J. (2017). The presence of contaminations in sewage sludge—The current situation. *Journal of Environmental Management*, 203, 1126–1136.

Fortuna, A.M., Sinha, S., Das, T.K. and Bezbaruah, A.N. (2019). Adaption of microarray primers for iron transport and homeostasis gene expression in *Pseudomonas fluorescens* exposed to nano iron. *MethodsX*, 6, 1181–1187.

Fukami, J., Cerezini, P., & Hungria, M. (2018). Azospirillum: Benefits that go far beyond biological nitrogen fixation. *Amb Express*, 4, 8.

Glick, B. R. (2012). Plant growth-promoting bacteria: mechanisms and applications. *Scientifica*, 2012, 963401.

Gupta, V. V. S. R., Neate, S. M., & Leonard, E. (1997). *Life in the soil: The relationship between agriculture and soil organisms*. Adelaide, Australia: Cooperative Research Centre for Soil & Land Management, The University of Adelaide.

He, X., Deng, H., & Hwang, H.-M. (2019). The current application of nanotechnology in food and agriculture. *Journal of Food and Drug Analysis*, 27(1), 1–21.

Herrera, H., Novotna, A., Ortiz, J., Soto, J., & Arriagada, C. (2020). Isolation and identification of plant growth-promoting bacteria from rhizomes of *Arachnitis uniflora*, a fully mycoheterotrophic plant in southern Chile. *Applied Soil Ecology*, 149, 103512.

Jacob, D. L., Borchardt, J. D., Navaratnam, L., Otte, M. L., & Bezbaruah, A. N. (2013). Uptake and translocation of Ti from nanoparticles in crops and wetland plants. *International Journal of Phytoremediation*, 15(2), 142–153.

Jentsch, P. (2020). *Historical perspectives on apple production: Fruit tree pest management, regulation and new insecticidal chemistries*. Highland, NY: Department of Entomology, Cornell University. http://web.entomology.cornell.edu/jentsch/assets/historical-perspectives-on-apple-production.pdf. (accessed May 2020)

Kah, M., Beulke, S., Tiede, K., & Hofmann, T. (2013). Nanopesticides: State of knowledge, environmental fate, and exposure modeling. *Critical Reviews in Environmental Science and Technology*, 43(16), 1823–1867.

Kah, M., & Hofmann, T. (2014). Nanopesticide research: Current trends and future priorities. *Environment International*, 63, 224–235.

Kah, M., Tufenkji, N., & White, J. C. (2019). Nano-enabled strategies to enhance crop nutrition and protection. *Nature Nanotechnology*, 14(6), 532–540.

Kim, M.-J., Kim, W., & Chung, H. (2020). Effects of silver-graphene oxide on seed germination and early growth of crop species. *Peerj*, 27, 8.

Kumar, S., Nehra, M., Dilbaghi, N., Marrazza, G., Hassan, A. A., & Kim, K.-H. (2019). Nano-based smart pesticide formulations: Emerging opportunities for agriculture. *Journal of Controlled Release*, 294, 131–153.

Kumar, S., Nehra, M., Dilbaghi, N., Marrazza, G., Tuteja, S. K., & Kim, K.-H. (2020). Nanovehicles for plant modifications towards pest- and disease-resistance traits. *Trends in Plant Science*, 25(2), 198–212.

Kumari, S., Choudhary, R. C., Kumaraswamy, R. V., Bhagat, D., Pal, A., Raliya, R., et al. (2019). Zinc-functionalized thymol nanoemulsion for promoting soybean yield. *Plant Physiology and Biochemistry*, 145, 64–74.

Lindstroem, K., & Mousavi, S. A. (2019). Effectiveness of nitrogen fixation in rhizobia. *Microbial Biotechnology*, 13(5), 1314–1335.

Liu, R. Q., & Lal, R. (2014). Synthetic apatite nanoparticles as a phosphorus fertilizer for soybean (*Glycine max*). *Scientific Reports*, 4(1), 1–6.

Liu, Y., Yue, L., Wang, Z., & Xing, B. (2019). Processes and mechanisms of photosynthesis augmented by engineered nanomaterials. *Environmental Chemistry*, 16(6), 430–445.

Mehmood, M. A., Zhao, H., Cheng, J., Xie, J., Jiang, D., & Fu, Y. (2020). Sclerotia of a phytopathogenic fungus restrict microbial diversity and improve soil health by suppressing other pathogens and enriching beneficial microorganisms. *Journal of Environmental Management*, 59, 109857.

Miranda-Villagomez, E., Iris Trejo-Tellez, L., Carlos Gomez-Merino, F., Sandoval-Villa, M., Sanchez-Garcia, P., & Angel, A.-M. M. (2019). Nanophosphorus fertilizer stimulates growth and photosynthetic activity and improves P status in rice. *Journal of Nanomaterials*, 18, 2019.

Mishra, J., Singh, R., & Arora, N. K. (2017). Plant growth-promoting microbes: Diverse roles in agriculture and environmental sustainability. In V. Kumar, M. Kumar, S. Sharma, & R. Prasad (Eds.), *Probiotics and plant health* (pp. 71–111). Singapore, Singapore: Springer.

Mortimer, M., Li, D., Wang, Y., & Holden, P. A. (2020). Physical properties of carbon nanomaterials and nanoceria affect pathways important to the nodulation competitiveness of the symbiotic N-2-fixing bacterium *Bradyrhizobium diazoefficiens*. *Small*, 16, 1906055.

Moya, P., Barrera, V., Cipollone, J., Bedoya, C., Kohan, L., Toledo, A., et al. (2020). New isolates of *Trichoderma* spp. as biocontrol and plant growth-promoting agents in the pathosystem *Pyrenophora teres*-barley in Argentina. *Biological Control*, 141, 104152.

Ouyang, K., Mortimer, M., Holden, P. A., Cai, P., Wu, Y., Gao, C., et al. (2020). Towards a better understanding of *Pseudomonas putida* biofilm formation in the presence of ZnO nanoparticles (NPs): Role of NP concentration. *Environment International*, 137, 105485–105485.

Puri, A., Padda, K. P., & Chanway, C. P. (2020). in vitro; and in vivo; analyses of plant-growth-promoting potential of bacteria naturally associated with spruce trees growing on nutrient-poor soils. *Applied Soil Ecology*, 149, 103538.

Rai, M., Ingle, A. P., Pandit, R., Paralikar, P., Shende, S., Gupta, I., et al. (2018). Copper and copper nanoparticles: Role in management of insect-pests and pathogenic microbes. *Nanotechnology Reviews*, 7(4), 303–315.

Rai, P. K., Kumar, V., Lee, S., Raza, N., Kim, K.-H., Ok, Y. S., et al. (2018). Nanoparticle-plant interaction: Implications in energy, environment, and agriculture. *Environment International*, 119, 1–19.

Rastogi, A., Tripathi, D. K., Yadav, S., Chauhan, D. K., Zivcak, M., Ghorbanpour, M., et al. (2019). Application of silicon nanoparticles in agriculture. *3 Biotech*, 9(3), 1–11.

Rodrigues, S. M., Trindade, T., Duarte, A. C., Pereira, E., Koopmans, G. F., & Romkens, P. (2016). A framework to measure the availability of engineered nanoparticles in soils: Trends in soil tests and analytical tools. *Trac-Trends in Analytical Chemistry*, 75, 129–140.

Rodriguez-Gonzalez, V., Terashima, C., & Fujishima, A. (2019). Applications of photocatalytic titanium dioxide-based nanomaterials in sustainable agriculture. *Journal of Photochemistry and Photobiology C-Photochemistry Reviews*, 40, 49–67.

Rudrappa, T., Biedrzycki, M. L., & Bais, H. P. (2008). Causes and consequences of plant-associated biofilms. *Fems Microbiology Ecology*, 64(2), 153–166.

Sabir, S., Arshad, M., & Chaudhari, S. K. (2014). Zinc oxide nanoparticles for revolutionizing agriculture: Synthesis and applications. *The Scientific World Journal*, 1–8.

Sammauria, R., Kumawat, S., Kumawat, P., Singh, J., & Jatwa, T. K. (2020). Microbial inoculants: potential tool for sustainability of agricultural production systems. *Archives of Microbiology*, 202(4), 677–693.

Shang, Y., Hasan, M. K., Ahammed, G. J., Li, M., Yin, H., & Zhou, J. (2019). Applications of nanotechnology in plant growth and crop protection: A review. *Molecules*, 24(14), 2558.

Sinha, S., Das, T. K., Bezbaruah, A. N., & Fortuna, A.-M. (2018). Reduced expression of iron transport and homeostasis genes in *Pseudomonas fluorescens* during iron uptake from nanoscale iron. *Nanoimpact*, 12, 42–50.

Swift, T. A., Oliver, T. A. A., Galan, M. C., & Whitney, H. M. (2019). Functional nanomaterials to augment photosynthesis: Evidence and considerations for their responsible use in agricultural applications. *Interface Focus*, 9(1), 20180048.

Tarafdar, J. C., Raliya, R., Mahawar, H., & Rathore, I. (2014). Development of zinc nanofertilizer to enhance crop production in pearl millet (*Pennisetum americanum*). *Agricultural Research*, 3, 257–262.

Tien, T. M., Gaskins, M. H., & Hubbell, D. H. (1979). Plant-growth substances produced by *Azospirillum-brasilense* and their effect on the growth of pearl millet (*Pennisetum-americanum* L). *Applied and Environmental Microbiology*, 37(5), 1016–1024.

USEPA (United States Environmental Protection Agency). Biosolids. https://www.epa.gov/biosolids/frequent-questions-about-biosolids, (accessed June 2020) 2020.

Weil, R. R., & Brady, N. C. (2017). *The nature and properties of soils* (15th ed.). London, UK: Pearson Education.

Zahedi, S. M., Karimi, M., & da Silva, J. A. T. (2020). The use of nanotechnology to increase quality and yield of fruit crops. *Journal of the Science of Food and Agriculture*, 100(1), 25–31.

Zhang, Y., & Zhuang, W.-Y. (2020). *Trichoderma brevicrassum* strain TC967 with capacities of diminishing cucumber disease caused by *Rhizoctonia solani* and promoting plant growth. *Biological Control*, 142, 104151.

Zhao, L., Lu, L., Wang, A., Zhang, H., Huang, M., Wu, H., et al. (2020). Nano-biotechnology in agriculture: Use of nanomaterials to promote plant growth and stress tolerance. *Journal of Agricultural and Food Chemistry*, 68(7), 1935–1947.

Zhu, Y., Xu, F., Liu, Q., Chen, M., Liu, X., Wang, Y., et al. (2019). Nanomaterials and plants: Positive effects, toxicity and the remediation of metal and metalloid pollution in soil. *Science of the Total Environment*, 662, 414–421.

# 11

# Biogeochemical Interactions of Bioreduced Uranium Nanoparticles

*S. Sevinç Şengör*[1,*] *and Rajesh K. Sani*[2,3]

[1] *Bobby B. Lyle School of Engineering, Southern Methodist University, Dallas, TX, USA*
[2] *Department of Environmental Engineering, Middle East Technical University, Dumlupınar Bulvarı No: 1, Ankara, Turkey*
[3] *Department of Chemical and Biological Engineering, South Dakota School of Mines and Technology, Rapid City, SD, USA*

## 11.1 Introduction

Uranium (U) is a common radionuclide found within historical and current mining-impacted regions, and exists in surface and groundwater, soil, and sediment at many sites worldwide. U in the subsurface has been a global problem in aquifers, water supplies, and related ecosystems because of its toxicity, which is primarily due to its chemical rather than radiological action. U-contaminated groundwater threatens ecosystem health and poses a real threat to human health due to the solubility and mobility properties of the oxidized form of U, which is U(VI). A promising strategy for in situ remediation of subsurface U contamination is through facilitating biostimulation of iron- and/or sulfate-reducing indigenous bacterial species resulting in dissimilatory reduction of soluble U(VI) to insoluble U(IV) – uraninite ($UO_2$) (Anderson et al., 2003; Finneran, Housewright, & Lovley, 2002; Lovley & Phillips, 1992). Therefore, the biological reduction of soluble U(VI) to insoluble U(IV) with subsequent precipitation of bioreduced U is considered as a viable option to remediate U because of its low solubility under reducing conditions, low cost, and minimal secondary waste generation (Akob et al., 2008; Brodie et al., 2006; Lovley & Phillips, 1992). Although most of the studies have considered the uraninite mineral ($UO_2$) as the primary product of U(VI) bioreduction, recent studies have shown the formation of distinct monomeric/mononuclear U(IV) phases of U bioreduction with appreciable mass fraction at the

*Current address: Department of Environmental Engineering, Middle East Technical University, Dumlupınar Bulvarı No: 1, Ankara, Turkey

*Microbial Interactions at Nanobiotechnology Interfaces: Molecular Mechanisms and Applications*, First Edition. Edited by R. Navanietha Krishnaraj and Rajesh K. Sani.

nanometer-particle scale and below, where the wealth of U(IV) chemistry is not fully understood (Bernier-Latmani et al., 2010; Boyanov et al., 2011; Fletcher et al., 2010; Sivaswamy et al., 2011). Our recent studies have also illustrated that a significant mass fraction of bioreduced U exists as a mobile nanoparticulate phase under growth conditions (Şengör et al., 2016). These results raise several fundamental questions regarding long-term U reclamation stewardship. These bioreduced U phase particles are vulnerable to be transported in the groundwater, possibly facilitated by association to other mobile colloids, where each distinct phase will likely have differential rates of transport and transformation in aquifers.

Over 25 dissimilatory metal-reducing bacteria (DMRB) have been reported to enzymatically reduce U(VI) to bioreduced U. Sulfate-reducing bacteria (SRB) are representative of the DMRB that can reduce U(VI) enzymatically and nonenzymatically, and are present in many metal-contaminated subsurface sites. Using SRB, our group's results (Şengör et al., 2016) show that a significant mass fraction of bioreduced U (35–60%) exists as a mobile phase. We have shown that SRB reduce U(VI) to nanometer-sized bioreduced U particles (1–5 nm), which reside both extracellularly and intracellularly (Sani, Peyton, & Dohnalkova, 2006). The accumulation of bioreduced U inside the cell (periplasm and cytoplasm) may physically shield it from subsequent oxidation processes, thus furthering the sequestering abilities of bioreduced and immobilized U(IV). However, extracellular reduced U has been shown to be oxidized quickly due to a large surface area-to-mass ratio of the nanoparticles (Singh et al., 2014; Spycher et al., 2011 and references therein). It has been shown using column experiments with natural materials that sulfate reduction can commence even though significant bioavailable Fe(III), a more energetically favorable electron acceptor, remains (Komlos, Mishra, Lanzirotti, Myneni, & Jaffe, 2008). Sulfate-reducing conditions are also often encountered in field settings, suggesting that sulfate-reducing conditions control the stability of bioreduced U, and the fate and transport of U reduced by SRB are very important as general components in understanding subsurface microbial reductive precipitation of U and other actinides. Therefore, for the successful application of long-term U bio-stabilization, it is vital to understand the structure, composition, and stabilities of these U(IV) phases, as these species may have significantly different characteristics, compositions, and responses to the biochemical conditions than the uraninite mineral, and these complex, interactive, and critical processes need to be incorporated into reactive transport models. This chapter aims to provide a review on biogeochemical interactions of bioreduced uranium nanoparticles, with the following chapter objectives:

- Review U contamination in subsurface settings and review the recent advances in U remediation strategies.
- Provide a review on overall biogeochemical interactions of bioreduced U nanoparticles.

- Provide an overall understanding of U mobility in the subsurface.
- Identify and characterize the fundamental processes that control the stability of reduced U, with a focus on re-oxidation processes and future directions of research.

## 11.2 Coupled Biogeochemical Mechanisms and Interactions of U in the Subsurface

U is the 49th most abundant element in the Earth's crust. It was discovered in 1789 in Germany and was first isolated in 1841 in France (Palmisano & Hazen, 2003). It was used for nuclear research, nuclear fuel production, and manufacturing of nuclear weapons, resulting in widespread contamination. U in the subsurface has become a significant problem in aquifers and related ecosystems because of its toxicity, which is primarily due to chemical rather than radiological action (Craft et al., 2004). In the United States, over 130 sites in 31 states, under the jurisdiction of the U.S. Department of Energy (DOE), comprise over 2 million acres of total contaminated area and approximately 6.4 billion $m^3$ of contaminated media. Responsibility for management, remediation, or monitored natural attenuation, and long-term stewardship of these sites falls under the DOE.

In South Dakota, USA, there have been many reports documenting soil, water, and sediment impacts associated with historical U mining operations, particularly from the Cave Hills region in northwestern South Dakota (Kipp, Stone, & Stetler, 2009; Larson & Stone, 2011; Stone, Stetler, & Schwalm, 2007a) and in the southern Black Hill region (Webb, Davis, & Hodge, 1995). At the former Edgemont Mill Site located in the east of Edgemont, South Dakota, there were evidences of high U concentrations within soil, groundwater, and pit areas (Rastogi, Osman, Andersen, & Sani, 2010). Similar U mining activities and associated air, water, and soil contamination incidents have also been reported in other parts of the United States and around the globe (e.g., Palmisano & Hazen, 2003).

U is chemically analogous to other actinides in terms of its redox properties and tendency to form complexes with inorganic and organic ligands, and thus serves as a valuable model for the behavior of actinides in complex natural systems (Suzuki & Suko, 2006). The most common form of U in surface and groundwaters is the oxidized U(VI) species, which exist as different ions depending on the alkalinity, pH, and complexants [e.g. $UO_2^{2+}$, $UO_2OH^+$, $UO_2(OH)_2$, $UO_3{\cdot}3H_2O$, $UO_2(CO_3)_2^{2-}$, $UO_2(CO_3)_2^{4-}$, and U(VI)-Ca-$CO_3$]. These oxidized species of U are highly soluble and mobile in aqueous environments. The oxidized U(VI) presents a significant health hazard due to its solubility and mobility properties. On the other hand, under reducing conditions, insoluble U(IV) formation is thermodynamically favored, and is commonly found in the form of mineral uraninite ($UO_2$). So in order to minimize risks associated with U(VI), the facilitation of abiotic or biotic reducing conditions is a plausible remediation strategy to biogeochemically precipitate U.

There is a significant interest in in situ solution leach mining (ISL) for mining many sandstone dominated U roll-front formations found throughout the west, including the southern Black Hills region. The most challenging and environmentally sensitive aspect of ISL mining is aquifer reclamation once the uranium ore body is "exhausted," since the groundwater will contain significantly elevated and highly soluble (and thus highly mobile) uranyl species. As part of commonly employed reclamation procedures utilized by solution mining companies, groundwater is flushed through the aquifer and the resulting U(VI)-laden water is treated ex situ using reverse osmosis, and subsequently pumped back to the aquifer up gradient of the reclamation zone. Sodium sulfate is further added during reclamation to promote sulfate-reducing conditions. To date, no current or former ISL mining operation in the United States has successfully reclaimed an aquifer to pre-mining baseline conditions. A promising strategy for in situ remediation of subsurface U contamination is through facilitating biostimulation of iron- and/or sulfate-reducing indigenous bacterial species resulting in dissimilatory reduction of soluble U(VI) to insoluble U(IV) – uraninite ($UO_2$) (Anderson et al., 2003; Finneran et al., 2002; Lovley & Phillips, 1992).

Bioreduced U is generally regarded as the most desirable product of this bioreductive process due to its inherent low solubility (hence stability) under reducing conditions (Ulrich et al., 2008). Recent results however indicate that in some cases a portion of reduced U forms molecular U(IV) (Alessi et al., 2012; Boyanov et al., 2010; Fletcher et al., 2010), that by virtue of its potential mobility is thought to be a less desirable end product. At least three types of bioreduced U(IV) have been reported in the literature: uraninite, nanoparticulate uraninite, and mononuclear U(IV) phases. Although uraninite and nanoparticulate uraninite have been reported to be more crystalline, mononuclear U(IV) has been considered to be more amorphous, posing challenges due to its mobile and unstable nature (Bernier-Latmani et al., 2010; Boyanov et al., 2011; Fletcher et al., 2010; Singh et al., 2014).

It has been established in liquid batch cultures that over 25 species of phylogenetically diverse prokaryotes can mediate enzymatic U(VI) reduction to insoluble U(IV) as uraninite $UO_2$ (Suzuki, Kemner, & Banfield, 2002). The most studied microbes are the dissimilatory metal- and sulfate-reducing bacteria (DMRB and SRB) including strains of the genera *Anaeromyxobacter*, *Geobacter*, *Shewanella*, and *Desulfovibrio* (e.g. Gorby & Lovley, 1992; Liu, Zachara, Fredrickson, Kennedy, & Dohnalkova, 2002; Lovley & Phillips, 1992; Lovley,

Phillips, Gorby, & Landa, 1991; Wu, Sanford, & Loffler, 2006), and are representative of strains present in many contaminated subsurface sites. Biostimulation of SRB for U remediation can be especially advantageous since SRB can reduce U(VI) by direct enzymatic mechanisms as well as indirectly via sulfide production (Beyenal et al., 2004; Lloyd & Lovley, 2001; Lovley, 1993; Lovley, Roden, Phillips, & Woodward, 1993a; Lovley, Widman, Woodward, & Phillips, 1993b), as illustrated in Figure 11.1a and b. SRB are present in many contaminated subsurface sites and carry out dissimilatory reduction of sulfate to sulfide (Abdelouas

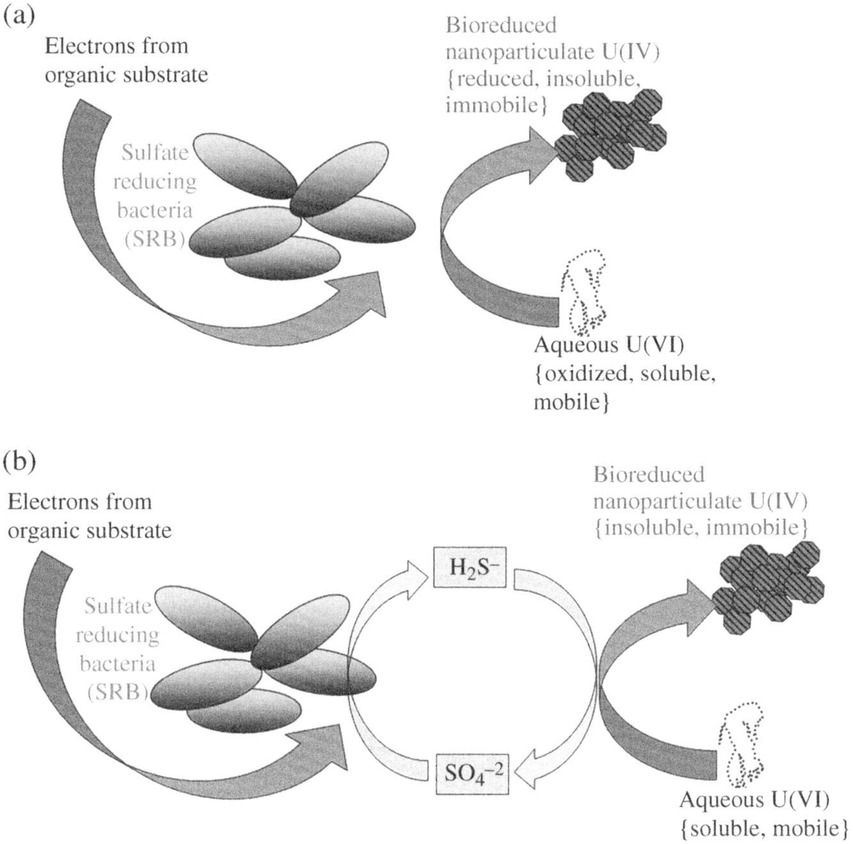


**Figure 11.1** Illustration of biologically induced redox reactions that lead to immobilization of U by SRB. (a) Direct enzymatic reduction of soluble U(VI) by SRB using electrons from organic compounds; (b) nonenzymatic reduction of soluble U(VI) by SRB, where the oxidation of organic compounds or hydrogen is coupled to the reduction of sulfate ($SO_4^{-2}$) to hydrogen sulfide ($H_2S^-$). This reduced form then interacts with multivalent metals, such as U to form insoluble metal precipitates.

et al., 2000; Abdelouas, Lu, Lutze, & Nuttall, 1998; Barnes et al., 1998). The resulting sulfide often reacts abiotically to form sulfide precipitates of many metals including Fe and lead (Pb) (Bhattacharya, Jumawan, Sun, Sund-Hegelburg, & Schwitzgebel, 1981; Poulson, Colberg, & Drever, 1997). Recent data from our labs demonstrate that sulfide can act as a potent reductant of U(VI) at environmentally relevant bicarbonate concentrations (5 mM) (Beyenal et al., 2004). Thus, SRB have great potential for transforming complex aqueous mixtures of heavy metals and radionuclides in the deep subsurface. Furthermore, reduction of Fe(III)-(hydr) oxides, either enzymatically (Coleman, Hedrick, Lovley, White, & Pye, 1993; Le Gall, Payne, Chen, Liu, & Xavier, 1994; Robert & Berthelin, 1986) or indirectly by sulfide production (Brim, Vankateswaren, Kostandarithes, Fredrickson, & Daly, 2003; Brooks et al., 2003; Buerge & Hug, 1997, 1999), can strongly influence the biogeochemistry of anoxic soil and sedimentary environments, as well as the persistence and mobility of various types of contaminants including U (Buerge & Hug, 1999; Caccavo Jr., Blakemore, & Lovley, 1992; Sani, Peyton, Amonette, & Geesey, 2004, 2005). In this regard, the reduction of Fe(III)-(hydr)oxides is of particular interest, given the abundance of Fe in most soils and sediments (Amonette, 2003).

It has also been shown using column experiments with natural materials that sulfate reduction can commence even though significant bioavailable Fe(III), a more energetically favorable electron acceptor, remains (Komlos et al., 2008). Sulfate-reducing conditions are also often encountered in field settings. From 2002 through 2007, bioremediation field experiments have been conducted at a Uranium Mill Tailings Remedial Action (UMTRA) site in Rifle, Colorado, using acetate amendment to stimulate microbially mediated immobilization of uranium in the unconfined aquifer (Anderson et al., 2003; Chang et al., 2005; Vrionis et al., 2005). These experiments involved the injection of acetate as an electron donor to react primarily with three competing electron acceptors: U(VI), Fe(III), and sulfate. The removal of U(VI) from solution was coincident with an accumulation of Fe(II) and enrichment of the iron-reducing bacteria, *Geobacteraceae*. The field observations were consistent with previous studies and showed that the stimulation of acetate-oxidizing metal-reducing bacteria (i.e. *Geobacter* sp.) concomitantly reduced and immobilized aqueous U(VI) (Finneran et al., 2002), and the acetate-oxidizing sulfate-reducing bacteria were not as effective at U(VI) immobilization (Anderson et al., 2003; Lovley, 1993; Ortiz-Bernad, Anderson, Vrionis, & Lovley, 2004; Yabusaki et al., 2007). Conversely, when ethanol was used to stimulate U(VI) reduction at a field site in Oak Ridge, TN, the reduction of U(VI), Fe(III), Mn(IV), and sulfate was observed to proceed concomitantly (Istok et al., 2004). At this site, several additions of electron donor at concentrations up to 300 mM were required, and these caused major shifts in rates of microbial activity, viable biomass, and community composition (Istok et al., 2004; North et al., 2004). Vrionis et al. (2005)

investigated the microbial and geochemical responses to a second acetate biostimulation using the same 2002 field experiment as described in Anderson et al. (2003). Enrichment of the iron-reducing bacteria (*Geobacteraceae)* increased levels of Fe(II) and decreases in U(VI) and Fe(III) minerals were observed. This was followed by a sulfate-reducing phase characterized by observance of sulfide, depleted Fe(III), and 16S rDNA sequences most closely related to *Desulfobacterales*. The study showed that the spatial variations in acetate delivery, as well as geochemical heterogeneity, at the site resulted in the spatial variability of the microbial processes. Each of these field studies showed that acetate stimulated Fe(III)-reducing bacteria (FeRB), which resulted in immobilization of U. However, upon transition from Fe(III)-reducing to sulfate-reducing conditions, U remobilization (Anderson et al., 2003) was also observed. These data suggest that sulfate-reducing conditions control the stability of bioreduced U; and the study of fate and transport of U reduced by SRB is very important in the understanding of subsurface microbial reductive precipitation of U and other actinides. Therefore, a better understanding of remediation or immobilization techniques involving natural indigenous bacteria (e.g. SRB) to promote biological bioreduced U production is warranted due to health risks associated with historical, current, and proposed U mining activities in the western United States and throughout the world. It is vital to understand further the complex biogeochemical processes that govern the fate and transport of U, especially with respect to immobilized U.

With respect to U fate and transport, microorganisms may influence radionuclide mobility through co-reduction of Fe(III) and mineral sorption (Jeon, Dempsey, Burgos, Barnett, & Roden, 2005; Langmuir, 1978; Lovley et al., 1991; Ortiz-Bernad et al., 2004). In anaerobic conditions, the mobility of U is dependent on its ability to form insoluble precipitates [U(IV), uraninite] or sorb to mineral phases such as Fe(III)(hydr)oxides (Figure 11.2). Microbial U(VI) reduction occurs much faster than abiotic reduction, and thus is a preferred mechanism for aquifer reclamation during ISL mining, assuming stimulation of microorganisms can be properly achieved. Thus, it is inherently important to understand the usefulness of biogenic U reduction, especially using SRB as part of ISL reclamation strategies. In addition to U, SRB have the potential to precipitate other heavy metals contaminants found in the region such as arsenic, copper, chromium, lead, and zinc through the formation of insoluble metal sulfides or hydroxides. Many of these heavy metals are present throughout the western United States due to both natural erosion processes and historical mining activities (Kipp et al., 2009; Larson & Stone, 2011; Rastogi et al., 2010; Stone, Stetler, & Schwalm, 2007a). The formation of metal sulfides, especially Fe-sulfides may also help in maintaining a low redox-potential barrier for in situ bioremediation of these metals by scavenging oxygen, thus hindering re-oxidation of reduced contaminant precipitates (Amonette, 2003).

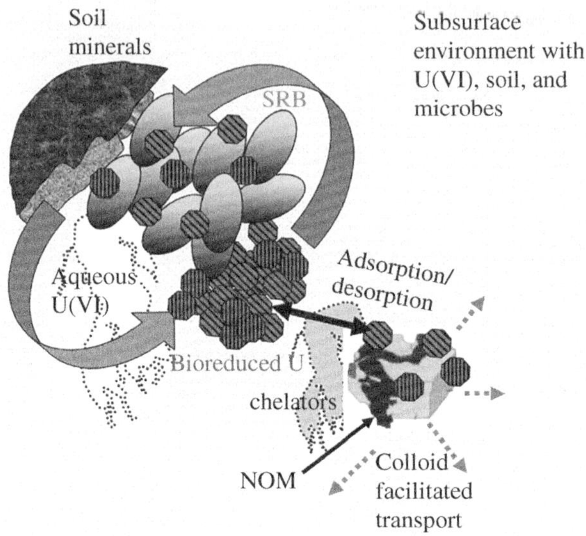


**Figure 11.2** Illustration of main biogeochemical interactions of soluble U(VI) and bioreduced U in natural subsurface settings. Fundamental research is needed to understand the structure, composition, and stabilities of bioreduced U(IV) phases as a colloid or colloid-facilitated transport, for long-term U bio-stabilization applications. These complex and interactive processes focusing on the mechanisms of bioreduced U(IV) reoxidation in addition to the impact of NOM and chelators need to be further examined for successful engineered U bioremediation techniques.

Chelators have been commonly used in the past as cleaning agents during industrial processing of nuclear fuels in order to stimulate metal–chelate complex formation in the subsurface (Palmisano & Hazen, 2003). Our previous work has also shown natural organic matter (NOM) to be an effective metal chelator and electron shuttle during dissimilatory iron reduction (Burgos et al., 2007; Stone, Burgos, Royer, & Dempsey, 2007b). Bank et al. (2008) investigated the impact of NOM on batch U(VI) sorption/reduction experiments and Singhal, Kumar, Rupali Karpe, Datta, and Hedge (2005) studied the speciation of U with NOM in groundwater. However, the knowledge of the mechanisms of NOM as electron shuttling agents and metal complexants in U and Fe bioreduction processes is rather sparse. Furthermore, our group's recent work on batch studies with chelating agents has shown EDTA and NTA promoted $UO_2$ re-oxidation with Fe(III) (hydr)oxide dissolution, indicating the significance of chelators found in U-contaminated sites in mobilizing U (Stewart, Girardot, Spycher, Sani, & Peyton, 2012). Therefore, *it is vital* to further examine the impact of NOM as well as chelators within these characterizations (Figure 11.2) for successful engineered U bioremediation techniques.

## 11.3 Biogenic Uraninite Precipitation and Its Nanoparticulate Forms

Biogenic uraninite arising from SRB stimulation comprises of 1–5 nanometer-sized particles of uraninite (e.g. Suzuki et al., 2002), the redox/dissolution characteristics of which after accounting for surface area do not depend significantly on microbial mediator (Sharpe et al., 2009). These characteristics are also similar to those of chemogenically synthesized $UO_2$; however, the chemogenic product is ~10 times larger with an average particle size of ~50 nm (Ulrich et al., 2009). Uraninite produced by the DMRB *Shewanella oneidensis* MR-1 is a near-stoichiometric crystal with a diameter of ~3 nm (Burgos et al., 2008; Schofield et al., 2008; Singer, Farges, & Brown, 2009), that may form irregular aggregates of 100 nm scale or larger. Although studies have shown that these small uraninite particles appear to conglomerate or aggregate after the formation of large U(IV) particles (>0.2 μm), thus minimizing the potential of further mobilization (Lovley et al., 1991), recent studies of molecular U(IV) show concerns for sparingly mobile nature of U(IV) in porous sediments, strengthening the need to investigate colloid-facilitated transport of biogenic U(IV) (Alessi et al., 2012; Boyanov et al., 2010).

Uraninite produced by the DSRB *Desulfovibrio desulfuricans* G20 grown in a metal toxicity medium (MTM, Sani, Geesey, & Peyton, 2001) containing bicarbonate buffer (30 mM, pH 7) forms particulate material that includes nanometer-sized particles (3–5 nm, Figure 11.3c) that are both extracellular and intracellular (periplasm and cytoplasm, Figure 11.3a, b). Figure 11.3a and b (Sani et al., 2006) show unstained G20 thin sections showing that ~80% of the cells had U precipitates associated with them, and 1/3 of those had U precipitates in the cytoplasmic spaces. Selected area electron diffraction pattern and crystallographic analysis of TEM lattice-fringe images (*d*-spacings 0.164, 0.193, 0.273, and 0.316 nm consistent with UO2, Joint Committee for Powder Diffraction Studies (JCPDS) 41-1442) confirmed the structure of precipitated U in the cytoplasm as being that of uraninite, with *d*-spacings consistent with those previously obtained for U reduced by *Desulfovibrio desulfuricans* and *Geobacter metallireducens* (Gorby & Lovley, 1992; Lovley & Phillips, 1992; Sharpe et al., 2009). HR-TEM images also showed that individual particle diameters of U precipitates formed in the cytoplasm were in the range of 3–5 nm and occurred as both discrete and aggregated particles. These results (with bicarbonate buffer) are consistent with two other

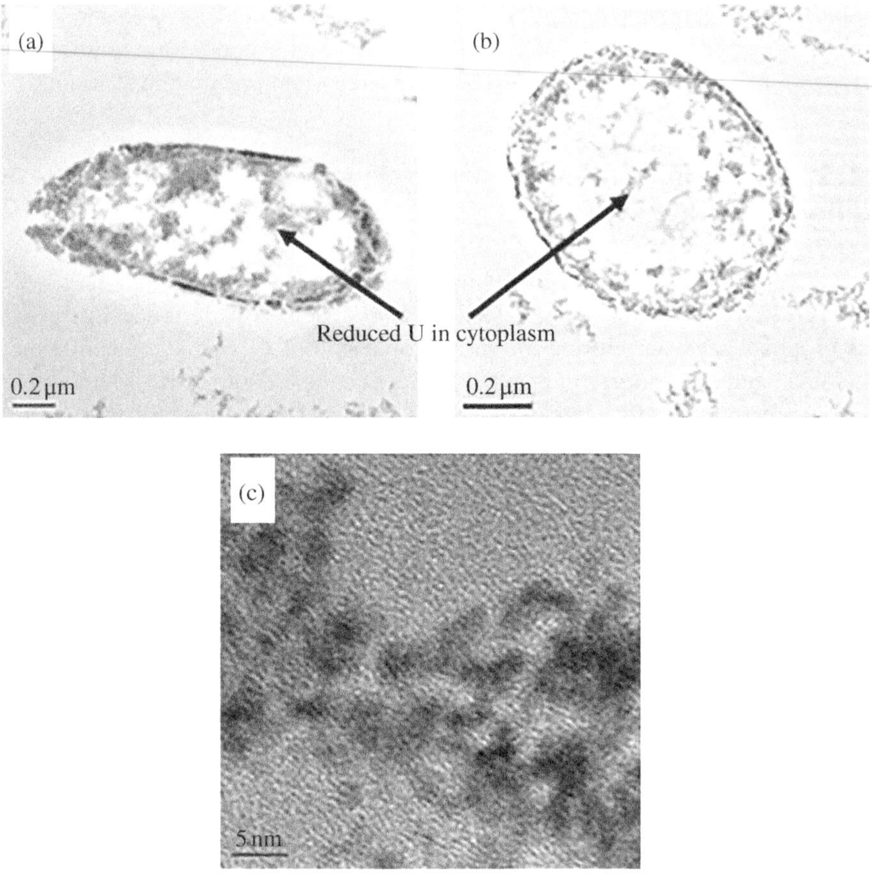


**Figure 11.3** Transmission electron microscopy images of *Desulfovibrio desulfuricans* G20 culture treated with U(VI) in MTM/bicarbonate buffer. *Source*: Sani et al., 2006.

reports where cytoplasmic uraninite deposit localizations were observed: (i) *Pseudomonas* sp. (McLean & Beveridge, 2001) and (ii) *Desulfovibrio äspöensis* isolated from near the Äspö Hard Rock Laboratory in Sweden (Merroun & Selenska-Pobell, 2008).

In contrast to the bicarbonate buffer system, the results obtained from TEM images of unstained thin sections of G20 culture treated with U(VI) in MTM containing piperazine-1,4-bis(2-ethanesulfonic acid) buffer (PIPES, 30 mM, pH 7) differed in at least two aspects. First, 50% of the cells had no reduced U associated with them. Second, in sharp contrast to the bicarbonate buffered system, there were no cells observed that had U precipitates in the cytoplasmic spaces (Figure 11.4a–c; Sani et al., 2006). Further experiments with different initial

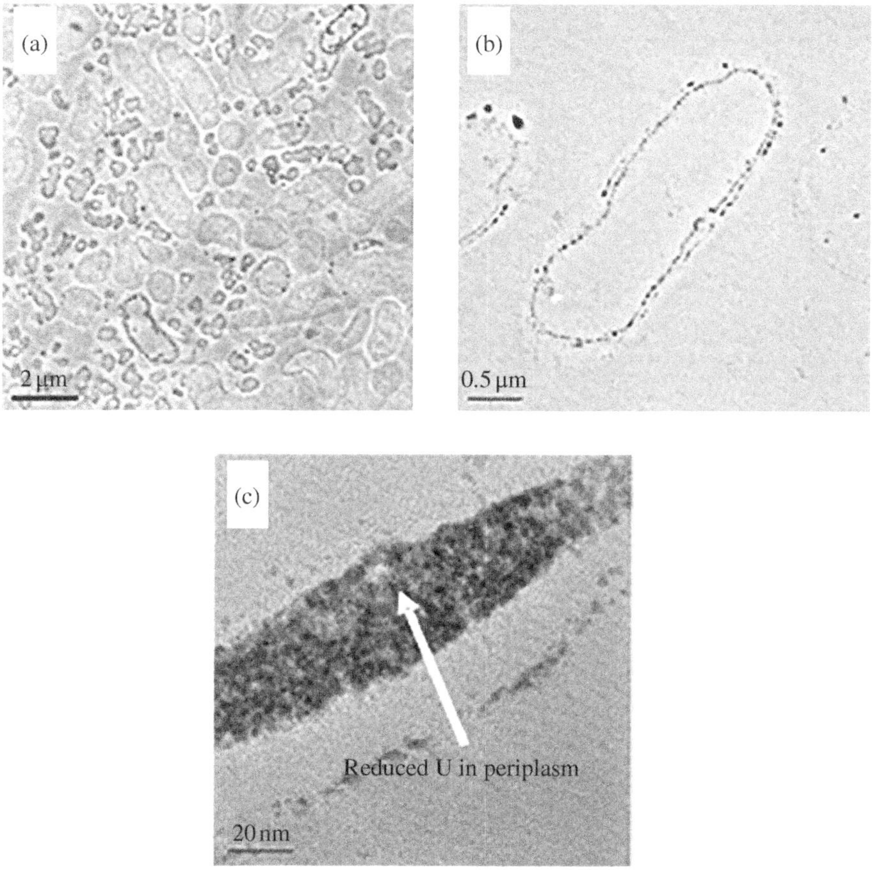


**Figure 11.4** Transmission electron microscopy images of *Desulfovibrio desulfuricans* G20 culture treated with U(VI) in MTM/PIPES buffer. *Source*: Sani et al., 2006.

U(VI) concentrations (25–900 μM) in MTM containing PIPES or bicarbonate buffer indicate that aggregation of uraninite depended on the initial concentration of U(VI) and on the type of buffer used in the growth medium (Şengör et al., 2016).

The reduced U produced exhibited a *range of sizes*. For example, when *D. desulfuricans* G20 was grown in MTM/PIPES solution, >60% of the reduced uranium (calculated using filtered and filtered oxidized U(VI) concentrations using an existing KPA, Figure 11.5a, yellow shaded area) passed through 0.2 μm filters (Gelman Acrodisc). 82% U was reduced but only 25% U (out of 82%) was precipitated out (Figure 11.5b). The remaining 57% reduced U was in aqueous phase and thus mobile. However, when G20 was grown in MTM/bicarbonate solution, >35%

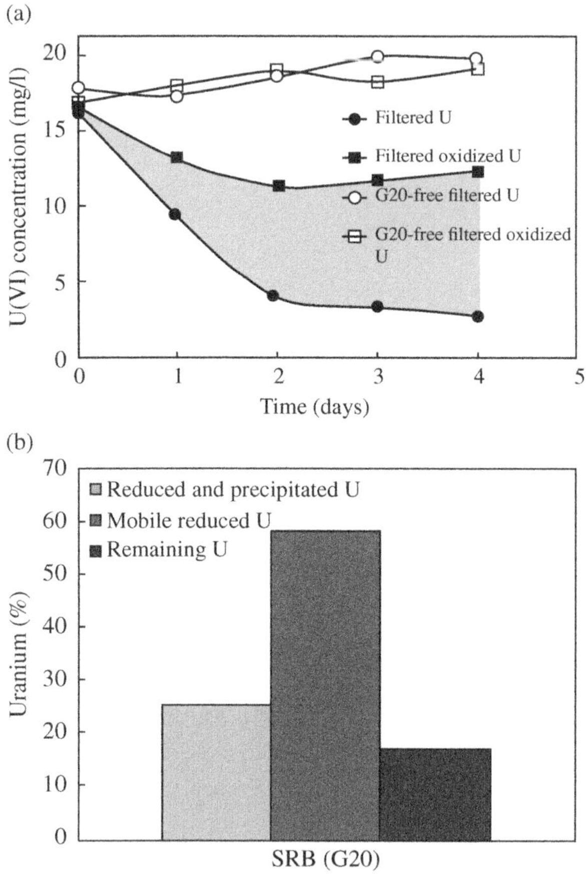


**Figure 11.5** (a) U reduction profiles of *Desulfovibrio desulfuricans* G20. Yellow highlighted region shows the reduction of U in aqueous phase; (b) comparative distribution of U in different phases.

of the U(IV) could pass through 0.2 μm filters (data not shown). These results show the presence of a significant amount (35–60%) of reduced U in the mobile phase. This U(IV) may range in size from nanoparticulate uraninite to molecular U(IV) as reported by Fletcher et al. (2010). Size fractionation results of the reduced U nanoparticles further demonstrated that most of the bioreduced U aggregates were in sizes between 18 and 200 nm and should be considered mobile (Şengör et al., 2016).

Most of the work on U reduction has been done in the absence of soil minerals using batch cultures (Ganesh, Robinson, Reed, & Sayler, 1997; Lovley et al., 1991; Lovley & Phillips, 1992; Sani et al., 2004, 2005; Sani, Peyton, Smith,

Apel, & Petersen, 2002; Spear, Figueroa, & Honeyman, 1999, 2000). To study this process in more realistic systems, experiments have also been performed to quantify the reduction of U(VI) in the presence of hematite and quartz under SRB growth conditions. Sorption of U(VI) to the mineral surfaces occurred over a nine day period before approaching equilibrium. After equilibration, inoculation with SRB yielded significant growth with a lag period (of 18 days in the case of 90 μM U(VI)), followed by concomitant reduction of sulfate and removal of U(VI) from solution (Sani et al., 2005). In these experiments, up to 98% of U(VI) was reduced when hematite was present. However, 42% of the reduced U passed through a 0.2 μm filter (Sani et al., 2004, 2005), indicating that while the U was reduced, it had the potential to be mobile in a natural groundwater due to the small particle and aggregate size.

Furthermore, additional assessments have also been performed using sediments collected from the Cave Hills U mining region of South Dakota that indicate the occurrence of both biotic and abiotic U reductive processes (Larson, Kipp, Mott, & Stone, 2012). Both laboratory and field results indicate a significant percentage (35–60%) of bioreduced U that exists at the sub-0.2 μm size fraction – which poses several fundamental questions regarding the long-term stability of these supposed immobile constituents. These results suggest that in general, a significant mass fraction of biogenic U(IV) remains in suspension with an appreciable mass fraction at the nanometer-particle scale and below. As such it will be subject to transport phenomena in groundwater, possibly facilitated by association to other mobile colloids.

## 11.4 Re-oxidation and Stability of Bioreduced Uranium

Microbially mediated U reduction is induced in the presence of a carbon source and electron donor, as discussed above. However, of critical importance is the long-term stability of nanometer-sized particles of bioreduced U. For effective long-term immobilization of uranium through bioreduction or chemical reduction, there should be no abiotic or biotic re-oxidation of biogenic U(IV). While bioreduced U is subject to oxidation in the presence of several oxidants including oxygen, nitrite, Fe(III), Mn(IV) etc. additional requirements aside from maintenance of anoxic conditions are needed for long-term stability. For example, nitrate reduction products ($NO_2^-$, $N_2O$, NO) were found to cause U(IV) re-oxidation, except when acetate, $H_2S$, aqueous Fe(II), and iron-sulfide minerals were present in excess relative to nitrate (Senko, Mohamed, Dewers, & Krumholz, 2005). Other experiments have shown rapid oxidation of U(IV) can occur through abiotic interactions with Fe(III), which subsequently generate U(VI) and Fe(II) (Senko, Istok, Suflita, & Krumholz, 2002), with the rate of oxidation dependent on mineralogy and surface area of Fe(III) minerals present. Others have shown that the presence of chelating agents (e.g. citrate, EDTA, and NTA; Stewart et al., 2012); humic substances (Burgos et al., 2007; Gu, Yan, Zhou, & Watson, 2005); and microbially produced bicarbonate (Wan et al., 2005) stimulated oxidation of $UO_2$(s) through the formation of stable U(VI) complexes. Further, pure cultures of DMRB (e.g. SRB – *Desulfovibrio desulfuricans* G20) were shown to re-oxidize U(IV) under electron donor limiting, sulfate-reducing conditions (Sani et al., 2004, 2005; Singh et al., 2014). In these studies, it has been shown that once the electron donor is consumed, oxidants even as weakly energetic as iron (III) (hydr)oxides can re-oxidize biogenically produced $UO_2$ (e.g. Sani et al., 2004). Re-oxidation of $UO_2$ by Fe(III) (hydr)oxides is troubling because of the ubiquitous presence of these subsurface minerals, and because the re-oxidation of $UO_2$ in this case has been shown to occur even under sulfate-reducing conditions (redox potentials of $\leq -200$ mV) (Sani et al., 2005). The thermodynamic viability of $UO_2$ oxidation by Fe(III) (hydr)oxides has been assessed (Ginder-Vogel, Criddle, & Fendorf, 2006), showing oxidation, even under mildly reducing conditions in the presence of ferrihydrite, and also depending on solution composition, particularly on Fe(II) and inorganic carbon and calcium concentrations. Ginder-Vogel, Criddle, and Fendorf (2010) demonstrate that biogenic uraninite oxidation by ferrihydrite appears to proceed through a soluble U(IV) intermediate and results in the production of Fe(II) and dissolved U(VI). It has also been shown that the re-oxidation of bioreduced U occurred only in the presence of growing cultures of SRB (G20). Hematite added to heat-killed cell controls, however, re-oxidized only a very small amount of U, whereas the hematite-free and G20-free controls did not result in any re-oxidation. When hematite was treated with bioreduced U under abiotic conditions, no bioreduced U re-oxidation was observed, thus indicating that the re-oxidation was somehow biocatalyzed (Sani et al., 2005).

In a natural multicomponent, multiphase setting, phase-association may exert primary control on U fate and transport. Thermodynamically, sulfide is preferentially oxidized by Fe(III) (hydr)oxides, compared to biogenic $UO_2$, and for this reason the relative rate of sulfide and $UO_2$ oxidation may play a key role In whether $UO_2$ re-oxidizes or not. The amount of Fe(II) in solution is another important factor, with precipitation of Fe(II) minerals lowering the activity of Fe(II) in solution and thus increasing the potential for $UO_2$ re-oxidation. The previously reported greater (and unintuitive) $UO_2$ re-oxidation by hematite compared to ferrihydrite may be explained by the greater reactivity of ferrihydrite toward sulfide, yielding higher Fe(II) in solution than when hematite is present (Spycher et al., 2011).

These findings suggest that while abiotic and biotic U(IV) formation appear to be effective processes for subsurface U contaminant immobilization, further investigations are needed to better understand the biogeochemical factors that control long-term stabilities of bioprecipitated U(IV).

## 11.5 Summary and Conclusions

Radionuclides including U constitute a significant environmental issue in the western United States as well as throughout the world, and have attracted the attention of many stakeholders. Because of its toxicity, which is primarily due to chemical rather than radiological action, U has been a priority pollutant in the subsurface, particularly for the DOE. Uranium-contaminated groundwater threatens ecosystem health, and poses a real threat to human health when sensitive receptors are affected. In situ bioreductive immobilization may be an attractive option at many sites for removal of U from the aqueous phase because of low cost and minimal secondary waste generation. Although the biogenic uraninite mineral ($UO_2$) has been considered as the primary product of U(VI) bioreduction, recent studies have shown the formation of distinct monomeric/mononuclear U(IV) phases. These bioreduced U(IV) particulates were observed to be at the nanometer-particle scale and below, raising concerns of re-mobilization and re-oxidation of the bioreduced U, once traveled in long distances in the subsurface environment.

In situ microbial reduction of soluble U(VI) to insoluble U(IV) is one potential mechanism of U immobilization. It has been shown that acetate stimulates Fe(III)-reducing bacteria (FeRB), mainly *Geobacter* species, resulting in immobilization of U. However, at the Old Rifle UMTRA Site, upon transition from Fe(III)-reducing to sulfate-reducing conditions, U remobilization was observed (Anderson et al., 2003). These data suggest that FeRB may initially dominate microbial activity, but once readily bioavailable Fe(III) is reduced, the systems will shift to sulfate-reducing conditions, which will then control the stability of bioreduced U. While SRB have been shown to reduce U(VI) in the absence of soil minerals under nongrowth conditions (Ganesh et al., 1997; Ganesh, Robinson, Chu, Kucsmas, & Reed, 1999; Lovley & Phillips, 1992; Lovley et al., 1991, Lovley, Roden, et al., 1993a, 1993b; Payne, Gentry, Rapp-Giles, Casalot, & Wall, 2002; Spear et al., 1999, 2000), data on U(VI) reduction under growth conditions with soil minerals are sparse (Sani et al., 2004, 2005). Recent evidence indicates that SRB-mediated uraninite nanocrystals (<5 nm diameter) can be re-oxidized in the presence of SRB and Fe(III)-(hydr)oxides, specifically hematite, goethite, and ferrihydrite (Sani et al., 2004, 2005). Thus, residual Fe(III) may be available to re-oxidize biologically reduced U, thereby significantly affecting U speciation and partitioning between aqueous and solid phases. Further fundamental research is needed to characterize the mechanism(s) of, and accurately model, these important geochemical reactions. It currently is not known how SRB activity affects the stability of bioreduced U in the presence of Fe(III)-(hydr)oxides.

The mobility and bioavailability of subsurface U is dependent on abiotic, biotic, and coupled reactions that determine oxidation state, aqueous complexation, and interactions with mineral surfaces, and thus, understanding how these complex processes proceed as part of U remediation stewardship strategies is paramount. Consequently, any serious investigation of U mobility must consider the coupled effects of aqueous complexation with various ligands, competitive sorption processes, and both biotic and abiotic reactions involving U and other redox sensitive species. This can only be achieved by closely integrating experimental studies with a comprehensive multicomponent biogeochemical modeling component. Field scale modeling, remediation system design, and monitoring of reductive immobilization schemes in situ need to rely on practical and accurate quantitative models that honor such dominant processes. This chapter provides a review of an overall understanding of biogeochemical interactions of nanoparticulate U in the subsurface, shedding light on the identification and characterization of fundamental processes that control the stability of reduced U. The chapter focuses on the current status of knowledge on these fundamental processes and interactions and also provides future directions of research and implementation for ultimate successful management of U and other similar radioactive contaminants in the subsurface environment.

## Questions and Answers

**1** What are the main forms of U in nature?
The most common form of U is the oxidized U(VI) species, which exist as different ions depending on the alkalinity, pH, and complexants [e.g. $UO_2^{2+}$, $UO_2OH^+$, $UO_2(OH)_2$, $UO_3{\cdot}3H_2O$, $UO_2(CO_3)_2^{2-}$, $UO_2(CO_3)_2^{4-}$ and U(VI)-Ca-$CO_3$]. These oxidized species of U are highly soluble and mobile in aqueous environments, and therefore present a significant health hazard due to their solubility and mobility properties. On the other hand, under reducing conditions, soluble U(VI) can be reduced to insoluble U(IV), which is commonly found in the form of mineral uraninite ($UO_2$).

**2** What is bioreduced U? What are the known forms of bioreduced U(IV) phases?
Bioreduced U is generally regarded as the most desirable product of biologically reduced U resulting from the bioreductive process, due to its inherent low solubility under reducing conditions. At least three types of bioreduced U(IV) have been reported in the literature: uraninite, nanoparticulate uraninite, and mononuclear U(IV) phases. Although uraninite and nanoparticulate uraninite have been reported to be more crystalline, mononuclear U(IV) has been considered to be more amorphous, posing challenges due to its mobile and unstable nature.

**3** What are the ways to remediate U contamination?
In order to minimize risks associated with U(VI), the facilitation of abiotic or biotic reducing conditions is a plausible remediation strategy to

biogeochemically precipitate U. Microbial U(VI) reduction is a preferred mechanism if stimulation of microorganisms can be properly achieved. A promising strategy for in situ remediation of subsurface U contamination is through facilitating biostimulation of iron- and/or sulfate-reducing indigenous bacterial species resulting in dissimilatory reduction of soluble U(VI) to insoluble U(IV).

4 What kind of microorganisms can reduce U(VI) to insoluble U(IV)?
More than 25 species of phylogenetically diverse prokaryotes can mediate enzymatic U(VI) reduction to insoluble U(IV) as uraninite $UO_2$. The most studied microbes are the dissimilatory metal-reducing bacteria (DMRB) and sulfate-reducing bacteria (SRB) including strains of the genera *Anaeromyxobacter, Geobacter, Shewanella,* and *Desulfovibrio.* Especially the role of SRB has been well-studied and reported for U(VI) bioreduction.

5 Describe enzymatic and nonenzymatic U(VI) reduction.
SRB can reduce U(VI) by direct enzymatic mechanisms as well as indirectly via sulfide production. Direct enzymatic reduction involves the use of electrons from organic compounds to directly reduce U(VI) to U(IV). In nonenzymatic reduction, the oxidation of organic compounds or hydrogen is coupled to the reduction of sulfate ($SO_4^{-2}$) to hydrogen sulfide ($H_2S^-$). This reduced form then interacts with multivalent metals, such as U to form insoluble metal precipitates.

6 What is re-oxidation? How does U(IV) reoxidation affect long-term stability of bioreduced U?
In the presence of several oxidants, biologically reduced U(IV) can be re-oxidized back to U(VI). This re-oxidation process significantly affects the U speciation and partitioning between aqueous and solid phases. For effective long-term immobilization of uranium through bioreduction or chemical reduction, there should be no abiotic or biotic reoxidation of biogenic U(IV). Therefore the long-term stability of nanometer-sized particles of bioreduced U is of critical importance.

7 Which oxidants can be responsible for U(IV) re-oxidation?
Bioreduced U may be re-oxidized in the presence of several oxidants including oxygen, nitrite, Fe(III), and Mn(IV). Nitrate reduction products ($NO_2^-$, $N_2O$, NO) were found to cause U(IV) re-oxidation, except when acetate, $H_2S$, aqueous Fe(II), and iron-sulfide minerals were present in excess relative to nitrate. The presence of chelating agents, humic substances, and microbially produced bicarbonate has also been shown to

oxidize U(IV) phases. Furthermore, pure cultures of SRB were shown to re-oxidize U(IV) under electron donor limiting, sulfate-reducing conditions. In these studies, it has been shown that once the electron donor is consumed, oxidants even as weakly energetic as iron(III) (hydr)oxides can re-oxidize biogenically produced $UO_2$.

8 What are chelators? What are their known roles in biogeochemical interactions of U in natural subsurface settings?
Chelators have been commonly used in the past as cleaning agents in order to stimulate metal–chelate complex formation in the subsurface. Natural organic matter (NOM) can also be used as an effective metal chelator and electron shuttle during dissimilatory iron reduction. Recent studies have shown that the chelating agents can promote $UO_2$ re-oxidation with Fe(III) (hydr)oxide dissolution, indicating the significance of chelators found in U-contaminated sites in mobilizing U. However, the knowledge of the mechanisms of NOM, electron shuttling agents and metal complexants in U and Fe bioreduction processes is not well known.

## References

Abdelouas, A., Lu, Y., Lutze, W., & Nuttall, H. E. (1998). Reduction of U(VI) to U(IV) by indigenous bacteria in contaminated ground water. *J. Contam. Hydrol.*, 35, 217–233.

Abdelouas, A., Lutze, W., Gong, W., Nuttall, H. E., Strietelmeier, B. A., & Travis, B. J. (2000). Biological reduction of uranium in groundwater and subsurface soil. *Sci. Total Environ.*, 250, 21–35.

Akob, D. M., Mills, H. J., Gihring, T. M., Kerkhof, L., Stucki, J. W., Anastacío, A. S., . . . Kostka, J. E. (2008). Functional diversity and electron donor dependence of microbial populations capable of U(VI) reduction in radionuclide-contaminated subsurface sediments. *Appl. Environ. Microbiol.*, 74, 3159–3170.

Alessi, D. S., Uster, B., Veeramani, H., Suvorova, E. I., Lezama-Pacheco, J. S., Stubbs, J. E., . . . Bernier-Latmani, R. (2012). Quantitative separation of monomeric U(IV) from $UO_2$ in products of U(VI) reduction. *Environ. Sci. Technol.*, 46(11), 6150–6157.

Amonette, J. E. (2003). Iron redox chemistry of clays and oxides: environmental applications. In A. Fitch (Ed.), *Electrochemical Properties of Clays*, CMS Workshop Lectures (Vol. 10, pp. 89–146). Aurora, CO: The Clay Minerals Society.

Anderson, R. T., Vrionis, H. A., Ortiz-Bernad, I., Resch, C. T., Long, P. E., Dayvault, R., . . . Lovley, D. R. (2003). Stimulating the in situ activity of

*Geobacter* species to remove uranium from the groundwater of a uranium-contaminated aquifer. *Appl. Environ. Microbiol.*, 69, 5884–5891.

Bank, T. L., Kukkadapu, R. K., Madden, A. S., Ginder-Vogel, M. A., Baldwin, M. E., & Jardine, P. M. (2008). Effects of gamma-sterilization on the physico-chemical properties of natural sediments. *Chem. Geol.*, 251, 1–7.

Barnes, S. P., Bradbrook, S. D., Cragg, B. A., Marchesi, J. R., Weightam, A. J., Fry, J. C., & Parkes, R. J. (1998). Isolation of sulfate-reducing bacteria from deep sediment layers of the Pacific Ocean. *Geomicrobiol. J.*, 15, 67–84.

Bernier-Latmani, R., Veeramani, H., Della, V. E., Junier, P., Lezama-Pacheco, J. S., Suvorova, E. I., . . . Bargar, J. R. (2010). Non-uraninite products of microbial U(VI) reduction. *Environ. Sci. Technol.*, 44, 9456–9462.

Beyenal, H., Sani, R. K., Peyton, B. M., Dohnalkova, A., Amonette, J. E., & Lewandowski, Z. (2004). Uranium immobilization by sulfate reducing biofilms. *Environ. Sci. Technol.*, 38, 2067–2074.

Bhattacharya, D., Jumawan, A., Sun, G., Sund-Hegelburg, C., & Schwitzgebel, K. (1981). Precipitation of heavy metals with sodium sulfide: bench-scale and full-scale experimental results. *AIChE Symp. Ser.*, 209, 31–38.

Boyanov, M. I., Fletcher, K. E., Kwon, M. J., Rui, X., O'Loughlin, E. J., Löffler, F. E., & Kemner, K. M. (2011). Solution and microbial controls on the formation of reduced U(IV) species. *Environ. Sci. Technol.*, 45(19), 8336–8344.

Boyanov, M. I., O'Loughlin, E. J., Kwon, M.-J., Mishra, B., Rui, X., Shibata, T., & Kemner, K. M. (2010). Goldschmidt abstracts 2010 – B. *Geochim. Cosmochim. Acta*, 74, A38–A134.

Brim, H., Vankateswaren, A., Kostandarithes, H. M., Fredrickson, J. K., & Daly, M. J. (2003). Engineering *Deinococcus geothermalis* for bioremediation of high-temperature radioactive waste environments. *Appl. Environ. Microbiol.*, 69, 4575–4582.

Brodie, E. L., Desantis, T. Z., Joyner, D. C., Baek, S. M., Larsen, J. T., Andersen, G. L., . . . Firestone, M. K. (2006). Application of a high-density oligonucleotide microarray approach to study bacterial population dynamics during uranium reduction and reoxidation. *Appl. Environ. Microbiol.*, 72, 6288–6298.

Brooks, S. C., Fredrickson, J. K., Carroll, S. L., Kennedy, D. W., Zachara, J. M., Plymale, A. E., . . . Fendorf, S. (2003). Inhibition of bacterial U(VI) reduction by calcium. *Environ. Sci. Technol.*, 37, 1850–1858.

Buerge, I. J., & Hug, S. J. (1997). Kinetics and pH dependence of chromium(VI) reduction by iron(II). *Environ. Sci. Technol.*, 31, 1426–1432.

Buerge, I. J., & Hug, S. J. (1999). Influence of mineral surfaces on chromium(VI) reduction by iron(II). *Environ. Sci. Technol.*, 33, 4285–4291.

Burgos, W. D., McDonough, J. T., Senko, J. M., Zhang, G., Dohnalkova, A. C., Kelly, S. D., . . . Kemner, K. M. (2008). Characterization of uraninite nanoparticles produced by *Shewanella oneidensis* MR-1. *Geochim. Cosmochim. Acta*, 72(20), 4901–4915.

Burgos, W. D., Senko, J. M., Dempsey, B. A., Roden, E. E., Stone, J. J., Kemner, K. M., & Kelly, S. D. (2007). Soil humic acid decreases biological uranium(VI) reduction by *Shewanella putrefaciens* CN32. *Environ. Eng. Sci.*, 24(6), 755–761.

Caccavo, F., Jr., Blakemore, R. P., & Lovley, D. R. (1992). A hydrogen-oxidizing, Fe(III)-reducing microorganism from the Great Bay Estuary, New Hampshire. *Appl. Environ. Microbiol.*, 58, 3211–3216.

Chang, Y. J., Long, P. E., Geyer, R., Peacock, A. D., Resch, C. T., Sublette, K., . . . White, D. C. (2005). Microbial incorporation of C-13-labeled acetate at the field scale: detection of microbes responsible for reduction of U(VI). *Environmental Science and Technology*, 39(23), 9039–9048.

Coleman, M. D., Hedrick, D. B., Lovley, D. R., White, D. C., & Pye, K. (1993). Reduction of Fe(III) in sediments by sulfate-reducing bacteria. *Nature*, 36, 436–438.

Craft, E., Abu-Qare, A., Flaherty, M., Garofolo, M., Rincavage, H., & Abou-Donia, M. (2004). Depleted and natural uranium: chemistry and toxicological effects. *J. Toxicol. Environ. Health Part B Crit. Rev.*, 7, 297–317.

Finneran, K. T., Housewright, M. E., & Lovley, D. R. (2002). Multiple influences of nitrate on uranium solubility during bioremediation of uranium-contaminated subsurface sediments. *Environ. Microbiol.*, 4, 510–516.

Fletcher, K. E., Boyanov, M. I., Thomas, S. H., Wu, Q., Kemner, K. M., & Löffler, F. E. (2010). U(VI) reduction to mononuclear U(IV) by *Desulfitobacterium* species. *Environ. Sci. Technol.*, 44(12), 4705–4709.

Ganesh, R., Robinson, K. G., Chu, L., Kucsmas, D., & Reed, G. D. (1999). Reductive precipitation of uranium by Desulfovibrio desulfuricans: evaluation of cocontaminant effects and selective removal. *Water Research*, 33(16), 3447–3458.

Ganesh, R., Robinson, K. G., Reed, G. D., & Sayler, G. (1997). Reduction of hexavalent uranium from organic complexes by sulfate- and iron-reducing bacteria. *Appl. Environ. Microbiol.*, 63, 4385–4391.

Ginder-Vogel, M., Criddle, C. S., & Fendorf, S. (2006). Thermodynamic constraints on the oxidation of biogenic UO2 by Fe(III) (hydr)oxides. *Environ. Sci. Technol.*, 40, 3544–3550.

Ginder-Vogel, M., Criddle, C. S., & Fendorf, S. (2010). Kinetic and mechanistic constraints on the oxidations of biogenic uraninite by ferrihydrite. *Environ. Sci. Technol.*, 44, 163–169.

Gorby, Y. A., & Lovley, D. R. (1992). Enzymatic uranium precipitation. *Environ. Sci. Technol.*, 26, 205–207.

Gu, B., Yan, H., Zhou, P., & Watson, D. B. (2005). Natural humics impact uranium bioreduction and oxidation. *Environ. Sci. Technol.*, 39, 5268–5275.

Istok, J. D., Senko, J. M., Krumholz, L. R., Watson, D., Bogle, M. A., Peacock, A., . . . White, D. C. (2004). In situ bioreduction of technetium and uranium in a nitrate-contaminated aquifer. *Environ. Sci. Technol.*, 38(2), 468–475.

Jeon, B. H., Dempsey, B. A., Burgos, W. D., Barnett, M. O., & Roden, E. E. (2005). Chemical reduction of U(VI) by Fe(II) at the solid–water interface using natural and synthetic Fe(III) oxides. *Environ. Sci. Technol.*, 39, 5642–5649.

Kipp, G., Stone, J. J., & Stetler, L. D. (2009). Metal transport in sediments near abandoned uranium mines in Harding County, South Dakota. *Appl. Geochem.*, 24, 2246–2255.

Komlos, J., Mishra, B., Lanzirotti, A., Myneni, S. C. B., & Jaffe, P. R. (2008). Real-time speciation of uranium during active bioremediation and U(IV) reoxidation. *J. Environ. Eng.*, 134(2), 78–86.

Langmuir, D. (1978). Uranium solution-mineral equilibria at low temperatures with applications to sedimentary ore deposits. *Geochim. Cosmochim. Acta*, 42(6), 547–569.

Larson, L., Kipp, G., Mott, H., & Stone, J. J. (2012). Sediment pore-water interactions associated with transport of arsenic and uranium within a watershed impacted by historical uranium mining activity, North Cave Hills region, South Dakota. *Appl. Geochem.*, 27(4), 879–891. doi:10.1016/j.apgeochem.2012.01.008

Larson, L., & Stone, J. J. (2011). Sediment-bound arsenic and uranium within the Bowman–Haley reservoir, North Dakota. *Water Air Soil Pollut.*, 219, 27–42.

Le Gall, J., Payne, W. J., Chen, L., Liu, M. Y., & Xavier, A. V. (1994). Localization and specificity of cytochromes and other electron transfer proteins from sulfate-reducing bacteria. *Biochimie*, 76, 655–665.

Liu, C., Zachara, J. M., Fredrickson, J. K., Kennedy, D. W., & Dohnalkova, A. (2002). Modeling the inhibition of the bacterial reduction of U(VI) by β-$MnO_2$(s). *Environ. Sci. Technol.*, 36, 1452–1459.

Lloyd, J. R., & Lovley, D. R. (2001). Microbial detoxification of metals and radionuclides. *Curr. Opin. Biotechnol.*, 12, 248–253.

Lovley, D. R. (1993). Dissimilatory metal reduction. *Annu. Rev. Microbiol.*, 47, 263–290.

Lovley, D. R., & Phillips, E. J. P. (1992). Bioremediation of uranium contamination with enzymatic uranium reduction. *Environ. Sci. Technol.*, 26, 2228–2234.

Lovley, D. R., Phillips, E. J. P., Gorby, Y. A., & Landa, E. R. (1991). Microbial reduction of uranium. *Nature*, 350, 413–416.

Lovley, D. R., Roden, E. E., Phillips, E. J. P., & Woodward, J. C. (1993a). Enzymatic iron and uranium reduction by sulfate-reducing bacteria. *Mar. Geol.*, 113, 41–53.

Lovley, D. R., Widman, P. K., Woodward, J. C., & Phillips, E. J. P. (1993b). Reduction of uranium by cytochrome c3 of *Desulfovibrio vulgaris*. *Appl. Environ. Microbiol.*, 59, 3572–3576.

McLean, J., & Beveridge, T. J. (2001). Chromate reduction by a *Pseudomonas* isolated from a site contaminated with chromated copper arsenate. *Appl. Environ. Microbiol.*, 67, 1076–1084.

Merroun, M. L., & Selenska-Pobell, S. (2008). Bacterial interactions with uranium: an environmental perspective. *J. Contam. Hydrol.*, 102, 285–295.

North, N. N., Dollhopf, S. L., Petrie, L., Istok, J. D., Balkwill, D. L., & Kostka, J. E. (2004). Change in bacterial community structure during in situ biostimulation of subsurface sediment cocontaminated with uranium and nitrate. *Appl. Environ. Microbiol.*, 70(8), 4911–4920.

Ortiz-Bernad, I., Anderson, R. T., Vrionis, H. A., & Lovley, D. R. (2004). Resistance of solid-phase U(VI) to microbial reduction during in situ bioremediation of uranium-contaminated groundwater. *Appl. Environ. Microbiol.*, 70, 7558–7560.

Palmisano, A., & Hazen, T. (2003). *Bioremediation of metals and radionuclides: What it is and how it works (No. LBNL-42595-(2003))*. Berkeley, CA: Lawrence Berkeley National Lab.(LBNL).

Payne, R. B., Gentry, D. A., Rapp-Giles, B. J., Casalot, L., & Wall, J. D. (2002). Uranium reduction by *Desulfovibrio desulfuricans* strain G20 and a cytochrome c3 mutant. *Applied and Environmental Microbiology*, 68(6), 3129–3132.

Poulson, S. R., Colberg, P. J. S., & Drever, J. I. (1997). Toxicity of heavy metals (Ni, Zn) to *Desulfovibrio desulfuricans. Geomicrobiol. J.*, 14, 41–49.

Rastogi, G., Osman, S., Andersen, G. L., & Sani, R. K. (2010). Microbial diversity associated with uranium mining-impacted soils as revealed by high-density 16S microarray and clone library. *Microb. Ecol.*, 59(1), 94–108.

Robert, M., & Berthelin, J. (1986). Role of biological and biochemical factors in soil mineral weathering. Chap 12. In P. M. Huang & M. Schnitzer (Eds.), *Interactions of Soil Minerals with Natural Organics and Microbes* (pp. 453–495). Madison, WI: Soil Science Society of America. (Special Publication No. 17)

Sani, R. K., Geesey, G., & Peyton, B. M. (2001). Assessment of Pb toxicity to *Desulfovibrio desulfuricans* G20: influence of components of lactate C medium. *Adv. Environ. Res.*, 5(3), 269–276.

Sani, R. K., Peyton, B. M., Amonette, J. E., & Dohnalkova, A. (2005). Reoxidation of uranium in the presence of iron(III)-(hydr)oxides under sulfate-reducing conditions. *Environ. Sci. Technol.*, 39, 2059–2066.

Sani, R. K., Peyton, B. M., Amonette, J. E., & Geesey, G. G. (2004). Reduction of uranium(VI) under sulfate reducing conditions in the presence of Fe(III)-(hydr) oxides. *Geochim. Cosmochim. Acta*, 68, 2639–2648.

Sani, R. K., Peyton, B. M., & Dohnalkova, A. (2006). Effect of uranium(VI) in aqueous media on *Desulfovibrio desulfuricans* G20. *Environ. Toxicol. Chem.*, 25, 1231–1238.

Sani, R. K., Peyton, B. M., Smith, W. A., Apel, W. A., & Petersen, J. N. (2002). Dissimilatory reduction of Cr(VI), Fe(III), and U(VI) by cellulomonas isolates. *Appl. Microbiol. Biotechnol.*, 60, 192–199.

Schofield, E. J., Veeramani, H., Sharp, J. O., Suvorova, E., Bernier-Latmani, R., Mehta, A., . . . BargarJR (2008). Structure of biogenic $UO_2$ produced by *Shewanella oneidensis* MR-1. *Environ. Sci. Technol.*, 42(21), 7898–7905.

Şengör, S. S., Singh, G., Dohnalkova, A., Spycher, N., Ginn, T. R., Peyton, B. M., & Sani, R. K. (2016). Impact of different environmental conditions on the aggregation of biogenic U (IV) nanoparticles synthesized by *Desulfovibrio alaskensis* G20. *BioMetals*, 29(6), 965–980.

Senko, J. M., Istok, J. D., Suflita, J. M., & Krumholz, L. R. (2002). In situ evidence for uranium immobilization and remobilization. *Environ. Sci. Technol.*, 36, 1491–1496.

Senko, J. M., Mohamed, Y., Dewers, T. A., & Krumholz, L. R. (2005). Role for Fe(III) minerals in nitrate-dependent microbial U(IV) oxidation. *Environ. Sci. Technol.*, 39, 2529–2536.

Sharpe, J. O., Schofield, E. L., Veeramani, H., Suvorova, E. I., Kennedy, D. W., Marshall, M. J., . . . Bernier-Latmani, R. (2009). Structural similarities between biogenic uraninites produced by phylogenetically and metabolically diverse bacteria. *Environ. Sci. Technol.*, 43, 8295–8301.

Singer, D. M., Farges, F., & Brown, G. E. (2009). Biogenic nanoparticulate $UO_2$: synthesis, characterization, and factors effecting surface reactivity. *Geochim. Cosmochim. Acta*, 73(12), 3593–3611.

Singh, G., Şengör, S. S., De, J., Bhalla, A., Kumar, S., Stewart, B., . . . Sani, R. K. (2014). Reoxidation of biogenic reduced uranium—a challenge towards bioremediation. *Crit. Rev. Environ. Sci. Technol.*, 44, 391–415.

Singhal, R. K., Kumar, A., Rupali Karpe, P. J., Datta, M., & Hedge, A. G. (2005). Association of uranium with colloids of natural organic matter in subsurface aquatic environment. *J. Radioanal. Nucl. Chem.*, 265, 405–408.

Sivaswamy, V., Boyanov, M. I., Peyton, B. M., Viamajala, S., Gerlach, R., Apel, W. A., . . . Borch, T. (2011). Multiple mechanisms of uranium immobilization by *Cellulomonas* sp. strain ES6. *Biotechnol. Bioeng.*, 108(2), 264–276.

Spear, J. R., Figueroa, L. A., & Honeyman, B. D. (1999). Modeling the removal of uranium U(VI) from aqueous solutions in the presence of sulfate-reducing bacteria. *Environ. Sci. Technol.*, 33, 2667–2675.

Spear, J. R., Figueroa, L. A., & Honeyman, B. D. (2000). Modeling reduction of uranium U(VI) under variable sulfate concentrations by sulfate-reducing bacteria. *Appl. Environ. Microbiol.*, 66, 3711–3721.

Spycher, N. F., Issarangkun, M., Stewart, B., Sengor, S., Belding, E., Ginn, T. M., . . . Sani, R. K. (2011). On modeling biogenic uraninite precipitation and reoxidation by iron(III)(hydr)oxides: thermodynamic and kinetic considerations. *Geochim. Cosmochim. Acta*, 75, 4426–4440.

Stewart, B. D., Girardot, C., Spycher, N., Sani, R. K., & Peyton, B. M. (2012). Influence of chelating agents on biogenic uraninite reoxidation by Fe(III)(hydr)oxides. *Environ. Sci. Technol.*, 47(1), 364–371.

Stone, J. J., Burgos, W. D., Royer, R. A., & Dempsey, B. A. (2007b). Effect of natural organic matter on zinc inhibition of biological Fe(III) and nitrate reduction by *Shewanella putrefaciens* CN32. *Environ. Sci. Technol.*, 41(15), 5284–5290.

Stone JJ, Stetler LD, and Schwalm A. 2007a. Final Report: North Cave Hills Abandoned Uranium Mines Impact Investigation. *Report No. 04-17*, Department of Civil and Environmental Engineering, South Dakota School of Mines and Technology. Prepared for USDA-USFS Northern Regional Office, Missoula, MT.

Suzuki, Y., Kemner, K. M., & Banfield, J. F. (2002). Nanometre-size products of uranium bioreduction. *Nature*, 419, 134.

Suzuki, Y., & Suko, T. (2006). Geomicrobiological factors that control uranium mobility in the environment: update on recent advances in the bioremediation of uranium-contaminated sites. *J. Mineral. Petrol. Sci.*, 101(6), 299–307.

Ulrich, K. U., Ilton, E. S., Veeramani, H., Sharp, J. O., Bernier-Latmani, R., Schofield, E. J., . . . Giammar, D. E. (2009). Comparative dissolution kinetics of biogenic and chemogenic uraninite under oxidizing conditions in the presence of carbonate. *Geochim. Cosmochim. Acta*, 73, 6065–6083.

Ulrich, K. U., Singh, A., Schofield, E. J., Bargar, J. R., Veeramani, H., Sharp, J. O., . . . Giammar, D. E. (2008). Dissolution of biogenic and synthetic $UO_2$ under varied reducing conditions. *Environ. Sci. Technol.*, 42(15), 5600–5606.

Vrionis, H. A., Anderson, R. T., Ortiz-Bernad, I., O'Neill, K. R., Resch, C. T., Peacock, A. D., . . . Lovley, D. R. (2005). Microbiological and geochemical heterogeneity in an in situ uranium bioremediation field site. *Appl. Environ. Microbiol.*, 71(10), 6308–6318.

Wan, J., Tokunaga, T. K., Brodie, E., Zheng, Z., Herman, D., Hazen, T. C., . . . Sutton, S. R. (2005). Reoxidation of bioreduced uranium under reducing conditions. *Environ. Sci. Technol.*, 39, 6162–6169.

Webb C, Davis A, and Hodge V. 1995. Rare earth elements at abandoned uranium mines in the southern Black Hills of South Dakota. *Preprints-Society of Mining Engineers of AIME.*

Wu, Q., Sanford, R. A., & Loffler, F. E. (2006). Uranium(VI) reduction by anaeromyxobacter dehalogenans strain 2CP-C. *Appl. Environ. Microbiol.*, 72(5), 3608–3614.

Yabusaki, S. B., Fang, Y., Long, P. E., Resch, C. T., Peacock, A. D., Komlos, J., . . . Anderson, R. T. (2007). Uranium removal from groundwater via *in situ* biostimulation: field-scale modeling of transport and biological processes. *J. Contam. Hydrol.*, 93(1), 216–235.

# 12

# Characterization and Quantification of Mobile Bioreduced Uranium Phases

*S. Sevinç Şengör[1,*] and Rajesh K. Sani[2,3]*

[1] *Bobby B. Lyle School of Engineering, Southern Methodist University, Dallas, TX, USA*
[2] *Department of Environmental Engineering, Middle East Technical University, Dumlupınar Bulvarı No: 1, Ankara, Turkey*
[3] *Department of Chemical and Biological Engineering, South Dakota School of Mines and Technology, Rapid City, SD, USA*

## 12.1 Introduction

Uranium (U) contamination is a widespread environmental concern in many surface water, groundwater, soil, and sediment environments due to its high toxicity and long-lives in the ecosystem. Biological reduction of soluble U(VI) to insoluble U(IV) with accompanying precipitation of uraninite nanoparticles has been considered as a viable option to remediate U. Under reducing conditions, sulfate-reducing bacteria (SRB), which are naturally present in many field settings, have been known to enzymatically and nonenzymatically reduce U(VI) and control the stability of bioreduced U (Sani, Geesey, & Peyton, 2001; Sani, Peyton, Amonette, & Geesey, 2004; Sani, Peyton, & Dohnalkova, 2006). In particular, SRB has been able to reduce U(VI) to nanometer-sized uraninite particles, observed both inside and outside the cells (Sani et al., 2006; Şengör et al., 2016). Although the accumulation of uraninite particles inside the cell (periplasm and cytoplasm) may physically prevent any transport and mobilization of U(IV) nanoparticles, the extracellular U(IV) are prone to be transported and subsequently oxidized back to U(VI) under relevant biochemical conditions (Singh et al., 2014).

Recent studies indicate that biogenic U(IV) can occur as a combination of nanoparticulate uraninite and mononuclear U(IV) phases with appreciable mass fraction to exist at nanometer particle scale and below. Each of these distinct phases may have significant different characteristics, compositions, and responses to the

*Current address: Department of Environmental Engineering, Middle East Technical University, Dumlupınar Bulvarı No: 1, Ankara, Turkey

*Microbial Interactions at Nanobiotechnology Interfaces: Molecular Mechanisms and Applications*, First Edition. Edited by R. Navanietha Krishnaraj and Rajesh K. Sani.

biochemical conditions and thus will likely have differential rates of transport and transformation in aquifers. It is vital to understand the structure, composition, and stabilities of these U(IV) phases for predicting long-term risk of U migration with or without re-oxidation in natural aquatic systems. Developing a fundamental, mechanistic understanding of the fate of uranium in complex systems requires quantifying the partitioning of U in these multiple distinctive phases. Therefore, the characterization of biogenic U(IV) has been crucial for its long-term stability and success of bioremediation strategies. The goal of this chapter is to provide an overview of the tools for characterization and quantification of bioreduced U phases. A numerical integration method is also presented for easy quantification of mobile bioreduced U(IV) using total dissolved U and U(VI) concentration data. The objectives of the chapter are listed as follows:

- Provide an overview of the tools for physicochemical analysis of reduced U(IV) nanoparticles.
- Present a numerical integration method for easy quantification of mobile bioreduced U(IV) phases.

## 12.2 Characterization of Biogenic U(IV)

Tools for physicochemical analysis of colloid phases include size exclusion chromatography, ultracentrifugation, dialysis, and flow field-flow fractionation based methods, each of which have been shown to be useful for investigating colloidal systems (Geckeis, Ngo Manh, Bouby, & Kim, 2003; Zhang, Bryan, Livens, & Jones, 1997; Zänker, Richter, & Huttig, 2003). Flow field-flow fractionation (Fl FFF) has been recognized as a fast, inexpensive technique capable of high-resolution separations of colloids from the particulate phase (Litzén, 1993). Fl FFF combined with inductively coupled plasma mass spectrometry (ICP-MS) is a well-established technique to study the sorption of metals and radionuclides to colloidal material (Amarasiriwardena, Siripinyanond, & Barnes, 2001; Geckeis et al., 2003; Ranville & Beckett, 2000). In general, Fl FFF-ICP-MS studies have focused on the characterization of metal-colloid size fractions rather than quantifying the degree to which metals and radionuclides are associated with each fraction. The usefulness of FFF-ICP-MS for determining the association of U with nanoparticles was demonstrated by Jackson, Ranville, Bertsch, and Sowder (2005a) by analyzing 0.45 micron filtrates from soil leaching experiments. Their results showed that in the operationally defined "dissolved" fraction, U was associated with 100 nm inorganic nanoparticles (based on co-elution of U with Al and Fe) and 2–5 nm organic matter (based on U co-elution with UV absorbance at 254 nm). In another experiment, the ability to perform adsorption isotherm experiments using very small sample volumes was demonstrated (Jackson, Ranville, & Neal, 2005b). In these experiments bacterial suspensions (3 ml) were spiked with variable amounts of U and a linear isotherm was produced. In a study of pore waters from a clay-rich aquitard, Ranville, Hendry, Thorsten, Qianli, and Bruce (2007) demonstrated that the FFF-ICP-MS technique was sufficiently sensitive to quantify U-organic matter size distributions in an uncontaminated aquifer without the need for pre-concentration. In a more quantitative framework, Lesher, Ranville, and Honeyman (2009) used Fl FFF-ICP-MS to develop a U-hematite sorption envelope, and validated the dataset by comparison to the same envelope created using the more traditional measurement technique of comparing the aqueous concentration in centrifuged and filtered samples with untreated samples.

Developing a fundamental, mechanistic understanding of the fate of uranium in complex systems requires quantifying the partitioning of U in colloidal, particulate, or nanoparticulate phases. Although most of the studies have considered the uraninite mineral ($UO_2$) as the primary product of U(VI) bioreduction, the formation of distinct monomeric/mononuclear U(IV) phases of U bioreduction has recently been documented, where their chemical characterization has not yet been fully understood (Bernier-Latmani et al., 2010; Boyanov et al., 2011; Fletcher et al., 2010; Sivaswamy et al., 2011). These bioreduced U phase particles are vulnerable to be transported in the subsurface environment, possibly facilitated by association to other mobile colloidal phases and these distinctive U phases will likely demonstrate differential rates of transport and transformation in impacted aquifers. Therefore, tracking these distinctive U phases is critical for successful U bio-stabilization applications in the long term. Mononuclear U(IV) can be measured as described by Fletcher et al. (2010). In addition to FFF-ICP-MS, size fractions of U can also be obtained by ultrafiltration technique using Nanosep devices following Alessi et al. (2010). Sequential filtration with filters ranging in pore size from 2 μm down 0.6 nm [(200 nm, 1 kDa (pore size, 0.6 nm), 5 kDa (pore size, 3 nm), 30 kDa (pore size, 18 nm), 100 kDa (pore size, 60 nm) can be used to separate U reduction products (and associated cellular mass) into different size fractions prior to measuring U concentrations. Samples of unfiltered, filtered oxidized, and unfiltered oxidized can be analyzed for U(VI) content on a kinetic phosphorescence analyzer (KPA). Additionally, all samples can be oxidized and reanalyzed on the KPA to obtain values for total U (UTOT). Uranium (IV) in each size fraction can then be calculated by the difference where U(IV) = UTOT − U(VI). These analyses can quantify bioreduced U that is (i) cell associated (unfiltered oxidized, [A] minus oxidized U in pellets of unfiltered [B]), (ii) in suspension, in aqueous phase, or associated with other colloids, e.g. organic compounds/by-products (filtered oxidized, [C] filtered minus unoxidized, [D]), aggregate/immobile form (unfiltered oxidized, [A] minus 0.2 μm filtered oxidized [C]), and particles <0.6 nm

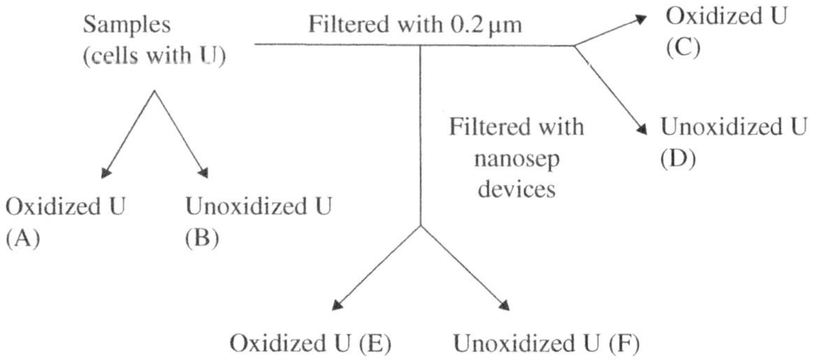


**Figure 12.1** Determination of various phases of bioreduced U.

in size (oxidized, [E] minus unoxidized, [F]) (Figure 12.1). Uranium within the cell space can also be quantified by mass balance, where intracellular U = UTOT − (adsorbed U + extracellular U). Therefore, a mass balance of the bioreduced U distribution among the colloidal, immobile, and extracellular phases can be determined.

## 12.3 Quantification of Mobile Bioreduced U(IV) Nanoparticles

A numerical integration method can also be used for easy quantification of mobile bioreduced U(IV) phases using total dissolved U and U(VI) concentration data, as presented in this section. The amount of the precipitated U(IV) nanoparticles by bioreduction can be calculated by the difference between the measured total dissolved U {i.e. total of U(VI) and U(IV) after oxidized to U(VI)} and U(VI) concentration measurements, in the filtrate samples by KPA-11 analyzer. In order to oxidize U(IV) to U(VI), 0.1 ml of filtered samples can be exposed to air for one hour, and 0.3 ml of concentrated $HNO_3$ can then be added, which would oxidize most of the U(IV) into U(VI) as described previously (Ganesh, Robinson, Reed, & Sayler, 1997; Lovley & Phillips, 1992). These measurements would correspond to the total U {i.e. U(VI) and oxidized (U(IV)}, which is referred to as "filter oxidized U(VI)" (Şengör et al., 2016).

Once the U(VI) and filter oxidized U(VI) concentrations are measured, several interpolation functions can be used for both the U(VI) and filter oxidized U(VI) data points, preferably with the aid of a computer software, e.g. Mathcad. Mathcad has several interpolation and regression functions, where *linterp* provides the easiest straight line interpolation between data points (Maxfield, 2009). Once the functions are defined for both U(VI) and filter oxidized U(VI) concentrations, trapezoidal rule can be used to approximate the region under these functions as a "trapezoid" and the corresponding areas can be calculated for the total experimental time periods, following Eq. 12.1 (Figure 12.2):

$$\int_{t_1}^{t_2} f(x) \approx \Delta t \left[ \frac{f(t_1) + f(t_2)}{2} \right] \tag{12.1}$$

The difference between the areas under the two functions {i.e. corresponding to U(VI) and filter oxidized U(VI) functions: *f*(U(VI)) and *f*(filter oxidized U(VI)), respectively} can be used to approximate the total amount of reduced U(IV) particles (Figure 12.2). As these reduced U(IV) particles would pass through a 0.2 μm filter, these would be characterized as "mobile" U(IV) nanoparticles, as discussed by Şengör et al. (2016). Therefore, the approximation of the total amount of reduced U(IV) particles from the difference between the areas under the curves (AUC) can be further extended to approximate the mobile % of U(IV) nanoparticles based on the initial concentration of U (VI) as follows (Eq. 12.2):

$$\%\text{ of mobile U(IV) nanoparticles} \approx 100 \times \frac{\text{difference between AUC's}}{C_0} \tag{12.2}$$

where $C_0$ is the initial U(VI) concentration.

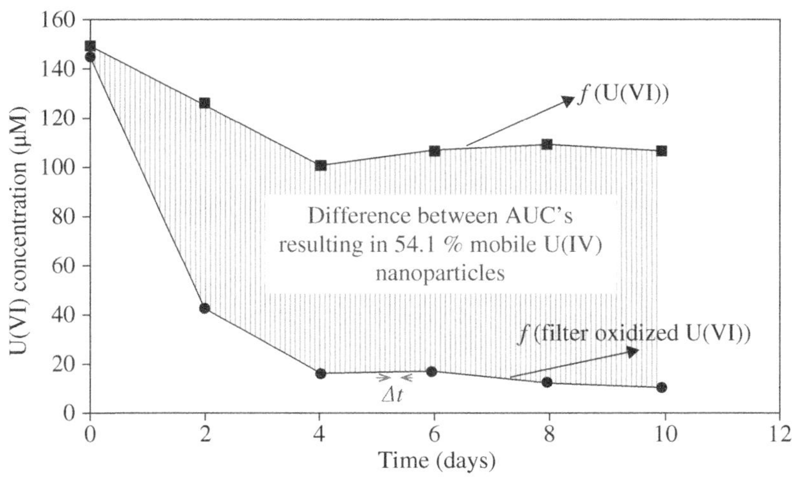


**Figure 12.2** Illustration of an example approximation of total amount of reduced U(IV) particles from the difference between the AUCs corresponding to *f*(U(VI)) and *f*(filter oxidized U(VI)) functions, using the trapezoidal rule of integration by Mathcad software. For an initial U(VI) concentration of 150 μM, the calculated difference between AUCs corresponds to 54.1% of mobile U(IV) nanoparticles.

Trapezoidal rule is one of the most commonly used approximations as part of Newton–Cotes formulas for numerical integration (Cruz-Uribe & Neugebauer, 2002). It is observed to result in extremely accurate integration of periodic functions and to have faster convergence than Simpson's rule for functions with weaker smoothness conditions (Rahman & Schmeisser, 1990; Weideman, 2002). Sufficiently small values of $\Delta t$ need to be used in order to maximize the precision of the approximation. An error analysis can also be performed using various techniques (Fourier series, Residue calculus, and Euler–Maclaurin summation) to determine the difference between the value of the integral and the numerical result (Atkinson, 1989; Weideman, 2002). It should be noted that the method provided here for the numerical integration of the AUCs is one of the simplest and easiest methods available among many others that could be used for the approximation of the mobile U(IV) nanoparticles. More sophisticated and advanced software can alternatively be used with higher precision and convergence.

## 12.4 Summary and Conclusions

This chapter aims to provide an overview of the tools for physicochemical characterization of reduced U(IV) nanoparticles in the aquatic environment. One approach involves serial ultrafiltration and one involves using FFF-ICP-MS. Sequential filtration with filters ranging in pore size from 2 μm down 0.6 nm (1 kDa) can be used to separate U reduction products (and associated cellular mass) into different size fractions prior to measuring U concentrations. Uranium (VI) concentrations can be measured on a kinetic phosphorescence analyzer (KPA) in both filtered and unfiltered samples. Additionally, all samples can be oxidized and reanalyzed on the KPA to obtain values for total U (UTOT). Uranium (IV) in each size fraction can then be calculated by difference where U(IV) = UTOT − U(VI). In a complimentary approach, more detailed information on U size distribution and phase association can come from analysis of subsamples using a FFF-ICP-MS, where the amount of U in each of the nanoparticulate and colloidal phases (i.e. uraninite as aggregate, associated with cells, or associated with iron [sulfide] minerals) can be determined.

A numerical integration method for easy quantification of mobile bioreduced U(IV) phases is also presented. The amount of the precipitated U(IV) nanoparticles by microbially mediated reduction can be approximated using the U(VI) concentration and total dissolved U {i.e. total of U(VI) and U(IV) after oxidation to U(VI)} measurements. An interpolation function can be used for both the U(VI) and filter oxidized U(VI) data points to define the corresponding functions for *f*(U(VI)) and *f*(filter oxidized U(VI)). Trapezoidal rule can be used to approximate the areas under these curves. The difference between the areas under the two functions can then be used to approximate the total amount of reduced U(IV) particles to approximate the mobile % of U(IV) nanoparticles based on the initial concentration of U(VI).

For successful application of long term U bio-stabilization, it is vital to understand the characterization, structure, and composition of distinctive U(IV) phases in multiple particulate, nanoparticulate, and colloidal forms, as these distinctive U phases will likely demonstrate differential rates of transport and transformation in impacted aquifers. Characterization of biogenic U(IV) phases is also critical to facilitate data interpretation in the flow systems and provide sufficient information for predictive modeling of biogenic U(IV) behavior in U-contaminated subsurface environments. Further research is also needed to significantly improve methods for tracking U in the aqueous and subsurface environments.

## Questions and Answers

**1** What is the significance of characterization and quantification of bioreduced U phases?

The distinct bioreduced U phases may have different characteristics, compositions, and responses to the biochemical conditions and thus will likely have differential rates of transport and transformation in aquifers. It is vital to characterize and quantify these U(IV) phases for predicting long-term risk of U migration with or without re-oxidation in natural aquatic systems. Characterization and quantification of these biogenic U(IV) phases are also critical to facilitate data interpretation in flow systems and to provide data that scientists, engineers, regulators, and stakeholders can use to help assess the behavior of biogenic U(IV) on a site-specific basis.

**2** What are the tools available for physicochemical analysis of U(IV) nanoparticles?

Size exclusion chromatography, ultracentrifugation, dialysis, flow field-flow fractionation (Fl FFF) based methods have been shown to be useful for investigating colloidal systems. Fl FFF combined with inductively coupled plasma mass spectrometry (ICP-MS) is also a well-established technique to study the sorption of metals and radionuclides to colloidal materials. In addition to FFF-ICP-MS, sequential filtration approach can be used to separate U reduction products into different size fractions.

**3** Describe sequential filtration technique to characterize various phases of bioreduced U.
Sequential filtration with filters ranging in pore size from 2 μm down 0.6 nm (1 kDa) can be used to separate U reduction products (and associated cellular mass) into different size fractions prior to measuring U concentrations. Uranium (VI) concentrations can be measured on a KPA in both filtered and unfiltered samples. Additionally, all samples can be oxidized and reanalyzed on the KPA to obtain values for total U (UTOT). Uranium (IV) in each size fraction can then be calculated by difference where U(IV) = UTOT − U(VI).

**4** What is trapezoidal rule? How can it be used to quantify bioreduced U phases?
Trapezoidal rule is one of the most commonly used approximations for numerical integration. Using U(VI) concentration and total dissolved U {i.e. total of U(VI) and U(IV) after oxidation to U(VI)}data, an approximate amount of precipitated U(IV) nanoparticles by microbially mediated reduction can be quantified. An interpolation function can be used for both the U(VI) and filter oxidized U(VI) data points to define the corresponding functions for *f*(*U*(*VI*)) and *f*(filter oxidized U(VI)). Trapezoidal rule can be used to approximate the areas under these curves. The difference between the areas under the two functions can then be used to approximate the total amount of reduced U(IV) particles to approximate the mobile % of U(IV) nanoparticles based on the initial concentration of U(VI).

**5** In a laboratory experiment, an initial amount of 500 μM U(VI) is used to determine the mobile percentage of bioreduced U(IV) using SRB. Daily measurements of U(VI) and filter oxidized U(VI) data points for 10 days are used to obtain linear interpolation functions between the data points, where the difference between the areas of these curves corresponding to *f*(U(VI)) and *f*(filter oxidized U(VI)) functions is approximated to be 105.20 μM. What percent of the bioreduced U(IV) from this experiment would be mobile?
Using Eq. 12.2, the mobile % of U(IV) phases can be approximated as:

$$\%\text{ of mobile U}\left(\text{IV}\right)\text{nanoparticles} \approx 100 \times \frac{\text{difference between AUC's}}{C_0}$$

$$\%\text{ of mobile U}\left(\text{IV}\right)\text{nanoparticles} \approx 100 \times \frac{105.20\,\mu\text{M}}{500\,\mu\text{M}}$$

$$\%\text{ of mobile U}\left(\text{IV}\right)\text{nanoparticles} \approx 21\%$$

## References


Alessi, D. S., Uster, B., Veeramani, H., Stubbs, J. E., Lezama-Pacheco, J. S., Bargar, J. R., & Bernier-Latmani, R. (2010). Goldschmidt abstracts 2010. *A Geochimica et Cosmochimica Acta*, 74, A1–A37.

Amarasiriwardena, D., Siripinyanond, A., & Barnes, R. M. (2001). Trace elemental distribution in soil and compost-derived humic acid molecular fractions and colloidal organic matter in municipal wastewater by flow field-flow fractionation-inductively coupled plasma mass spectrometry (flow FFF-ICP-MS). *Journal of Analytical Atomic Spectrometry*, 16, 978–986.

Atkinson, K. E. (1989). *An introduction to numerical analysis* (2nd ed.). New York, NY: Wiley. ISBN: 978-0-471-50023-0

Bernier-Latmani, R., Veeramani, H., Della Vecchia, E., Junier, P., Lezama-Pacheco, J. S., Suvorova, E. I., . . . Bargar, J. R. (2010). Non-uraninite products of microbial U(VI) reduction. *Environmental Science & Technology*, 44, 9456–9462.

Boyanov, M. I., Fletcher, K. E., Kwon, M. J., Rui, X., O'Loughlin, E. J., Löffler, F. E., & Kemner, K. M. (2011). Solution and microbial controls on the formation of reduced U(IV) species. *Environmental Science & Technology*, 45(19), 8336–8344.

Cruz-Uribe, D., & Neugebauer, C. J. (2002). Sharp error bounds for the trapezoidal rule and Simp-son's rule. *Journal of Inequalities in Pure and Applied Mathematics*, 3(4), 1–22.

Fletcher, K. E., Boyanov, M. I., Thomas, S. H., Wu, Q., Kemner, K. M., & Löffler, F. E. (2010). U(VI) reduction to mononuclear U(IV) by *Desulfitobacterium* species. *Environmental Science & Technology*, 44(12), 4705–4709.

Ganesh, R., Robinson, K. G., Reed, G. D., & Sayler, G. (1997). Reduction of hexavalent uranium from organic complexes by sulfate- and iron-reducing bacteria. *Applied and Environmental Microbiology*, 63, 4385–4391.

Geckeis, H., Ngo Manh, T., Bouby, M., & Kim, J. I. (2003). Aquatic colloids relevant to radionuclide migration: Characterization by size fractionation and ICP-mass spectrometric detection. *Colloids and Surfaces A: Physicochemical and Engineering Aspects*, 217(1-3), 101–108.

Jackson, B. P., Ranville, J. F., Bertsch, P. M., & Sowder, A. G. (2005a). Characterization of colloidal and humic-bound Ni and U in the "dissolved" fraction of contaminated sediment extracts. *Environmental Science & Technology*, 39(8), 2478–2485.

Jackson, B. P., Ranville, J. F., & Neal, A. L. (2005b). Application of flow field flow fractionation-ICPMS for the Study of Uranium Binding in Bacterial Cell Suspensions. *Analytical Chemistry*, 77(5), 1393–1397.

Lesher, E., Ranville, J. F., & Honeyman, B. D. (2009). Analysis of pH dependent uranium (VI) sorption to nanoparticulate hematite by flow field-flow fractionation – inductively coupled plasma mass spectrometry. *Environmental Science & Technology*, 43(14), 5403–5409.

Litzén, A. (1993). Separation speed, retention, and dispersion in asymmetrical flow field-flow fractionation as functions of channel dimensions and flow rates. *Analytical Chemistry*, 65, 461–470.

Lovley, D. R., & Phillips, E. J. P. (1992). Bioremediation of uranium contamination with enzymatic uranium reduction. *Environmental Science & Technology*, 26, 2228–2234.

Maxfield, B. (2009). *Essential Mathcad for engineering, science, and math*. Oxford, UK: Academic Press, Elsevier.

Rahman, Q. I., & Schmeisser, G. (1990). Characterization of the speed of convergence of the trapezoidal rule. *Numerische Mathematik*, 57(1), 123–138. doi:10.1007/BF01386402

Ranville, J. F., & Beckett, R. (2000). Aquatic colloids. In M. E. Schimpf, K. Caldwell, & J. C. Giddings (Eds.), *Field flow fractionation handbook*. New York, NY: Wiley.

Ranville, J. F., Hendry, M. J., Thorsten, N. R., Qianli, X., & Bruce, D. H. (2007). Quantifying uranium complexation by groundwater. *Journal of Contaminant Hydrology*, 233–246.

Sani, R. K., Geesey, G., & Peyton, B. M. (2001). Assessment of Pb toxicity to *Desulfovibrio desulfuricans* G20: Influence of components of lactate C medium. *Advances in Environmental Research*, 5(3), 269–276.

Sani, R. K., Peyton, B. M., Amonette, J. E., & Geesey, G. G. (2004). Reduction of uranium(VI) under sulfate reducing conditions in the presence of Fe(III)-(hydr) oxides. *Geochimica et Cosmochimica Acta*, 68, 2639–2648.

Sani, R. K., Peyton, B. M., & Dohnalkova, A. (2006). Effect of uranium(VI) in aqueous media on *Desulfovibrio desulfuricans* G20. *Environmental Toxicology and Chemistry*, 25, 1231–1238.

Şengör, S. S., Singh, G., Dohnalkova, A., Spycher, N., Ginn, T. R., Peyton, B. M., & Sani, R. K. (2016). Impact of different environmental conditions on the aggregation of biogenic U(IV) nanoparticles synthesized by *Desulfovibrio alaskensis* G20. *Biometals*, 29(6), 965–980.

Singh, G., Şengör, S. S., De, J., Bhalla, A., Kumar, S., Stewart, B., . . . Sani, R. K. (2014). Reoxidation of biogenic reduced uranium—A challenge towards bioremediation. *Critical Reviews in Environmental Science and Technology*, 44, 391–415.

Sivaswamy, V., Boyanov, M. I., Peyton, B. M., Viamajala, S., Gerlach, R., Apel, W. A., . . . Borch, T. (2011). Multiple mechanisms of uranium immobilization by *Cellulomonas* sp. strain ES6. *Biotechnology and Bioengineering*, 108(2), 264–276.

Weideman, J. A. C. (2002). Numerical integration of periodic functions: A few examples. *The American Mathematical Monthly*, 109(1), 21–36. doi:10.2307/2695765

Zänker, H., Richter, W., & Huttig, G. (2003). Scavenging and immobilization of trace contaminants by colloids in the waters of abandoned ore mines. *Colloids and Surfaces A: Physicochemical and Engineering Aspects*, 217(1-3), 21–31.

Zhang, Y. J., Bryan, N. D., Livens, F. R., & Jones, M. N. (1997). Selectivity in the complexation of actinides by humic substances. *Environmental Pollution*, 96(3), 361–367.

# Index



*Microbial Interactions at Nanobiotechnology Interfaces: Molecular Mechanisms and Applications*, First Edition. Edited by R. Navanietha Krishnaraj and Rajesh K. Sani.

Printed and bound by CPI Group (UK) Ltd, Croydon, CR0 4YY